MATH AND SCIENCE FOR YOUNG CHILDREN

Rosalind Charlesworth
Louisiana State University

Karen K. Lind
University of Louisville

Delmar Publishers Inc.®

NOTICE TO THE READER

Delmar Staff
Associate Editor: Jay Whitney
Production Manager: Gerry East
Project Editor: Christopher Chien
Production Supervisor: Karen Seebald
Production Coordinator: Sandra Woods
Art Manager: John Lent
Design Coordinator: Susan C. Mathews

For information address Delmar Publishers Inc.
2 Computer Drive West, Box 15-015
Albany, New York 12212-5015

Printed in the United States of America
Published simultaneously in Canada
by Nelson Canada,
a division of The Thomson Corporation

10 9 8 7 6 5 4 3

Library of Congress Cataloging-in-Publication Data:

Charlesworth, Rosalind.
Math and science for young children / Rosalind Charlesworth.
Karen K. Lind.
p. cm.
ISBN 0-8273-3402-8
1. Mathematics—Study and teaching (Primary) 2. Science—Study and teaching (Primary) I. Lind, Karen. II. Title.
QA135.5.C463 1990
372.3′5044—dc20 89-25629
CIP

Contents

SECTION IV SYMBOLS AND HIGHER-LEVEL ACTIVITIES

SECTION V MATHEMATICS CONCEPTS AND OPERATIONS FOR THE PRIMARY GRADES

SECTION VI USING SKILLS, CONCEPTS, AND ATTITUDES FOR SCIENTIFIC INVESTIGATIONS IN THE PRIMARY GRADES

SECTION VII THE MATH AND SCIENCE ENVIRONMENT

APPENDICES

Preface

Math and Science for Young Children is designed to be used by students in training and teachers in service in early childhood education. To the student, it introduces the excitement and extensiveness of math and science experiences in programs for young children. For teachers in the field, it presents an organized, sequential approach to creating a developmentally appropriate math and science curriculum for preschool and primary children.

Activities are presented in a developmental sequence designed to support young children's construction of the concepts and skills essential to a basic understanding of mathematics and science. A developmentally appropriate approach to assessment is stressed to have an individualized program in which each child is presented at each level with tasks that can be accomplished successfully before moving on to the next level.

A further emphasis is placed on three types of learning: naturalistic, informal, and structured. Much learning can take place through the child's natural exploratory activities if the environment is designed to promote such activity. The adult can reinforce and enrich this naturalistic learning by careful introduction of information and structured experiences.

The back-to-basics and pressure-cooker instructional practices of the eighties have produced a widespread use of inappropriate instructional practices with young children. Mathematics for preschoolers has been taught as "pre-math," apparently under the assumption that math learning only begins with addition and subtraction in the primary grades. It has also been taught in both preschool and primary as rote memory material using abstract paper and pencil activities. Science, on the other hand, has been largely ignored with the excuse that teaching the basics precluded allowing time for science. This text is designed to counteract these developments and to bring to the attention of early childhood educators the interrelatedness of math and science and the necessity of providing young children with opportunities to explore concretely these domains of early concept learning.

Rosalind Charlesworth is a professor in the Department of Curriculum and Instruction of the College of Education at Louisiana State University, Baton Rouge, Louisiana. Working in collaboration with faculty in the LSU School of Human Ecology, Dr. Charlesworth has been instrumental in developing the graduate programs in Early Childhood Education at LSU.

Karen Lind is an assistant professor in the Department of Early and Middle Childhood Education at the University of Louisville, Louisville, Kentucky. Dr. Lind is a veteran of the primary classroom and teaches courses in science methods for undergraduate and graduate students.

Dr. Charlesworth's career in early childhood education has included experiences with both normal and special needs young children in laboratory schools, public schools, and day care and through research in social and cognitive development and behavior. She also taught courses in early education and child development at other universities before joining the faculty at Louisiana State University. She is the author of the popular Delmar text, *Understanding Child Development*, has published many articles in professional journals, and gives presentations regularly at major professional meetings. Dr. Charlesworth has provided service to the field through active involvement in professional organizations. She is a past president of the Louisiana Association on Children Under Six, was a member of the Editorial Board of the Southern Association on Children Under Six journal *Dimensions*, and is a member of the Board of the National Association of Early Childhood Teacher Educators.

Dr. Lind is president of the Council for Elementary Science International (CESI). She is the early childhood column editor of *Science & Children*, a publication of the National Science Teacher's Association (NSTA). Her research and inservice programs focus on integrating science into preschool and primary classroom settings.

Acknowledgments

The authors wish to express their appreciation to the following individuals and Early Childhood and Development Centers:

- Dee Radeloff, for her collaboration in the writing of *Experiences in Math for Young Children*, which served as the starting point for this book.
- Dr. Mark Malone of the University of Colorado at Colorado Springs, who contributed to the planning of this text. Dr. Malone also demonstrated great patience while introducing Dr. Charlesworth to the mysteries of word processing on a personal computer.
- Jerry Bergman, for many of the photographs used from *Experiences in Math for Young Children*.
- Photographers Robert L. Knaster (University of Louisville) and Mike Ogburn (Jefferson County Public Schools) for their patience and expertise.
- Artist Bonita S. Carter for the care and accuracy taken in her original art and photographs.
- Kate Charlesworth, for her tolerance of her mother's writing endeavors.
- Eugene F. Lind and Paul and Marian Kalbfleisch for their encouragement during the writing process.
- The children and teachers who were photographed in the following Early Childhood Centers and Elementary schools:

  Louisiana State University College of Education Laboratory Elementary School, Baton Rouge, Louisiana
  Louisiana State University School of Human Ecology Laboratory Preschool, Baton Rouge, Louisiana
  Plan-Do-Talk Day Care Center, Bowling Green, Ohio
  Dunn's Kiddie Kare, Bowling Green, Ohio
  Bowling Green State University Child Development Center, Bowling Green, Ohio
  Walden School, Louisville, Kentucky
  Anchorage Elementary School, Anchorage, Kentucky
  Jefferson County Public Schools, Louisville, Kentucky
- The following teachers who provided a place for observation and/or cooperated with our efforts to obtain photographs:

  Lois Rector, Kathy Tonore, Lynn Morrison, and Nancy Crom (LSU Laboratory Elementary School), Joan Benedict (LSU Laboratory Preschool), Nancy Miller, and Candy Jones (East Baton Rouge Parish Public Schools). Jenny Flackler (Walden School), Maureen Awbrey (Anchorage Schools), and Jenny Fackler (Walden School). Elizabeth Beam (Zachary Taylor) and Dr. Anna Smythe (Cochran), Jefferson County Public Schools.
- Anchorage Schools computer teacher Sharon Campbell who provided recommendations for using computers with young children.
- University of Louisville graduate student Christy McGee and Louisiana State University graduate students Sr. Paula Marie Blouin, Carole Johnson, Rebecca Frederick, Robin Jarvis, Patricia Diez, and Ann Duffel, who provided firsthand examples of teachers and young children experiencing mathematics.

- Phyllis Marcuccio, Editor of *Science and Children* and Director of Publications for the National Science Teachers Association for generously facilitating the use of articles appearing in *Science and Children* and other NSTA publications.
- The staff of Delmar Publishers for their patience and understanding throughout this project.

DEDICATION

This book is dedicated to the memory of a dear friend

ADA DAWSON STEPHENS

SECTION I

Concept Development in Mathematics and Science

UNIT 1 How Concepts Develop

OBJECTIVES

After studying this unit, the student should be able to:

- Define concept development
- Identify children developing concepts
- Describe the commonalities between math and science
- Label examples of Piaget's developmental stages of thought
- Identify conserving and nonconserving behavior and state why conservation is an important developmental task
- Explain how young children acquire knowledge

Early childhood is a period when children actively engage in aquiring basic concepts. Concepts are the building blocks of knowledge; they allow people to organize and categorize information. As we watch children in their everyday activities we can observe concepts being constructed and used. For example:

- One-to-one correspondence: passing apples, one to each child at the table; putting pegs in pegboard holes; putting a car in each garage built from blocks.
- Counting: Counting the pennies from the penny bank, the number of straws needed for the children at the table, the number of rocks in the rock collection.
- Classifying: Placing square shapes in one pile and round shapes in another; putting cars in one garage and trucks in another.
- Measuring: Pouring sand, water, rice, or other materials from one container to another.

As you proceed through this text, you will see that young children begin to construct many concepts during the preprimary period; they then apply them to the problem-solving tasks that are the beginnings of scientific inquiry.

During the preprimary period children learn and begin to apply concepts basic to both mathematics and science. As children enter the primary period (grades one through three), they apply these early basic concepts when exploring more abstract inquiries in science and to help them understand more complex concepts in mathematics such as addition, subtraction, multiplication, division, and the use of standard units of measurement.

Figure 1–1 The infant learns about the world through his senses.

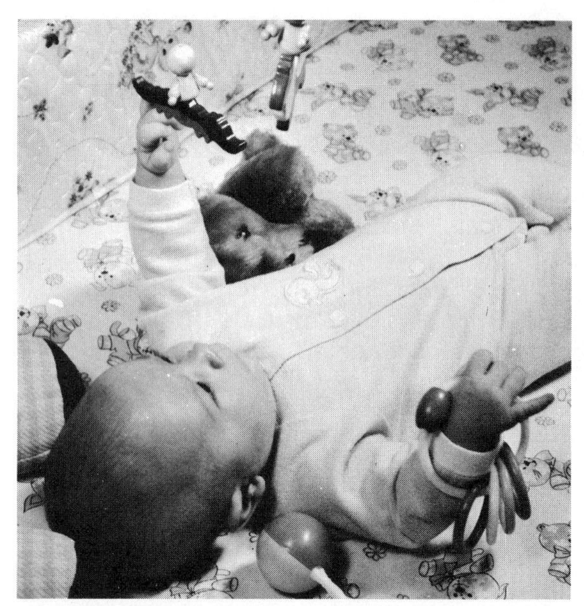

Figure 1–2 The infant learns about distance as he reaches and grasps.

Figure 1–3 The infant learns the shape concept of roundness.

As young children grow and develop physically, socially, and mentally, their concepts grow and develop as well. *Development* refers to changes that take place due to growth and experience. It follows an individual timetable for each child. Development is a series or sequence of steps that each child reaches one at a time. Different children of the same age may be weeks, months, or even a year or two apart in reaching certain stages and still be within the normal range of development. This text examines concept development in math and science from birth through the primary grades. For an overview of this developmental sequence, see Figure 1–11 (page 13).

Concept growth and development begins in infancy. The baby explores the world with his senses. He looks, touches, smells, hears, and tastes. The child is born curious. He wants to know all about his environment. The baby begins to learn ideas of size, weight, shape, time, and space. As he looks about, he senses his relative smallness. He grasps things and finds that some fit his tiny hand and others do not. The infant learns about weight when items of the same size cannot always be lifted. He learns about shape. Some things stay where he puts them while others roll away. He learns time sequence. When he wakes up, he feels wet and hungry. He cries. Mom comes. He is changed and then fed. Next he plays, gets tired, and goes to bed to sleep. As the infant begins to move, he develops an idea of space. He is placed in a crib, in a playpen, or on the floor in the center of the living room. As the baby first looks and then

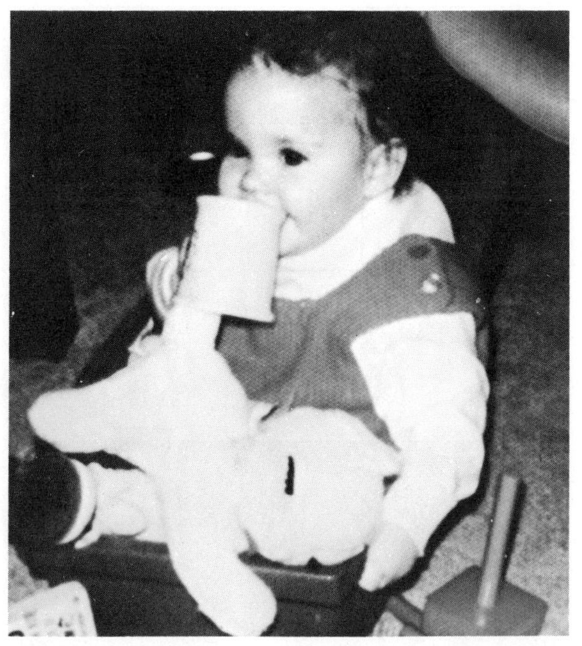

Figure 1–4 The toddler learns about size, shape, and space.

moves, he dicovers space. Some spaces are big. Some spaces are small.

As the child learns to crawl, to stand, and to walk, he is free to discover more on his own and learns to think for himself. He holds and examines more things. He goes over, under, and in large objects and discovers his size relative to them. The toddler sorts things. He puts them in piles—of the same color, the same size, the same shape, or with the same use. The young child pours sand and water into containers of different sizes. He piles blocks into tall structures and sees them fall and become small parts again. He buys food at a play store and pays with play money. As the child cooks imaginary food, he measures imaginary flour, salt, and milk. He sets the table in his play kitchen, putting one of everything at each place just as is done at home. The free exploring and experimentation of the first two years is the opportunity for the development of muscle coordination and the senses of taste, smell, sight, and hearing. The child needs these skills as a basis for future learning.

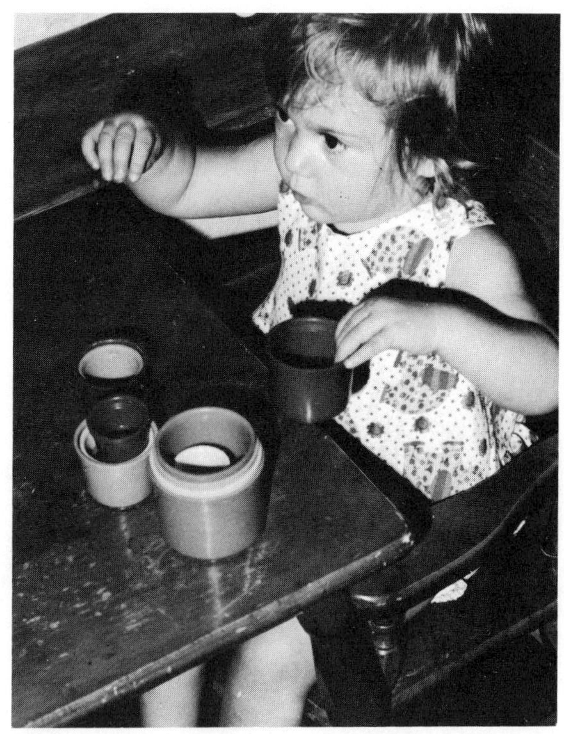

Figure 1–5 Sorting and matching are important toddler activities.

As young children leave toddlerhood and enter the preschool and kindergarten levels of the preprimary period, exploration continues to be the first step in dealing with new situations; at this time, however, they also begin to apply basic concepts to collecting and organizing data to answer a question. Collecting data requires skills in observation, counting, recording, and organizing. For example, for a science investigation, kindergartners might be interested in the process of plant growth. Supplied with lima bean seeds, wet paper towels, and glass jars, the children place the seeds in the jars where they are held against the sides with wet paper towels. Each

Figure 1–6 The young child learns about measurement.

Figure 1–7 The young child learns that there is one buttonhole for each button.

Name Mary

How many days until I see green sprouting up?

1	2	3	4	5	6	7	8	9	10	11	12
X	X	X									

Figure 1–8 Mary records each day that passes until her bean seed sprouts.

Figure 1–9 This second grader learns through concrete activity in the first stages of the learning cycle.

day they add water as needed and observe what is happening to the seeds. They dictate their observations to their teacher, who records them on a chart. Each child also plants some beans in dirt in a small container such as a paper or plastic cup. The teacher supplies each child with a chart for his or her bean garden. The children check off each day on their charts until they see a sprout (Figure 1–8). Then they count how many days it took for a sprout to appear; they compare this number with those of the other class members, as well as with the time it takes for the seeds in the glass jars to sprout. The children have used the concepts of number and counting, one-to-one correspondence, time, and comparison of the number of items in two groups. Primary children might attack the same problem but can operate more independently and record more information, use standard measuring tools (i.e., rulers), and do background reading on their own (Figure 1–9).

COMMONALITIES IN MATH AND SCIENCE IN EARLY CHILDHOOD

The young child's understanding of math and science grows from the development of some of the same basic concepts during early childhood. Much of our understanding of how and when this development takes place comes from research based on Jean Piaget's theory of how concept development occurs. Piaget's theory is briefly described in the next part of the unit. First, the commonalities that tie math and science together are examined.

Math and science are interrelated in that the basic math concepts of comparing, classifying, and measuring are basic process skills of science (see Unit 5 for a more in-depth explanation). That is, basic math concepts are needed in order to solve problems in science. The other science process skills (observing, communicating, inferring, hypothesizing, and defining and controlling variables) are equally important for solving problems in both science and mathematics. For example, consider the principle of the ramp, a basic concept in physics. Suppose a two-foot-wide plywood board is leaned against a large block so that it becomes a ramp. The children are given a number of balls of different sizes and weights to roll down the ramp. Once they have the idea of the game through free exploration, the teacher might insert some questions such as, "What do you think would happen if two balls started to roll at exactly the same time from the top of the ramp?" or, "What would happen if you changed the height of the ramp or had two ramps of different heights? Of different lengths?" The students could guess, explore what happens, using ramps of varying steepness and length and balls of various types, observe what happens, communicate their observations, and describe commonalities and differences. They might observe differences in speed and distance traveled contingent on the size or weight of the ball, the height and length of the ramp, or other variables. In this example, children could use math concepts of speed, distance, height, length, and counting (how many blocks are propping each ramp?) while engaged in scientific observation. For another example, suppose the teacher brings several pieces of fruit to class: one red apple, one green apple, two oranges, two grapefruit, and two bananas. The children examine the fruit to discover as much about it as possible. They observe size, shape, color, texture, taste, and composition (juicy or dry, segmented or whole, seeds, and so on). Observations may be recorded using counting and classification skills (How many of each fruit type? Of each color? How many are spheres? How many are juicy? and so on). The fruit can be weighed and measured, prepared for eating, and divided equally among the students.

As with these two examples, it will be seen throughout the text that math and science concepts and skills can be acquired as children engage in traditional early childhood activities such as playing with blocks, water, sand, and manipulative materials, as well as during dramatic play, cooking, and outdoor activities.

PIAGETIAN PERIODS OF CONCEPT DEVELOPMENT AND THOUGHT

Jean Piaget contributed enormously to understanding the development of children's thought. Piaget identified four periods of cognitive, or mental, growth and development. Early childhood educators are concerned with the first two periods and the first half of the third.

The first period identified by Piaget, called the *sensorimotor period* (from birth to about age two), is described in the first part of the unit. It is the time when children begin to learn about the world. They use all their sensory abilities—touch, taste, sight, hearing, smell, and muscular. They also use growing motor abilities—to grasp, to crawl, to stand, and, eventually, to walk. Children in this first period are explorers and need opportunities to use their sensory and motor abilities to learn basic skills and concepts. Through these activities the young child *assimilates* (takes into the mind and comprehends) a great deal of information. By the end of this period, children have developed the concept of object permanence. That is, they realize that objects exist even when they are out of sight. They also develop the ability of object recognition. They learn to identify objects using the information they have acquired about features such as color, shape, and size. As children near the end of the sensorimotor period, they reach a stage where they can engage in *representational thought*; that is, instead of acting impetuously, they can think through a solution before attacking a problem. They also enter into a time of rapid language development.

The second period, called the *preoperational period*, extends from about ages two to seven. During this period children begin to develop concepts that are more like those of adults, but these are still incomplete in relation to what they will be like at maturity. These concepts are often referred to as *preconcepts*. During the early part of the preoperational period, language continues to undergo rapid growth, and speech is used increasingly to express concept knowledge. Children begin to use concept terms such as big and small (size), light and heavy (weight), square and round (shape), late and early (time), long and short (length), and so on. This ability to use language is one of the symbolic behaviors that emerges during this period. Children also use symbolic behavior in their representational play, where they may use sand to represent food, a stick to represent a spoon, or another child to represent father, mother, or baby. Play is a major arena in which children develop an understanding of symbolic functions that underlie the later understanding of abstract symbols such as numerals, letters, and written words.

An important characteristic of preoperational children is *centration*. When materials are changed in form or arrangement in space, children may see them as changed in amount as well. This is because preoperational children tend to *center* on the most obvious aspects of what is seen. For instance, if the same amount of liquid is put in both a tall, thin glass and a short, fat glass, preoperational children say there is more in the tall glass "because it is taller." If clay is changed in shape from a ball to a snake, they say there is less clay "because it is thinner." If a pile of coins is placed close together, preoperational children say there are fewer coins than they would if the coins were spread out. When the physical arrangement of material is changed, preoperational children seem to be unable to hold the original picture of its shape in mind. They lack *reversibility*: That is, they cannot reverse the process of change mentally. The ability to hold or save the original picture in the

Original	Physical Change	Question	Non-Conserving Answer	Conserving Answer
Same amount of drink. 		Is there still the same amount of drink?	No, there is more in the tall glass.	Yes, you just put the drink in different size glasses.
Same amount of clay. 	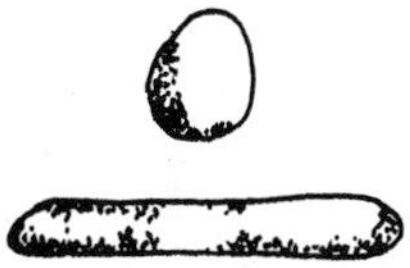	Is there still the same amount of clay?	No, there is more clay in the snake because it is longer.	Yes, you just rolled it out into a different shape.
Same amount of pennies. 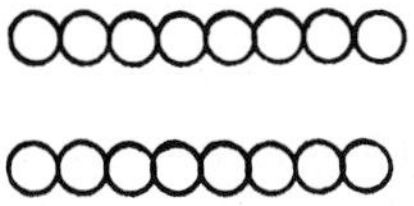	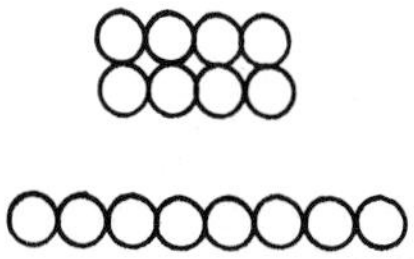	Are there still the same number of pennies?	No, there are more in the bottom row because it is longer.	Yes, you just moved the pennies closer together (points to top row).

Figure 1–10 Physical changes in conservation tasks

mind and reverse physical change mentally is referred to as *conservation*. The inability to conserve is a critical characteristic of preoperational children. During the preoperational period children work with the precursors of conservation such as counting, one-to-one correspondence, shape, space, and comparing. They also work on seriation (putting items in a logical sequence, such as fat to thin or dark to light) and classification (putting things in logical groups according to some common criteria such as color, shape, size, use, and so on).

During the third period, called *concrete operations* (usually from ages seven to eleven), children are becoming *conservers*. That is, they are becoming more and more skilled at retaining the original picture in mind and making a mental reversal when appearances are changed. The time between ages five and seven is one of transition to concrete operations. Each child's thought processes are changing at their own rate. During this time of transition, therefore, a normal expectation is that some children are already conservers and others are not. This is a critical consideration for kindergarten and primary teachers because the ability to conserve number (the pennies problem) is a good indication that children are ready to deal with abstract symbolic activities. That is, they will be able to mentally manipulate groups that are represented by number symbols with a real understanding of what mathematical operations mean. Section II of this text covers the basic concepts that children have to understand and integrate in order to conserve.

Piaget's final period is called *formal operations* (ages eleven through adulthood). During this period, children can learn to use the scientific method independently. That is, they learn to solve problems in a logical and systematic manner. They begin to understand abstract concepts and to attack abstract problems. They can imagine solutions before trying them out. For example, suppose a person who has reached the formal operations level is given samples of several colorless liquids and is told that some combination of these liquids will result in a yellow liquid. A person at the formal operations level would plan out how to systematically test to find the solution; a person still at the concrete operational level might start combining without considering all the parameters of the problem, such as labeling each liquid, keeping a record of which combinations have been tried, and so on. Note that this period may be reached as early as age eleven; however, it may not be reached at all by many adults.

PIAGET'S VIEW OF HOW CHILDREN ACQUIRE KNOWLEDGE

According to Piaget's view, children acquire knowledge by constructing it through their interaction with the environment. Children do not wait to be instructed to do this; they are continually trying to make sense out of everything they encounter. Piaget divides knowledge into three areas:

- **Physical knowledge** is the type that includes learning about objects in the environment and their characteristics (color, weight, size, texture, and other features that can be determined through observation and are physically within the object).
- **Logico-mathematical knowledge** is the type that includes relationships each individual constructs (such as same and different, more and less, number, classification, and so on) in order to make sense out of the world and to organize information.
- **Social (or conventional) knowledge** is the type that is created by people (such as rules for behavior in various social situations).

Physical and logico-mathematical knowledge depend on each other and are learned simultaneously. That is, as the physical characteristics of objects are learned, logico-mathematical categories are constructed to organize information. For example, in the popular story *Goldilocks and the Three Bears*, papa bear is big, mama bear is middle sized, and baby bear is the smallest (seriation), but all three (number) are bears because they are covered with fur and have a certain body shape with a certain combination of features common only to bears (classification).

Constance Kamii, a student of Piaget's, has actively translated Piaget's theory into practical applications for the instruction of young children. Kamii emphasizes that according to Piaget, autonomy (independence) is the aim of education. Intellectual autonomy develops in an atmosphere where: children feel secure in their relationships with adults; where they have an opportunity to share their ideas with other children; and where they are encouraged to be alert and curious, come up with interesting ideas, problems and questions, use initiative in finding out the answers to problems, have confidence in their abilities to figure out things for themselves, and speak their minds with confidence. Young children need to be presented with problems to be solved through games and other activities that challenge their minds. They must work with concrete materials and real problems such as the examples provided earlier in the unit.

The Curriculum Standards for School Mathematics, published by The National Council of Teachers of Mathematics (NCTM, 1988), emphasizes five goals for students:

1. Learning to value mathematics
2. Becoming confident in one's own ability
3. Becoming a mathematical problem solver
4. Learning to communicate mathematically
5. Learning to reason mathematically

These goals can be achieved for young children through the concrete approach to integrated science and mathematics presented in this text.

THE LEARNING CYCLE

The authors of the Science Curriculum Improvement Study program designed a Piagetian-based teaching strategy called the *learning cycle.* The learning cycle approach is designed to include four factors which Piaget felt interacted in concept learning: physical maturation, physical experience, social interaction, and self-regulation. The *physical growth* of the body, including the nervous system, follows a sequence (the periods of mental growth, which is an aspect of this sequence, have already been described in this unit); *physical experience* refers to the interaction with real objects and events which enables children to construct concepts; *social interaction* refers to the exchange of ideas and the communication between children as they engage in active learning; *self-regulation* refers to the mental activity that goes on as concepts are constructed. The processes underlying concept learning are described in Unit 6. In this unit the learning cycle will be described as an example of an instructional strategy inspired by Piaget's theory of learning.

Charles E. Barman has updated and clarified the learning cycle approach in his paper *An expanded view of the learning cycle: New ideas about an effective teaching strategy*. Barman labels the three phases of the cycle as follows:

1. exploration
2. concept introduction
3. concept application

During the *exploration phase* the teacher remains in the background, observing and occasionally inserting a comment or question (see Unit 2, naturalistic and informal learning). The students actively manipulate materials and interact with each other. The teacher's knowledge of child development guides the selection of materials and how they are placed in the environment so that they provide a developmentally appropriate setting in which young children can explore and construct concepts.

During the *concept introduction* phase the teacher provides direct instruction; this begins with a discussion of the information the students have discovered. The teacher helps the children record their information. During this phase the teacher clarifies and adds to what the children have found out for themselves by using explanations, print materials, films, guest speakers, and other available resources (see Unit 2, structured learning experiences).

The third phase of the cycle is the *application phase*. The teacher (or the children themselves) suggest a new problem to which the information learned in the first two phases can be applied. Again, the children are actively involved in concrete activities and exploration.

These phases can be applied to the ramp-and-ball example described earlier. During the first phase, the ramp and the balls are available to be examined. The teacher inserts some suggestions and questions as the children work with the materials. In the second phase, the teacher communicates with the children regarding what they have observed. The teacher might also provide explanations, label the items being used, and otherwise assist the children in organizing their information; at this point books and/or films about simple machines could be provided. For the third phase, the teacher poses a new problem and challenges the children to apply their concept of the ramp and how it works to the new problem. For example, some toy vehicles might be provided to use with the ramp(s).

Barman describes three types of *Learning cycle* lessons which vary according to the way data is collected by the students and the type of reasoning the students engage in. The three types of lessons are 1) descriptive, 2) empirical-inductive, and 3) hypothetical-deductive. Most young children will be involved in descriptive lessons, where they mainly observe, interact, and then describe their observations (descriptive lesson). They may begin to generate "guesses" regarding the reasons for what they observed, but serious hypothesis generation requires concrete operational thought (empirical-inductive lesson). In the third type of lesson, students observe, generate hypotheses, and design experiments to test their hypotheses (hypothetical-deductive lesson). This type of lesson requires formal operational thought. However, this does not mean that preoperational and concrete operational children should be discouraged from generating ideas on how to find out if their guesses will prove to be true; quite the contrary. They need to be encouraged to take the risk. Often they will come up with a viable solution, even though they may not yet have reached the level of mental maturation necessary to understand the underlying physical or logico-mathematical reasons.

THE ORGANIZATION OF THE TEXT

This text is divided into seven sections. The sequence is both integrative and developmental. Section I is an integrative section that sets the stage for instruction. Development, acquisition, and promotion of math and science concepts is described. A plan for assessing developmental levels is provided. Finally, the basic concepts of science and their application are described.

Sections II, III, and IV encompass the developmental mathematics and science program for sensorimotor- and preoperational-level children. Section II describes the fundamental concepts basic to both math and science along with suggestions for instruction and materials. Section III focuses on applying these fundamental concepts, attitudes, and skills at a more advanced level. Section IV deals with higher level concepts and activities.

Sections V and VI encompass the acquisition of concepts and skills for children at the concrete operations level. At this point, the two subject areas conventionally become more discrete in terms of instruction. However, they should continue to be integrated because science explorations can enrich children's math skills and concepts through concrete applications, whereas mathematics is used to organize and interpret data collected through observation.

In Section VII, parallel units provide suggestions of materials and resources and descriptions of math and science in action in the classroom and math in the home. Finally, the appendices include concept assessment tasks, lists of children's books that contain math and science concepts as well as fingerplays, games and songs, and food requirements for animals.

As Figure 1–11 illustrates, concepts are not acquired in a series of quick, short-term lessons; development begins in infancy and continues throughout early childhood and, of course, beyond. As you read each unit, keep referring back to Figure 1–11; it can help you relate each section to periods of development.

SUMMARY

Concept development begins in infancy and grows through four periods throughout a lifetime. The exploratory activities of the infant and toddler during the sensorimotor period are the basis of later success. As they use their senses and muscles, children learn about the world. During the preoperational period, concepts grow

Period	Concepts and Skills			
	Section II Fundamental	Section III Applied	Section IV Higher Level	Section V Primary
Sensorimotor (Birth to age two)	Observation Problem-solving One-to-one correspondence Number Shape Space			
Preoperational (Two to seven years)	Sets and classifying Comparing Counting Parts and wholes Language	Ordering Informal measurement: Weight Length Temperature Volume Time Sequence	Number symbols Sets and symbols	
Transitional (Five to seven years)		Graphing	Concrete addition and subtraction	
Concrete operations (Seven to eleven years)				Whole number operations Fractions Number facts Place value Geometry Measurement with standard units

Figure 1–11 The development of math and science concepts

rapidly, and children develop the basic concepts and skills of science and mathematics, moving toward intellectual autonomy through independent activity, which serves as a vehicle for the construction of knowledge. Sometime between ages five and seven, children enter the concrete operations period and learn to apply abstract ideas and activities to concrete knowledge of the physical and mathematical world. The *Learning cycle* lesson is described as an example of a developmentally inspired teaching strategy. This text presents the concepts, skills, and attitudes fundamental to math and science for young children.

FURTHER READING AND RESOURCES

Barman, C. R. 1989. An expanded view of the learning cycle: New ideas about an effective teaching strategy. *Council of Elementary Science International Monograph IV*. Indianapolis, IN: Indiana University.

Charlesworth, R. 1987. *Understanding child development*. Albany, NY: Delmar.

Forman, G. E. and Hill, F. 1984. *Constructive play*. Menlo Park, CA: Addison-Wesley.

Inhelder, B. and Piaget, J. 1969. *The early growth of logic in the child*. NY: Norton.

Kamii, C. 1986. Cognitive learning and development. *Today's kindergarten*. NY: Teacher's College Press.

Karplus, R. and Thier, H. D. 1967. *A new look at elementary school science—Science curriculum improvement study*. Chicago: Rand McNally.

Labinowicz, E. 1985. *Learning from children: New beginnings for teaching numerical thinking*. Menlo Park, CA: Addison-Wesley.

Lawson, A. E. and Renner, J. W. 1975. Piagetian theory and biology teaching. *American Biology Teacher*. 37 (6): 336–343.

National Council of Teachers of Mathematics 1988. *Curriculum and evaluation standards for school mathematics*. Reston, VA: Author.

Piaget, J. 1965. *The child's conception of number*. NY: Norton.

Sprung, B., Froschl, M., and Campbell, P.B. 1985. *What will happen if . . .* Mt. Ranier, MD: Gryphon House.

Sunal, C.S. 1982. Philosophical bases for science and mathematics in early childhood education. *School Science and Mathematics*. 82 (1): 2–10.

SUGGESTED ACTIVITIES

1. Using the descriptions in the unit, prepare a list of behaviors that would indicate that a young child at each of Piaget's first three periods of development is engaged in behavior exemplifying the acquisition of math and science concepts. Using your list, observe four young children at home or at school. One child should be six to eighteen months old, one 18 months to 2½ years old, one age three to five, and one age six to seven. Record everything each child does that is on your list. Note any similarities and differences observed among the four children.

2. Interview three mothers of children ages two to eight. Using your list from Activity #1 as a guide, ask them which of the activities each of their children do. Ask them if they realize that these activities are basic to the construction of math and science concepts, and note their responses. Did you find that these mothers appreciated the value of children's play activities in math and science concept development?

3. Interview two or three young children. Present the conservation of number problem illustrated in Figure 1–10 (see Appendix A for detailed instructions). Tape their responses. Listen to the tape and describe what you learn. How did the children's responses differ? How were the answers similar?

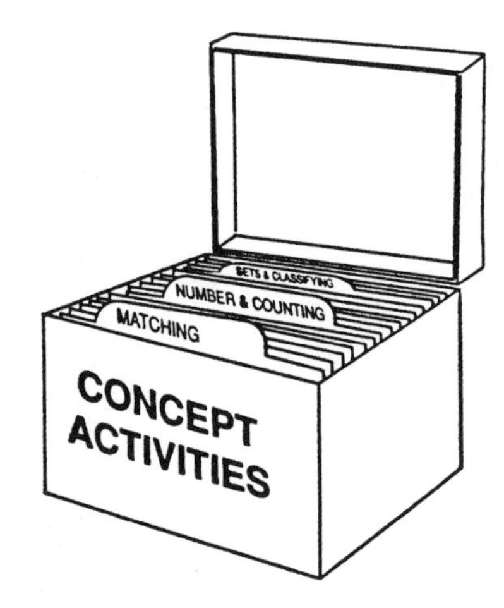

Figure 1–12 Start a math/science activity file now so you can keep it up to date.

4. You should begin to record on 5½″ × 8″ file cards each math and science activity that you learn about. Buy a package of cards, some dividers, and a file box. Label your dividers with the titles of Units 8 through 44. Figure 1–12 illustrates how your file should look.

REVIEW

A. Define the term *concept development*.

B. Match each of the listed names of concepts and skills with a correct description of child behavior from those that follow.

Concepts and skills:

1. money
2. size
3. sorting
4. space
5. time

Description of child behavior:

a. The child examines objects of different sizes or compares his size with that of another person or thing.

b. The child goes through a sequence of regular events: arrives at school, has free play, eats a snack, paints or hears a story, toilets, washes, eats lunch, takes a nap, has a snack, plays outdoors, goes home.

c. The child discovers he can crawl under the table but not under the couch.

d. Two children are playing store and one exchanges some pieces of green paper for some food.

e. A child is putting blocks in different piles by color and size.

C. Describe the commonalities between math and science.

D. Decide which of the following descriptions describes a child in the sensorimotor (SM), preoperational (P), or concrete operational (CO) Piagetian stages.

1. Mary watches as her teacher makes two balls of clay of the same size. The teacher then rolls one ball into a snake shape and asks, "Mary, do both balls still have the same amount or does one ball have more clay?" Mary laughs, "They are still the same amount. You just rolled that one out into a snake."

2. Michael shakes his rattle and then puts it in his mouth and tries to suck on it.

3. John's mother shows him two groups of pennies. One group is spread out and one group is stacked up. Each group contains ten pennies. "Which bunch of pennies would you like to have, John?" John looks carefully and then says, "I'll take these because there are more," as he picks up the pennies, that are spread out.

E. In review question D, which child, Mary or John, is a conserver? How do you know? Why is it important to know that a child is or is not a conserver?

F. Explain how young children acquire knowledge. Include the place of the learning cycle in knowledge acquisition. Provide examples from your observations.

UNIT 2 How Concepts Are Acquired

OBJECTIVES

After studying this unit, the student should be able to:

- List and define the three types of learning experiences described in the unit
- Recognize examples of each of the three types of learning experiences
- State possible responses to specific opportunities for the child to learn concepts

Concepts are acquired through children's active involvement with the environment. As they explore their surroundings, they actively construct their own knowledge. Specific learning experiences can be characterized as naturalistic (or spontaneous), informal, or structured. These experiences differ in terms of who controls the choice of activity: the adult or the child.

Naturalistic experiences are those in which the child controls choice and action; informal where the child chooses the activity and action, but at some point there is adult intervention; and structured where the adult chooses the experience for the child and gives some direction to the child's action (Figure 2–1).

NATURALISTIC EXPERIENCES

Naturalistic experiences are those initiated spontaneously by children as they go about their daily activities. These experiences are the major mode of learning for children during the sensorimotor period. Naturalistic experiences can be a valuable mode of learning for older children also (Figure 2–2).

The adult's role is to provide an interesting and rich environment. That is, there should be many things for the child to look at, touch, taste, smell, and hear. The adult should observe the

TYPES OF ACTIVITY	INTERACTION EMPHASIZED
Naturalistic	Child/Environment
Informal	Child/Environment/Adult
Structured	Adult/Child/Environment

Figure 2–1 Concepts are learned through three types of activity.

Figure 2–2 The child learns about measurement through a naturalistic activity.

child's activity and note how it is progressing and then respond with a glance, a nod, a smile, or a word of praise to encourage the child. The child needs to know when he is doing the appropriate thing.

Some examples of naturalistic experiences are listed:

- Kurt hands Dad two pennies saying, "Here's your two dollars!"
- Tamara takes a spoon from the drawer—"This is big." Mom says, "Yes."
- Roger is eating orange segments. "I got three." (Holds up three fingers.)
- Nancy says, "Big girls are up to here," as she stands straight and points to her chin.
- Cindy (age four) sits on the rug sorting colored rings into plastic cups.
- Tanya and Tim (both age four) are having a tea party. Tim says, "The tea is hot."
- Sam (age five) is painting. He makes a dab of yellow. Then he dabs some blue on top. "Hey! I've got green now."
- Trang Fung (age six) is cutting her clay into many small pieces. Then she squashes it together into one big piece.
- Sara (age six) is restless during the after-lunch rest period. As she sits quietly with her head on her desk, her eyes rove around the room. Each day she notices the clock. One day she realizes that when Mrs. Red Fox says, "One-Fifteen, time to get up," that the short-hand is always on the one and the long hand is always on the three. After that she knows how to watch the clock for the end of rest time.
- Theresa (age seven) is drawing with markers. They are in a container that has a hole to hold each one. Theresa notices that there is one extra hole. "There must be a lost marker," she comments.
- Vanessa (age eight) is experimenting with cup measures and containers. She notices that each cup measure holds the same amount even though each is a different shape. She also notices that you cannot always predict how many cups of liquid a container holds just by looking at it. The shape can fool you.

INFORMAL LEARNING EXPERIENCES

Informal learning experiences are initiated by the adult as the child is engaged in a naturalistic experience. These experiences are not preplanned for a specific time. They occur when the adult's experience and/or intuition indicates it is time to act. This might happen for various reasons—for example, the child might need help or is on the right track in solving a problem but needs a cue or encouragement. It might also happen because the adult has in mind some concepts that should be reinforced and takes advantage of a *teachable moment*. Informal learning experiences occur when an opportunity for instruction presents itself by chance. Some examples are (Figure 2–3):

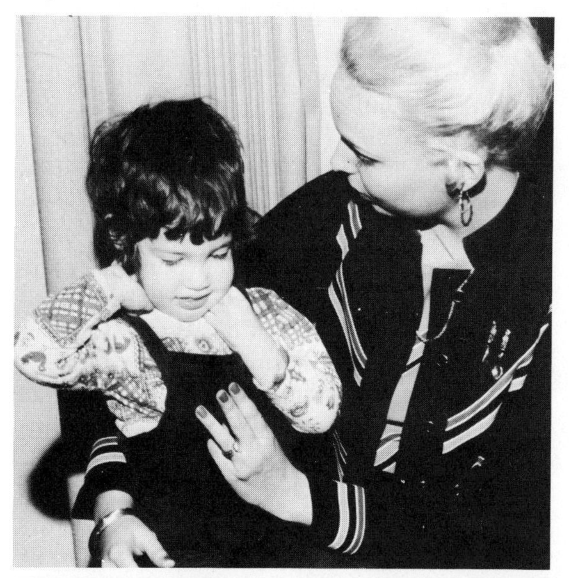

Figure 2–3 "One, two, three." An informal learning experience.

- "I'm six years old," says three-year-old Kate while holding up three fingers. Dad says, "Let's count those fingers. One, two, three fingers. You are three years old."
- Bob (age four) is setting the table. He gets frustrated because he does not seem to have enough cups. "Let's check," says his teacher. "There is one placemat for each chair. Let's see if there is one cup on each mat." They move around the table checking. They come to a mat with two cups. "Two cups," says Bob. "Hurrah!" says his teacher.
- With arms outstretched at various distances, Tim (age four) asks, "Is this big? Is this big?" Mr. Brown says, "What do you think? What *is* 'this' big?" Tim looks at the distance between his hands with his arms stretched to the fullest. "This is a big person." He puts his hands about eighteen inches apart. "This is a baby." He places his thumb and index finger about half an inch apart. "This is a blackberry." Mr. Brown watches with a big smile on his face.
- Juanita (age four) has a bag of cookies. Mrs. Ramirez asks, "Do you have enough for everyone?" Juanita replies, "I don't know." Mrs. R. asks, "How can you find out?" Juanita says, "I don't know." Mrs. R. tells her, I'll help you. We'll count them."
- Kindergartners George and Sam are playing with some small rubber figures called "Stackrobats."® George links some together horizontally, while Sam joins his vertically. The boys are competing to see who can make the longest line. When George's line reaches across the diameter of the table, he encounters a problem. Miss Jones suggests that he might be able to figure out another way to link the figures together. He looks at Sam's line of figures and then at his. He realizes that if he links his figures vertically he can continue with the competition.
- Dean, a first grader, runs into Mrs. Red Fox's classroom on a spring day after a heavy rainstorm. He says, "Mrs. Red Fox! I have a whole bunch of worms." Mrs. Red Fox asks Dean where he found the worms and why there are so many out this morning. She suggests he put the worms on the science table where everyone can see them. Dean follows through and places a sign next to the can: "Wrms fnd by Dean."
- Second grader Liu Pei is working with blocks. She shows her teacher, Mr. Wang, that she has made three stacks of four blocks. "She asks, "When I have three stacks of four, is that like when my big brother says 'three times four'?" "Yes," responds Mr. Wang. "When you have three stacks of four that is three times four."
- Jason notices that each time he feeds Fuzzy the hamster, Fuzzy runs to the food pan before Jason opens the cage. He tells his teacher, who uses the opportunity to discuss anticipatory responses, why they develop, and their significance in training animals. He asks Jason to consider why this might happen so consistently and to think about other times he has noticed this type of response in other animals or humans. Several other children join the discussion. They decide to keep individual records of any anticipatory responses they observe for a week, compare observations, and note trends.

STRUCTURED LEARNING EXPERIENCES

Structured experiences are preplanned lessons or activities. They can be done with individuals or small or large groups at a special time or an opportune time. The following are examples of some of these structured activities (Figure 2–4):

- With an individual at a specific time. Cindy is four years old. Her teacher de-

Figure 2–4 A structured learning situation. "Put them in order from smallest to largest."

cides that she needs some practice counting. She says, "Cindy, I have some blocks here for you to count. How many are in this pile?"

- With a small group at a specific time. Mrs. Red Fox is sitting with a group of six in a semicircle in front of her. She says, "I have some balls in this basket. Look at them and tell me what you see." She places a basket of balls in front of the group. In the basket are about a dozen balls that range in size from a table tennis ball to a basketball. After the children examine the balls and discuss their characteristics, Mrs. Red Fox picks up the basketball and says, "Find a ball that is smaller." After the children respond, she puts the basketball and the table tennis ball aside. She picks up a tennis ball and says, "Find a ball in the basket that is larger."
- With an individual at an opportune time. Mrs. Flores knows that Tanya needs help with the concept of shape. Tanya is looking for a game to play. Mrs. Flores says, "Try this shape-matching game, Tanya. There are squares, circles, and triangles on the big card. You find the little cards with the shapes that match."
- With a group at an opportune time. Mrs. Raymond has been working with the children on the concepts of light and heavy. They ask her to bring out some planks to make ramps for the packing boxes and the sawhorses. She brings out the planks and explains to the group, "These are the heavy planks. Put them on the packing boxes. These are the light planks. Put them on the sawhorses."
- With a large group at a specific time. Ms. Hebert realizes classification is an important concept that should be applied throughout the primary grades. It is extremely important in organizing science data. For example, to study skeletons, students brought bones from home. Ms. Hebert puts out three large sheets of construction paper and has the students explore the different ways bones can be classified (such as chicken, turkey, duck, cow, pig, deer) or placed in subcategories (such as grouping chicken bones into wings, backs, legs, and so on).

As a final example, consider how the same concepts might be constructed at all three levels with the same materials.

- Naturalistic experience. Mr. Flores places lids from various sized containers in a plastic tub on the rug where the children can examine them. At first, the children examined the lids one by one, then put them in the tub, and spilled them out. Recently, Mr. Flores noticed that some children separate the lids into groups by color; others

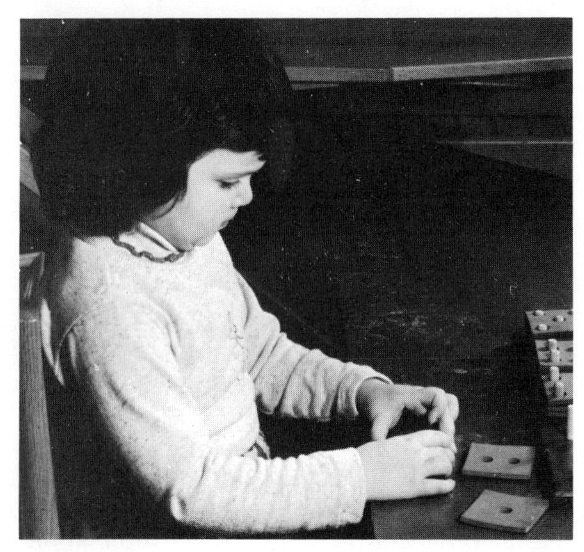

Figure 2–5 Self-correcting materials lend themselves to naturalistic, informal, and structured activities.

sort them by size; and others line them up in order from large to small.

- Informal learning experience. One day Mr. Flores sits down on the rug with the children. He says, "Tell me about these lids and what you can do with them". The children tell him how some are the same color, some are the same size, and so on. As they talk, they show him examples. Then Mr. Flores asks some specific questions. "Bob, how many red lids do you have?" "Juanita, do you have more red lids or green lids?" He holds up a lid and says, "Everyone find a lid that is bigger than this one."
- Structured learning experience. Mr. Flores has the tub of lids, some large pieces of poster board, and an assortment of marking pens. The poster boards are marked off in squares like graph paper. He says, "Today we are going to make some graphs using information from this lid collection." He and the children then discuss what criteria they might use.

Note that throughout the examples in this unit the adults ask a variety of questions and provide different types of directions for using the materials. Questions and instructions can be divergent or convergent. Divergent questions and instructions do not have one right answer but provide an opportunity for creativity, guessing, and experimenting. Questions that begin "Tell me about . . .", "What do you think . . .?", "What have you found out . . .?", "What can we do with . . .?" and directions such as "You can examine these . . ." or "You may play with these . . ." are divergent.

Convergent questions or directions ask for a specific response or activity. There is a specific piece of information called for, such as "How many . . .?", "Tell me the names of the parts of a plant.", "Find a ball smaller than this one.", and so on. Adults often ask only convergent questions and give convergent directions. Remember that children need time to construct their ideas. Divergent questions and directions encourage them to think and act for themselves. Convergent questions and directions can provide the adult with specific information regarding what the child knows, but too many of these questions tend to make the child think that there might be only one right answer to all problems. This can squelch creativity and the willingness to guess and experiment.

SUMMARY

Three types of learning experiences have been described and defined. The teacher and parent learn through practice how to make the best use of naturalistic, informal, and structured experiences so that the child has a balance of free exploration and specific planned activities.

FURTHER READING AND RESOURCES

Baratta-Lorton, M. 1976. *Math their way.* Menlo Park, CA: Addison-Wesley.

Benham, N.B., Hosticka, A., Payne, J.D., and Yeotis. C. 1982. Making concepts in science and mathematics visible and viable in the early childhood curriculum. *School Science and Mathematics.* 82 (1): 29–37.

Forman, G.E., and Kuschner, D.S. 1983. *The child's construction of knowledge.* Washington, D.C.: National Association for the Education of Young Children.

Kamii, C., and DeVries, R. 1978. *Physical knowledge in preschool education.* Englewood Cliffs, NJ: Prentice-Hall.

Kamii, C. 1985. *Children reinvent arithmetic.* New York: Teachers College Press.

Smith, R.F. 1987. Theoretical framework for preschool science experiences. *Young Children.* 42 (2): 34–40.

SUGGESTED ACTIVITIES

1. Observe a prekindergarten, kindergarten, and primary classroom. Keep a record of concept learning experiences that are naturalistic, informal, and structured. Compare the differences in the numbers of each type of experience observed in each of the classrooms.
2. During your observations, also note any times you think opportunities for naturalistic or informal learning experiences were missed.

REVIEW

A. List the three types of learning experiences and write your own definition or description of each.

B. Indicate whether the examples which follow are naturalistic, informal, or structured.

1. "Mama, I'll cut this donut in half and give you part." "Good idea," says Mom.
2. Eighteen-month-old Brad has lined up four small dishes and is putting a toy dog in each one.
3. Teacher and four children are sitting at a table. Each child has a pile of colored chips in front of him. "Line up three red chips." Children follow directions. "Put a blue chip next to each red chip."
4. "I have three cookies," says Leroy. Teacher notices that Leroy has four cookies. "Let's count those cookies, Leroy, just to be sure."
5. Three children are pouring water. They have many sizes and shapes of containers. "I need a bigger bowl, please," says one to another.
6. The children are learning about the human body. Ms. Jones has a model skeleton. She says, "This is the ankle bone. Show me your ankle bone."
7. Trang Fung brings her pet mouse to school. Each child observes the mouse and asks Trang Fung questions about his

habits. Several children draw pictures and write stories about the mouse.

8. Children in Ms. Hebert's class are playing on the jungle gym. They are trying to find out who can hang upside down by the knees the longest. Several complain of being dizzy. After they rest for a few minutes, Mrs. Hebert discusses with them why the human body might feel dizzy after hanging upside down too long.

9. Mrs. Red Fox introduces her class to LOGO through structured floor games. They take turns pretending to be a turtle and try to follow commands given by the teacher and the other students.

C. Tell how you would react in the following situations. Would you respond with a naturalistic, an informal, or a structured learning experience?

1. Richard and Diana are playing house. They are setting the table for dinner. They carefully place each place setting in front of each chair.

2. Most of the children in the class seem to have trouble telling squares from rectangles.

3. Sam says, "I have more crayons than you have, George." "No, you don't." "Yes, I do!"

4. Pete is trying to put a round peg in a square hole. He is beginning to look upset.

5. The children need some help in understanding and using time words such as *yesterday*, *today*, and *tomorrow* and *early* and *late*.

6. The children in Mr. Wang's class are discussing the show they must put on for the students in the spring. Some children want to do a show with singing and dancing, others do not. Brent suggests that they vote. The others agree. Derrick and Theresa count the votes. They agree that there are seventeen in favor of a musical show and ten against.

7. One of your students brings in a large crate of oranges, lemons, and grapefruit that his family purchased during a trip to Florida.

8. When you arrive at school you discover that Lollipop the mouse is giving birth.

UNIT 3 Promoting Young Children's Concept Development

OBJECTIVES

After studying this unit, the student should be able to:

- List in order and define the six steps in choosing concept objectives and activities
- Identify definitions of assessment and evaluation
- Discuss the advantages of using the six-step method
- Identify examples of each of the six steps
- Describe the choices the teacher must make after evaluation
- Evaluate whether a teacher uses the six steps

A teacher must know students well to help them learn to their fullest capacities. Objectives and activities must be chosen with care so that the children can move as fast and go as far as possible. The steps for planning concept experiences are the same as those used for any subject. Six questions must be answered (Figure 3–1):

- Where is the child now? **Assess**
- What should he learn next? **Choose objectives**

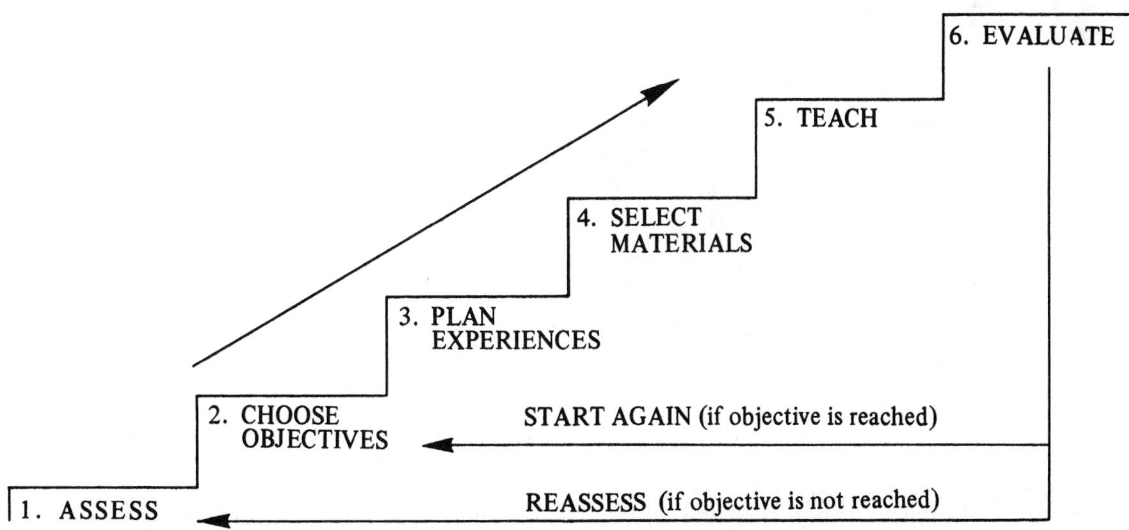

Figure 3–1 What should be taught and how?—FOLLOW THE STEPS.

- What should the child do in order to accomplish these objectives? **Plan experiences**
- Which materials should be used to carry through the plan? **Select materials**
- Do the plan and the materials fit? **Teach** (do the planned experiences with the child)
- Has the child learned what was taught (reached objectives)? **Evaluate**

ASSESSING

Each child should be individually assessed. Two methods for this are used most frequently. Children can be interviewed individually using specific tasks, and they can be observed during their regular activities. The purpose of assessment is to find out what children know and what they can do before instruction is planned. The topic of assessment is covered in detail in unit 4 (Figures 3–2a and 3–2b).

Specific Task Assessment

The following are examples of some specific tasks that can be given to a child:

Figure 3–2a Assessment may be done with an individual interview: "Show me the block that is bigger."

Figure 3–2b A primary child is interviewed individually: "How many cubes did I hide? How many are left?"

- Present the child with a pile of ten counters (buttons, coins, poker chips, or other small things) and say, "Count these for me."
- Show the child two groups of chips; a group of three and a group of six. Ask, "Which group has more chips?"
- Show the child five cardboard dolls, each one a half inch taller than the next. Say: "Which is the tallest?" "Which is the smallest?" "Line them up from the smallest to the tallest."
- Show the child cards with one shape drawn on each card: triangle, circle, rectangle, diamond, and square. Say, "Find the square," or "Tell me the name of each shape."
- Give a six-year-old a simple addition problem. Say, "You have three yellow cars," and Place three yellow cars in front of the child. "Now you buy two blue cars," and place the blue cars next to the yellow cars. Say, "Write in numbers how many yellow cars you have, how many blue cars you have, and how many cars you have altogether."

- Put thirty counting chips in front of a seven-year-old. Say, "Here are thirty chips. Show me how many groups of ten there are."
- Place a pile of counting chips in front of an eight-year-old and ask, "Show me two times three using these chips."

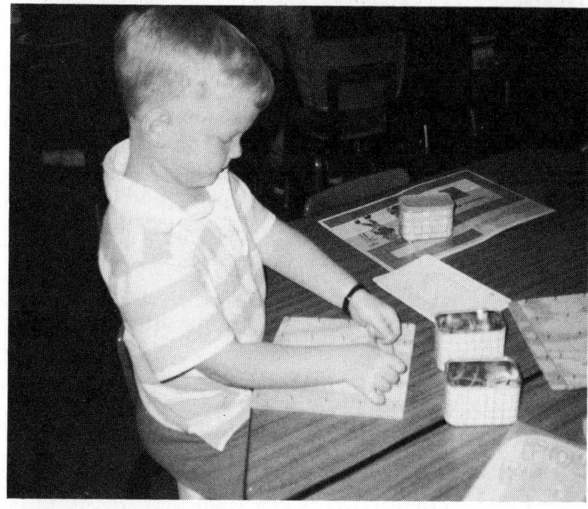

Figure 3–3a and 3–3b Assessment may also be done through observing children as they do their classroom activities.

Assessment by Observation

The following are examples of observations that can be made as children play and/or work:

- Does the one-year-old show an interest in experimenting by pouring things in and out of containers of different sizes?
- Does the two-year-old spend time sorting objects and lining them up in rows?
- Does the three-year-old show an interest in understanding size, age, and time by asking how big he is, how old he is, and when will . . . questions?
- Does the four-year-old set the table correctly? Does he ask for help to write numerals, and does he use them in his play activities?
- Can the five-year-old divide a bag of candy so that each of his friends receives an equal share?
- If there are five children and three chairs, can a six-year-old figure out how many more chairs are needed so everyone will have one?
- If a seven-year-old is supposed to feed the hamster two tablespoons of pellets each day, can he or she decide how much food should be left for the weekend?
- Four eight-year-olds are making booklets. Each booklet requires four pieces of paper. Can the children figure out how many pieces of paper will be needed to make the four booklets?

Through observation the teacher can find out if the child can apply concepts to real life problems and activities. By keeping a record of these observations, the teacher builds up a more complete picture of the child's strengths and weakness (Figures 3–3a and 3–3b).

CHOOSING OBJECTIVES

Once the child's level of knowledge is identified, objectives can be selected. That is, a decision can be made as to what the child should be

able to learn next. For instance, look at the first task example in the previous section. Suppose a five-year-old child counts fifty objects correctly. The objective for this child would be different than the one for another five-year-old who can count only seven objects accurately. The first child does not need any special help with object counting. A child who counts objects at this level at age five can probably figure out how to go beyond fifty alone. The second child might need some help and specific activities with counting objects beyond groups of seven.

Suppose a teacher observes that a two-year-old spends very little time sorting objects and lining them up in rows. The teacher knows that this is an important activity for a child of this age, one most two-year-olds engage in naturally without any special instruction. The objective selected might be that the child would choose to spend five minutes each day sorting and organizing objects. Once the objective is selected the teacher then decides how to go about helping the child reach it.

PLANNING EXPERIENCES

Remember that young children construct concepts through naturalistic activities as they explore the environment. As they grow and develop, they feel the need to organize and understand the world around them. Children have a need to label their experiences and the things they observe. They notice how older children and adults count, use color words, label time, and so on. An instinctive knowledge of math and science concepts develops before an abstract understanding. When planning, it is important for adults to keep in mind the following:

- Naturalistic experiences should be emphasized until the child is into the preoperational period.
- Informal instruction is introduced during the sensorimotor period and increases in frequency during the preoperational period.
- Structured experiences are used sparingly during the sensorimotor and early preoperational periods and are brief and sharply focused.

Abstract experiences can be introduced gradually during the preoperational and transitional periods and increased in frequency as the child reaches concrete operations, but they should always be preceded by concrete experiences. Keep these factors in mind when planning for young children. These points are covered in detail in the section on selecting materials.

Planning involves deciding the best way for each child to accomplish the selected objectives. Will naturalistic, informal, and/or structured experiences be best? Will the child acquire the concept best on his/her own? With a group of children? One to one with an adult? In a small group directed by an adult? Once these questions have been answered, the materials can be chosen. Sections II, III, IV, V, and VI tell how to plan these experiences for the concepts and skills that are acquired during the early years.

SELECTING MATERIALS

Three things must be considered when selecting science and math materials. First, there are some general characteristics of good materials. They should be sturdy, well made, and constructed so that they are safe for children to use independently. They should also be useful for more than one kind of activity and for teaching more than one concept.

Second, the materials must be designed for acquisition of the selected concepts. That is, they must fit the objective(s).

Third, the materials must fit the children's levels of development. As stated, acquiring a concept begins with concrete experiences with real things. For each concept included in the curriculum, materials should be sequenced from concrete to abstract and from three-dimensional (real objects), to two-dimensional (cutouts), to

pictorial, to paper and pencil. Too often, however, the first steps are skipped and children are immersed in paper and pencil/workbook and ditto sheet activities without the prerequisite concrete experiences and before they have developed the perceptual and motor skills necessary to handle a writing implement with ease. Six steps to be followed from concrete materials to paper and pencil are described as follows. Note that step one is the first and last step during the sensorimotor period; during the preoperational period, the children move from step one to step five; and during concrete operations, they move into step six.

- Step 1. Real objects are used for this first step. Children are given time to explore and manipulate many types of objects such as blocks, chips, stones, sticks, and materials such as sand, water, mud, clay, and playdough. Whether instruction is naturalistic, informal, or structured, concrete materials are used.
- Step 2. Real objects are used along with pictorial representations. For example, blocks can be matched with printed or drawn patterns. When cooking, each implement to be used (measuring spoons and cups, bowls, mixing spoons, and so on) can be depicted on a pictorial sequenced recipe chart. Children can draw pictures each day showing the height of their bean sprouts.
- Step 3. Cutouts, which can be motorically manipulated, are introduced. For example, cardboard cutouts of different sizes, colors, and shapes can be sorted. Cutout dogs can be matched with cutout doghouses. Cutout human body parts can be put together to make a whole body. Although the materials have moved into two dimensions, they can still be manipulated.
- Step 4. Pictures are next. Commercially available pictorial materials, teacher-created or magazine pictures, and cutup workbook pages can be used to make card games as well as sequencing, sorting, and matching activities. For example, pictures of people in various occupations might be matched with pictures of their equipment. Pictures of a person at different ages can be sequenced from baby to old age. Groups of objects drawn on a card can be counted and matched with the appropriate numeral.

STOP HERE IF CHILDREN HAVE NOT YET REACHED THE TRANSITION STAGE.

- Step 5. Wipeoff folders are the next step. Activities that normally involve the use of a writing implement are mounted on cardboard or glued into a file folder and laminated or covered with clear plastic. Teacher artwork or workbook pages with good quality artwork can be used. Children can begin by marking answers with a chip or other concrete marker and move on to practice marking answers with wipeoff crayons. For example, a basic graph pattern might be available. Children can make graphs using concrete objects as markers or by making marks with wipeoff crayons. Children can be introduced to using writing implements and printed workbook pages with the opportunity to correct mistakes and experiment with responses.
- Step 6. At this level, paper and pencil activities are introduced. When the teacher observes that the children understand the concept with materials at the first five levels, this level is introduced. If the materials are available, children usually start experimenting when they feel ready.

An example of sequencing materials using the six steps follows. Suppose one of the objectives for children in kindergarten is to compare differences in dimensions. One of the dimensions to be compared is length. Materials can be sequenced as follows:

- Step 1. Real objects. Children explore the properties of Unifix® Blocks and Cuisinaire® Rods. They fit Unifix Blocks together into groups of various lengths. They compare the lengths of the Cuisinaire® rods. They do measurement activities such as comparing how many Unifix® Cubes fit across the short side of the table versus the long side of the table.
- Step 2. Real objects with pictures. The Unifix® Cubes are used to construct rows that match pictured patterns of various lengths. Sticks are used to measure pictured distances from one place to another.
- Step 3. Cutouts. Unifix® and Cuisinaire® cutouts are used to make rows of various lengths. Cutouts of snakes, fences, and so on are compared.
- Step 4. Pictures. Cards with pictures of pencils of different lengths are sorted and matched. A picture is searched for the long and the short path, the dog with long ears, the dog with short ears, the long hose, the short hose, and so on.

STOP HERE IF THE CHILDREN HAVE NOT YET REACHED THE TRANSITION STAGE.

- Step 5. Wipeoff folders. Children use objects or wipeoff crayons to mark drawings of long and short things.
- Step 6. Paper and pencil activities are introduced. Students might draw long and short things. If they are into the transitional stage they might do workbook pages if required by the school.

At the early steps, children might be able to make comparisons of materials with real objects

PERIODS OF DEVELOPMENT	HOW CONCEPTS ARE ACQUIRED Naturalistic	Informal	Structured
Sensorimotor	Real objects Objects and pictures Pictures	Real objects Objects and pictures Pictures	
Preoperational	Real objects Objects and pictures Cutouts Pictures	Real objects Objects and pictures Cutouts Pictures	Real objects Objects and pictures Cutouts Pictures
Transitional	Real objects Objects and pictures Cutouts Pictures	Real objects Objects and pictures Cutouts Pictures Wipeoffs Paper and pencil	Real objects Objects and pictures Cutouts Pictures Wipeoffs
Concrete Operations	Real objects Objects and pictures Cutouts Pictures	Real objects Objects and pictures Cutouts Pictures Wipeoffs Paper and pencil	Real objects Objects and pictures Cutouts Pictures Wipeoffs Paper and pencil

Figure 3–4 Two dimensions of early childhood concept instruction with levels of materials used.

and even with cutouts and picture cards, but they might fail if given just paper and pencil activities. In this case it would be falsely assumed that they do not understand the concept when, in fact, it is the materials that are inappropriate.

The chart in Figure 3–4 depicts the relationship between the cognitive developmental periods—naturalistic, informal, and structured ways of acquiring concepts—and the six levels of materials (Figure 3–4). Each unit of this text has examples of various types of materials. Section VII contains lists and descriptions of many that are excellent.

TEACHING

Once the decision has been made as to what the child should be able to learn next and in what context the concept acquisition will take place, the next step is teaching. Teaching occurs when the planned experiences using the selected materials are put into operation. If the first four steps have been performed with care, the experience should go smoothly. The children will be interested and will learn from the activities because they match their level of development and style of learning. They might even acquire a new concept or skill or extend and expand one already learned.

The time involved in the teaching step might be a few minutes or several weeks, months, or even years depending on the particular concept being acquired and the age and ability of the child. For instance, time sequence is initially learned through naturalistic activity. From infancy, children learn that there is a sequence in their daily routine: sleeping, waking up wet and hungry, crying, being picked up, cleaned, fed, and played with; and sleeping again. In preschool, they learn a daily routine such as coming in, greeting the teacher, hanging up coat, eating breakfast, playing indoors, having a group activity time, snacking, playing outdoors, having a quiet activity, lunch, playing outdoors, napping, having a small group activity time, and going home. Time words are acquired informally as children hear terms such as yesterday, today, tomorrow, o'clock, next, after, and so on. In kindergarten, special events and times are noted on a calendar. Children learn to name the days of the week and months of the year and to sequence the numerals for each of the days. In first grade, they might be given a blank calendar page to fill in the name of the month, the days of the week, and the number for each day. Acquiring the concept of time is a very complex experience and involves many prerequisite concepts that build over many years. Some children will learn at a fast rate, others at a slow pace. One child might learn that there are seven days in a week the first time this idea is introduced; another child might take all year to acquire this information. Some children need a great deal of structured repetition; others learn from naturalistic and informal experiences. Teaching developmentally involves flexible and individualized instruction (Figure 3–5).

Even with careful planning and preparation, an activity might not work well the first time. When this happens, analyze the situation by asking the following questions:

- Was the child interested?
- Was the task too easy or too hard?
- Did the child understand what he was asked to do?
- Were the materials right for the task?
- Were they interesting?
- Is further assessment needed?
- Was the teacher enthusiastic?
- Was it just a "bad" day for the child?

You might try the activity again using the same method and the same materials or with a change in the method and/or materials. In some cases, the child might have to be reassessed to be sure the activity is appropriate for his or her developmental level.

Figure 3–5 Teaching: "This shape is a square. Feel it. Now find another square."

EVALUATING

The sixth step is evaluation. What has the child learned? What does he know and what can he do after the concept experiences have been presented? The assessment questions are asked again. If the child has reached the objective, a new one can be chosen. The steps of planning, choosing materials, teaching, and evaluating are repeated. If the child has not reached the objective, the same activities can be continued or a new method may be tried. For example, a teacher wants a five-year-old to count out the correct number of objects for each of the number symbols from zero to ten. She tries many kinds of objects for the child to count and many kinds of containers in which to place the things he counts, but the child is just not interested. Finally she gives him small banks made from baby food jars and real pennies. The child finds these materials are exciting and goes on to learn the task quickly and with enthusiasm.

Evaluation may be done using formal, structured questions and tasks and specific observations as will be presented in Unit 4. Informal questions and observations of naturalistic experiences can be used for evaluation also. For example, when a child sets the table in the wrong way, it can be seen without formal questioning that he has not learned from instruction. He needs some help. Maybe organizing and placing a whole table setting is more than he can do now. Can he place one item at each place? Does he need to go back to working with a smaller number (such as a table for two or three)? Does he need to work with simpler materials which have more structure (such as pegs in a pegboard)? To look at these more specific skills, the teacher would then return to the assessment step. At this point she would assess not only the child but also the types of experiences and materials she has been using. Sometimes assessment leads the teacher to the right objective but the experience and/or materials chosen are not (as in the example given) the ones that fit the child.

Figure 3–6 Evaluation: "Do both rows have the same number of pennies?"

Frequent and careful evaluation helps both teacher and child avoid frustration. An adult must never take it for granted that any one plan or any one material is the best choice for a specific child. The adult must keep checking to be sure the child is learning what the experience was planned to teach him.

SUMMARY

This unit has described six steps that provide a guide for what to teach and how to teach it. Following these steps can minimize guesswork. The steps are (1) assess, (2) choose objectives, (3) plan experiences, (4) choose materials, (5) teach, and (6) evaluate.

FURTHER READING AND RESOURCES

Baratta-Lorton, M. 1976. *Math their way*. Menlo Park, CA: Addison-Wesley.

Benham, N.B., Hosticka, A., Payne, J.D., and Yeotis, C. 1982. Making concepts in science and mathematics visible and viable in the early childhood curriculum. *School Science and Mathematics*. 82 (1): 45–64.

Charlesworth, R. 1987. *Understanding child development*. Albany, NY: Delmar.

Charlesworth, R. 1986. Kindergarten mathematics. *Dimensions*. 14 (2): 4–6.

Charlesworth, R. 1981. Math experiences for young children. *Dimensions*. 9 (4): 168–169.

Ginsburg, H. 1988. *Children's arithmetic: How they learn it and how you teach it*. Austin, TX: Pro Ed.

Knight, M.E., and Graham, T.L. 1984. What's so hard about teaching science? *Day Care and Early Education*. 12 (2): 14–16.

Richardson, K. 1984. *Developing number concepts using Unifix Cubes*. Menlo Park, CA: Addison-Wesley.

Williams, C.V., and Kamii, C. 1986. How do children learn by handling objects? *Young Children*. 42 (1): 23–26.

SUGGESTED ACTIVITIES

1. Interview two early childhood teachers. Ask them to tell you what kinds of math and science experiences they include in their programs. Find out how they decide what to teach, to whom, and which materials to use. Go through their responses later and try to evaluate whether they use any or all of the steps described in this unit.

2. Go to the library and look through recent issues of professional publications for teachers of young children. For your Activities File make a card for each article you find that gives ideas for using assessment and evaluation to help in choosing objectives, planning, and choosing materials. Summarize on each card the ideas which you feel will be helpful to you.

3. Spend a morning in an early childhood center. Note all the math and science experiences. Go through your notes and evaluate what you observed. What steps did you see? Did you see any incidents where you felt that the teacher needed to evaluate her own teaching or the child's learning? Why?

REVIEW

A. Place the following list of steps for choosing objectives and activities in the correct order.

1. Evaluate
2. Plan experiences
3. Select materials
4. Teach
5. Assess
6. Choose objectives

B. Write the names of the steps that go with the definitions below.

1. Finding out what the child knows and can do *before* you present a concept experience.
2. Finding out what the child knows and can do *after* you present a concept experience.
3. Deciding what the child should learn next.
4. Making decisions as to the best way for each child to arrive at the chosen objectives.
5. Doing the planned experiences using the selected materials.
6. Selecting materials that help the child acquire the chosen objectives.

C. Read the following descriptions and label them with the correct step name.

1. The teacher goes to her Activities File and looks in the section on materials. She takes out the cards labeled *Measurement*, looks through them, and takes out two.
2. The teacher looks over the results of the assessment tasks she has just given to Jimmie. She sees he has difficulty with *more* and *less*. He can tell when the differences are large but not when they are small. There has to be at least four more in one group for him to be able to label the groups correctly. She tries to decide on the next step for Jimmie.
3. The teacher is seated at a table with Cindy. On the table are many different objects. "Cindy, we've been learning about how different things belong together. You put the things from this pile into smaller piles of things which belong together." After Cindy finishes, the teacher asks, "Tell me why these belong together." She points to each pile in turn.
4. It is the beginning of the year. Before planning her math program Mrs. Ramirez questions each child individually to find out exactly which skills and ideas he has and at what stage he is.
5. Mr. Brown is having the children learn how to use measuring cups and spoons in preparation for making a chocolate cake. They are using the cups with sand and water in order to discover the relationships between the different size cups.
6. The next objective for Kate is to be able to correctly organize a three-part-sequence story. Mrs. Raymond goes to her Activity File and looks under *Time and Sequence*.
7. Mr. Wang looks through the library of computer software for programs that can be used to reinforce addition and subtraction facts.

D. Discuss the advantages of following the steps presented in this unit.

E. After evaluation what choices does the teacher have?

F. Read in the following paragraph about Miss

Collins' way of choosing objectives and activities. Analyze and evaluate her approach in terms of what you have learned in this unit.

Miss Collins believes that all children are about the same unless they have an extreme handicap. Her math and science programs are the same from year to year. She assumes that all children entering her class need to learn the same things with the same materials. She believes that she does a fine job and that when her students leave, they are all ready for the next year's work, although she never questions them at the end of the year to find out.

UNIT 4 Assessing the Child's Developmental Level

OBJECTIVES

After studying this unit, the student should be able to:

- Explain how to find the child's level of concept development
- Explain the value of commercial assessment instruments for concept assessment
- Make a developmental assessment task file
- Recognize tasks which might be used at different developmental levels
- Be able to assess the concept development level of young children.

Children's levels of concept development are found by seeing which concept tasks they are able to do. The first question in teaching is "Where is the child now?" To find the answer to this question the teacher assesses. The teacher gives the child tasks to solve (such as those described in Unit 3). She observes what the child does as he solves the problems and records the answers he gives. This information is used to guide the next steps in teaching. The long-term objective for young children is to be sure that they have a strong foundation in basic concepts that will take them through the transition into the concrete operational stage when they begin to deal seriously with abstract symbols in math and independent investigations in science. Following the methods and sequence in this text helps reach this goal and at the same time achieves some further objectives:

- Builds a positive feeling in the child toward math and science
- Builds confidence in the child that he can do math and science activities
- Builds a questioning attitude in response to his curiosity regarding math and science problems

ASSESSMENT METHODS

Observation and interview are assessment methods the teacher uses to find out the child's level of development. Examples of both of these methods were included in Unit 3. More are provided in this unit. Assessment is appropriately done through observations and interviews using teacher-developed assessment tasks. Commercial instruments used for initial screening may also supply useful information but are limited in scope for the everyday assessment needed for planning. Initial screening instruments usually cover a broad range of areas and provide a profile that indicates overall strengths and weaknesses. These strengths and weaknesses can be looked at in more depth by the classroom teacher for information needed to make normal instructional decisions or by a diagnostic specialist (i.e., school psychologist or speech and language therapist) where an initial screening indicates some serious developmental problem. Individually administered screening instruments should be the only type used with young children. Child responses should require the use of concrete materials and/or pictures, verbal answers, or motoric responses such as pointing or rearranging some

Figure 4–1 The teacher learns about the child's concept of her body in space.

objects. Paper and pencil should be used only for assessment of perceptual motor development (i.e., tasks such as name writing, drawing a person, or copying shapes). Booklet-type paper and pencil tests administered to groups or individuals are inappropriate until children are well into concrete operations, can deal with abstract symbols, and have well-developed perceptual motor skills (Figure 4–1).

Observational Assessment

Observation is used to find out how children use concepts during their daily activities. The teacher has in mind the concepts the children should be using. Whenever she sees a concept reflected in a child's activity, she writes down the incident and places it in the child's record folder. This helps her plan future experiences.

Throughout this book, suggestions are made for behaviors which should be observed. The following are examples of behaviors as the teacher would write them down for the child's folder:

- Brad (eighteen months old) dumped all the shape blocks on the rug. He picked out all the circles and stacked them up. Shows he can sort and organize.
- Cindy (four years old) carefully set the table for lunch all by herself. She remembered everything. Understands one-to-one correspondence.
- Chris (three years old) and George (five years old) stood back to back and asked Cindy to check who was taller. Good cooperation—first time Chris has shown an interest in comparing heights (Figure 4–2).
- Mary (five years old), working on her own, put the right number of sticks in juice cans marked with the number symbols zero through twenty. She is ready for something more challenging.
- Trang Fung and Sara (six-year-olds), on their own decided to find out how many cups of water are needed to fill containers of various sizes and shapes. Each time they filled a container, they wrote down its name and how many cups of water it held. They learned how to set up an investigation.
- Derrick and Brent (seven-year-olds) compare their baseball card collections and figure out how many more or less cards each has. They understand the concepts of more and less.
- Ann and Jason (eight-year-olds) argue about which materials will float and sink. They asked their teacher if they could test their theories. They got the water, collected

Figure 4–2 The teacher observes as the children compare heights.

some objects, and set up a chart to record their predictions and then the names of the items that sink and those that float. This demonstrates understanding of how to develop an investigation to solve a problem.

Observational information may also be recorded on a checklist. For example, concepts can be listed and each time the child is observed demonstrating one of the behaviors the date can be put next to that behavior. Soon there will be a profile of the concepts the child demonstrates spontaneously, (Figure 4–3).

Interview Assessment

The individual interview is used to find out specific information in a direct way. The teacher can present a task to the child and observe and record the way the child works on the task and the solution he arrives at for the problem presented by the task (Figure 4–4). The rightness and wrongness of the answer is not as important as how the child arrives at the answer. Often a child starts out on the right track but gets off somewhere in the middle of the problem. For example, Kate (age three) is asked to match four saucers with four cups. This is an example of one-to-one correspondence. She does this task easily. Next she is asked to match five cups with six saucers, "Here are some cups and saucers. Find out if there is a cup for every saucer." She puts a cup on each saucer. Left with an extra saucer, she places it under one of the pairs. She smiles happily. By observing the whole task, the teacher can see that Kate does not feel comfortable with the concept of "one more than." This is normal for a preoperational three-year-old. She finds a way to do away with the problem by putting two saucers under one cup. She understands the idea of matching one to one but cannot have things out of balance. Only by observing the whole task can the teacher see the reason for what appears to be a "wrong" answer to the task.

For another example, Tim, who is just four and a half years old, is given the following task. First he is shown cards with the number symbols zero to six. He is asked to name each number symbol and does so correctly. He is then asked to place the correct number of chips by each number symbol. Tim's responses tell the teacher that he recognizes and can name number symbols but that he cannot yet match the symbols with the right number of chips. He can recognize groups up to four but does not yet have the idea of groups of more than four. He tried to count out five chips and six chips but lost track after four. As will be seen in Unit 6, Tim's behavior is normal for a four-year old.

Finally, Theresa, a second grader, is given several math problems to complete:

5 + 4   9 + 1   6 − 2   3 + 6

CONCEPT ACTIVITY OBSERVATION CHECKLIST

Child's Name ______________________ Birth Date __________

School Year ______________________ Grade/Group __________

Concept Activities *(Concepts and activities are described in the text)*	**Dates Observed**
Selects Math Center	
Selects Science Center	
Selects Cooking Center	
Selects Math Concept Book	
Selects Science Book	
Selects sand or water	
Sets the table correctly	
Counts spontaneously	
Sorts play materials into logical groups	
Uses comparison words (i.e., *bigger, fatter*, etc.)	
Builds with blocks	
Works with part/whole materials	
Demonstrates an understanding of order and sequence	
Points out number symbols in the environment	
Demonstrates curiosity by asking questions, exploring the environment, and making observations	
Uses concept words	

Figure 4–3 Concept observation checklist

Figure 4–4 The child enjoys the individual interview.

She writes:

5+4=9 9+1=10 6−2=8 3+6=9

Her teacher asks her to show her with Unifix Cubes how she did the problems. Theresa counts out five cubes and four cubes for the first problem, nine cubes and one cube for the second, six cubes and two cubes for the third, and three cubes and six cubes for the fourth. He decides Theresa needs help with noticing signs and knowing which type of operation is called for.

If Kate's, Tim's, and Theresa's answers were observed only at the end point and recorded as right or wrong, the crux of their problems would be missed. Only the individual interview offers the opportunity to observe a child solve a problem from start to finish without distractions or interruptions.

An important factor in the one-to-one interview is that it must be done in an accepting manner by the adult. She must value and accept the child's answers whether they are right or wrong from the adult point of view. If possible, the interview should be done in a quiet place where there are no other things which might take the child's attention off the task. The adult should be warm, pleasant, and calm. Let the child know that he is doing well with smiles, words ("Good," "Fine," "You're a good worker," "Keep trying hard"), and gestures (nod of approval, pat on the shoulder).

If persons other than one of the teachers do the assessment interviews, the teacher should be sure that they spend time with the children before the interviews (Figure 4–5). Advise a person doing an interview to sit on a low chair or on the floor next to where the children are playing. Children are usually curious when they see a new person. One may ask, "Who are you? Why are you here?" The children can be told, "I am Ms. X. Someday I am going to give each of you

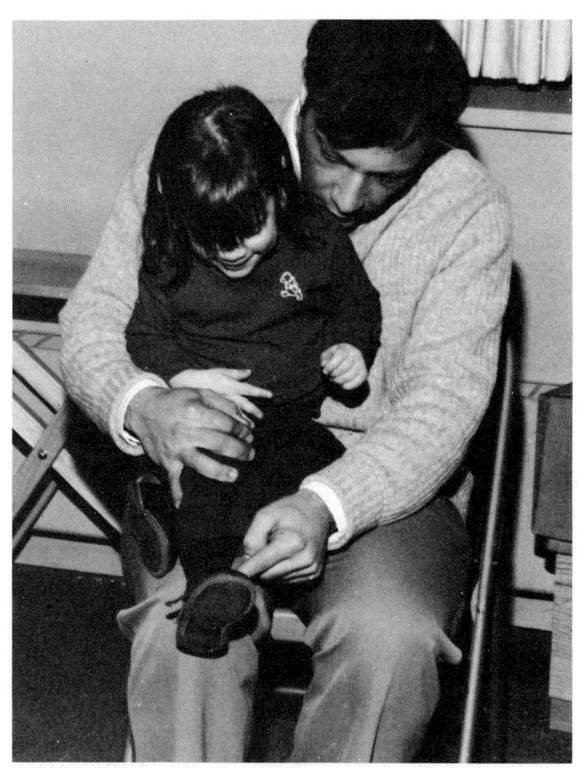

Figure 4–5 A parent volunteer gets acquainted with the children before doing any interviews.

a turn to do some special work with me. It will be a surprise. Today I want to see what you do in school and learn your names." If the interviewer pays attention to the children and shows an interest in them and their activities, they will feel comfortable and free to do their best when the day comes for their assessment interview.

If the teacher does the assessment herself, she also should stress the special nature of the activity for her and each child, "I'm going to spend some time today doing some special work with each of you. Everyone will get a turn."

ASSESSMENT TASK FILE

Each child and each group of children is different. The teacher needs to have on hand questions to fit each age and stage she might meet in individual young children. She also needs to add new tasks as she discovers more about children and their development. A card file of assessment tasks should be set up. Such a file has three advantages:

- The teacher has a personal involvement in creating her own assessment tasks and is more likely to use them, understand them, and value them.
- The file card format makes it easy to add new tasks and revise or remove old ones.
- There is room for the teacher to use her own creativity to add new questions and make materials.

Use the tasks in Appendix A to begin the file. Other tasks can be developed as the student proceeds through the units in this book and through her future career with young children. Directions for each task can be put on five-by-eight-inch plain white file cards. Most of the tasks will require the use of concrete materials and/or pictures. Concrete materials can be items found around the home and center. Pictures can be purchased or cut from magazines and readiness type workbooks and glued on cards.

The basic materials needed are: 5″ × 8″ file card box, 5″ × 8″ unlined file cards, 5″ × 8″ file dividers, black pen, set of colored markers, ruler, scissors, glue, clear Contac® or laminating material, and preschool/kindergarten readiness workbooks with artwork (such as the Golden Readiness Workbooks published by Western Publishing Company).

In Appendix A each assessment task is set up as it would be on a five-by-eight-inch card. Note that on each card what the adult says to the child is always printed in CAPITAL LETTERS so the instructions can be found and read easily. The tasks are set up developmentally from the sensorimotor level (birth to age two) to the preoperational level (ages two to seven) to early concrete operations (ages six to eight). The ages are flexible relative to the stages and are given only to serve as a guide for selecting the first tasks to present to each child.

Each child is at his own level. If the first tasks are too hard, the interviewer should start at a lower level. If the first tasks are quite easy for the child, the interviewer should start at a higher level. Figure 4–6 is a sample recording sheet format that could be used to keep track of each child's progress. Some teachers prefer an individual sheet for each child, others a master sheet for the whole class. The names and numbers of the tasks to be assessed are entered in the first column. Several columns are provided for entering the date and the level of progress (+, accomplished; **v**, needs some help; –, needs a lot of help) for children who need repeated periods of instruction. The column on the right is for comments on the process used by the child that might give some clues as to specific instructional needs.

ASSESSMENT TASKS

The assessment tasks included in Appendix A represent the concepts that must be acquired by young children from birth through the primary grades. Most of the tasks require an indi-

DEVELOPMENTAL TASKS RECORDING SHEET

Child's Name ____________________ Birth Date ____________

School Year ____________ School ____________ Teacher ____________

Grade/Group ____________ Person Doing Assessment ____________

Levels: +, accomplishes; v partial; −, cannot do task

Task	Levels			Comments
	Date	Date	Date	

Comments:

Figure 4–6 Recording sheet for developmental tasks

vidual interview with the child. Some tasks are observational and require recording of activities during playtime. The infant tasks and observations assess the development of the child's growing sensory and motor skills. As was discussed in the first unit, these sensory and motor skills are basic to all later learning.

The assessment tasks are divided into nine developmental levels. *Levels One and Two* are tasks for the child in the sensorimotor stage (Figure

Figure 4–7 Even the toddler works hard on the interview tasks.

4–7). *Levels Three through Five* include tasks of increasing difficulty for the prekindergarten child (Figures 4–8 and 4–9). The *Level Six* tasks are those things which the child can usually do when he enters kindergarten between the ages of five and six (Figure 4–10). This is the level he is growing toward during his prekindergarten years. Some children will be able to accomplish all these tasks by age five; others not until six or over. *Level Seven* summarizes the math words which are usually a part of the child's natural speech by age six. *Level Eight* is included as an assessment for advanced prekindergartners and for children in centers that have a kindergarten program. The child about to enter first grade should be able to accomplish the tasks at *Level Six* and *Level Eight*. He should also be using most of the concept words correctly. *Level Nine* includes tasks to be accomplished during the primary grades.

EXAMPLE OF AN INDIVIDUAL INTERVIEW

The following is a part of the *Level Five* assessment interview as given to Bob (four and one-half years old). A corner of the storage room has been made into an assessment center. Mrs.

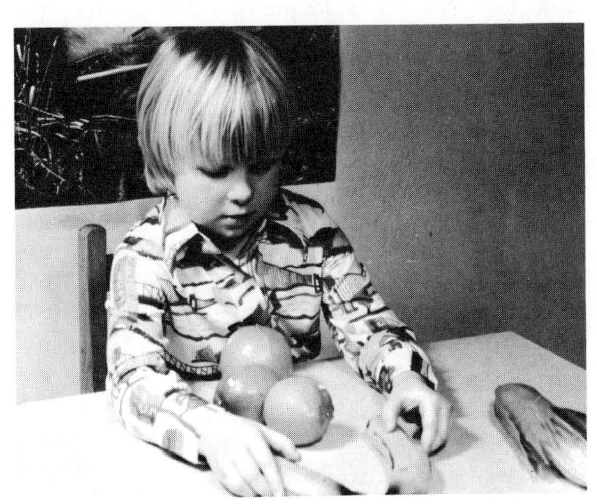

Figure 4–8 For a Free Sort the three-year-old child is given a collection of objects and told, "Put the things together that belong together."

Figure 4–9 The four-year-old finds that the Free Sort is a challenging task.

Figure 4–10 The five-year-old tries to find out if there are the same number of players on each team.

Figure 4–11 "This one is the triangle."

Ramirez comes in with Bob, "You sit there and I'll sit here, Bob. We have some important things to do." They both sit down at a low table and Mrs. Ramirez begins.

An interview does not have to include any special number of tasks. For the preoperational child, the teacher can begin with matching and proceed through the ideas and skills one at a time so that each interview can be quite short if necessary.

If the person doing the interviewing has the time for longer sessions and the children are able to work for a longer period of time, the following can serve as suggested maximum amounts of time (Figures 4–11, 4–12, 4–13)

- Fifteen to twenty minutes for two-year-olds
- Thirty minutes for three-year-olds
- Forty-five minutes for four-year-olds
- Up to an hour with five-year-olds and older

Mrs. Ramirez:	Bob's Response:
HOW OLD ARE YOU?	"I'm four." (He holds up four fingers.)
COUNT TO TEN FOR ME, BOB. (Mrs. Ramirez nods her head up and down.)	"One, two, three, four, five, six, seven, eight, nine, ten,. . .I can go some more. Eleven, twelve, thirteen, twenty!"
HERE ARE SOME BLOCKS. HOW MANY ARE THERE? (She puts out ten blocks.)	(He points, saying) "One, two, three, four five, six, seven, eight, nine, ten, eleven, twelve." (He points to some more than once.)
GOOD, BOB. NOW COUNT THESE. (Five blocks)	(He counts, pushing each one he counts to the left) "One, two, three, four, five."

She puts the blocks out of sight and brings up five plastic horses and five plastic cowboys. FIND OUT IF EACH COWBOY HAS A HORSE.

(Bob looks over the horses and cowboys. He lines up the horses in a row and then puts a cowboy on each.) "Yes, there are enough."

FINE, BOB. (She puts the cowboys and horses away. She takes out some inch cube blocks. She puts out two piles of blocks: five yellow and two orange.)

DOES ONE GROUP HAVE MORE?

"Yes." He points to the yellow.

GOOD. She puts out four blue and three green.

DOES ONE GROUP HAVE LESS?

He points to the green blocks.

WELL DONE.

She takes out five cutouts of bears of five different sizes. FIND THE BIGGEST BEAR.

"Here it is." (He picks the right one.)

FIND THE SMALLEST BEAR.

(He points to the smallest.)

PUT ALL THE BEARS IN A ROW FROM BIGGEST TO SMALLEST.

(Bobby works slowly and carefully.) "All done." (Two of the middle bears are reversed.)

(Mrs. Ramirez smiles.)

GOOD FOR YOU, BOB. YOU'RE A HARD WORKER.

Figure 4–12 "Put these in a row from the skinniest to the fattest."

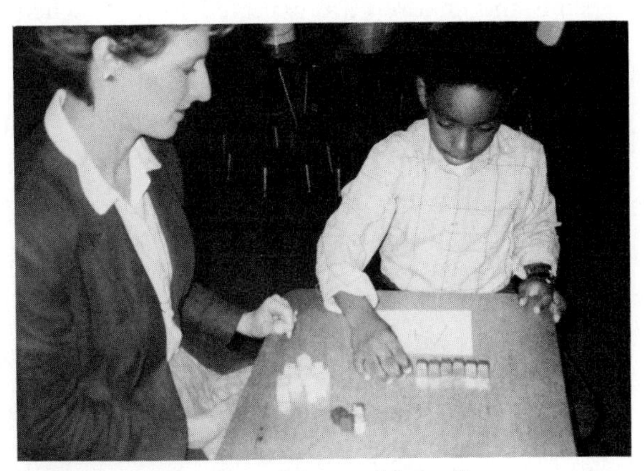

Figure 4–13 "Do this problem for me using these Unifix cubes." A primary child enjoys the one-to-one interview.

SUMMARY

There are two ways children can be assessed to find their developmental level. They can be observed and they can be interviewed.

Observation is most useful when looking at how children use concepts in their everyday activities. The interview with one child at a time gives the teacher an opportunity to look at very specific ideas and skills.

Guidelines are given for doing an interview. There is a summary of the nine levels of developmental tasks which are included in the Appendix. A sample of part of an interview shows how the exchange between interviewer and child might progress.

FURTHER READING AND RESOURCES

Meisels, S.J. 1987. Uses and abuses of developmental screening and school readiness tests. *Young Children*. 42 (2): 4–6, 68–74.

Richardson, K. 1984. *Developing number concepts using unifix cubes*. Menlo Park, CA: Addison-Wesley.

SUGGESTED ACTIVITIES

1. Interview the kindergarten supervisor (or if your system starts at grade one, the primary supervisor) in your local public school system. Find out what kinds of math knowledge and skills the children are expected to have when they enter school (kindergarten or grade one). Compare their list with the tasks in *Level Six* (and *Seven* if the children start at grade one): Are they the same? What differences are there?
2. Contact the office which supervises school psychological services in your local school district. Find out if there is preschool testing (sometimes called *screening*) for children entering the system for the first time. If there is a preschool testing program, find out what kinds of tasks are used. Compare the concept questions used by the school district with the ones suggested in the Appendix of this text. Are the questions about the same? Do they cover the same topics? Would the child who succeeds at *Levels Six*, *Seven* (and *Eight* for grade one entrance) be ready for the kindergarten (or first grade) programs of your system? What changes might you have to make in the tasks from the text?
3. Find at least two children at two different age levels that you could have permission to assess. Make up the assessment cards and gather the materials needed. Try out the tasks with the children. Discuss with the other students in your class any problems you discovered. As a group, work out improvements in the method and list suggestions for making the assessment run smoothly.
4. Based on the results recorded for the assessments done in the preceding activity, write out some teaching objectives for the children in the areas where they had the most difficulty. Look ahead in the text and find the units which tell how to teach in the areas assessed. Pick out some activities that fit your objectives.
5. Invite a kindergarten and/or a primary teacher to visit your class and describe how they assess and evaluate the levels of their students in math and science.
6. Go to the library. Find five articles in periodicals which discuss assessment and/or evaluation of young children. List the main ideas you read. What did you learn that will help you in the future?

REVIEW

A. Explain how to find a child's level of concept development. Why is it important to make assessment the first step in teaching? what is the value of commercial assessment instruments?

B. Read incidents 1–4 which follow. What is being done wrong in each situation? What should be done?

1. Mr. Brown is interviewing a child in the teachers' lounge. Other teachers are coming in and going out. There is a lot of talking among the adults. The child keeps looking away from the assessment materials to see who is in the room.

2. Ms. Collins has a volunteer parent come in to do some of her assessment tasks. She tells the parent, "Just go ahead and take the children in order from the attendance list. The faster the better."

3. Mr. Flores is interviewing Jimmy. Mr. Flores puts a marble and a ping-pong ball on the table.

WHAT IS THIS? (Points to marble)	"A marble."
WHAT IS THIS? (Points to ball)	"A ping-pong ball."
IS ONE HEAVIER THAN THE OTHER OR ARE THEY BOTH THE SAME?	"The ping pong ball is heavier because it is bigger." (Jimmy looks closely at Mr. F.'s face which has a very serious expression.)

4. Mrs. Raymond is interviewing Cindy. Mrs. R places a nickel, a dime, a penny, and a dollar bill on the table.

TELL ME THE NAME OF EACH OF THESE.	"Penny." (Points to penny.)
OKAY, RIGHT.	
WRONG, YOU HAVE THOSE MIXED UP.	"Dime." (Points to nickel.) "Nickel." (Points to dime.)

C. Decide at which developmental level each of the following tasks would be placed. Use the tasks in the Appendix as a guide.

1. The child is shown a clock: WHAT IS THIS? WHAT DOES IT TELL US?

2. Between the child and the adult on the table are two rows of poker chips matched one to one. The adult says, NOW, WATCH WHAT I DO. She moves the chips in the row nearest her so they are touching each other in the middle of the row rather than being an inch apart.

3. The adult watches as the child places plastic beads in a coffee can and then dumps them out again.

4. The adult notes that the child has just begun to line his blocks and toy cars up in rows.

5. The child is shown drawings of a triangle, a square, and a circle. He is asked to point to each as it is named.

6. The child holds a rattle for two or three seconds.

7. The child is asked to use Unfix® Cubes to show how fifteen pieces of candy can be divided up equally among five children.

8. The adult notes that several children are able to set up their own graphs for depicting data.

D. The teacher is doing some concept assessments. What should she do in each of the following situations?

1. She has a group of infants ranging in age from three months to twelve months. Another teacher tells her that these children are too young to learn math. What could the first teacher say to show that concept learning begins in infancy?

2. She is starting out with a group of children ages four to five years. She knows nothing about their level of concept development.

3. The teacher is giving the *Level Three* tasks to Pete who is two and one-half years old. She finds that he can give his age correctly and hold up two fingers. He can count five objects and can state which of two dolls is the big one. He can follow directions and put a block in, on, under, next to, and over a cup. The teacher has noticed that he makes trains when playing with blocks and puts people on each block to ride the train.

4. Mary is four years old. When asked how old she is, she holds up three fingers and says, "I'm six." When asked to count to ten she says, "One, two, six, four, ten." The teacher gives her ten blocks to count. Mary points in a disorganized way as she says, "One, two, five, eight, six!"

5. The teacher gives five-year-old Bobby two groups of chips: ten red and ten blue. FIND OUT IF THERE IS THE SAME NUMBER IN EACH GROUP. Without hesitation Bobby counts each group, "They are the same—ten in each group. I can prove it to you." He stacks a red on each blue and quickly counts the ten stacks correctly. "That was easy."

E. The best review of this unit is to do the third and fourth activities in the suggested activities section. Begin to make an assessment kit and find out what it is like to use the materials with children.

UNIT 5 The Basics of Science

OBJECTIVES

After studying this unit, the student should be able to:

- Define the relative importance of science content, processes, and attitudes in teaching young children
- Explain why science should be taught to young children
- Identify the major areas of science instruction
- List the science attitudes and process skills appropriate to preschool and primary grades
- Select appropriate science topics for teaching science to young children

SCIENCE AND WHY WE TEACH IT TO YOUNG CHILDREN

When people think of science they generally first think of the content of science. Science is often viewed as an encyclopedia of discoveries and technological achievements. Formal training in science classes often promotes this view by requiring memorization of seemingly endless science concepts. Science has been compiling literally millions of discoveries, facts, and data over thousands of years. We are now living in the age that is sometimes described as the "Knowledge Explosion." Consider the fact that the amount of scientific information created between the years 1900 and 1950 equals that which was learned from the beginning of recorded history until the year 1900. Since 1950 the rate of production of scientific information has increased even further. Some now estimate that the total amount of scientific information produced now doubles every two to five years.

If you tried to teach all that has been learned in science in preschool and continued daily straight through high school, you would make only a small dent in the body of knowledge. It is simply impossible to learn everything. Despite this, however, far too many teachers approach the task of teaching children science as if it were a body of information that anyone can memorize. The fact is that it is nearly impossible to predict what specific information taught to primary age students today will be of use to them as they pursue a career through the next century (Figure 5–1).

It is entirely possible that today's body of science knowledge will change before a child graduates from high school. Scientists are constantly looking at data in different ways and coming to new conclusions. Thus, it cannot be predicted with any certainty which facts will be the most important for students to learn for life in the twenty-first century. What is known is that people in the next century will have to face new problems that they will attempt to solve. Life, in a sense, is a series of problems. The people who are most successful in future decades will be those who are best equipped to solve the problems they encounter.

This discussion is intended to put the nature of science in perspective. Science in pre-

Figure 5–1 Exploring science

school through college should be viewed more as a verb than a noun. It is not so much a body of knowledge as it is a way of thinking and acting. Science is a way of trying to discover the nature of things. The attitudes and skills that have moved science forward through the centuries are the same attitudes and skills that enable individuals to solve the problems that they encounter in everyday life.

An approach to science teaching that emphasizes the development of thinking and the open-minded attitudes of science would seem to be most appropriate to the instruction of young children. This unit covers processes, attitudes, content, and the importance of science in language arts and reading.

TEACHING SCIENCE PROCESSES

Children discover the content of science by applying the processes of science. This can be done through science activities, class discussions, reading, and a variety of other teaching strategies. These are the thinking skills necessary to learn science.

Process Skills

1. *Observing*. Using the senses to gather information.
2. *Comparing*. Looking at similarities and differences in real objects. In the primary grades, students begin to compare and contrast ideas and concepts.
3. *Classifying*. Grouping and sorting according to categories, such as size, shape, color, use, and so on.
4. *Measuring*. Quantitative descriptions made by an observer either directly through observation or indirectly with an instrument.
5. *Communication*. Communicating ideas, directions, and descriptions orally or in written form such as pictures, maps, graphs, or journals so others can understand what you mean.
6. *Inferring*. Based on observations but suggests more meaning about a situation than can be directly observed. When children infer, they recognize patterns and expect these patterns to recur under similar circumstances.
7. *Predicting*. Making reasonable guesses or estimations based on observations or data.
8. *Hypothesizing*. Formal conditional statements about a phenomenon being investigated. The typical form of a hypothesis is that: if . . . then . . . . For a primary child an example would be, if water is put in the freezer overnight then it freezes.
9. *Defining and controlling variables*. Determining which variables in an investigation should be studied or should be controlled to conduct a controlled experiment. For example, when we find out if a plant grows in the dark, we must also grow a plant in the light.

Figure 5–2 Process skills

Process skills are those that allow students to process new information through concrete ex-

Figure 5–3 Classifying keys

periences (Figure 5–2). The skills most appropriate for preschool and primary students are *observing*, *comparing*, *measuring*, *classifying*, and *communicating* (Figure 5–3). Sharpening these skills is essential for coping with daily life as well as for future study in science and mathematics. As students move through the primary grades, mastery of these skills will enable them to perform intermediate process skills that include *gathering* and *organizing* information, *inferring*, and *predicting*. If students have a strong base of primary and intermediate process skills, they will be prepared by the time they reach the intermediate grades to apply those skills to the more sophisticated and abstract skills, such as *forming hypotheses* and *separating variables*, which are required in experimentation.

Grade level suggestions for introducing specific science process skills are given as a general guide for their appropriate use. Since students

Figure 5–4 Graphing our family

vary greatly in experience and intellectual development, you may find that your early childhood students are ready to explore higher level process skills sooner. In such cases you should feel free to stretch their abilities by encouraging them to work with more advanced science process skills. For example, four- and five-year-olds can begin with simple versions of intermediate process skills, such as making a guess about a physical change (What will happen when the butter is heated?) as a first step toward predicting. They can gather and organize simple data (such as counting the days until the chicks hatch) and make simple graphs like those described in unit 20 of this text (Figure 5–4).

PROCESS SKILLS

Knowledge and concepts are developed through the use of process skills. It is with these skills that individuals think through and study problems.

Observing

Observation is the ability to describe something using the five senses (sight, smell, sound, touch, taste). It is a fundamental skill upon which all other scientific skills are based.

Teaching strategies that reinforce observation skills require children to watch carefully to note specific phenomena that they might ordinarily overlook. For example, when Mr. Wang's class observes an aquarium, he guides them by asking, "Which fish seems to spend the most time on the bottom of the tank? Do the fish seem to react to things like light or shadow or an object in their swimming path?"

Storybooks and informational books can also encourage the use of process skills. In the popular book *Bubbles*, Bernie Zubrowski asks children to note the colors they see and to describe the different patterns that bubbles make when viewed from several angles.

Comparing

As children develop skills in observation, they will naturally begin to compare and contrast and to identify likenesses and differences. This procedure, which sharpens their observation skills, and is the first step toward classifying.

Teachers can encourage children to find likenesses and differences throughout the school day. A good example of this strategy can be seen when, after a walk through a field, Mrs. Red Fox asks her first graders, "Do you have any seeds sticking to your clothes? Do any of them have wings to carry them through the air?"

Classifying

Classifying begins when children group and sort real objects. To group, children need to compare objects and develop subsets. A *subset* is a group that shares a common characteristic unique to that group. For example, the jar may be full of buttons, but children are likely to begin classifying by sorting the buttons into subgroups of red buttons, yellow buttons, blue buttons, and other colors.

Mrs. Jones has her kindergarten children collect many kinds of leaves. They place individual leaves between two squares of wax paper. Mrs. Jones covers the wax paper squares with a piece of smooth cloth and presses the cloth firmly with a warm steam iron. The leaf is now sealed in and will remain preserved for the rest of the year.

Once the leaves are prepared, the children choose a leaf to examine, draw, and describe. They carefully observe and compare leaves to discover each leaf's unique characteristics. Then, the children place the leaves into subgroups of common characteristics.

Measuring

Measuring is the skill of quantifying observations. This can involve numbers, distances, time, volumes, and temperature, which may or may not be quantified with standard units. Nonstandard units are involved when children say that they have used two "shakes" of salt while cooking or a "handful" of rice and a "couple" of beans when creating their collage.

Measuring involves placing objects in order, such as an ordered sequence (seriation), or it can be ordering according to length or shade. Children can also invent units of measure. For example, when given beans to measure objects, Vanessa may say, "The book is twelve beans long," but Ann finds that the same book is eleven beans long. Activities such as this help children see a need for a standard unit of measure—like an inch.

Communicating

In science, communicating refers to the skill of describing a phenomenon. A child communicates ideas, directions, and descriptions orally or in written form, such as in pictures, dioramas, maps, graphs, journals, and reports. Communication requires that information be collected, arranged, and presented in a way that helps others understand your meaning.

Teachers encourage communication when they ask children to keep logs, draw diagrams or graphs, or otherwise record an experience they have observed. Children respond well to tasks such as recording daily weather by writing down the date, time of day, and drawing pictures of the weather that day. They will enjoy answering questions such as, "What was the temperature on Tuesday? Was the sun out on Wednesday?"

Inferring

When children infer, they make a series of observations, categorize them, then try to give them some meaning. An inference is arrived at indirectly (not directly, like a simple observation). For example, you look out the window and see the leaves moving on the trees. You infer that the wind is blowing. You have not experienced the wind directly, but based on your observations you know that the wind is blowing. In this case, your inference can be tested simply by walking outside.

Predicting

Predicting is closely related to inferring. When you predict, you say what you expect to happen. You make a reasonable guess or estimation based on observations of data. Children enjoy simple prediction questions. After reading *Science in a Vacant Lot* by Seymour Simon, children can count the number of seeds in a seed package and then predict how many of the seeds will grow into plants. And as they prepare to keep a record of how two plants grow (one has been planted in topsoil, the other in subsoil), they are asked, "Which plant do you think will grow better?"

HYPOTHESIZING AND CONTROLLING VARIABLES = *EXPERIMENT*

To be called an experiment, an investigation must contain a hypothesis and control variables. A hypothesis is a more formal operation than the investigative questions that young children explore in the preschool and primary grades. A hypothesis is a statement of a relationship that might exist between two variables. A typical form of a hypothesis is: if . . . , then . . . .

In a formal experiment, variables are defined and controlled. Although experiments can be attempted with primary age children, experimental investigations are most appropriate in the middle grades.

DEVELOPING SCIENTIFIC ATTITUDES

In some ways, attitudes toward a subject or activity can be as important as the subject itself. Some examples are: individuals who know that cigarette smoking may kill but continue to smoke, or people who know that wearing a seatbelt greatly improves their chances of surviving an accident and preventing injury but choose not to wear them. The same is true with scientific attitudes.

The scientific attitudes of curiosity, skepticism, positive self-image, and positive approach to failure are highlighted and other relevant attitudes are listed in Figure 5–5.

Curiosity

Preschool and primary students are obviously not mentally developed to a point where they can think consciously about forming attitudes for systematically pursuing problems, but they can practice behaviors that will create lifelong habits that reflect scientific attitudes.

Curiosity	Checking evidence
Withholding judgment	Positive approach to failure
Skepticism	Positive self-image
Objectivity	Willingness to change
Open-mindedness	Positive attitude toward change
Avoiding dogmatism	Avoiding superstitions
Avoiding gullibility	Integrity
Observing carefully	Humility
Making careful conclusions	

Figure 5–5 Scientific attitudes

Curiosity is thought to be one of the most valuable attitudes that can be possessed by anyone. It takes a curious individual to look at something from a new perspective, question something long believed to be true, or look more carefully at an exception to the rule. This approach that is basic to science is natural to young children. They use all their senses and energies to "find out" about the world around them. Often, years of formalized experiences in school, which allow little time for exploration and questioning, squelch this valuable characteristic. Educational experiences that utilize firsthand inquiry experiences like learning cycle make use of a child's natural curiosity rather than trying to suppress it (Figure 5–6).

Figure 5–6 "Look what I made!"

Skepticism

Do you believe everything that you see? Are you skeptical about some things that you hear? Good! This attitude reflects the healthy skepticism required by both science and the child's environment. Children need to be encouraged to question, wonder, ask "why," and be cautious about accepting things at face value. Experiences designed around direct observation of phenomena and gathering data naturally encourage children to explore new situations in an objective and open-minded fashion. This type of experience can do much toward developing confidence and a healthy skepticism.

Positive Approach to Failure and Self-Image

A positive approach to failure and a positive self-image are closely related attitudes. Students need the opportunity to ask their own questions and seek their own solutions to problems. At times this may mean that they will pursue dead ends, but often much more is learned in the pursuit than in the correct answer. If children are conditioned to look to adult authority figures to identify and solve problems, they will have a difficult time approaching new problems both as students and as adults.

In the last twenty years, some educators believed that children should not be allowed to experience failure. Educational situations were structured so that every child could be successful nearly all the time. It was reasoned that the experience of failure would discourage students from future study. In the field of science, however, it is as important to find out what does not work as it is to find out what does. In fact, real growth in science tends to happen when solutions do not fit what was predicted. Although students should not be constantly confronted with frustrating learning situations, a positive attitude toward failure may better serve them in developing problem-solving skills. After all, in much of science inquiry, there are no "right" or "wrong" answers.

The remaining science attitudes, *willingness to change*, *positive attitude toward change*, *withholding judgment*, *avoiding superstitions*, *integrity*, and *humility*, are additional important attitudes both to science and functioning as a successful adult. These can be encouraged in science teaching both through the teacher's exhibiting these behaviors and acknowledging students when they demonstrate them. All of these attitudes that support the enterprise of science are also quite valuable tools for young students in approaching life's inevitable problems.

SCIENCE AND THE DEVELOPMENT OF LITERACY

An often-heard question is, "Why take time to teach science to young children?" Many teachers believe that they have too much to do in a day and they cannot afford to take the time to teach science. The answer to why science should be taught is, "You cannot afford NOT to teach science." Piaget's theory leaves no question as to the importance of learning through activity. The Council for Basic Education reports that there is impressive evidence that hands-on science programs aid in the development of language and reading skills. Some evidence indicates that achievement scores increase as a result of such programs. This statement is supported by many researchers. The following are possible explanations for this improvement:

In its early stages, literacy can be supported by giving children an opportunity to manipulate familiar and unfamiliar objects. During science experiences children use the thinking skills of science to match, discriminate, sequence, describe, and classify objects. These perceptual skills are among those needed for reading and writing (Figure 5–7). A child who is able to make fine discriminations between objects will be prepared to discriminate between letters and words of the alphabet. As children develop conventional reading and writing skills they can apply their knowledge to facilitate their explorations in science by reading background material and recording hypotheses, observations, and interactions.

What better way to allow for children to develop communication skills than to participate in the "action plus talk" of science! Depending on the age level, children may want to communicate what they are doing to the teacher and other students. They may even start talking about themselves. Communication by talking, drawing, painting, modeling, constructing, drama, puppets, and writing should be encouraged. These are natural communication outcomes of hands-on science.

Reading and listening to stories about the world is difficult when you do not have a base of experience. If story time features a book about a hamster named Charlie, it might be difficult to understand what is happening if you have not seen a hamster. However, the communication gap is bridged if children have knowledge about small animals. Once a child has contact with the object represented by the written word, meaning can be developed. Words do not make a lot of sense when you do not have the background experience to understand what you read (Figure 5–8).

The relationship of language developed in the context of the direct experiences of science is explored in Unit 7 of this text. Ideas for working with experience charts, tactile sensations, listen-

EXAMPLES OF PROBLEM-SOLVING SKILLS IN SCIENCE	CORRESPONDING READING SKILLS
Observing	Discriminating shapes Discriminating sounds Discriminating syllables and accents
Identifying	Recognizing letters Recognizing words Recognizing common prefixes Recognizing common suffixes Recognizing common base words Naming objects, events, and people
Describing	Isolating important characteristics Enumerating characteristics Using appropriate terminology Using synonyms
Classifying	Comparing characteristics Contrasting characteristics Ordering, sequencing Arranging ideas Considering multiple factors
Designing investigations	Asking questions Looking for potenial relationships Following organized procedures Reviewing prior studies Developing outlines
Collecting data	Taking notes Surveying reference materials Using several parts of a book Recording data in an orderly fashion Developing precision and accuracy
Interpreting data	Recognizing cause and effect relationships Organizing facts Summarizing new information Varying rate of reading Inductive and deductive thinking
Communicating results	Using graphic aids Logically arranging information Sequencing ideas Knowledge of technical vocabulary Illuminating significant factors Describing with clarity
Formulating conclusions	Generalizing Analyzing critically Evaluating information Recognizing main ideas and concepts Establishing relationships Applying information to other situations

Figure 5–7 Science and reading connection. *(Reprinted with permission from the National Science Foundation*, Carter, G.S. and Simpson, R.D. Science and Reading: A Basic Duo. *The Science Teacher*, March 1989, p. 20.)

Figure 5–8 Children prepare for animal visits

ing, writing, and introducing words are included in curriculum integration.

Science also provides various opportunities to determine cause-and-effect-relationships. A sense of self-esteem and control over their lives develops when children discover cause-and-effect relationships and when they learn to influence the outcome of events. For example, a child can experience a causal relationship by deciding whether to add plant cover to an aquarium. Predicting the most probable outcome of actions gives children a sense of control, which is identified by Mary Budd Rowe as "fate control." She found that problem-solving behaviors seem to differ according to how people rate on fate control measures. Children scoring high on these measures performed better at solving problems.

Keep in mind that the child who is academically advanced in math and reading is not always the first to solve a problem or assemble the most interesting collection. If sufficient time to work with materials is provided, children with poor language development may exhibit good reasoning. It is a mistake to correlate language skills with mental ability.

APPROPRIATE SCIENCE CONTENT

The science content for preschool and primary education is not greatly different from that of any other elementary grade level. As already mentioned, the way the science is taught is probably far more important than the science content itself. The four main areas of science emphasis that are common in the primary grades are life science, physical science, earth science and health science. Ideally, each of the four main areas should be given a balanced coverage.

Life Science

Science teaching at this level is traditionally dominated by life science experiences. This is not because it is most appropriate but rather because of tradition. Teaching at early elementary grades has its roots in the Nature Study and Garden School movements of the first half of the twentieth century. Many programs and materials for young children concentrate much time on life science to the exclusion of other science content. Although life science should not be the entire curriculum for young children, it can be an important part of the curriculum. Children are natural observers and enjoy finding out about the living world around them (Figure 5–9).

Life science investigations lend themselves quite readily to simple observations, explorations, and classifications. As with all science content at this level, hands-on experiences are essential to development of relevant concepts, skills, and attitudes. The areas of content typically covered with young children are plants, animals, and ecology. These experiences should build a foundation for students' understanding of environmental problems and solutions in higher grade levels and in adult life. Intelligent decision mak-

Figure 5–9 Chris examines a worm

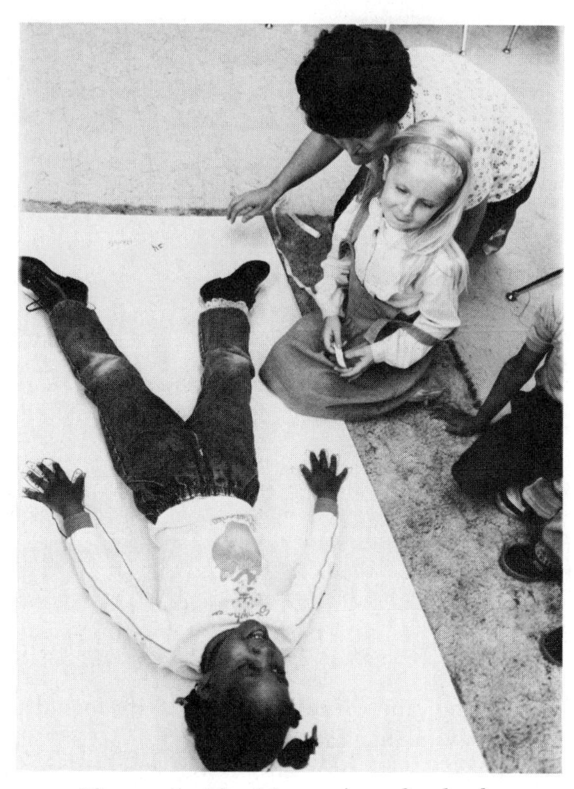

Figure 5–10 Measuring the body

ing regarding the interaction of science, technology, and the environment may well be a critical factor for survival in the next century.

Health Science

The study of health and the human body is receiving increased emphasis in elementary education. Recent concerns about problems of drug abuse, communicable diseases, and the relationship of nutrition and health have given rise to education in both factual information and "refusal skills." These learnings will help children take action to prevent the spread of disease, maintain a healthy body, and ask the types of questions that will ensure informed decisions.

Young children are curious about their bodies and are eager to learn more about themselves. They will enjoy exploring body parts and their relationships, body systems, foods, and nutrition. Misconceptions and worries that children have can be clarified by learning "all about me" in a variety of hands-on experiences (Figure 5–10).

Physical Science

Young children enjoy pushing on levers, making bulbs light, working with magnets, using a string-and-can telephone, and changing matter. This is the study of physical science—forces, motion, energy, and machines. Teachers will enjoy watching a child assemble an assortment of blocks, wheels, and axles into a vehicle that really works. Physical science activities are guaran-

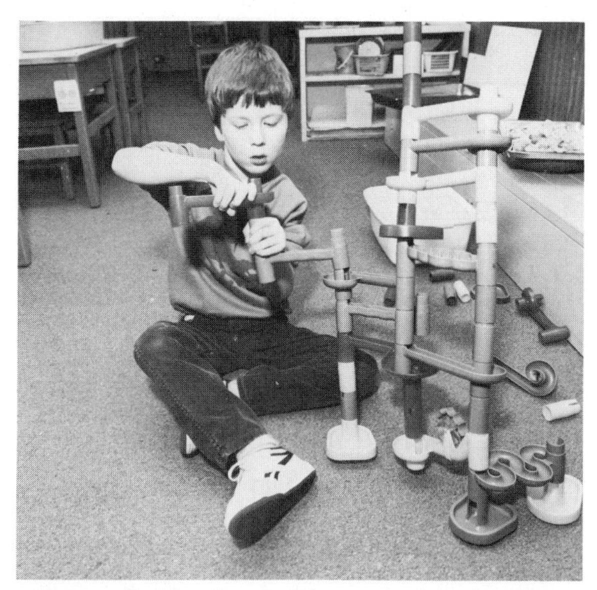

Figure 5–11 Controlling speed and action

teed to make a child's face light up or ask, "How did you do that?" (Figure 5–11).

Sometimes the content of this area is overlooked, which is unfortunate because physical science lends itself quite well to the needs of young children. One advantage of physical science activities is that they are more foolproof than many other activities. For example, if a young child is investigating the growth of plants, many things can go wrong that will destroy the investigation. Plants can die, get moldy, or take so long to give the desired effect that the children lose interest. Physical science usually "happens" more quickly. If something damages the investigation, it can always be repeated in a matter of minutes. Repeatability of activities is a significant advantage in developing a process orientation to science.

Keep in mind that children are growing up in a technological world. They interact daily with technology. It is likely that future life-styles and job opportunities may depend on skills related to the realm of physical science.

Earth Science

The study of earth science also allows many opportunities to help children develop process skills. Children are eager to learn about weather and how soil is formed. Air, land, water, rocks, and the sun, moon, and stars are all a part of earth science. Although these topics are attention grabbers, the teacher of young children must be certain to make the phenomena concrete for them to be effective. Hands-on experiences need not be difficult. Try making fossil cookies, weather and temperature charts, parachutes, and rock and cloud observation to teach a concept. Unit 37 gives a number of examples of how the earth sciences can be made appropriate for young children.

SUMMARY

Our major goal in science education is to develop scientifically literate people who can think critically. In order to teach science to tomorrow's citizens, process skills and attitudes must be established as major components of any science content lesson. Facts alone will not be sufficient for children who are born into a technological world. Children interact daily with science. Their toasters pop, their can openers whir, and their televisions, VCRs, and computers are commonplace. Preparation to live in a changing world as productive individuals should begin early in a child's life.

The manipulation of science materials, whether initiated by the child and/or the teacher, creates opportunities for language and literacy development. Hands-on experiences are essential if the child is to receive the maximum benefits from science instruction. The intent of this book is to tell how to plan and teach these kinds of experiences for young children.

FURTHER READING AND RESOURCES

Forman, G. E., and Hill, F. 1984. *Constructive play: Applying Piaget in the classroom.* Menlo Park, Calif.: Addison-Wesley.

Forman, G. E., and Kuschner, D.S. 1983. *The child's construction of knowledge.* Washington, D.C.: National Association for the Education of Young Children.

Harlan, J. 1984. *Science experiences for the early childhood years.* 4th ed. Columbus, Ohio: Merrill.

Kamii, C., and Devries, R. 1978. *Physical knowledge in preschool education.* Englewood Cliffs, N.J.: Prentice-Hall.

Kamii, C., and Lee-Katz, L. 1979. Physics in preschool education: A Piagetian approach. *Young Children.* 34: 4–9.

Lind, K. K., and Milburn, M.J. 1988. Mechanized childhood. *Science and Children.* 25 (5): 32–33.

Mechling, K. R., and Oliver, D. L. 1983. *Science teaches basic skills.* Washington, D.C.: National Science Teachers Association.

Rowe, M. B. 1973. *Teaching Science as Continuous Inquiry.* New York, N.Y.: McGraw-Hill.

Sunal, C. S. 1982. Philosophical bases for science and mathematics in early childhood education. *School Science and Mathematics.* 82 (1): 2–10.

Tipps, S. 1982. Making better guesses: A goal in early childhood science. *School Science and Mathematics.* 82 (1): 29–37.

SUGGESTED ACTIVITIES

1. The results of science are everywhere. Make a list of all of the science examples that you encountered on your way to class.

2. Recall a typical primary school day from your past. What type of science activities were introduced? How were the experiences introduced? Compare your experience with that of a child in a classroom in which you have observed. What are the similarities and differences? Do technological advancements make a difference?

3. Examine the teacher's edition of a recent elementary science textbook at the primary grade level. What science attitudes are claimed to be taught in the text? Decide if there is evidence of these attitudes being taught in either the student or teacher edition of the text.

4. Interview teachers of preschool and primary children to find out what types of science activities they introduce to students. Is the science content balanced? Which activities include the exploration of materials? Record your observations for class discussion.

5. If you were to teach science to urban children, what type of activities would you provide? Apply the question to rural and suburban environments. Would you teach science differently? Why or why not?

6. Pick one of your favorite science topics. Plan how you would include process, attitude, and content.

REVIEW

A. Discuss how the traditional view of science as a body of knowledge differs from a contemporary view of science as a process.

B. How does the knowledge explosion affect the science that is taught to young children?

C. List process skills that should be introduced to all preschool and primary age students.

D. What is meant by the term *healthy skepticism*?

E. What are the four major areas of science content appropriate for early childhood education?

F. Why has health education received increased attention in early childhood education?

UNIT 6 How Young Scientists Use Concepts

OBJECTIVES

After studying this unit, the student should be able to:

- Develop lessons using a variety of science process skills such as observing, comparing, measuring, classifying, and predicting
- Apply problem-solving strategies to lessons designed for young students
- Use data collecting and analysis as a basis for designing and teaching science lessons
- Design experiences for young children that enrich their experience at the preoperational level and prepare them for the concrete operational level
- Describe the process of self-regulation
- Describe the proper use of a discrepant event in teaching science

CONCEPT FORMATION IN YOUNG CHILDREN

Young children try very hard to explain the world around them. Do any of the following statements sound familiar?

> "Thunder is the sound of the angels bowling."
>
> "Chickens lay eggs and pigs lay bacon."
>
> "Electricity comes from a switch on the wall."
>
> "The sun follows me when I take a walk."

These are the magical statements of intuitive thinkers. These children use their senses or intuition to make judgments. Their logic is unpredictable, and they frequently prefer to use "magical explanations" to explain what is happening in their world. Clouds become the "smoke of angels" and rain falls because "it wants to help the farmers." These comments are typical of the self-centered view of intuitive children. They think that the sun rises in the morning just to shine on them. It never occurs to them that others might also benefit. They also have a difficult time remembering more than one thing at a time.

Statements and abilities such as these inspired Jean Piaget's curiosity about young children's beliefs. His search for answers about how children think and learn has contributed to our understanding that learning is an internal process. In other words, it is the child who brings meaning to the world and not vice versa. These misconceptions of a child are normal. This is what the child believes, thus, this is what is real to the child (Figure 6–1).

The temptation is to try and move children out of their magical stage of development. This is a mistake. Although some misconceptions can be corrected, others have to wait for more advanced thinking to develop. Students should not be pushed, pulled, or dragged through developmental stages. Instead, the goal is to enhance the development of young children at their present level of operation. In this way, they have the richness of experiences to take with them when the next level of development naturally occurs.

Figure 6–1 Children have misconceptions

Enhancing Awareness

When teaching concepts that are too abstract for children to fully understand, try to focus on aspects of the concept that can be understood. This type of awareness can be enhanced by the use of *visual depictions, observing, drawing*, and *discussion*.

Although some aspects of weather might remain a mystery, understanding can be developed by recording the types of weather that occur in a month. Construct a large calendar and put it on the wall. Every day, have a student draw a picture that represents the weather for the day. When the month is completed, cut the calendar into individual days. Then, glue the days that show similar weather in columns to create a bar graph. Ask the children to form conclusions about the month's weather from the graph. In this way, children can relate to weather patterns and visualize that these patterns change from day to day (Figure 6–2).

In a similar way, children will be better prepared for later studies of nutrition if they have some understanding of their own food intake. Graphing snacks is a way for children to visually compare and organize what they eat. On posterboard make five columns and mount pictures from the basic food groups at the bottom of each column. Discuss which food group each child's favorite snacks belong in. Have the children write their initials or attach a picture of themselves in the appropriate food column. Ask, "What do most of us like for snacks?" Discuss healthy snacks and make a new chart at the end of the week. Is there a difference in choices? (Figure 6–3).

For observations to be effective, they should be done in ten minutes or less, conducted with a purpose, and brought together by discussions. Unconnected observations do not aid in concept formation. For example, if a purpose is not given observing two flickering, different size candles, children will lose interest within minutes. Instead of telling children to "go look" at two burning candles, ask them to look and to find the difference between the candles. Children will become excited with the discovery that the candles are not the same size (Figure 6–4).

Discussions that follow observations heighten a child's awareness of that observation. A group of children observing a fish tank to see where the fish spends most of its time will be prepared to share what they saw happen in the tank. However, there may not be agreement.

Differences of opinion about observations stimulate interest and promote discussion. The children are likely to return to the tank to see for themselves what others say they saw. This would probably not happen without the focusing effect of discussion.

Drawing can provide excellent opportunities for observation and discussion. An effective use of drawing to enhance concept development would be to have children draw a tiger from memory before going to the zoo. Strategies like this usually reveal that more information is

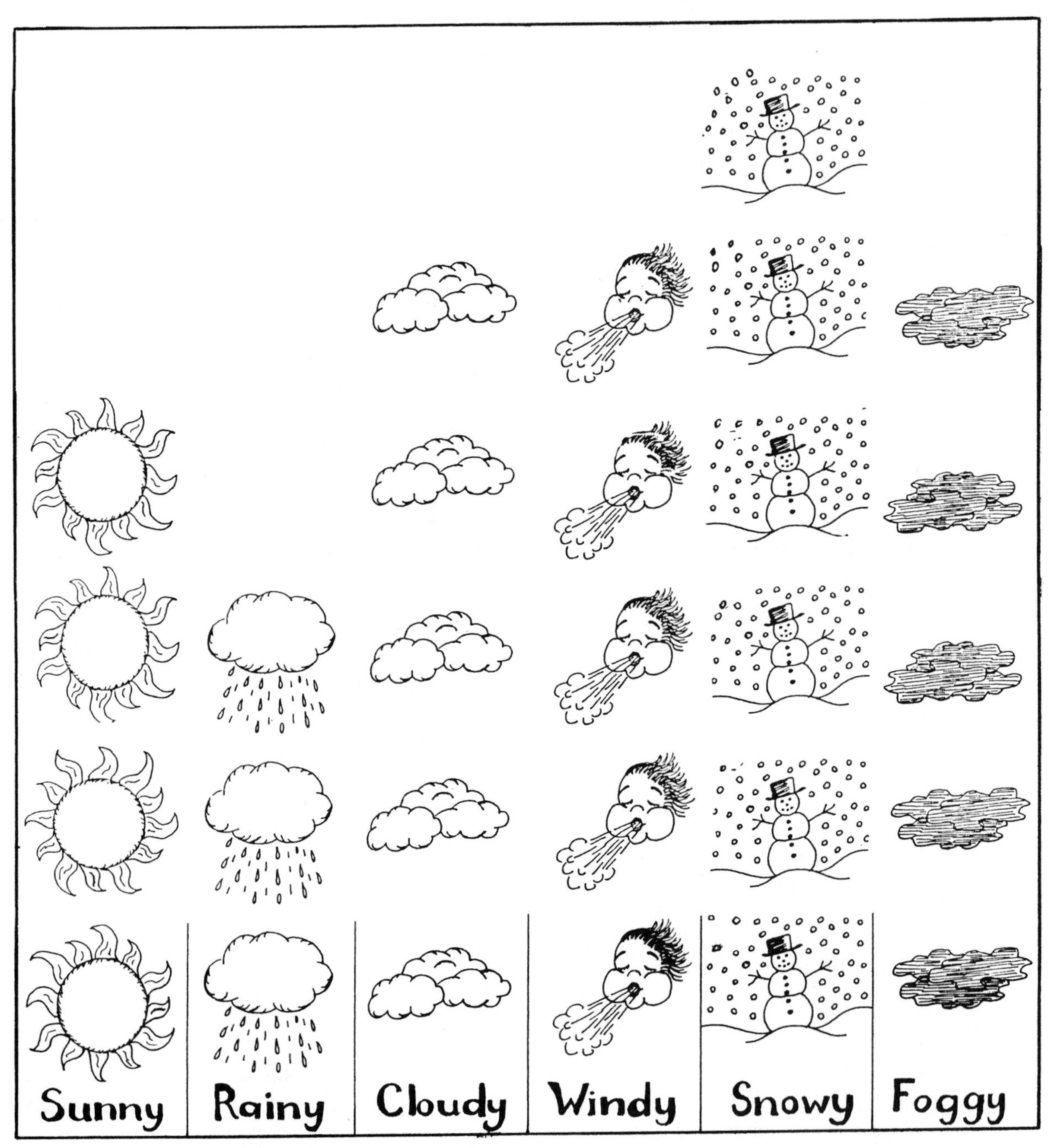

Figure 6–2 How many cloudy days were there this month?

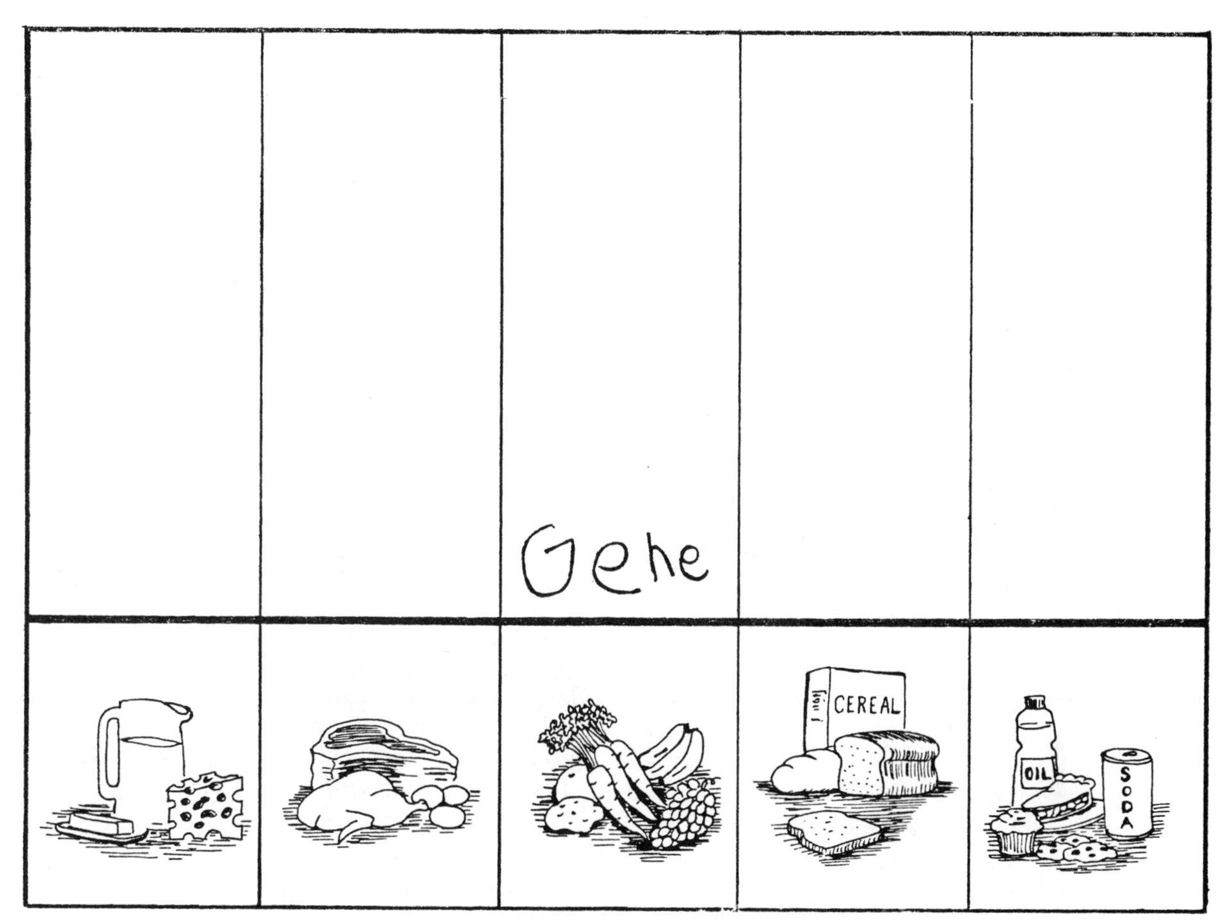

Figure 6–3 Find your favorite snacks.

needed to construct an accurate picture. Children will be eager to observe details that they might not otherwise notice about the tigers in the zoo (Figure 6–5).

Teacher Magic and Misconceptions

Children need time to reflect and absorb ideas to fully understand a concept. Misconceptions can occur at any stage. Be sure to give them plenty of time to manipulate and explore. For example, when a teacher mixes yellow and blue paints to create green, children might think that the result is teacher magic. However, if given the paints and the opportunity to discover green, they might remember that yellow and blue mixed together make green. Because children cannot carry out most operations mentally, they need to manipulate materials to develop concepts (Figure 6–6).

Misconceptions can occur at any stage of development. Some children in the primary grades may be in a transitional stage or have

Figure 6–4 Children notice that the candles are not the same size.

Figure 6–5 Robin's drawing of a tiger after a visit to the zoo

Generate a list of topics and construct questions to ask children.

For example:

Rain	What is rain? Where does rain come from?
Thunder	What do you think thunder is? Where does the loud sound come from?
Grass	What makes the grass green?
Night	Why does it get dark at night?
Seasons	Why are there seasons? Why do the leaves fall off the trees?
River	Why are there rivers in some places but not in others?
Sun	Why does the sun move?

Figure 6–6 Common misconceptions and questions

moved into the concrete operation stage of development. Although these students will be able to do much logical thinking, the concepts they work with must still be tied to concrete objects that they can manipulate. Firsthand experiences with materials continue to be essential for learning.

The child in this developmental stage no longer looks at the world through "magical eyes." Explanations for natural events are influenced by other natural objects and events. For example, a child may now say, "The rain comes from the sky" rather than "It rains to help farmers." This linking of physical objects will first appear in the early concrete development stages. Do not be misled by an apparent new awareness. The major factor in concept development is still contingent upon children's need to manipulate, observe, discuss, and visually depict things to understand what is new and different about them. This is true in all of the Piagetian developmental stages discussed in this book.

SELF-REGULATION AND CONCEPT ATTAINMENT

Have you ever seen someone plunge a turkey skewer through a balloon? Did you expect the balloon to burst? What was your reaction when it stayed intact? Your curiosity was probably peaked—you wanted to know why the balloon did not burst. Actually, you had just witnessed a *discrepant event* and entered the process of self-regulation. This is when your brain responds to interactions between you and your environment. Gallagher and Reid describe self-regulation as the active mental process of forming concepts. Knowing a little bit about how the brain functions in concept development will help in an explanation of this process. Visualize the human mind with thousands of concepts stored in various sections of the brain, sort of like a complex system of mental pigeonholes, much like a postal sorting system. As children move through the world and encounter new objects and phenomena, they assimilate and accommodate new information and store it in the correctly labeled mental category in their minds.

The brain, functioning like a postal worker, naturally classifies and stores information into the appropriate pigeonholes. New information is always stored close to all of the related information that has been previously stored. This grouping of closely related facts and phenomena related to a concept is called a *cognitive structure*. In other words, all that we know about the color red is stored in the same area of the brain. Our cognitive structure of red is developed further each time we have a color-related experience. The word *red* becomes a symbol for what we understand and perceive as the color red (Figure 6–7).

Our understanding of the world is imperfect because sooner or later, there is some point at which true understanding ends and misconceptions exit but go unquestioned. This is because incorrect interpretations of the world are stored alongside correct ones.

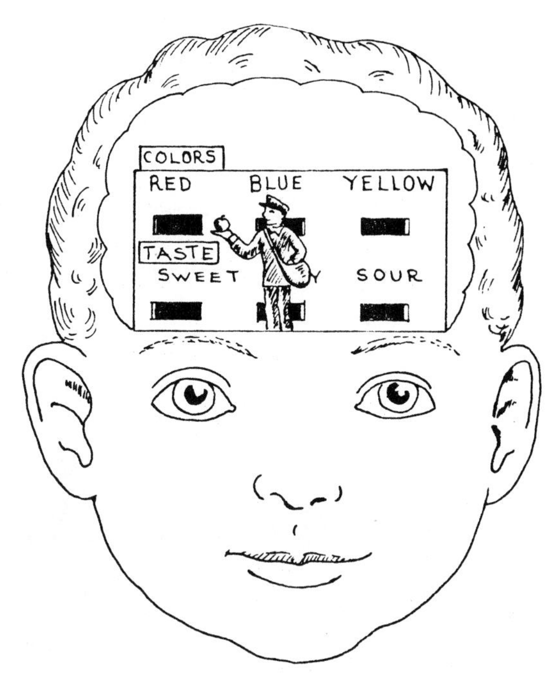

Figure 6–7 The brain functions like a postal worker.

Continuing the postal worker analogy, if the information doesn't quite fit into an existing pigeonhole, it is stored somewhere else. The postal workers probably get frustrated trying to find a suitable pigeonhole.

A point can also be reached where new information conflicts with older information stored in a given cognitive structure. When children realize that they do not understand something they previously thought they understood, they are said to be in what Piaget calls a state of *disequilibrium*. This is where you were when the balloon did not behave as you expected. The balloon did not behave as you expected and things no longer fit neatly together.

This is the *teachable moment*. When children are perplexed, their minds will not rest until they can find some way to make the new information fit. Since existing structures are inade-

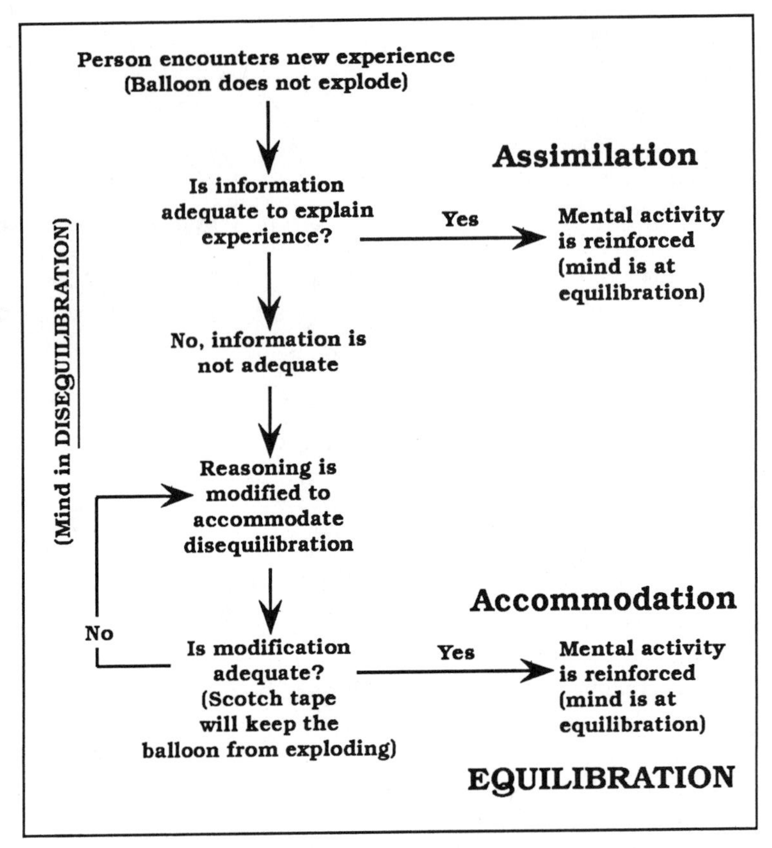

Figure 6–8 The self-regulation process *(Courtesy Indiana University).*

quate to accommodate all of the existing information, they must continually modify or replace with new cognitive structures. When in this state, children actively seek out additional information to create the new structure. They ask probing questions, observe closely, and inquire independently into the materials at hand. In this state, they are highly motivated and very receptive to learning. When children have had enough information to satisfy their curiosity and to create a new cognitive structure that explains most or all of the facts, they return to a state of *equilibrium* where everything appears to fit together. As children move from disequilibrium, to equilibrium two mental activities take place. When confronted with something that they do not understand, children fit it into a scheme, something they already know. If this does not work for them, they modify the scheme, or make a new one. This is called *accommodation*. Assimilation and accommodation work together to help students learn concepts.

To make use of the process of self-regulation in your classroom, find out at what point your students misunderstand or are unfamiliar with the topic you are teaching. This will allow you to present information contrary or beyond their existing cognitive structure and thus put them in a state of disequilibrium, where learning occurs.

Finding out what children know can be done in a number of ways. In addition to referring to the assessment units in this book, listen to children's responses to a lesson or question, or simply ask them to describe their understandings of a concept. For example, before teaching a lesson on animals, ask, "What does an animal look like?" You would be surprised at the number of young children who think that a life form has to have legs to be considered an animal.

DISCREPANT EVENTS

A *discrepant event* puts students in disequilibrium and prepares them for learning. They are curious and want to find out what is happening. It is recommended that you take advantage of the natural learning process to teach children what you want them to understand. The following scenes might give you some ideas for discrepant events to improve lessons you plan to teach.

- Mr. Wang's second grade class is working on a unit on the senses. His students are aware of the function of the five senses, but they may not know that the sense of smell plays as large a role in appreciating food as the sense of taste. His students work in pairs with one child blindfolded. The blindfolded students are asked to pinch their noses shut and taste several foods such as bread, raw potatoes, or apples to see if they can identify them.
- Students switch roles and try the same investigation with various juices such as apple, orange, tomato, and so on. Most students cannot identify juices. Having experienced this discrepant event, the students will be more interested in finding out about the structures of the nose related to smell. They may be more motivated to conduct an investigation about how the appearance of food affects its taste (Figure 6–9).

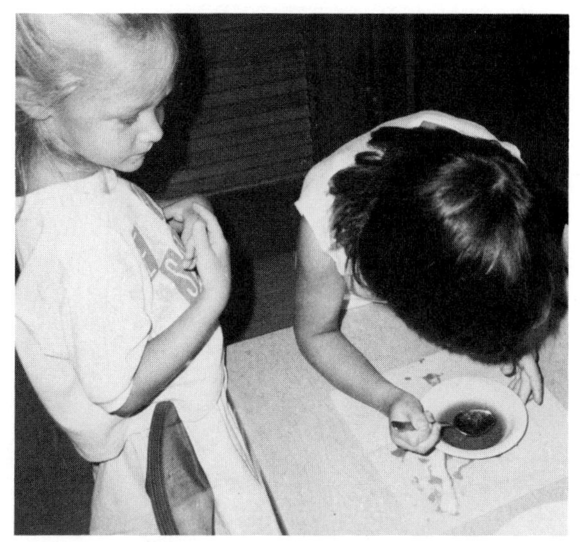

Figure 6–9 "Honey tastes sweet."

- Mrs. Fox fills two jars with water while her first grade class watches. She fills one jar to the rim and leaves about an inch of space in the other jar. She puts a lid on each jar and places it on a tray in the school yard on a very cold winter day. Her students return later in the day and find that both have frozen, but the completely filled jar has burst. Students are eager to find out why.
- Kindergarten students Ann and Vanessa, have been instructed to place two ice cubes in a glass and fill it to the brim with water. The ice cubes float on the water and extend about half an inch above the edge of the glass. Ms. Hebert asks them what will happen when the ice cubes melt. Vanessa thinks the water will overflow because there will be more water. Ann thinks the water level will drop because the ice cubes will contract when they melt. They watch in puzzlement as they realize that the water level stays the same as the cubes melt (Figure 6–10).

Figure 6–10 Ann and Vanessa check their predictions.

USING THE LEARNING CYCLE TO BUILD CONCEPTS

You can assist your students in creating new cognitive structures by designing learning experiences in a manner congruent with how children learn naturally. One popular approach is the application of the learning cycle. The learning cycle, described in Unit 1, is based upon the cycle of equilibration originally described by Piaget. Learning cycle, which is used extensively in elementary science education, combines aspects of naturalistic, informal, and structured activity methods—suggested elsewhere in this book—into a method of presenting a lesson.

As you have learned, the discrepant event is an effective device for motivating students and placing them in disequilibrium. Learning cycle is a useful approach to learning for many of the same reasons. Learning begins with a period of free exploration. Exploration can be as simple as giving students the materials to be used in a day's activity at the beginning of a lesson so they

Figure 6–11 Making the machine work.

can play with them for a few minutes. Minimal or no instructions should be given other than those related to safety, breaking the materials, or logistics in getting the materials. By letting students manipulate the materials, they will explore and very likely discover something they did not know before or something other than what they expected to happen. The following example utilizes all three steps of the learning cycle (Figure 6–11).

MAKING THE BULB LIGHT

EXPLORATION PHASE

Mr. Wang placed a wire, a flashlight bulb, and a size D battery on a tray. Each group of three students was given a tray of materials and told to try to figure out a way to make the bulb light. As each group successfully lit the bulb, he asked them if they could find another arrangement of materials that would make the bulbs light. After about ten minutes, most groups had found at least one way to light the bulb.

CONCEPT INTRODUCTION PHASE

After playing with the wires, batteries, and bulbs, Mr. Wang had the students bring their trays and form a circle on the floor. He asked the students if they had found out anything interesting about the materials. Jason showed the class one arrangement that worked to light the bulb. To introduce the class to the terms *open*, *closed*, and *short circuit*, Mr. Wang explained to the class that an arrangement that lights is called a closed circuit. Ann showed the class an arrangement that she thought would work but did not. Mr. Wang explained that this is an open circuit. Chad showed the class an arrangement that did not light the bulb but made the wire and battery very warm. Mr. Wang said this was a short circuit. Then he drew an example of each of the three types of circuit on the board and labeled them.

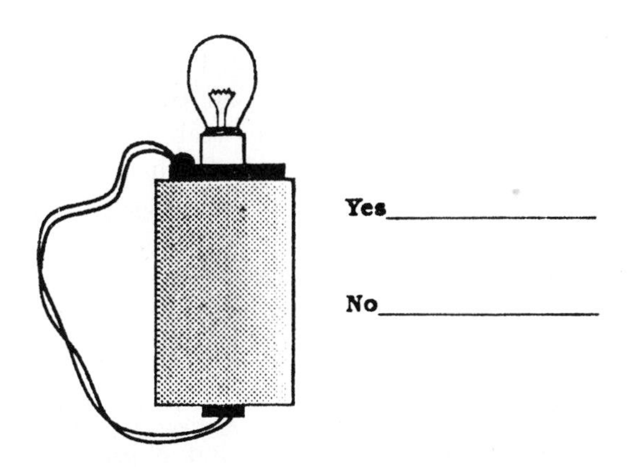

Figure 6–12 Battery prediction sheet

CONCEPT APPLICATION PHASE

Next, Mr. Wang showed the class a worksheet with drawings of various arrangements of batteries and bulbs. He asked the children to predict which arrangements would light and form a closed circuit, which would form an open circuit by lighting, and which would heat up the battery and wire, forming a short circuit. After recording their predictions independently, the children returned to their small groups to test each of the arrangements. They recorded the actual answers beside their predictions (Figure 6–12).

After the students completed the task, Mr. Wang called them together. They discussed the results of their investigation, sharing which arrangements of batteries, wire, and bulbs they predicted correctly and incorrectly. Then Mr. Wang reviewed the terms learned during the lesson and asked the students to read a short section of their science text that talked about fire and electrical safety. The reading discussed the dangers of putting metal objects in wall sockets, and fingers in light sockets, and suggested precautions when flying kites near power lines.

After reading the section aloud, Mr. Wang asked the class if they could see any relationship between the reading and the day's activity. Chad responded that flying a kite into an electrical power line or sticking a dinner fork in a wall socket were similar to what he did when he created a short circuit with the battery and wire. Other students found similar relationships between the reading activity and their lives.

In the preceding activity, you have seen a functioning model of the theory of cognitive development described by Piaget implemented in the classroom. The exploration phase invites assimilation and disequilibrium, the concept development phase provides for accommodation, and as the concept application phase expands the concept, reinforcement strategies for retaining the new concept are provided.

USING PART OF THE LEARNING CYCLE TO BUILD CONCEPTS

Although a formal investigation using learning cycle is an excellent way to present lessons to children, teaching all lessons in this manner may not be possible or desirable. At times, exploration and observation might be the full lesson. Giving students an opportunity to practice their skills of observation is often sufficient for them to learn a great deal about unfamiliar objects or phenomena. In the following scenarios, teachers made use of exploration observations to create lessons.

- Mr. Brown constructed a small bird feeder and placed it outside the window of his prekindergarten room. One cold winter day his efforts were rewarded. Students noticed and called attention to the fact that there were several kinds of birds at the feeder. Brad said that they all looked the same to him. Leroy pointed out different characteristics of the birds to Brad. Diana and Cindy noticed that the blue jay constantly chased other birds away from the feeder and that a big cardinal moved away from the feeder as soon as any other bird approached. Students wondered why the blue jay seemed to scare all the other birds and the cardinal seemed to be afraid of even the small birds. They spent several minutes discussing the possibilities.
- On Richard's birthday, his preschool teacher, Miss Collins, decided that the class should make a microwave cake to celebrate the occasion. The students helped Miss Collins mix the ingredients and commented on the sequence of events as the cake cooked for seven minutes in the microwave. Richard was the first to observe how bubbles started to form in small patches. Then George commented that the whole cake was bubbling and getting bigger. After removing the cake from the microwave, many children noticed how the cake shrank, got glossy, then lost its gloss as it cooled (Figure 6–13).

Some lessons can be improved by having children do more than just observing and exploring. These exploration lessons include *data collection* as an instructional focus. Data collection and interpretation are important to real science and real problem solving. Although firsthand observation will always be important, most breakthroughs in science are made in the analysis of carefully collected data. Scientists usually spend much more time searching through stacks of data than peering down the barrel of a microscope or through a telescope.

Data collection for young children is somewhat more abstract than firsthand observation. Therefore, it is important that students have sufficient practice in making predictions, specula-

Figure 6–13 The children help mix the ingredients.

tions, and guesses with firsthand observations before they begin to collect and interpret data. Nevertheless, young students can benefit from early experience in data collection and interpretation. Initial data collections are usually pictorial in form. Long-term patterns and changes that children cannot easily observe in one setting are excellent beginnings for data collection.

- Weather records such as those discussed earlier in the unit can expose children to patterns during any time of the year. After charting the weather with drawings or attaching pictures that represent changing conditions, have children decide which clothing is most appropriate for a particular kind of weather. Drawing clouds, sun, rain, lightning, snow, and so on that correspond to the daily weather and relating that information to what is worn can give students a sense of why data collection is useful.
- Growing plants provides excellent opportunities for early data collection. Mrs. Fox's first grade class charted the progress of bean plants growing in paper cups on the classroom windowsill. Each day students cut a strip of paper the same length as the height of their plant and glued the strips to a large sheet of newsprint. Over a period of weeks, students could see how their plants grew continuously even though they noticed few differences by just watching them. After pondering the plant data, Dean asked Mrs. Fox if the students could measure themselves with a strip of paper and chart their growth for the rest of the year. Thereafter, Mrs. Fox measured each student once a month. Her students were amazed to see how much they had grown during the year.

Another technique for designing science lessons is to allow students to have input into the process of problem solving and designing investigations. This might be called a *concept introduction lesson* because it utilizes the concept introduction phase of the learning cycle as the basis for a lesson. Although initial investigation and problem-solving experiences may be teacher designed, students eventually will be able to contribute to planning their own investigations. Most students probably will not be able to choose a topic and plan the entire investigation independently until they reach the intermediate grades, but their input into the process of planning gives them some ownership of the lesson and increases their confidence to explore ideas more fully. When solving real problems, identifying the problem is often more critical than the skills of attacking it. Students need practice in both aspects of problem solving.

The following examples depict students giving input into the problem to be solved and then helping to design how the problem should be approached.

- Mr. Wang's second grade class had had previous experience in charting the growth of plants. He told his class, "I'd like us to design an investigation about how fast plants grow. What things do you think could affect how fast a plant grows?" As Mr. Wang listed factors on the chalkboard, his students suggested a variety of factors including the amount of water, fertilizer, sunshine, temperature, type of seed, how much they are talked to, if they are stepped on, and so on. Students came up with possibilities that had never occurred to Mr. Wang. Next, he broke the class into small groups and told each group that they could have several paper cups, seeds, and some potting soil. They were asked to choose a factor from the list that they would like to investigate. Then, Mr. Wang helped each group plan their investigation.
- Derrick and Brent decided to study the effect of light on their plant. Mr. Wang asked them how they would control the amount of light their plants receive. Brent suggested that they bring a light bulb to place near the plants so that they could leave the light on all day. Derrick said that he would put the plants in a cardboard box for the hours they were not supposed to receive light. Mr. Wang asked them to think about how many plants they should use. They decided to use three: one plant would receive light all day, one plant would receive no light, and one plant would receive only six hours of light.

SUMMARY

In all of the preschool and primary developmental stages described by Piaget, keep in mind that a child's view of the world and concepts are not the same as yours. Their perception of phenomena is from their own perspective and experiences. Misconceptions arise. So, explore the world to expand thinking and be ready for the next developmental stage. Teach children to observe with all of their senses, classify, predict, and communicate to discover other viewpoints.

There are many possible methods for designing science instruction for young children. Learning cycle is an application of the theory of cognitive development described by Piaget. Learning cycle can incorporate a number of techniques into a single lesson, or each of the components of learning cycle can be used independently to develop lessons. Other effective methods for designing lessons include discrepant events, data collection and analysis, problem solving, and cooperative planning of investigations. All of these methods emphasize science process skills that are important to the development of concepts in young children.

FURTHER READING AND RESOURCES

Barman, C. 1989. *New ideas about an effective teaching strategy.* Council for Elementary Science International. Indianapolis: Indiana University.

Benham, N. B., Hosticka, A., Payne, J. D., and Yeotis, C. 1982. Making concepts in science and mathematics visible and viable in the early childhood curriculum. *School Science and Mathematics.* 82 (1): 45–64.

Charlesworth, R. 1987. *Understanding Child Development.* Albany, N.Y.: Delmar.

Gallagher, J. M., and Reid, D. K. 1981. The Learning Theory of Piaget and Inhelder. Monterey, Calif.: Brooks/Cole.

Harlen, W., and Symington, D. 1988. Helping children to observe. *Primary Science Taking the Plunge.* New Hampshire: Heinemann.

Price, G. G. 1982. Cognitive learning in early

childhood education: mathematics, science, and social studies. *Handbook of Research in Early Childhood Education*. New York: The Free Press.

Renner, J. W., and Marek, E. A. 1988. *The Learning Cycle*. New Hampshire: Heinemann.

Shaw, J. M., and Owen, L. L. 1987. Weather—beyond observation. *Science and Children*. 25 (3): 27–29.

Sprung, B., Froschl, M., and Campbell, P.B. 1985. *What will Happen If. . . .?* New York: Educational Equity Concepts.

Tipps, S. 1982. Making better guesses: A goal in early childhood science. *School Science and Mathematics*. 82 (1): 29–37.

SUGGESTED ACTIVITIES

1. Interview children to determine their understanding of cause-and-effect relationships. Questions such as those in Figure 6–6 will be effective in gaining insight into the perceptions and misconceptions of young children. Compare remarks with classmates.

2. Brainstorm at least three strategies to prepare children to understand topics such as rain, lightning, clouds, mountains, and night. Pick a topic and prepare a learning experience that will prepare children for future concept development.

3. Pretend that you are presenting two lessons about seeds (seeds grow into plants) to a class of first graders. What types of learning experiences will you design for these students? How will you apply Piaget's theory of development to this topic? In what way would you adapt the experiences for a different developmental level?

4. Present a discrepant event to your classmates. After discovering how the event was achieved, analyze the process of equilibration that the class went through. Design a discrepant event that focuses on senses that would be effective for kindergarten age children.

5. Learning cycle has been presented as an effective way to build concepts in young children. What is your concept of the learning cycle and how will you use it with children? Select a concept and prepare a learning cycle to teach it.

6. Reflect on your own primary science experiences. What types of learning strategies did you encounter? Were they effective? Why or why not?

REVIEW

Briefly answer each of the following:

A. Describe what is meant by a discrepant event.

B. Why are discrepant events used in science education?

C. List the three major parts of the learning cycle.

D. Describe how cognitive structures change over time.

E. Why is data collection important with young children?

F. Create an example of a discrepant event that you might use with young children on a topic of your choice.

UNIT 7 Planning for Science

OBJECTIVES

After studying this unit, the student should be able to:

- Develop science concepts with subject area integrations
- Explain and use the strategy of webbing in unit planning
- Identify and develop science concepts in a lesson; design lesson plans for teaching science to children
- Construct evaluation strategies

INTEGRATING SCIENCE INTO THE CURRICULUM

Concepts will be more likely to be retained by children if presented in a variety of ways and extended over a period of time. For example, after a trip to the zoo, extend the collective experience by having children dictate a story about their trip. Other activities could focus on following up pre-visit discussions, directing children to observe specifics at the zoo, such as differences in animal noses. Children might enjoy comparing zoo animal noses to those of their pets by matching pictures of similar noses on a bulletin board. In this way, concepts can continue to be applied and related to past experiences as the year progresses.

Additional integrations might include drawing favorite animal noses, creating plays about animals with specific types of noses, writing about an animal and its nose, and creating "smelling activities." You might even want to introduce reasons an animal has a particular type of nose. One popular idea is to purchase plastic animal noses, distribute them to children, and play a "Mother-May-I-Like" game. Say, "If you have four legs and roar, take two giant steps. If you have two legs and quack, take three steps forward" (Figure 7–1).

Think of how much more science can be learned if we make connections between it and other subjects. This requires preparing planned activities and taking advantage of every teachable moment that occurs in your class to introduce children to science.

Opportunities abound for teaching science in early childhood. Consider actively involving children with art, blocks, dramatic play, woodworking, language arts, math, and creative movement. Learning centers are one way to provide excellent integrations (Figure 7–2). Centers are discussed in Unit 40 of this book. The following ideas may encourage your thinking.

- *Painting.* Finger painting helps children learn to perceive with their fingertips and demonstrates the concept of color diffusion as they clean their hands. Shapes can be recognized by painting with fruit and familiar objects.
- *Water center.* Concepts such as volume and conservation begin to be grasped when children measure with water and sand. Buoyancy can be explored with boats and sinking and floating objects.
- *Blocks.* Blocks are an excellent way to introduce children to friction, gravity, and simple machines. Leverage and efficiency can be reinforced with *woodworking*.
- *Books.* Many books introduce scientific concepts while *telling a story*. Books with pic-

Figure 7–1 "If you have two legs and quack, take three steps forward."

tures give views of unfamiliar things and an opportunity to explore detail and infer and discuss.

- *Music and rhythmic activities.* These let children experience the movement of air against their bodies. Air resistance can also be demonstrated by dancing with a scarf.
- *Playground.* The playground can provide an opportunity to predict weather, practice balancing, and experience friction. The concrete world of science integrates especially well with reading and writing. Basic words, object guessing, experience charts, writing stories, and working with tactile sensations all encourage early literacy development (Figures 7–3 and 7–4).

Children Learn in Different Ways

It is important to provide children with a variety of ways to learn science. Even very young children have developed definite patterns in the way they learn. Observe a group of children engaged in free play: some prefer to work alone quietly, others do well in groups. Personal learning style also extends to a preference for visual or auditory learning. The teacher in the

Figure 7–2 Children enjoy reading about bears.

Figure 7–3 "How do we make chocolate chip cookies?"

following scenario keeps the wide range of learning styles in mind when planning science experiences.

Ms. Hebert knows that the children in her kindergarten class exhibit a wide range of learning styles and behaviors. For example, Ann wakes up slowly. She hesitates to jump in and explore and prefers to work alone. On the other hand, Vanessa is social, verbal, and ready to go first thing in the morning. As Ms. Hebert plans activities to reinforce the observations of flamingos that her class made at the zoo, she includes experiences that include both group discussion and individual work. Some of the visual, auditory, and small and large group activities include flamingo number puzzles, drawing and painting flamingos, and acting out how flamingos eat, rest, walk, and honk. In this way, Ms. Hebert meets the diverse needs of the class and integrates concepts about flamingos into the entire week (Figure 7–5). Organizing science lessons with subject matter correlations in mind insures integration. The following section outlines ways to plan science lessons and units.

ORGANIZING FOR TEACHING SCIENCE

In order to teach effectively, teachers must organize what they plan to teach. How they organize depends largely on their teaching situation. Some school districts require that a textbook series or curriculum guide be used when teaching science. Some have a fully developed program to follow, and others have no established guidelines. Regardless of district directives, the strategies discussed in this chapter can be adapted to a variety of teaching situations.

Planning for Developing Science Concepts

The first questions to ask when organizing for teaching is, "What do I want the children to know?" and "What is the best way to do this?"

Figure 7–4 Putting peanut claws on bear claw cookies is a tasty tactile activity.

You might have a general topic in mind, such as air, but do not know where to go from there. One technique that might help organize your thoughts is *webbing*, a strategy borrowed from literature. A web depicts a variety of possible concepts and curricular experiences that you might use to develop concepts. By visually depicting your ideas, you will be able to tell at a glance the concepts covered in your unit. As the web emerges, projected activities can be balanced by subject area, e.g., social studies, movement, art, drama, and math, and by a variety of naturalistic, informal, or structured activities.

Start your planning by selecting what you want children to know about a topic. For example, the topic "air" contains many science concepts. Four concepts about air that are commonly taught to first graders begin the web depicted in Figure 7–6.

1. Air is all around us.
2. Air takes up space.

Figure 7–5 A flamingo lacing card

3. Air can make noise.
4. Air can be hot and cold.

After selecting the concepts that you plan to develop, begin adding appropriate activities to achieve your goal. Look back at Unit 6 and think of some of these strategies that will best help you teach about air. Remember, "messing around" time and direct experience are vital for learning.

The developed web in Figure 7–7 shows at a glance the main concepts and activities that will be included in this unit. You may not want to teach all of these activities but you will have the advantage of flexibility when you make decisions.

Next, turn your attention to how you will evaluate children's learning. Remember, preschool and primary age children will not be able

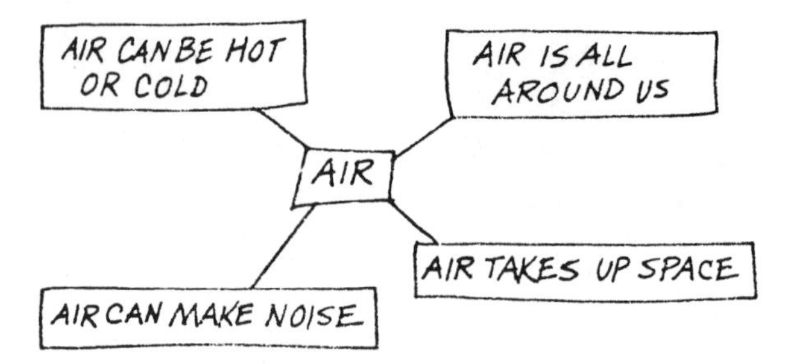

Figure 7–6 Begin by making a web of the science concepts you want to teach.

to verbalize their true understanding of a concept. They simply have not advanced to the formal stages of thinking. Instead, have students show their knowledge in ways that can be observed. Have them explain, predict, show, tell, draw, describe, construct, and so on (Figure 7–8). These are the verbs that indicate an action of some kind. For example, as students explain to you why they have grouped buttons together in a certain way, be assured that the facts are there and so are the concepts, and one day they will come together in a fully developed concept statement. Concept development takes time and cannot be rushed.

The webbed unit you have developed is a long-term plan for organizing science experiences around a specific topic. Formal units usually contain overall goals and objectives, a series of lessons, and an evaluation plan. Goals are the broad statements that indicate where you are heading with the topic. Objectives state how you plan to achieve your goals. Practical teaching direction is

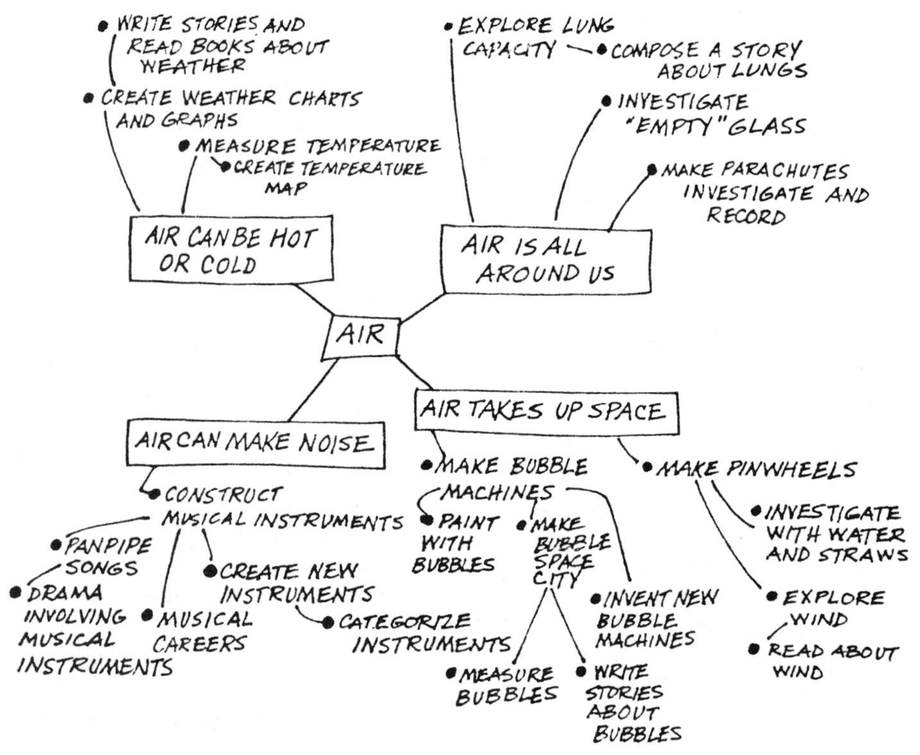

Figure 7–7 Example of a webbed unit

ACTION VERBS

Simple Action Words

arrange	design	map
attempt	discriminate	match
categorize	distinguish	measure
chart	explain	name
circle	formulate	order
classify	gather	organize
collect	graph	place
compare	identify	point
compile	include	recall
complete	indicate	report
construct	infer	select
contrast	label	sort
count	list	state
define	locate	tell
describe		

Figure 7–8 Verbs that indicate action

provided by daily lesson plans. An evaluation plan is necessary to assess student learning and your own teaching. There is a variety of ways these components can be organized, but Figure 7–9 outlines the essential ingredients.

Lesson Planning

The lesson plan is a necessary component of the unit. It helps you plan the experiences that will aid in concept development. The following lesson plan is adaptable and focuses on developing a science concept, manipulating materials, and extending and reinforcing the concept with additional activities and subject area integrations. Refer to Figure 7–10, The Bubble Machine, for an example of this lesson plan format.

Basic Science Lesson Plan Components

Concept. Concepts are usually the most difficult part of a lesson plan. The temptation is to write an objective, or topic title. However, in order to really focus your teaching on the major concept

COMPONENTS IN A UNIT PLAN

TOPIC	What is the topic or concept?
GOAL	Where are you heading?
OBJECTIVES	How do you plan to achieve your goals?
LESSONS	What will you teach?
EVALUATION	Did the students learn what you wanted them to learn?

Figure 7–9 Components in unit planning

to be developed, you must find the science in what you intend to teach. For example, ask yourself, "What do I want the children to learn about air?"

Objective. Then ask, "What do I want the children *to do* to help them understand that air takes up space?" When you have decided on the basic experience, be sure and identify the process skills that children will use. In this way, you will be aware of content and process.

Define the teaching process in behavioral terms. State what behavior you want the children to exhibit. This will make evaluation easier because you have stated what you expect the children to accomplish. Although many educators state behavioral objectives with conditions, most teachers find that beginning a statement with "the child should be able to" followed by an action verb is an effective way to state objectives. Some examples might be:

The child should be able to describe the parts of a flower.

The child should be able to construct a diorama of the habitat of a tiger.

The child should be able to draw a picture that shows different types of animal noses.

Materials. In order for children to manipulate materials, you must decide which materials should be organized in advance of the lesson.

CONCEPT: "Air takes up space. Bubbles have air inside of them."

OBJECTIVE: The child will construct a bubble-making machine by manipulating materials and air to produce bubbles. The child will observe and describe the bubbles. The child will infer that air is inside of bubbles.

MATERIALS: Mix a basic bubble solution of 8 tablespooons liquid detergent and 1 quart water (expensive detergent makes stronger bubbles).
Large pails, measuring spoons, quart jar or milk carton, cups, straws

PROCEDURE:

Initiating Activity: Demonstrate an assembled Bubble Machine. Have children observe the machine and tell what they think is happening.

How to do it: Help children assemble bubble machines. Insert straw into the side of the cup. Pour the bubble mixture to just below the hole in the side of the cup. Give children 5 minutes to explore blowing bubbles with the bubble machine. Then ask children to see how many bubbles they can blow. Ask: What do your bubbles look like? Describe your bubbles. Add food coloring for more colorful bubbles.

What happens to your bubbles? Do they burst? How can you make them last longer?

What do you think is in the bubbles? How can you tell? What did you blow into the bubble? Can you think of something else that you blow air into to make larger? (balloon)

EXTENSIONS:

1. Have students tell a story about the bubble machine as you record it on chart paper.
2. Encourage children to make bubble books with drawings that depict their bubbles, bubble machines, and the exciting time they had blowing bubbles. Encourage the children to write, or pretend to write, about their pictures. Threes and fours enjoy pretending to write; by five or six children begin to experiment with inventing their own spellings. Be sure to accept whatever they produce. Have children read their books to the class. Place them in the library center for browsing.
3. Make a bubbles bulletin board. Draw a cluster of bubbles and have students add descriptive words about bubbles.
4. Challenge students to invent other bubble machines. (Unit 36 of this book contains activities that teach additional concepts of air and bubbles.)

Figure 7–10 The bubble machine lesson

Ask, "What materials will I need to teach this lesson?

Procedure. When planning the lesson ask, "How will this experience be conducted?" You must decide how you will initiate the lesson with children, present the learning experience, and relate the concept to the childrens' past experiences. Questions that encourage learning should be considered and included in the lesson plan.

It is recommended that an initiating experience begin the lesson. This experience could be the "messing around" with materials stage of the learning cycle, a demonstration or discrepant event, a question sequence that bridges what you intend to teach with a previous lesson or experience. The idea is that you want to stimulate and interest the children about what they are going to do in the lesson (Figure 7–11).

Extension. To ensure maximum learning of the concept, ways to keep the idea going must be planned. This can be done by extending the concept with additional learning activities, integrating the concept into other subject areas, preparing learning centers, and so on.

Assessment Strategies

You cannot teach effectively if you do not plan for evaluation. To continue with another lesson before you know what students understand from the current lesson seems pointless. One reason for avoiding evaluation might be the tendency to associate the term *evaluation* with the term *grading*. This is a misconception. Grading takes place when a symbol that stands for a level of achievement is assigned to learning. Evaluation, on the other hand, is finding out what students understand. Evaluation takes place *before*, *during*, and *after* teaching.

Figure 7–11 A bubble machine

Evaluation done before teaching is diagnostic in nature and takes place when you assess what children know about a topic and determine which stage of development they are in. For example, when you ask children where they think rain comes from, you are evaluating and discovering misconceptions.

As teachers evaluate student progress during teaching, changes in teaching strategies are decided. If one strategy is not working, try something else.

As children work on projects, you will find yourself interacting with them on an informal basis. Listen carefully to children's comments and watch them manipulate materials. You cannot help but assess how things are going. You might even want to keep a record of your observations or create a chart that reflects areas of concern to you, such as attitudes or interaction between students. When observations are written down in an organized way, they are called *anecdotal records* (Figure 7–12).

If you decide to use anecdotal records, be sure to write down dates and names and to tell students you are keeping track of things that happen. A review of your records will be valuable when you complete the unit. Recording observations can become a habit and provide an additional tool in assessing children's learning and your teaching strategies. Anecdotes are also invaluable resources for parent conferences.

Responses to oral questions can be helpful in evaluating while teaching. Facial expressions are especially telling. Everyone has observed a blank look that usually indicates a lack of understanding. This look may be because the question asked was too difficult. So, ask an easy question or present your question in a different manner for improved results.

One way to evaluate during the after teach-

EVALUATION

Things to record when observing, discussing, or keeping anecdotal records

Cognitive:

- How well do the children handle the materials?
- Do the children cite out-of-school examples of the science concept?
- Is the basic concept being studied referred to as the children go about their day?

Attitude and skill:

- Do the children express like or dislike of the topic?
- Are there any comments that suggest prejudice?
- Do children evidence self-evaluation?
- Are ideas freely expressed in the group?
- Are there any specific behaviors that need to be observed each time science is taught?

Figure 7–12 Keeping anecdotal records

ing component is to ask questions about the activity in a lesson review. Some teachers write main idea questions on the chalkboard. Then, they put a chart next to the questions and label it "What We Found Out." As the lesson progresses, the chart is filled in by the class as a way of showing progress and reviewing the lesson or unit.

Another strategy is to observe children applying the concept. For example, as "The Three Billy Goats" is being read, Joyce comments, "The little billy goat walks just like the goat that we saw at the zoo." You know from this statement that Joyce has some idea of how goats move.

Sometimes students have difficulty learning because an activity doesn't work. One basic rule when teaching science is, "Always do the activity first." This includes noting questions or possible problems you may encounter. If you have trouble, students will too (Figure 7–13).

Evaluating the Unit

How well did you design your unit? Ask yourself some questions, such as the following, to help evaluate your work. These questions pull together the major points of this unit.

1. Have you related the unit to the children's past experiences?
2. Are a variety of process skills used in the activities?
3. Have you integrated other subjects with the unit?
4. When you use reading and writing activities, do they follow hands-on experiences?
5. Do you allow for naturalistic, informal, and directed activities?
6. Is a variety of teaching strategies included?
7. Are both open-ended and narrow questions asked?
8. Will the evaluation strategy provide a way to determine if children can apply what they have learned?
9. What local resources are included in the unit?

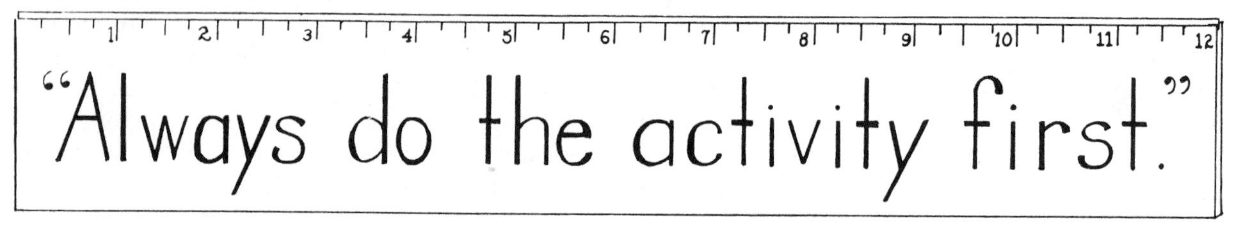

Figure 7–13 The golden rule

Three Basic Types of Units

Some teachers like to develop *resource units*. The resource unit is an extensive collection of activities and suggestions focusing on a single science topic. The advantage of a resource unit is the wide range of appropriate strategies available to meet the needs, interests, and abilities of the children. As the unit is taught, additional strategies and integrations are usually added. For example, Mrs. Jones knows that she is going to teach a unit on seeds in her kindergarten class. She collects all of the activities and teaching strategies that she can find. When she is ready to teach the seed unit, she selects the activities that she believes are most appropriate.

Teachers who design a teaching unit plan to develop a science concept, objectives, materials, activities; and evaluation procedures for a specific group of children. This unit is less extensive than a resource unit and contains exactly what will be taught, a timeline, and the order of activities. Usually, general initiating experiences begin the unit and culminating experiences end the unit. The specific teaching unit has value and may be used again with other classes after appropriate adaptations have been made. For example, Mr. Wang has planned a two-week unit on batteries and bulbs. He has decided on activities and planned each lesson period of the two weeks.

Extending the textbook in a textbook unit is another possibility. The most obvious limitation of this unit is a school district change in textbooks. A textbook unit is designed by outlining the science concepts for the unit and checking the textbook for those already covered in the book or teacher's manual. Additional learning activities are added for concepts not included in the text or sometimes instead of those in the text. Initiating activities to arouse interest in the topic might be needed. One advantage of this unit is using the textbook to better advantage. For example, after doing animal activities, use the text to confirm or extend knowledge.

Open-Ended and Narrow-Questions

Asking questions can be likened to driving a car with a stick shift. When teaching the whole class, start in low gear with a narrow question that can be answered yes or no, or with a question that has an obvious answer. This usually puts students at ease. They are happy; they know something. Then, try an open-ended question that has many answers. Open-ended questions stimulate discussion and offer opportunities for thinking. However, if the open-ended question is asked before the class has background information, the children might just stare at you, duck their heads, or exhibit undesirable behavior. Do not panic; quickly shift gears and ask a narrow question. Then, work your way back to what you want to find out.

Teachers who are adept at shifting between narrow and open-ended questions are probably excellent discussion leaders and have little trouble with classroom management during these periods.

Figure 7–14 "Can you make a piece of clay float?"

Open-ended questions are excellent interest builders when used effectively. For example, consider Ms. Hebert's initiating activity for a lesson about buoyancy.

Ms. Hebert holds a rock over a pan of water and asks a narrow question, "Will this rock sink when dropped in water?" (yes, no) Then, she asks an open-ended question, "How can we keep the rock from sinking into the water?" The children answer, "Tie string around it," "Put a spoon under it," and "Grab it."

As the discussion progresses, the open-ended question leads the children into a discussion about boats. After talking about how boats are used, give each child a piece of clay and instruct them to design a boat that will float (Figure 7–14).

SUMMARY

Children are more likely to retain science concepts if they are integrated with other subject areas. Making connections between science and other aspects of a child's school day requires that opportunities for learning be well planned and readily available to children. Learning science in a variety of ways encourages personal learning style and insures subject integrations.

The key to effective teaching is organization. Unit and lesson planning provide a way to plan what you want children to learn and how you want them to learn. A planning web is a useful technique for depicting ideas, outlining concepts, and integrating content.

The three basic types of planning units are resource units, teaching units, and textbook units. Teachers utilize whichever unit best suits their classroom needs. By asking open-ended and narrow questions, teachers develop science concepts and encourage higher-order thinking skills in their students.

FURTHER READING AND RESOURCES

Gega, P. 1986. *Science in Elementary Education.* New York: John Wiley & Sons, Inc.

Harlan, J. 1988. Science experiences for early childhood. Columbus, Ohio: Merrill.

Jacobson, W. J., and Bergman, A. B. 1987. *Science for children.* Englewood Cliffs, N.J.: Prentice-Hall.

Rice, K. Soap films and bubbles. *Science and Children 23* (8), 4–9.

Schwartz, J. J. 1988. *Encouraging Early Literacy.* New Hampshire: Heinemann.

Townsend, J., and Schiller, P. 1984. *More Than A Magnet, Less Than A Dollar.* Houston: Evergreen Publishing Company.

Victor, E. 1985. *Science for the Elementary School.* New York: Macmillan.

Wassermann, S., and Ivany, J.W.G. 1988. *Teaching Elementary Science.* Cambridge, Mass.: Harper & Row.

SUGGESTED ACTIVITIES

1. Choose a topic and construct a web of the major concepts that are important to know about the topic. Add the learning activities you plan to use to teach these concepts and suggest appropriate evaluation techniques. Work in teams to brainstorm ideas.
2. Select a lesson from a primary level textbook. Compare how the book presents the concept with what you know about how children learn science.
3. Interview a teacher to determine his or her

approach to planning; for example, whether a textbook series or curriculum guide is used, how flexibility is maintained, and what assessment techniques are found to be most effective. Ask teachers to describe how they plan units and lessons and to compare strategies used the first year of teaching with those used now. Discuss responses to your questions in a group.

4. Reflect on your past experiences in science classes. Which were the most exciting units? Which were the most boring? Do you remember what made a unit exciting or dull? Or, do you primarily remember a specific activity?
5. Can you integrate a unit on seeds with language arts? Then, integrate drama/movement, art, math, and social studies with the seed unit. Try designing a puppet show that would help teach the concept.
6. Practice writing questions that will help you evaluate children's understanding of a lesson on taste.

REVIEW

1. What is webbing and how can it be applied in designing a science unit?
2. Which process skills are involved in the Bubble Machine lesson, including extensions?
3. Identify learning cycle components in the Bubble Machine lesson.
4. What strategies could you use to integrate subject areas into science? Give an example.
5. List the major components of a science lesson plan.
6. Identify three ways of evaluating students other than paper and pencil tests.
7. What is a unit? List three basic types of units.
8. Explain the difference between narrow and open-ended questions and how each is used.

SECTION II

Fundamental Concepts and Skills

UNIT 8 One-to-One Correspondence

OBJECTIVES

After studying this unit, the student should be able to:

- Define one-to-one correspondence
- Identify naturalistic, informal, and structured one-to-one correspondence activities
- Describe five ways to vary one-to-one correspondence activities
- Assess and evaluate children's one-to-one correspondence skills

One-to-one correspondence is the most fundamental component of the concept of number. It is the understanding that one group has the same number of things as another. For example, each child has a cookie; each foot has a shoe; each person wears a hat. It is preliminary to counting and basic to the understanding of equivalence and to the concept of conservation of number described in Unit 1.

ASSESSMENT

To obtain information of an informal nature, note the children's behavior during their work, play, and routine activities. Look for one-to-one correspondence that happens naturally. For example, when the child plays train, he may line up a row of chairs so there is one for each child passenger. When he puts his mittens on, he shows that he knows that there should be one for each hand; when painting, he checks to be sure he has each paintbrush in the matching color of paint. Tasks for formal assessment are given in Appendix A. The following are examples:

SAMPLE ASSESSMENT TASK

Preoperational Ages 2–3

METHOD: Observation, individuals or groups

SKILL: Child demonstrates one-to-one correspondence during play activities.

MATERIALS: Play materials that lend themselves to one-to-one activities such as small blocks and animals, dishes and eating utensils, paint containers and paintbrushes, pegs and pegboards, sticks and stones, and so on.

PROCEDURE: Provide the materials and encourage the children to use them.

EVALUATION: Note if the children match items one to one such as putting small peg dolls in each of several margarine containers or on top of each of several blocks that have been lined up in a row.

INSTRUCTIONAL RESOURCE(S):

Charlesworth, R. and Lind, K. (1990) *Math and Science for Young Children*. Albany, New York: Delmar. Unit 8

SAMPLE ASSESSMENT TASK

Preoperational Ages 5–6

METHOD: Interview
SKILL: The child can place two groups of ten items each in one-to-one correspondence.
MATERIALS: Two groups of objects of different shapes and/or color (such as pennies and cube blocks or red chips and white chips). Have at least ten of each type of object.
PROCEDURE: Place two groups of ten objects in front of the child. Find out if there is the same amount (number) in each bunch (pile, group, set). If the child cannot do the task, go back and try it with two groups of five.
EVALUATION: The children should arrange each group so as to match the objects one-to-one, or they might count each group to determine equality.
INSTRUCTIONAL RESOURCE(S):
Charlesworth, R. & Lind, K. (1990) *Math and Science for Young Children*. Albany, New York: Delmar. Unit 8

NATURALISTIC ACTIVITIES

One-to-one correspondence activities develop from the infant's early sensorimotor activity. He finds out that he can hold one thing in each hand, but he can put only one object at a time in his mouth (Figure 8–1). The toddler discovers that five peg dolls will fit one each in the five holes in his toy bus. Quickly he learns that one person fits on each chair, one shoe goes on each foot, and so on. The two-year-old spends a great deal of his playtime in one-to-one correspondence activities. He lines up containers such as margarine cups, dishes, or boxes and puts a small toy animal in each one. He pretends to set the table for lunch. First he sets one place for himself and one for his bear with a plate for each. Then he gives each place a spoon and a small cup and saucer. He plays with his large plastic shapes and discovers there is a rod that will fit through the hole in each shape (Figure 8–2).

Figure 8–1 The small child finds he can hold one thing in each hand, but he can put only one object at a time in his mouth.

INFORMAL ACTIVITIES

There are many opportunities for informal one-to-one correspondence activities each day. There are many times when things must be

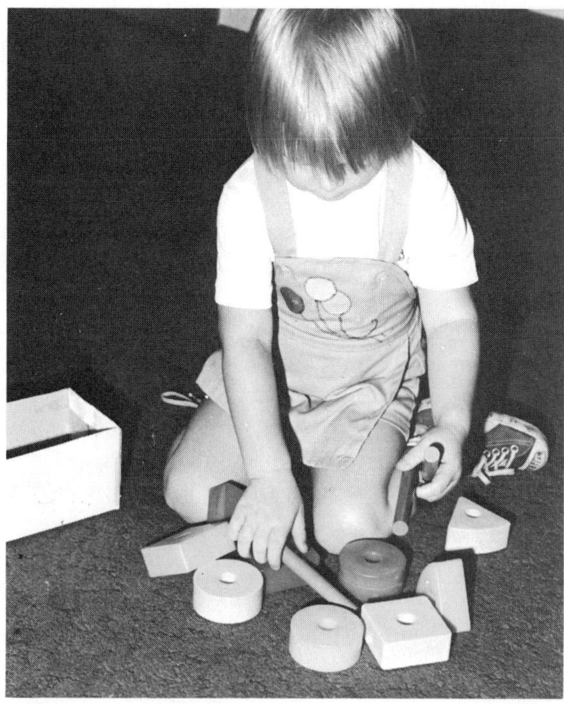

Figure 8–2 "Which rod goes with which shape?"

passed out to a group: food items, scissors, crayons, paper, napkins, paper towels, or notes to go home. Each child should do as many of these things as possible.

Checking on whether everyone has accomplished a task or has what they need is another chance for informal one-to-one correspondence. Does everyone have a chair to sit on? Does everyone have on two boots or two mittens? Does each person have on his coat? Does each person have a cup of milk or a sandwich? A child can check by matching: "Larry, find out if everyone has a pair of scissors, please."

One-to-one correspondence helps to solve difficulties. For instance, the children are washing rubber dolls in soap suds. Jeanie is crying, "Petey has two dolls and I don't have any." Mrs. Carter comes over. "Petey, more children want to play here now so each one can have only one baby to wash." One-to-one correspondence is often the basis for rules such as "Only one person on each swing at a time." "Only one piece of cake for each child today."

Other informal activities occur when children pick out materials made available during free play. These kinds of materials would include pegboards, felt shapes on a flannel board, bead and inch-cube block patterns, shape sorters, formboards, lotto games, and other commercial materials. Materials can also be made by the teacher to serve the same purposes. Most of the materials described in the next section can be made available for informal exploration both before and after they have been used in structured activities.

STRUCTURED ACTIVITIES

The extent and variety of materials that can be used for one-to-one correspondence activities is almost endless (Figure 8–3). Referring to Unit 3, remember the six steps from concrete to abstract materials. These steps are especially relevant when selecting one-to-one correspondence

Figure 8–3 One block is matched with each square.

materials. Five characteristics must be considered when selecting materials:

- Perceptual characteristics
- Number of items to be matched
- Concreteness
- Physically joined or not physically joined
- Groups of the same or not the same number

The teacher can vary or change one or more of the five characteristics and can use different materials. In this way, more difficult tasks can be designed (Figure 8–4).

Perceptual qualities are very important in matching activities. The way the materials to be matched look is important in determining how hard it will be for the child to match them. Materials can differ a great deal on how much the same or how much different they look. Materials are easier to match if the groups are different. To match animals with cages or to find a spoon for each bowl is easier than making a match between two groups of blue chips. In choosing objects, the task can be made more difficult by picking out objects that look more the same.

The number of objects to be matched is important. The more objects in each group, the more difficult it is to match. Groups with less than five things are much easier than groups with five or more. In planning activities, start with small groups (less than five) and work up

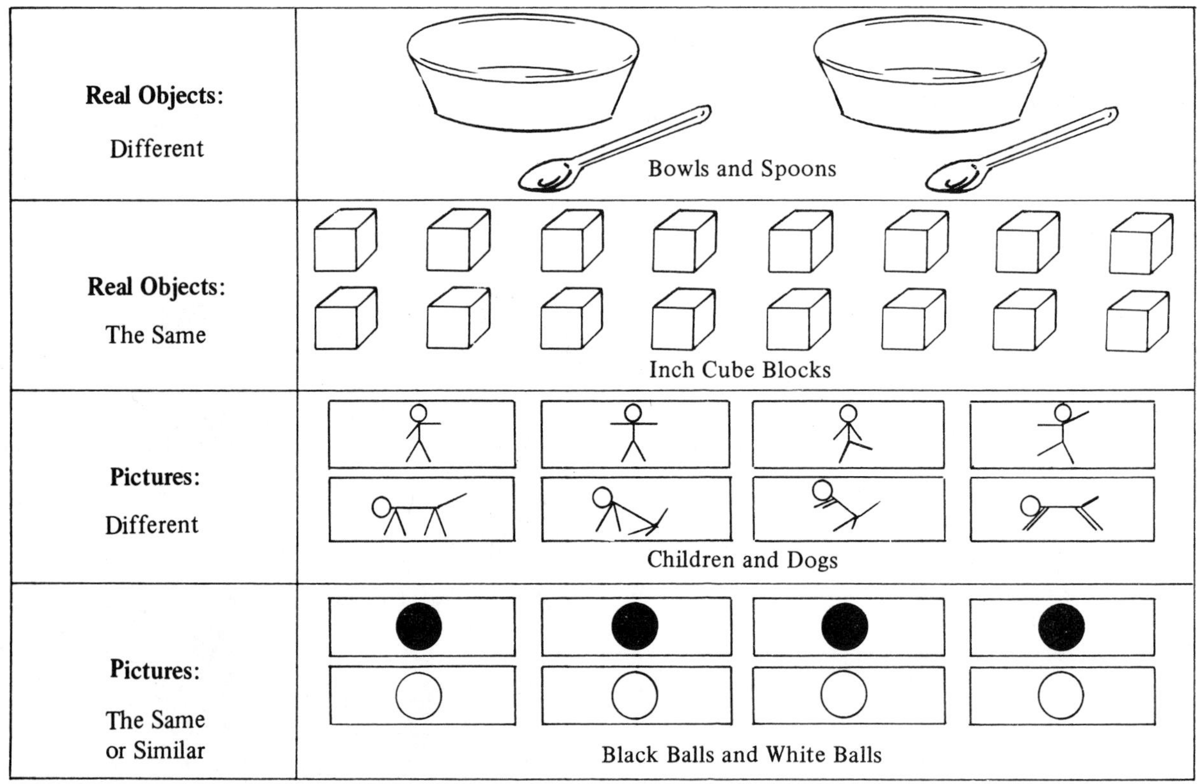

Figure 8–4 Examples of groups with different perceptual difficulty levels

step by step to groups of nine. When the child is able to place groups of ten in one-to-one correspondence, he or she has a well-developed sense of the concept.

How close to real materials are is referred to as *concreteness* (Figure 8–5). Remember from Unit 3 that instruction always should begin with concrete real objects. The easiest and first one-to-one correspondence activities should involve the use of real things such as small toys and other familar objects. Next, less familiar, more similar objects such as cube blocks, chips, and popsicle sticks can be used. The next level would be cutout shapes such as circles and squares, cowboys and horses, or dogs and doghouses. Next would come pictures of real objects and pictures of shapes. Real objects and pictures could also be used. Computer software can be used by young children who need practice in one-to-one correspondence. Programs that serve this purpose are:

Learning with Leeper. Sierra On-Line, Inc., Coarsegold, Calif. (Atari—400, 800; Atari Tape 16K; Atari Disk 24K; C64).

Let's go fishing. Learning Technologies, Inc. Dallas.

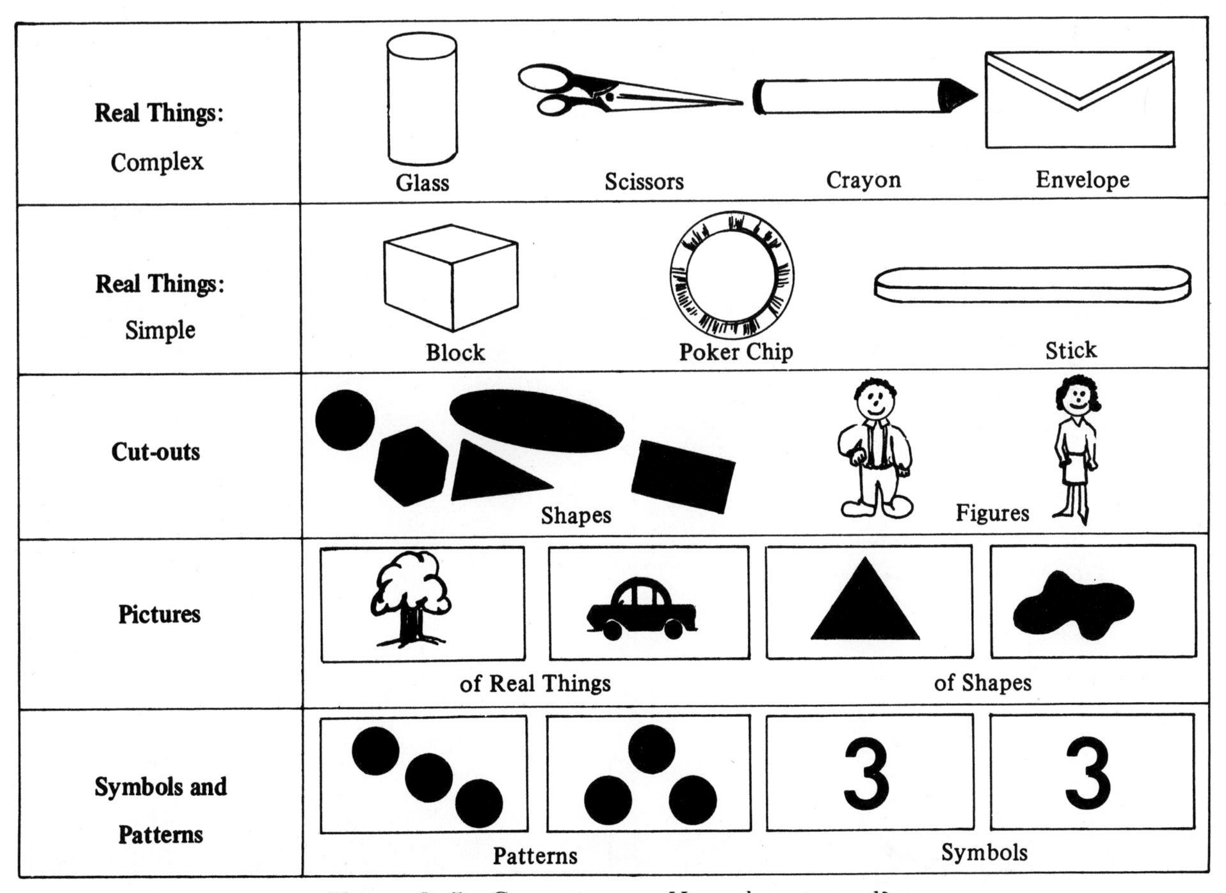

Figure 8–5 Concreteness: How close to real?

Match-up! Hayden Software Co. Lowell, Mass. (Apple II family; C64; IBM PC, PCJr; 48K).

Learning to hit the computer keys one at a time with one finger is a one-to-one experience in both the kinesthetic and perceptual-motor domains.

It is easier to tell if there is one-to-one correspondence if the objects are joined than if they are not joined. For example, it is easier to tell if there are enough chairs for each child if the children are sitting in them than if the chairs are on one side of the room and the children on the other. In beginning activities, the objects can be hooked together with a line or a string so that the children can more clearly see that there is or isn't a match. In Figure 8–6, each foot is joined to a shoe and each animal to a bowl. The hands and mittens and balls and boxes are not joined.

Placing unequal groups in one-to-one correspondence is harder than placing equal groups. When the groups have the same number, the child can check to be sure he has used all the items. When one group has more, he does not have this clue, Figure 8–7.

The sample lessons that are given illustrate some basic types of one-to-one correspondence activities. Note that they also show an increase in difficulty by varying the characteristics just described. The lessons are shown as they would be put on cards for the Idea File, Figures 8–9 through 8–14.

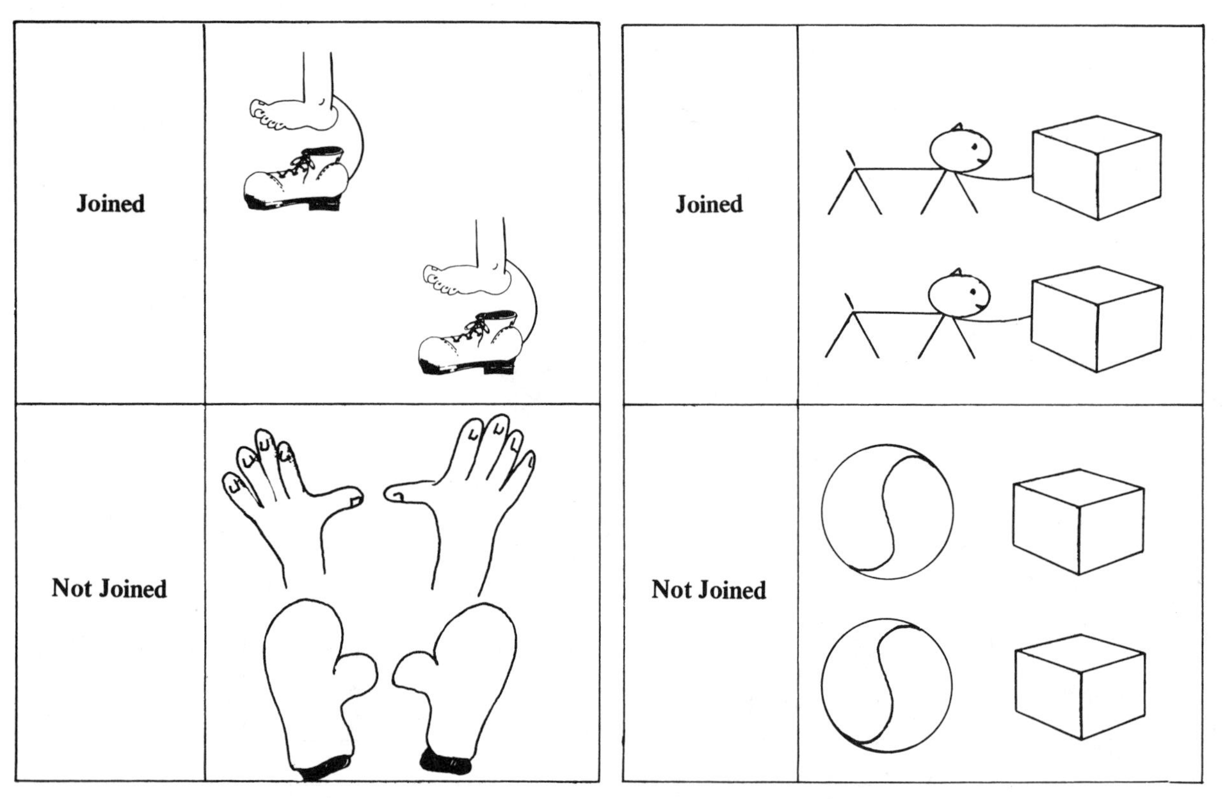

Figure 8–6 Joined groups and not joined groups

Figure 8–7 Matching equal and unequal groups

Figure 8–8 These children are learning the one-to-one concept and developing social co-operation skills.

EVALUATION

Informal evaluation can be done by noticing each child's response during structured activities. Also observe each child during free play to see whether he can pass out toys or food to other children, giving one at a time. On the shelves in the housekeeping area, paper shapes of each item (such as dishes, cups, tableware, purses, etc.) may be placed on the shelf where the item belongs. Hang pots and pans on a pegboard with the shapes of each pot drawn on the board. Do the same for blocks and other materials. Notice which children can put materials away by making the right match.

Using the same procedures as in the assessment tasks, a more formal check can be made regarding what the children have learned. Once

MATCHING – DOGS AND PEOPLE

Objective: To match joined groups of three objects.

Materials: Two sets of three objects which normally would go together. For example, doll people holding toy dogs on leashes.

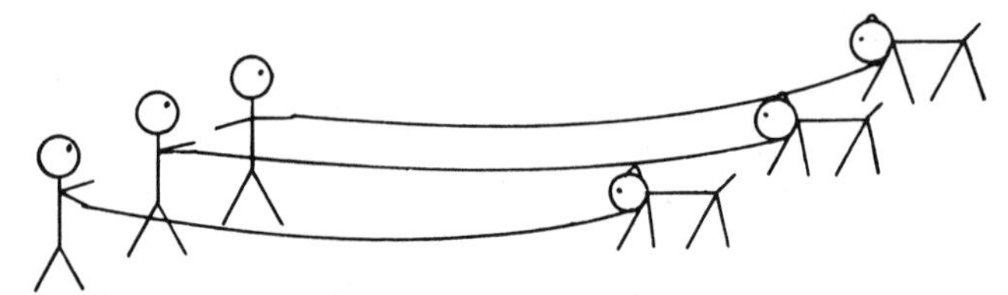

Activity: HERE ARE SOME PEOPLE AND SOME DOGS. THE DOGS ARE ON LEASHES. DOES EACH PERSON HAVE A DOG? SHOW ME HOW YOU CAN TELL. Note if the children can show or explain that the leashes connect the dogs and people.

Follow-up: Use other groups of objects such as cats and kittens, cups and saucers, houses and roofs, etc. Increase number of items in each group as the three to three task becomes easy.

Figure 8–9 One-to-one correspondence activity card—Dogs and people: Matching objects that are perceptually different.

MATCHING – THE THREE PIGS

Objective: To match joined groups of three items.

Materials: Two sets of cut-outs for the bulletin board. Three pieces of yarn. Make three pig cut-outs and three house cut-outs: straw, sticks, and bricks. Put them on bulletin board and connect each pig to his house with thick yarn:

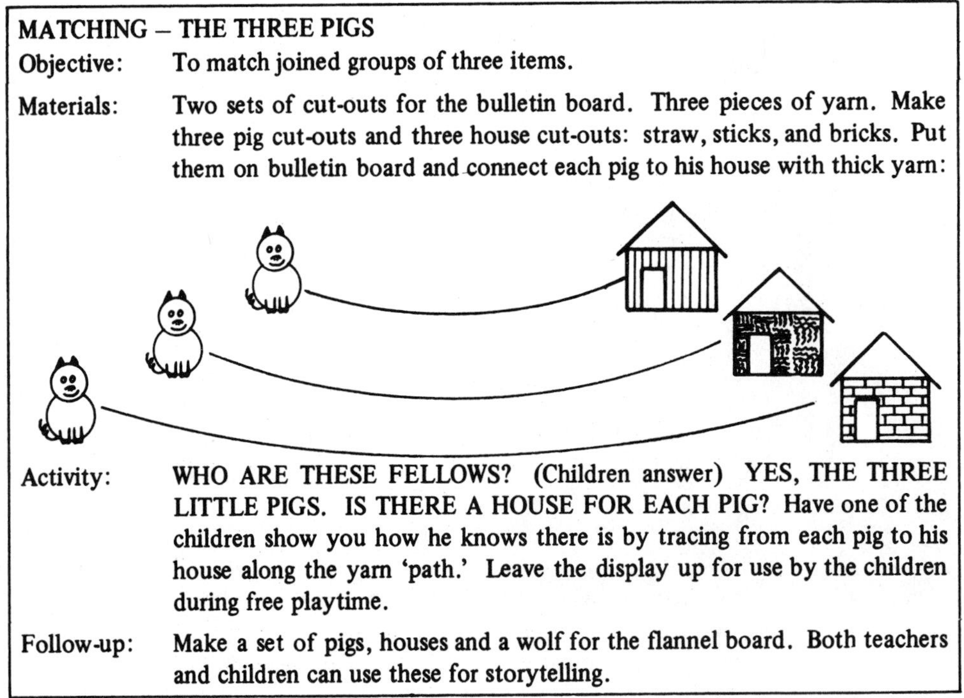

Activity: WHO ARE THESE FELLOWS? (Children answer) YES, THE THREE LITTLE PIGS. IS THERE A HOUSE FOR EACH PIG? Have one of the children show you how he knows there is by tracing from each pig to his house along the yarn 'path.' Leave the display up for use by the children during free playtime.

Follow-up: Make a set of pigs, houses and a wolf for the flannel board. Both teachers and children can use these for storytelling.

Figure 8–10 One-to-one correspondence activity card—The three little pigs matching cutouts that are perceptually different.

MATCHING – PENNIES FOR TOYS	
Objective:	To match groups of two and more objects.
Materials:	Ten pennies and ten small toys (for example, a ball, a car, a truck, three animals, three peg people, a crayon).
Activity:	LET'S PRETEND WE ARE PLAYING STORE. HERE ARE SOME PENNIES AND SOME TOYS. Show the child(ren) two toys. Place two pennies near the toys. DO I HAVE ENOUGH PENNIES TO BUY THESE TOYS IF EACH ONE COSTS ONE PENNY? SHOW ME HOW YOU CAN FIND OUT.
Follow-up:	Use more toys and more pennies as the children can match larger and larger groups.

Figure 8–11 One-to-one correspondence activity card—Pennies for toys: Matching real objects

children can do one-to-one with ten objects, try more—see how far they can go.

SUMMARY

The most basic number skill is one-to-one correspondence. Starting in infancy, children learn about one-to-one relationships. Sensorimotor and early preoperational children spend much of their playtime in one-to-one correspondence activities.

Many opportunities for informal one-to-one correspondence activities are available during play and daily routines. Materials used for structured activities with individuals and/or small

MATCHING – PICTURE MATCHING	
Objective:	To match groups of pictured things, animals, or people.
Materials:	Make or purchase picture cards which show items familiar to young children. Each set should have two groups of ten. Pictures from catalogues, magazines, or readiness workbooks can be cut out, glued on cards, and covered with clear Contac or laminated. For example, pictures of ten children should be put on ten different cards. Pictures of ten toys could be put on ten other cards.
Activity:	Present two people and two toys. DOES EACH CHILD HAVE A TOY? SHOW ME HOW YOU CAN FIND OUT. Increase the number of items in each group.
Follow-up:	Make some more card sets. Fit them to current science or social studies units. For example, if the class is studying jobs, have pilot with plane, driver with bus, etc.

Figure 8–12 One-to-one correspondence activity card—Picture matching

MATCHING – SIMILAR OR IDENTICAL OBJECTS

Objective: To match two to ten similar and/or identical objects.

Materials: Twenty objects such as poker chips, inch cube blocks, coins, cardboard circles, etc. There may be ten of one color and ten of another or twenty of the same color (more difficulty perceptually).

Activity: Begin with two groups of two and increase the size of the groups as the children are able to match the small groups. HERE ARE TWO GROUPS (BUNCHES, SETS) OF CHIPS (BLOCKS, STICKS, PENNIES, ETC.). DO THEY HAVE THE SAME NUMBER OR DOES ONE GROUP HAVE MORE? SHOW ME HOW YOU KNOW. Have the children take turns using different sizes of groups and different objects.

Follow-up: Glue some objects to a piece of heavy cardboard or plywood. Set out a container of the same kinds of objects. Have this available for matching during free playtime.

Figure 8–13 One-to-one correspondence activity card—Similar or identical objects

MATCHING – OBJECTS TO DOTS

Objective: To match 0–9 objects with 0–9 dots.

Materials: Ten frozen juice cans with dots (filled circles) painted a dark color on each from zero to nine:

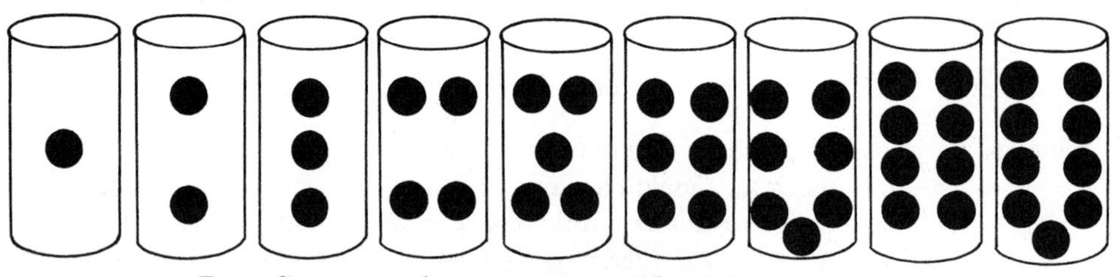

Forty-five tongue depressors or popsicle sticks.

Activity: Give each child a can. Put all the sticks in a container where the children can reach them. LOOK AT THE DOTS ON YOUR CAN. PUT THE SAME NUMBER OF STICKS IN YOUR CAN AS THERE ARE DOTS ON YOUR CAN. Have the children check with each other.

Follow-up: Put the cans and sticks out during free play. Encourage children who have had a hard time in the group to practice on their own.

Figure 8–14 One-to-one correspondence activity card—Matching objects to dots

groups should be made available for free exploration also.

Materials and activities can be varied in many ways to make one-to-one correspondence fun and interesting. Once children have a basic understanding of the one-to-one concept, they can apply the concept to higher-level activities involving equivalence and more and less.

FURTHER READING AND RESOURCES

Burton, G. 1985. *Towards a good beginning*. Menlo Park, Calif.: Addison-Wesley.

Connolly, A. J. 1982. *Early steps math*. Circle Pines, Minn.: American Guidance Service.

Scott, L. B., and Garner, J. 1978. Mathematical experiences for young children. St. Louis: McGraw-Hill.

Shaw, J. 1981. IDEAS! Math. *Dimensions*, 10 (1). 15–18.

SUGGESTED ACTIVITIES

1. Give a two-year-old some small containers and some safe objects that will fit in the containers. Observe his play. Note any evidence of one-to-one correspondence.
2. Select two or more of the structured activities from this unit. Prepare the materials needed and do the activities with two four-year-olds and two five-year-olds. Report the results to the class.
3. Create a one-to-one correspondence game. Try it out with a child. Show it to the class. Describe any improvements you would make the next time.
4. Copy the sample structured activities and place them in your Idea File. Add two more activities using different kinds of materials.
5. Discuss in a small group what you would do if you had a five-year-old in school who could not do one-to-one correspondence of two groups of ten objects.

REVIEW

A. Define one-to-one correspondence. Give at least one example.

B. Decide whether each of the following one-to-one correspondence activities is naturalistic, informal, or structured.

1. Mrs. Carter has some pictures. There are six dogs and six bones. "Petey, does each dog have a bone?"
2. Kate lines up some square blocks in a row. Then she puts a wooden doll on each block.
3. Cindy puts one shoe on each of her doll's feet.
4. Mary carefully passes one cup of milk to each child.
5. Mrs. Carter shows five-year-old George two groups of ten chips. "Find out if each group has the same number."

C. Describe several ways to vary one-to-one correspondence activities. Give examples.

D. Which of each of the following groups (a or

b) would be harder to place in one-to-one correspondence?

1. a. 4 blue chips to 4 white chips
   b. 10 blue chips to 10 white chips
2. a. 5 shoes and 5 socks
   b. 5 blobs and 5 squares
3. a. groups which are separated
   b. groups which can have each pair joined with a string
4. a. two groups of six
   b. a group of five and a group of six
5. a. cards with pictures of cups and saucers
   b. real cups and saucers

UNIT 9 Number and Counting

OBJECTIVES

After studying this unit, the student should be able to:

- Describe the concept of *number* and its relationship to counting
- Define *rote* and *rational counting* and explain their relationship
- Identify examples of rote and rational counting
- Teach counting using naturalistic, informal, and structured activities appropriate to each child's age and level of maturity

The development of the concept of number is closely tied to the acquisition of counting skills. *Number* is the understanding of what quantity means. It is the understanding of the "oneness" of one, "twoness" of two, and so on. Quantities from one to four or five are the first to be recognized. Infants can perceive the difference between these small quantities, and children as young as two-and-one-half or three years may recognize these small amounts so easily that they seem to do so without counting. The concept of number is constructed bit by bit from infancy through the preschool years, and gradually becomes a tool that can be used in problem solving.

Number's partner, counting, includes two operations—rote counting and rational counting. *Rote* counting involves reciting the names of the numerals in order from memory. That is, the child who says, "One, two, three, four, five, six, seven, eight, nine, ten" has correctly counted in a rote manner from 1 to 10. *Rational* counting involves attaching each numeral name in order to a series of objects in a group. For example, the child has some pennies in his hand. He takes them out of his hand one at a time and places them on the table. Each time he places one on the table he says the next number name in sequence. "One," places first penny. "Two," places another penny. "Three," places another penny. He has successfully done rational counting of three objects. Rational counting is a higher level of one-to-one correspondence (Figure 9–1).

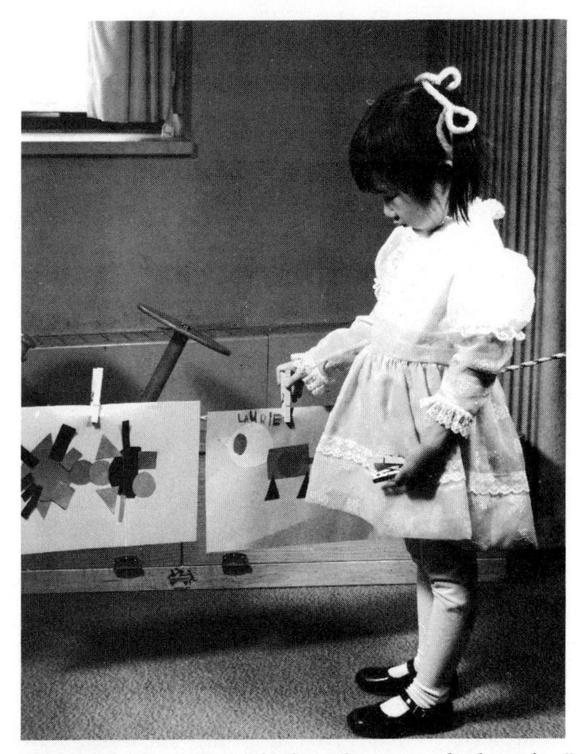

Figure 9–1 "One clothespin, two clothespins" is an example of rational counting.

A basic understanding of rote counting and one-to-one correspondence is the foundation of rational counting. The ability to rational count assists children in understanding the concept of number by enabling them to check their labeling of quantities as being a specific amount. It also helps them to compare equal quantities of different things, such as two apples, two children, and two chairs, and to realize that the quantity two is *two*, regardless of what makes up a group. In the long term, number, counting, and one-to-one correspondence all serve as the basis for developing the concept of number conservation, which is usually mastered by age six or seven. Too often the preprimary mathematics program centers on counting with repeated teacher-directed drill and practice. Children need repeated and frequent practice to develop counting skills, but this practice should be of short duration and center on naturalistic and informal instruction. Structured activities should include many applications, such as the following examples of data collection:

- How many children in the class have brothers? Sisters? Neither?
- How many days will it be until the first seed sprouts? Let's make some guesses, and then we'll keep track and see who comes close.
- How many days did we mark off before the first egg hatched?
- How many carrots do we need so that each rabbit will get one?
- How many of the small blocks will balance the big block on the balance scale?

It is a normal expectation that rote counting develops ahead of rational counting. For example, a two- or three-year-old who has a good memory might rote count to ten but only be able to rational count one or two or three objects accurately. When given a group of more than three to count, a young child might perform as described in the following example. Six blocks are placed in front of a two and one-half year old, and he is asked, "How many blocks do you have? Using his pointer finger, he "counts" and points:

- "One, two, three, four, five, six, seven, eight" (pointing at some blocks more than once and some not at all)
- "One, two, four, six, three, ten" (pointing to each block only once but losing track of the correct order)

Counting things is much more complicated than reciting number names in order from memory. The child must coordinate eyes, hands, speech, and memory to correctly do rational counting. This is difficult for the two- or three-year-old because he is still in a period of rapid growth in coordination. He is also limited in his

Figure 9–2 Rational counting involves coordination of memory, eyes, and hands.

ability to stick to a task. The teacher should not push a child to count more things than he can count easily and with success. Most rational counting experiences should be naturalistic and informal.

By age four or five the rate of physical growth is slowing. Coordination of eyes, hands, and memory is maturing. Rational counting skills should begin to catch up with rote counting skills (Figure 9–2). Structured activities can be introduced. At the same time naturalistic and informal activities should continue.

Kamii, in her book *Number in Preschool and Kindergarten* (see Further Reading), particularly emphasizes the necessity of being aware of the coordination of one-to-one correspondence and counting in the development of the concept of number. Four levels of development in counting have been identified by asking children to put out the same number of items as an adult put out using groups of sizes four to eight:

1. Children cannot understand what they are supposed to do.
2. They do a rough visual estimation or copy (that is, they attempt to make their group look like the model).
3. Children do a methodical one-to-one correspondence. It is usually past age five-and-one-half before children reach this stage.
4. Children count. That is, the child counts the items in the model and then counts out another group with the same amount. Children usually reach this stage at about age seven.

To develop the coordination of the two concepts, it is essential that children count and do one-to-one correspondence using movable objects. It becomes obvious that, among other weaknesses, the use of workbook pages precludes moving the objects to be counted and/or matched. In addition, opportunities should be provided for the children to work together so they can discuss and compare ideas. As the children work individually and/or with others, watch closely and note the thinking process that appears to take place as they solve problems using counting and one-to-one correspondence.

ASSESSMENT

The adult should note the child's regular activity. Does he recognize groups of zero to four without counting: "Mary has no cookies"; "I have two cookies"; "John has four cookies." Does he use rational counting correctly when needed: "Here, Mr. Black, six blocks for you." (Mr. Black has seen him count them out on his own.) For formal assessment, see the tasks in the Appendix. Following are sample tasks.

SAMPLE ASSESSMENT TASK

Preoperational Ages 3–6

METHOD: Interview, individual

SKILL: Demonstrates the ability to rote count

MATERIALS: None

PROCEDURE: COUNT FOR ME. COUNT AS FAR AS YOU CAN. If the child hesitates or looks puzzled, ask again. If the child still doesn't respond, say, ONE, TWO, WHAT'S NEXT?

EVALUATION: Note how far the child counts and the accuracy of the counting. Young children often lose track (i.e., "One, two, three, four, five, six, ten, seven, . . .") or miss a number name. Two's and three's may just count their ages, whereas four's usually can count accurately to ten and might try the teens and even beyond. By age five or six, children will usually begin to understand the commonalities in the twenties and beyond and move on toward counting to 100. Young children vary a great deal at each age level, so it is important to find where each individual is and move along from there.

INSTRUCTIONAL RESOURCE(S):

Charlesworth, R. & Lind, K. (1990) *Math and Science for Young Children*. Albany, NY: Delmar. Unit 9

SAMPLE ASSESSMENT TASK

Preoperational Ages 3–6

METHOD: Interview, individual or small group

SKILL: Child demonstrates ability to rational count

MATERIALS: Thirty or more objects such as cube blocks, chips, or Unifix® Cubes

PROCEDURE: Place a pile of objects in front of the child (about ten for a three-year-old, twenty for a four-year-old, thirty for a five-year-old, and as many as 100 for older children). COUNT THESE FOR ME. HOW MANY CAN YOU COUNT?

EVALUATION: Note how accurately the child counts and how many objects are attempted. In observing the process, note:

1. Does the child just use his eyes or does he actually touch each object as he counts.
2. Is some organizational system used, such as lining the objects up in rows, moving the ones counted to the side, and so on?
3. Compare accuracy on rational counting with rote counting.

INSTRUCTIONAL RESOURCE(S):

Charlesworth, R. & Lind, K. (1990) *Math and Science for Young Children*. Albany, NY: Delmar. Unit 9

NATURALISTIC ACTIVITIES

Number is an idea and counting is a skill which is used a great deal by young children in their everyday activities. Once these are in the child's thoughts and activity he will be observed often in number and counting activities. He practices rote counting often. He may run up to the teacher or parent saying, "I can count—one, two, three." He may be watching a TV program and hear "one, two, three, four. . . ." He may then repeat "one, two, ..." At first he may play with the number names saying to himself, "one, two, five, four, eight, ..." in no special order. Listen carefully and note that gradually he gets more of the names in the right order.

Number appears often in the child's activities once he has the idea in mind. A child is eating crackers, "Look at all my crackers. I have two crackers." One and two are usually the first amounts used by children two and three years old. They may use one and two for quite a while before they go on to larger groups. Number names are used for an early form of division. That is, a child has three cookies which he divides with his friends, "One for you, one for you, and one for me." The child is looking at a picture book, "There is one daddy and two babies." The child wants another toy, "I want one more little car, Dad."

INFORMAL ACTIVITIES

The alert adult can find a multitude of ways to take advantage of opportunities for informal instruction. For example, the child is watching a children's television program. The teacher is sitting next to him. A voice from the TV rote counts by singing, "One, two, three, four, five, six." The teacher says to the child, "That's fun, let's count too." They then sing together. "One, two, three, four, five, six." Or, the teacher and children are waiting for the school bus to arrive. "Let's count as far as we can while we wait. One, two, three, . . ." Since rote counting is learned through frequent but short periods of practice, informal activities should be used most for teaching.

Everyday activities offer many opportunities for informal rational counting and number activities. For instance, the teacher is helping a child get dressed after his nap. "Find your shoes and socks. How many shoes do you have? How many socks? How many feet?" Some children are meeting at the door. The teacher says, "We are going to the store. There should be five of us. Let's count and be sure we are all here."

Table setting offers many chances for rational counting (Figure 9–3). "Put six placemats on each table." "How many more forks do we need?" "Count out four napkins." Play activities also offer times for rational counting. "Mary, please give Tommy two trucks." A child is looking at his hands which are covered with fingerpaint: "How many red hands do you have, Joey?"

Figure 9–3 Setting the table is an opportunity for informal rational counting and one-to-one correspondence.

A more challenging problem can be presented by asking an open-ended question such as, "Get enough napkins for everyone at your table." or "Be sure everyone has the same number of carrot sticks." In these situations, children don't have a clue to help them decide how many they need or how many each person should get but have to figure out for themselves how to solve the problem. Often children will forget to count themselves. This presents an excellent opportunity for group discussion to try to figure out what went wrong. The teacher could follow-up such an incident by reading the book *Six Foolish Fishermen* (Elkin 1964, see Appendix B) in which the fishermen make the same mistake.

STRUCTURED ACTIVITIES

Rote counting is learned mostly through naturalistic and informal activities. However, there are short, fun things which can be used to help children learn the number names in the right order. There are many rhymes, songs, and fingerplays. Songs include those such as *This Old Man*, *Ten Little Indians*, and *The Twelve Days of Christmas*.

A favorite old rhyme is:

> One, two, buckle your shoe.
> Three, four, shut the door.
> Five, six, pick up sticks.
> Seven, eight, shut the gate.
> Nine, ten, a big fat hen.

A finger play can be used:

> **Five Little Birdies**
>
> *(Hold up five fingers. As each bird leaves fly your hand away and come back with one less finger standing up.)*
>
> Five little birdies sitting by the door
> One flew away and then there were four.
> Four little birdies sitting in a tree
> One flew away and then there were three.
> Three little birdies sitting just like you
> One flew away and then there were two.
> One little birdie sitting all alone
> He flew away and then there were none.

More direct ways of practicing rote counting are also good. Clapping and counting at the same time teaches number order and gives practice in rhythm and coordination. With a group, everyone can count at the same time, "Let's count together. One, two, three, ..." Individual children can be asked, "Count as far as you can."

Groups which have zero to four items are special in the development of rational counting skills. The number of items in groups this size can be perceived without counting. For this reason these groups are easy for children to understand. They should have many experiences and activities with groups of size zero to four before they work with groups of five and more. With structured activities it is wise to start with groups of size two because, as mentioned before, there are so many which occur naturally. For example, the child has two eyes, two hands, two arms, and two legs. Two pieces of bread are used to make a sandwich, and bikes with two wheels are signs of being big. For this reason, activities using two are presented first in the following examples. The activities are set up so that they can be copied onto activity cards for the file (Figure 9–4).

STRUCTURED ACTIVITIES

NUMBER: GROUPS OF TWO

OBJECTIVE: To learn the idea of two

MATERIALS: The children's bodies, the environment, a flannel board and/or a magnetic board, pairs of objects, pictures of pairs of objects

ACTIVITIES:

1. Put several pairs of felt pieces (such as hearts, triangles, or bunnies, for example) on the flannel board (or magnets on the magnet board). Point to each group in turn: WHAT ARE THESE? HOW MANY ARE THERE?
2. Have the children check their bodies and the other children's bodies for groups of two.
3. Have the children, one at a time, find groups of two in the room.
4. Using rummy cards, other purchased picture cards, or cards you have made, make up sets of cards with identical pairs. Give each child a pack with several pairs mixed up. Have them sort the pack and find the groups of two.

5. Have a container with many objects. Have the children sort out as many groups of two as they can find.

FOLLOW-UP: Have the materials available during free play.

NUMBER: GROUPS OF THREE

OBJECTIVE: To learn the idea of three

MATERIALS: Flannel board and/or magnet board, objects, picture cards

ACTIVITIES: Do the same types of activities using groups of three. Emphasize that three is one more than two.

FOLLOW-UP: Have the materials available during free play.

NUMBER: GROUPS OF ONE

OBJECTIVE: To learn the idea that one is a group

MATERIALS: Flannel board and/or magnet board, objects, picture cards

ACTIVITIES: Do the same types of activities using groups of one as were done for groups of two and three.

FOLLOW-UP: Have the materials available during free play.

NUMBER: ZERO

OBJECTIVE: To understand the idea that a group with nothing in it is called "zero"

MATERIALS: Flannel board, magnet board, and objects

ACTIVITIES:

1. Show the children groups of things on the flannel board, magnet board, and/or groups of objects. SEE ALL THESE THINGS? Give them a chance to look and respond. NOW I TAKE THEM AWAY. WHAT DO I HAVE NOW? They should respond with "nothing," "all gone," and/or "no more."
2. Put out a group of flannel pieces, magnet shapes, or objects of a size the children all know (such as one, two, three, or four). Keep taking one away. HOW MANY NOW? When none are left say: THIS AMOUNT IS CALLED ZERO. Repeat until they will answer "zero" on their own.
3. Play a Silly Game. Ask HOW MANY REAL LIVE TIGERS DO WE HAVE IN OUR ROOM? (continue with other things which obviously are not in the room).

FOLLOW-UP: Work on the concept informally. Ask questions: HOW MANY CHILDREN ARE HERE AFTER EVERYONE GOES HOME? (after snack if all the food has been eaten) HOW MANY COOKIES (CRACKERS, PRETZELS, etc.) DO YOU HAVE NOW?

Figure 9–4 When the child understands the number concept, then quantities can be associated with symbols (see Units 23 and 24).

After the children have the ideas of groups of zero, one, two, and three then go on to four. Use the same kinds of activities. When they have four in mind then do activities using groups of all five amounts. Emphasize the idea of *one more than* as you move to each larger group.

When the children have the idea of groups from zero to four they can then go on to groups larger than four. Some children are able to perceive five without counting just as they perceive zero through four without actually counting. Having learned the groups of four and less, children can be taught five by adding on one more to groups of four. When the children understand five as four with one more and six as five with one more and so on, then more advanced rational counting can begin. That is, children can work with groups of objects where they can only find the number by actually counting each object. Before working with a child on counting groups of six or more, the adult must be sure the child can do the following:

SAMPLE ACTIVITIES

NUMBER: USING SETS ZERO THROUGH FOUR

OBJECTIVE: To understand groups of zero through four

MATERIALS: Flannel board, magnet board, and/or objects

ACTIVITIES:

1. Show the children several groups of objects of different amounts. Ask them to point to sets of one, two, three, and four.
2. Give the children a container of many objects. Have them find sets of one, two, three, and four.
3. Show the children containers of objects (pennies, buttons, etc.). Ask them to find the ones with groups of zero, one, two, three, and four.
4. Give each child four objects. Ask each one to make as many different groups as he can.
5. Ask the children to find out HOW MANY _____ ARE IN THE ROOM? (Suggest things for which there are four or less.)

- Recognize groups of zero to four without counting
- Rote count to six or more correctly and quickly
- Recognize that a group of five is a group of four with one more added

The following are activities for learning about groups larger than four.

STRUCTURED ACTIVITIES

NUMBER/RATIONAL COUNTING: INTRODUCING FIVE

OBJECTIVE: To understand that five is four with one more item added

MATERIALS: Flannel board, magnet board, and/or objects

ACTIVITIES:

1. Show the children a group of four. HOW MANY IN THIS GROUP? Show the children a group of five. HOW MANY IN THIS GROUP? Note how many already have the idea of five. Tell them YES, THIS IS A GROUP OF FIVE. Have them make other groups with the same amount using the first group as a model.
2. Give each child five objects. Ask them to identify how many obejcts they have.
3. Give each child seven or eight objects. Ask them to make a group of five.

FOLLOW-UP: Have containers of easily counted and perceived objects always available for the children to explore. These would be items such as buttons, poker chips, Unifix® cubes and inch cubes.

NUMBER/RATIONAL COUNTING: GROUPS LARGER THAN FIVE

OBJECTIVE: To be able to count groups of amounts greater than five

MATERIALS: Flannel board and/or magnet board, objects for counting, pictures of groups on cards, items in the environment

ACTIVITIES:

1. One step at a time present groups on the flannel board and magnet board and groups made up of objects such as buttons, chips, popsicle sticks, inch cube blocks, etc. Have the children take turns counting them—together and individually.
2. Present cards with groups of six or more, showing cats, dogs, houses, or similar figures. Ask the children as a group or individually to tell how many items are pictured on each card.
3. Give the children small containers with items to count.
4. Count things in the room. HOW MANY TABLES (CHAIRS, WINDOWS, DOORS, CHILDREN, TEACHERS)? Have the children count all at the same time and individually.

FOLLOW-UP: Have the materials available for use during free play. Watch for opportunities for informal activities.

NUMBER/RATIONAL COUNTING: FOLLOW-UP WITH STORIES
OBJECTIVE: To be able to apply rational counting to fantasy situations.
MATERIALS: Stories which will reinforce the ideas of groups of numbers and rational counting skills: Some examples are *The Three Pigs*, *The Three Bears*, *The Three Billy Goats Gruff*, *Snow White and the Seven Dwarfs*, *Six Foolish Fishermen*.
ACTIVITIES: As these stories are read to the younger children, take time to count the number of characters who are the same animal or same kind of person. Use felt cutouts of the characters for counting activities and for one-to-one matching (as suggested in Unit 8). Have older children dramatize the stories. Get them going with questions such as:
HOW MANY PEOPLE WILL WE NEED TO BE BEARS? HOW MANY PORRIDGE BOWLS, (SPOONS, CHAIRS, BEDS) DO WE NEED? JOHN, YOU GET THE BOWLS AND SPOONS.
FOLLOW-UP: Have the books and other materials available for the children to use during free playtime.

NUMBER/RATIONAL COUNTING: FOLLOW-UP WITH COUNTING BOOKS
OBJECTIVE: To strengthen rational counting skills
MATERIALS: Counting books (see list in Appendix)
ACTIVITIES: Go through the books with one child or small group of children. Discuss the pictures as a language development activity and count the items pictured on each page.

RATIONAL COUNTING: FOLLOW-UP WITH ONE-TO-ONE CORRESPONDENCE
OBJECTIVE: To combine one-to-one correspondence and counting
MATERIALS: Flannel board and/or magnet board and counting objects
ACTIVITIES: As the children work with the counting activities, have them check their sets which they say are the same number by using one-to-one correspondence. See activities for Unit 8.
FOLLOW-UP: Have materials available during free playtime.

Four- to six-year-olds can play simple group games that require them to apply their counting skills. For example, a bowling game requires them to count the number of pins they knock down. A game in which they try to drop clothespins into a container requires them to count the number of clothespins that land in the container. They can compare the number of pins knocked down or the number of clothespins dropped into the containers by each child. By age six or seven children can keep a cumulative score using tic marks (lines) such as:

Student	Score
Derrick	/ / / / / /
Liu Pei	/ / / / / / / /
Brent	/ / / /
Theresa	/ / / / / /

Not only can they count, they can compare amounts to find out who has the most and if any of them have the same amount. Older children (see Units 23 and 24) will be interested in writing numerals and might realize that instead of tic marks they can write down the numeral that represents the amount to be recorded.

Quite a bit of computer software has been designed to reinforce counting skills and the number concept. Five-year-old George sits at the computer deeply involved with *Stickybear Numbers* (Weekly Reader Family Software). Each time he presses the space bar, a group with one less appears on the screen. Each time he presses a number, a group with that amount appears. His friend Kate joins him and comments on the pictures. They both count the figures in each group and compare the results. Activity #8 includes a list of software the reader might wish to review.

EVALUATION

Informal evaluation can be done by noting the answers given by the children during teaching sessions. The teacher should also observe the children during free play and notice whether they apply what they have learned. When they choose to explore materials used in the structured lessons during the free play period, questions can be asked. For instance, Kate is at the flannel board arranging the felt shapes in rows. As her teacher goes by, she stops and asks, "How many bunnies do you have in that row?" Or, four children are playing house, and the teacher asks, "How many children in this family?" Formal evaluation can be done with one child by using tasks such as those in the Appendix.

SUMMARY

The number concept involves the understanding of quantities. Counting and one-to-one correspondence work together to develop this concept. The number concept involves an understanding of "oneness," "twoness," and so on. Counting includes two types of skills: rote counting and rational counting. Rote counting is saying from memory the names of the numerals in order. Rational counting is attaching the number names in order to items in a group to find out the total number of items in the group.

Rote counting is mastered before rational counting. Rational counting begins to catch up with rote counting after age four or five. The number concept is learned simultaneously. Quantities greater than four are not identified until the child learns to rational count beyond four. Counting is learned for the most part through naturalistic and informal activities supported by structured lessons.

FURTHER READING AND RESOURCES

Beaty, J. J., and Tucker, W. H. 1987. *The computer as a paintbrush.* Columbus, Ohio: Merrill.

Baratta-Lorton, M. 1976. *Math their way.* Menlo Park, Calif.: Addison-Wesley.

Barson, A., and Barson, L. 1987. IDEAS for teachers: Dot Bingo. *Arithmetic Teacher.* 35 (1): 27, 29.

Clements, D. H., and Callahan, L. G. 1983. Number or prenumber foundational experiences for young children: Must we choose? *Arithmetic Teacher*. 31 (3): 34–37.

Gelman, R., and Gallistel, C. R. 1986. *The child's understanding of number*. Cambridge, Mass.: Harvard University Press.

Hoot, J. L. 1986. *Computers in early childhood education*. Englewood Cliffs, N.J.: Prentice-Hall.

Kamii, C. 1982. *Number in preschool and kindergarten*. Washington, D.C.: National Association for the Education of Young Children.

Richardson, K. 1984. *Developing number concepts using Unifix Cubes*. Menlo Park, Calif.: Addison-Wesley.

Scott, L. B., and Garner, J. 1978. *Mathematical experiences for young children*. St. Louis: McGraw-Hill.

SUGGESTED ACTIVITIES

1. Try the sample assessment tasks included in this unit with three or four children at different ages between two-and-one-half and six-and-one-half years. Record the results and compare the answers received at each age level.

2. Go to the library. Find five number books (such as those listed in the Appendix). Go through each one. Write a description of each one. Tell whether you would buy it for your own school and how you would use it.

3. Design a counting lesson of three or four activities. Try it out with individuals or small groups of children. Evaluate its success in a few sentences.

4. Add five cards with rote counting activities and five cards with rational counting activities to your card file of activities.

5. Collect small objects (such as buttons, toothpicks, etc.) and containers (such as empty pharmacists' pill bottles, egg cartons, etc.) and make your own counting kit.

6. Visit a classroom. Write down a brief description of each example of a counting activity observed. Label each as to whether it involved rote or rational counting and whether it was naturalistic, informal, or structured teaching.

7. View some children's educational television programs. Describe how counting is taught. Develop a plan for following up on what the child views in order to strengthen his learning.

8. Try to locate some of the computer software listed below and review it. Use the following criteria suggested by Spencer and Baskin (in Hoot, see Further Reading and Resources):

   a. Read the instructions that come with the program and then try it out. The first time through, take note of your impressions. Was the program interesting? Easy to use?

   b. Go through the program two more times. Are the prompts, questions, and responses logical and clear? Are the graphics, sound, and/or animation of good quality?

   c. If possible, try out each program you reviewed with children. Did they enjoy it? Could they work independently? Would they like to use it again?

   d. Obtain the opinions of teachers and/or parents who have used the program with children.

   e. Write an evaluation of each program including the following information:

      1) Program name, manufacturer, and price

      2) Concept or skill taught

      3) Brief description

4) Your evaluation

Suggested Software

Charlie Brown's 123s. Random House, New York: Apple II.

Finger Abacus. Edutek Corp., Palo Alto, Calif.: Apple II with Applsoft ROM, 32K.

Getting Ready to Read and Add. Sunburst, Pleasantville, N.Y.: 48K Apple II; 48K Atari; C64; 64K IBM PC/128K PCJr.

Introduction to Counting. Edu-Ware, Services, Agoura, Calif.: Apple II, IIe, II+; Franklin Ace, Atari 800.

Kinder Concepts. Midwest Software, Farmington, Mich.: C64/128; Apple II+.

Number Farm. Developmental Learning Materials, Allen, Tex.

Sequence. Spinnaker, Cambridge, Mass.: Apple, IBM-PC & PCJr, Commodore.

Stickybear Numbers. Xerox Education Publications, Middletown, Conn.: Apple IIs, 48K; Atari, 48K; Commodore 64.

REVIEW

A. Discuss each of the following and their relationship:

1. Number concept
2. Rote counting
3. Rational counting
4. One-to-one correspondence

B. Select the correct statements:

1. Rational counting develops before rote counting.
2. Rote counting is learned mainly through frequent, short periods of practice.
3. Most rote counting activities are naturalistic and informal.
4. Rote counting usually develops ahead of rational counting until about age four or five when rational counting starts to catch up.
5. Rational counting should be taught by having daily structured lessons beginning at age two.
6. When teaching rational counting, it is important to give the child groups of real objects to count.
7. Counting supports the development of the number concept but is not the only underlying skill necessary for this development.

C. Decide which of the following incidents would involve rote counting and which would involve rational counting.

1. Mary says to her teacher, "I am going to get six napkins to put on the table."
2. Johnny says, "I can count! Listen to me: one, two, three!"
3. The children are singing, "This old man he played one. . . ."
4. The teacher says to Joyce, "Please bring three more boxes of crayons."
5. The teacher hands Peter a string of colored beads, "How many beads on the string, Peter?"
6. "Let's count and clap—one (clap), two (clap). . . ."

7. "Juanita, please give everyone at your table one pencil and one piece of paper."

D. What should the adult do in the following situations?

1. Randy (age 3 1/2) looks up with a big smile and says, "One, two, three, four, five, six, seven, eight, nine, ten."
2. Tanya (age 2 1/2) says, "I have two eyes, you have two eyes."
3. Mary is five. She cannot count out eight forks for setting the table without making a mistake.
4. Bobby is 4 1/2. He is counting a stack of five blocks: "One, two, four, eight, seven—seven blocks."
5. The teacher is having lunch at a table with five three- and four-year-olds. She wants to have some informal counting experiences.
6. Some of the children who are of kindergarten age seem to be having trouble with counting activities.

UNIT 10 Sets and Classifying

OBJECTIVES

After studying this unit, the student should be able to:

- Describe features of sets
- Describe the activity of classifying
- Identify five types of criteria that children can use when classifying
- Assess, plan, and teach classification activities appropriate for young children

In mathematics and science, an understanding of sets and classifying is essential. The term *set* refers to things that are put together in a group based on some common criteria (such as color, shape, size, use). A set can contain from zero (an empty set) to an endless number of things (or members). However, most sets which children work with have some observable limit. A set of dishes is usually for a certain number of place settings such as service for eight or service for twelve. A set of tires for a car is usually four plus a spare which equals five. A set of tires for a large truck consists of more than five tires.

To add is to put together or join sets. The four tires on the wheels plus the spare in the trunk equals five tires. To subtract is to separate a set into smaller sets. For example, one tire goes flat and is taken off. It is left for repairs, and the spare is put on. A set of one (the flat tire) has been taken away or subtracted from the set of five. There are now two sets: four good tires on the car and one flat tire being repaired.

Before doing any formal addition and subtraction, the child needs to learn about sets and how they can be joined and separated (Figure 10–1). That is, children must practice sorting (separating) and grouping (joining). This type of activity is called *classification*. The child does tasks where he separates and groups things because they belong together for some reason. Things may belong together because they are the same color, or the same shape, do the same work, are the same size, are always together, and so on. For example, a child may have a box of wooden blocks and a box of toy cars (wood, metal, and plastic). The child has two sets: blocks and cars. He then takes the blocks out of the box and separates them by grouping them into four piles: blue blocks, red blocks, yellow blocks, and green blocks. He now has four sets of blocks. He builds a garage with the red blocks. He puts some cars in the garage. He now has a new set of toys. If he has put only red cars in the garage, he now has a set of red toys. The blocks and toys could be grouped in many other ways by using shape and material (wood, plastic, and metal) as the basis for the groups. Figure 10–2 illustrates some possible groups using the blocks and cars.

Young children spend much of their playtime in just such classification activities. As children work busily at these sorting tasks, they simultaneously learn words that label their activity. This happens when another person tells him the names and makes comments: "You have made a big pile of red things." "You have a pile of blue blocks, a pile of green blocks,. . . ." "Those are plastic and those are wood." As the child learns

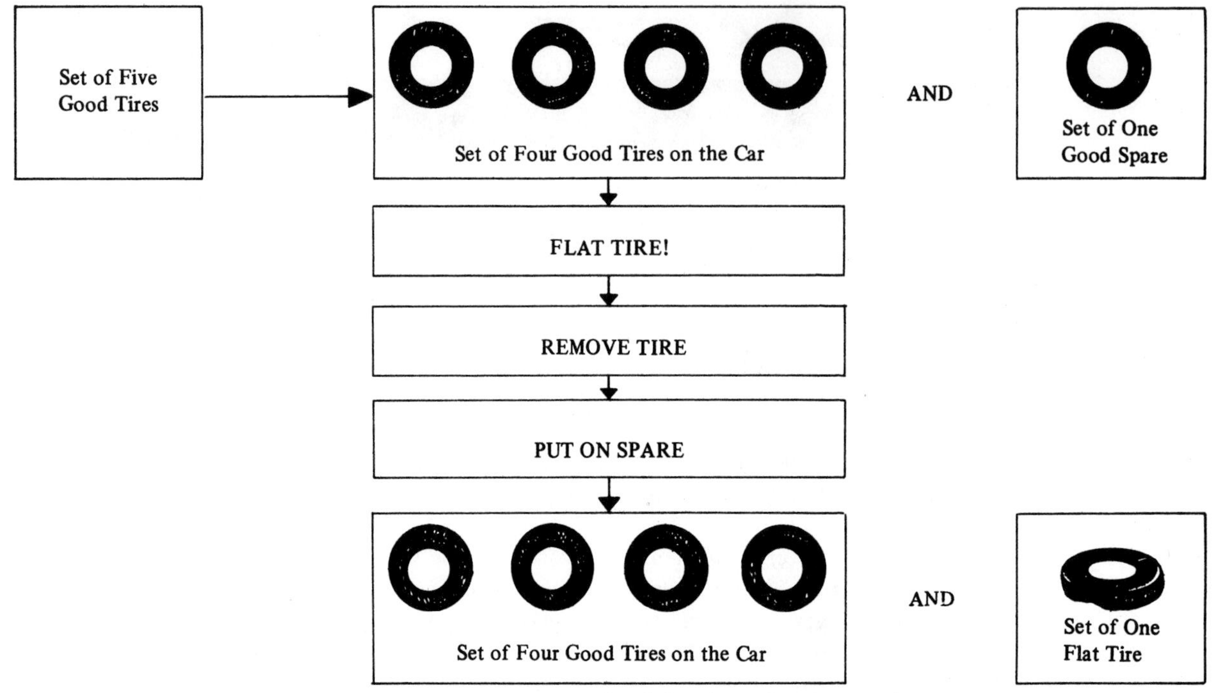

Figure 10–1 Sets can be joined and separated.

to speak, the adult questions him, "What color are these? Which ones are plastic?"

The child learns that things may be grouped together using a number of kinds of common features:

- **Color:** Things can go together that are the same color.
- **Shape:** Things may all be round, square triangular, and so on.
- **Size:** Some things are big and some are small; some are fat and some are thin; some are short and some are tall.
- **Material:** Things are made out of different materials such as wood, plastic, glass, paper, cloth, and metal.
- **Pattern:** Things have different visual patterns such as stripes, dots, flowers, or may be plain (no design).
- **Texture:** Things feel different from each other (smooth, rough, soft, hard, wet, dry).
- **Function:** Some items do the same thing or are used for the same thing (all are for eating, writing, playing music, for example).
- **Association:** Some things do a job together (candle and match, milk and glass, shoe and foot) or come from the same place (bought at the store or seen at the zoo) or belong to a special person (the hose, truck, and hat belong to the fire fighter).
- **Class name:** There are names which may belong to several things (people, animals, food, vehicles, weapons, clothing, homes).
- **Common features:** All have handles or windows or doors or legs or wheels, for example.

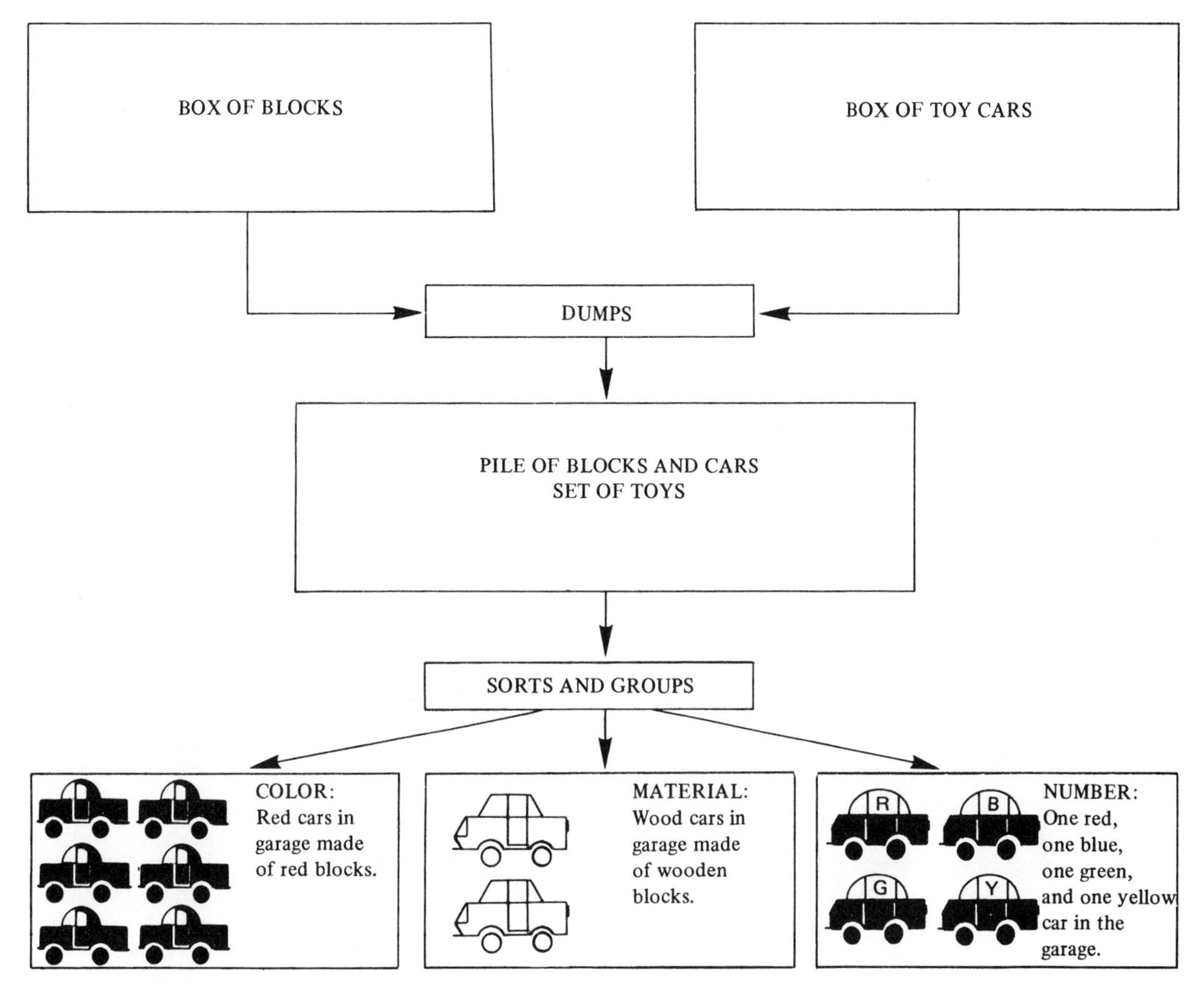

Figure 10–2 Classification (forming sets) may be evident in children's play.

- **Number:** All are groups of specific amounts (see Unit 6) such as pairs, groups of three, four, five, and so on.

ASSESSMENT

The adult should note and record the child's play activities. Does he sort and group his play materials? For example, he might play with a pegboard and put each color peg in its own row; build two garages with big cars in one and small cars in another: when offered several kinds of crackers for snack, he might pick out only triangle shapes: he might say, "Only boys can be daddies—girls are mothers."

More formal assessment can be done using the tasks in the Appendix. Two examples are shown:

SAMPLE ASSESSMENT TASK

Preoperational Ages 4–6

METHOD: Interview

SKILL: Child is able to classify and form sets using verbal and/or object clues

MATERIALS: Twenty to twenty-five objects (or pictures of objects or cutouts) that can be grouped into several possible sets by criteria such as color, shape, size, or category (i.e., animals, plants, furniture, clothing, or toys).

PROCEDURE: Set all the objects in front of the child in a random arrangement. Try the following types of clues:

1. FIND SOME THINGS THAT ARE ______________ (Name a specific color, shape, size, material, pattern, function, or class).
2. Hold up one object, picture, or cutout. Say, FIND SOME THINGS THAT BELONG WITH THIS. After the choices are made, ask, WHY DO THESE THINGS BELONG TOGETHER?

EVALUATION: Note if the child can make a logical looking group and provide a logical reason for the choices. That is, "Because they are cars," ("They are all green," "You can eat with them," etc.).

INSTRUCTIONAL RESOURCE(S):

Charlesworth, R., and Lind, K. 1990. *Math and Science for Young Children*. Albany, N.Y.: Delmar.

SAMPLE ASSESSMENT TASK

Preoperational Ages 4–6

METHOD: Interview

SKILL: Child is able to classify and form sets in a free sort

MATERIALS: Twenty to twenty-five objects (or pictures of objects or cutouts) that can be grouped into several possible sets by criteria such as color, shape, size, or category (i.e., animals, plants, furniture, clothing, or toys).

PROCEDURE: Set all the objects in front of the child in a random arrangement. Say, PUT THE THINGS TOGETHER THAT BELONG TOGETHER. If the child looks puzzled, backtrack to the previous task, hold up one item, and say FIND SOME THINGS THAT BELONG WITH THIS. When a set is completed, say, NOW FIND SOME OTHER THINGS THAT BELONG TOGETHER. Keep on until all the items are grouped. Then point to each group and ask, WHY DO THESE BELONG TOGETHER?

EVALUATION: Note if the child can make logical looking groups and provide a logical reason for each one. That is, "Because they are cars," ("They are all green," "You can eat with them," etc.).

INSTRUCTIONAL RESOURCE(S):

Charlesworth, R., and Lind, K. 1990. *Math and Science for Young Children.* Albany, N.Y.: Delmar.

NATURALISTIC ACTIVITIES

Sorting and grouping is one of the most basic and natural activities for the young child. Much of his play is organizing and reorganizing the things in his world. The infant learns the set of people who take care of him most of the time (day care provider, mother, father, and/or relatives and friends) and others are put in the set of "strangers" He learns that some objects when pressed on his gums makes the pain of growing teeth less. These are his set of teething things.

As soon as the child is able to sit up, he finds great fun in putting things in containers and dumping them out. He can never have too many boxes, plastic dishes, and coffee cans along with safe items such as large plastic beads, table tennis balls, or teething toys (just be sure the items are too large to swallow). With this type of activity, children have their first experiences making sets.

By age three the child sorts and groups to help organize his play activities. He sorts out from his things those which he needs for what he wants to do. He may pick out wild animal toys for his zoo; people dolls for his family play; big blocks for his house; blue paper circles to paste on paper; girls for friends, and so on.

The adult provides the free time, the materials (junk is fine as long as it is safe), and the space. The child does the rest.

INFORMAL ACTIVITIES

Adults can let children know that sorting and grouping activities are of value in informal ways by showing that they approve of what the children are doing. This can be done with a look, smile, nod, or comment.

Adults can also build children's classification vocabulary in informal ways. They can label the child's product and ask questions about what the child has done: "You have used all blue confetti in your picture." "You've used all the square blocks." "You have the pigs in this barn and the cows in that barn." "You painted green and purple stripes today." "Can you put the wild animals here and the farm animals here?" "Separate the spoons from the forks." "See if any of the cleaning rags are dry." "Put the crayons in the can and the pencils in the box." "Show me which things will roll down the ramp." "Which seeds are from your apple? Which are from your orange?" "Put the hamsters in the silver cage and the mice in the brown cage." As the children's vocabularies increase, they will be able to label and describe how and why they are sorting and grouping. In addition, words give them shortcuts for labeling sets.

STRUCTURED ACTIVITIES

Sorting and grouping, which form the basis of classifying sets of things, lend themselves to many activities with many materials (Figure 10–3). As already mentioned, real objects are used first, then pictures and objects, then cutouts, and then pictures. One-to-one correspondence skills are needed to sort and group. Classification takes the child into higher levels of grouping that go beyond one-to-one correspondence. Several

Figure 10–3 "Put the fruit in one bowl and the vegetables in the other."

things may be classified into any one set. The child must keep in mind the basis for his group as he sorts through all available things. Remembering that when given three pigs and three houses in separate piles that he must find a pig for each house is easier than being given a mixed pile of pigs and houses which he must first sort into a pile of pigs and a pile of houses before he can find if there is a house for each pig.

The following activities help children learn the idea of sets:

STRUCTURED ACTIVITIES

SETS AND CLASSIFICATION: COLOR

OBJECTIVE: To sort and group by color

MATERIALS: Several different objects that are the same color and four objects each of a different color; for example, a red toy car, a red block, a red bead, a red ribbon, a red sock, and so on, and one yellow car, one green ribbon, one blue ball, and one orange piece of paper

ACTIVITIES:

1. Hold up one red object, FIND THE THINGS THAT ARE THE SAME COLOR AS THIS. After all the red things have been found: THESE THINGS ARE ALL THE SAME COLOR. TELL ME THE NAME OF THE COLOR. If there is no correct answer: THE THINGS YOU PICKED OUT ARE ALL RED THINGS. Ask: WHAT COLOR ARE THE THINGS THAT YOU PICKED OUT?
2. Put all the things together again: FIND THE THINGS THAT ARE *NOT* RED.

FOLLOW-UP: Repeat this activity with different colors and different materials. During center time put out a container of brightly colored materials. Note if the children put them into groups by color. If they do, ask, "Why did you put those together?" Accept any answer they give but note whether they give a color answer.

SETS AND CLASSIFICATION: ASSOCIATION

OBJECTIVE: To form sets of things that go together by association

MATERIALS: Picture card sets may be bought or made. Each set can have a theme such as one of the following:

1. Pictures of people in various jobs and pictures of things that go with their job:

Worker	Things that go with the worker's job
letter carrier	letter, mailbox, stamps, hat, mailbag, mail truck
airplane pilot	airplane, hat, wings
doctor	stethoscope, thermometer, Band-Aid
trash collector	trash can, trash truck
police officer	handcuffs, pistol, hat, badge, police car
fire fighter	hat, hose, truck, boots and coat, hydrant, house on fire
grocer	various kinds of foods, bags, shopping cart

Start with about three sets and keep adding more.

2. Things that go together for use:

Item	Goes with
glass tumbler	carton of milk, pitcher of juice, can of soda pop
cup and saucer	coffee pot, teapot, steaming teakettle
match	candle, cigarette, campfire
paper	pencil, crayon, pen
money	purse, wallet, bank
table	four chairs

Start with three sets and keep adding more.

3. Things that are related, such as animals and their babies

ACTIVITIES:

1. One at a time show the pictures of people or things which are the main clue (the workers for example) and ask: WHO (WHAT) IS THIS? When they have all been named, show the "go with" pictures one at a time: WHO (OR WHAT) DOES THIS BELONG TO?
2. Give each child a clue picture. Hold each "go with" picture up in turn: WHO HAS THE PERSON (OR THING) THIS BELONGS WITH? WHAT DO YOU CALL THIS?
3. Give a deck of cards to one child: SORT THESE OUT. FIND ALL THE WORKERS AND PUT THE THINGS WITH THEM THAT THEY USE. Or, HERE IS A GLASS, A CUP AND SAUCER, AND SOME MONEY. LOOK THROUGH THESE PICTURES AND FIND THE ONES THAT GO WITH THEM.

FOLLOW-UP: Have sets of cards available for children to use during free playtime. Note whether they use them individually or make up group games to play. Keep introducing more sets.

SETS AND CLASSIFICATION: SIMPLE SORTING

OBJECTIVE: To practice the act of sorting

MATERIALS: Small containers such as margarine dishes filled with small objects such as buttons of various sizes, colors, and shapes, or with dried beans, peas, corn; another container with smaller divisions in it (such as an egg carton)

ACTIVITIES:

1. Have the sections of the larger container marked with a model such as each kind of button or dried bean. The children match each thing from their container with the model until everything is sorted and grouped into new sets in the egg carton (or other large container with small sections).
2. Use the same materials but do not mark the sections of the sorting container. See how the child will sort on his own.

FOLLOW-UP: Have these materials available during playtime. Make up more sets using different kinds of things for sorting.

SETS AND CLASSIFICATION: CLASS NAMES, DISCUSSION

OBJECTIVE: To discuss sets of things which can be put in the same class and decide on the class name

MATERIALS: A set of things which can be put in the same group on the basis of class name, such as

1. animals: several toy animals
2. vehicles: toy cars, trucks, motorcycles
3. clothing: a shoe, a shirt, a belt, and so on
4. things to write with: pen, pencil, marker, crayon, chalk

ACTIVITIES: The same plan can be followed for any group of things.

1. Bring the things out one at a time until three have been discussed. Ask about each.
   a. WHAT CAN YOU TELL ME ABOUT THIS?
   b. Five specific questions:
      WHAT DO YOU CALL THIS (WHAT IS ITS NAME?)
      WHAT COLOR IS IT?
      WHAT DO YOU DO WITH IT? or (WHAT DOES IT DO?) or (WHO USES THIS?)
      WHAT IS IT MADE OUT OF?
      WHERE DO YOU GET ONE?
   c. Show the three things discussed: WHAT DO YOU CALL THINGS LIKE THIS? THESE ARE ALL (ANIMALS, VEHICLES, CLOTHING, THINGS TO WRITE WITH, AND SO ON.)

2. Put two or more groups of things together that have already been discussed. Have the children sort them into sets and tell the class name for each set.

FOLLOW-UP: Put together sets like the above that include things from science and social studies.

SETS AND CLASSIFICATION: LEARNING THE NAME *SET*

OBJECTIVE: To learn the meaning of the term *set*

MATERIALS: After the children have had many sorting and grouping experiences, use materials that are familiar (that they have used in sorting and grouping activities).

ACTIVITIES:

1. Show the children groups of things they have already used, such as crayons, cups, buttons, toy cars, blocks, and so on. Tell them THIS IS A SET OF (OBJECT NAME). Show three or four sets of different things. When you have introduced several groups of things with THIS IS A SET OF (NAME OF OBJECTS), then point to each and ask: WHAT IS THIS? Always answer with YES, THIS IS A SET OF (OBJECT NAME) whether they say set or not. Next point to each set and ask, WHAT IS THE NAME OF THIS SET?
2. As soon as the children use the name set to refer to groups, present bunches of objects or pictures of objects as done before, ask him to find the sets. Use bunches of objects which can be sorted into at least three different sets.

FOLLOW-UP: Use the term *set* whenever you use classification activities.

SETS AND CLASSIFICATION: BOOKS

OBJECTIVE: To learn and discuss characteristics of sets using books

MATERIALS: Books with themes centering around a group of things with a unifying feature (Many of Golden Press' Golden Shape Books are excellent and inexpensive.)

Animals

1. The Dog Book, 1964.
2. Jungle Babies, 1969.
3. The Cat Book, 1964.
4. The Bunny Book, 1965.
5. The Nest Book, 1968.

Vehicles

1. The Car Book, 1968.
2. The Truck and Bus Book, 1966.
3. The Boat Book, 1965.

Christmas
1. The Christmas Tree Book, 1966.
People
1. People in Your Neighborhood, 1971.
2. People in My Family, 1971.
Everyday Things
1. The Shopping Book, 1975 (stores).
2. My House Book, 1966 (rooms in a house).
3. The Telephone Book, 1968 (things in a house).
4. The Sign Book, 1968.
5. The Hat Book, 1965.
6. The Snowman Book, 1965 (winter).
7. The Apple Book, 1964 (fruit).

ACTIVITIES:

1. Younger children enjoy looking at the pictures and labeling the items.
2. Read the books to young listeners.
3. Discuss the categories in each book. Children can compare with their own experience.
4. Bring real items that are pictured in the book.
5. Use as parts of broader units.
6. When the child has learned the term *set* the teacher can ask him to look through the books and tell if they contain sets.

FOLLOW-UP: Have the books available in the book center for the children to look at. Read and discuss the following books: *Brian Wildsmith's Wild Animals* (Watts 1967) and *Brian Wildsmith's Fishes* (Watts 1968). These books introduce the children to some other words for *set* or *group* such as school, flock, herd, and so on.

Classification is one of the most important fundamental skills in science. The following are examples of how classification might be used during science activities:

STRUCTURED ACTIVITIES

SETS AND CLASSIFICATION: SORTING A NATURE WALK COLLECTION

OBJECTIVE: To sort items collected during a nature walk

MATERIALS: The class has gone for a nature walk. They have collected leaves, stones, bugs, etc. They have various types of containers (i.e., plastic bags, glass jars, plastic margarine containers).

ACTIVITIES:

1. Have the children spread out pieces of newspaper on tables or on the floor.
2. Have them dump their plants and rocks on the table. Animals and insects purposely collected are in a separate container.
3. LOOK AT THE THINGS YOU HAVE COLLECTED. PUT THINGS THAT BELONG TOGETHER IN GROUPS. TELL ME WHY THEY BELONG TOGETHER. Let the children explore the materials and identify leaves, twigs, flowers, weeds, smooth rocks, rough rocks, light and dark rocks, etc. After they have grouped the plant material and the rocks, have them sort their animals and insects into different containers. See if they can label their collections (i.e., earthworms, ants, spiders, beetles, ladybugs).
4. Help them organize their materials on the science table. Encourage them to write labels or signs using their own spellings, or help them with spelling if needed. If they won't attempt to write themselves, let them dictate labels to you.

FOLLOW-UP: Encourage the children to examine all the collections and discuss their attributes. Have some plant, rock, insect, and animal reference books on the science table. Encourage the children to find pictures of items like theirs in the books. Read what the books tell about their discoveries.

SETS AND CLASSIFICATION: SORTING THINGS THAT SINK AND FLOAT

OBJECTIVE: To find out which objects in a collection sink and which float

MATERIALS: A collection of many objects made from different materials. You might ask each child to bring one thing from home and then add some items from the classroom. Have a large container of water and two empty containers labeled *sink* and *float*. Make a large chart with a picture/name of each item where the results of the explorations can be recorded (Figure 10–5).

ACTIVITIES:

1. Place the materials on the science table and explain to everyone what the activity is for.
2. During center time let individuals and/or groups of two or three experiment by placing the objects on the water and then in the appropriate container when they float or sink.
3. When the things are sorted, the children can record their names at the top of the next vacant column on the chart and check off which items sank and which floated.
4. After the items have been sorted several times have the students compare their lists. Do the items float and/or sink consistently? Why?

FOLLOW-UP: The activity can continue until everyone has had an opportunity to explore it. New items can be added. Some children might like to make a boat in the carpentry center.

Figure 10–4 "The rake goes with the leaves."

Sam and Mary are sitting at the computer using the program *Gertrude's Secrets.* This is a game designed to aid in basic classification skills of matching by specific common criteria. Mary hits a key that is the correct response and both children clap their hands as a tune plays and Gertrude appears on the screen in recognition of their success. A list of software that helps the development of classification is included in Activity #5.

EVALUATION

As the children play, note whether each one sorts and groups as part of his play activities. There should be an increase as they grow and have more experiences with sets and classification activities. They should use more feature names when they speak during work and play. They should use color, shape, size, material, pattern, texture, function, association words, and class names.

Things	Float	Sink
cotton ball	/////	///
cork	////////	
paper	////	////
foil	/////	///
wood	////////	
rock		////////

Figure 10–5 The students can record the results of their exploration of the floating and sinking properties of various objects.

1. Tim has a handful of colored candies. "First I'll eat the orange ones." He carefully picks out the orange candies and eats them one at a time. "Now, the reds." He goes on in the same way until all the candies are gone.
2. Diana plays with some small wooden animals. "These farm animals go here in the barn. Richard, you build a cage for these wild animals."
3. Mr. Flores tells Bob to pick out from a box of toys some plastic ones to use in the water table.
4. Mary asks the cook if she can help sort the clean tableware and put it away.

5. George and Sam build with blocks. George tells Sam, "Put the big blocks here, the middle-sized ones here, and the small blocks here."
6. Richard and Diana take turns reaching into a box that contains items that are smooth or rough. When they touch an item, they say whether it is smooth or rough, guess what it is, and then remove it and place it on the table in the smooth or the rough pile.
7. Tanya is working with some containers that contain substances with either pleasant, unpleasant, or neutral odors. On the table are three pictures: a happy face, a sad face, and a neutral face. She puts each of the containers on one of the three faces according to her feelings about each odor.

For more structured evaluation, the sample assessment tasks and the tasks in the Appendix may be used.

SUMMARY

Sets are composed of things that are put together in a group based on one or more common criteria. The act of putting things into groups by sorting out things which have some feature that is the same is called classification.

Classifying is a part of a child's normal play. He builds skills he will need later to add and subtract. He also adds to his store of ideas and words as he learns more features to be used to sort and to group.

Naturalistic, informal, and structured classification activities can be done following the sequence of materials from objects, to objects and pictures, to cutouts to picture cards. Books are another excellent pictorial mode for learning classes. Computer games that reinforce classification skills and concepts are also available. Classification is another essential math and science component.

FURTHER READING AND RESOURCES

Anselmo, S. 1981. Children develop classification skills. *Day Care and Early Education*, *8*(3), 31–33.

Baratta-Lorton, M. 1979. *Workjobs II.* Menlo Park, Calif.: Addison-Wesley.

Baratta-Lorton, M. 1976. *Math their way*. Menlo Park, Calif.: Addison-Wesley.

Burton, G. M. 1985. *Towards a good beginning.* Menlo Park, Calif.: Addison-Wesley.

Caballero, J. 1981. *Vanilla manilla folder games for young children*. Atlanta, GA: Humanics Ltd. Community Helpers sorting game.

Carson, P. and Dellosa, J. 1977. *All aboard for readiness skills*. Akron, Ohio: Carson-Dellosa Publishing. Little Larry sorting game.

Isenberg, J. and Jacobs, J. 1981. Classification: Something to think about. *Childhood Education*, *57*(5).

Isaak, B. 1982. *Classifying cat*. Santa Barbara, Calif.: The Learning Works, Inc.

Scott, L. B. and Garner, J. 1978. *Mathematical experiences for young children*. St. Louis: McGraw-Hill.

Seefeldt, C. 1981. Math in a button box. *Day Care and Early Education*, *8*(3), 53–57.

SUGGESTED ACTIVITIES

1. Put together a collection of objects of different sizes shapes colors, classes, materials, and uses. Put them in a large box or plastic dishpan. Present them in turn to a small group (two to six children each) of one-year-olds, two-year-olds, three-year-olds, four-year-olds,

and five-year-olds, six-year-olds, and seven-year-olds. Tell the children, HERE ARE SOME THINGS YOU CAN PLAY WITH. Write a description of what each group does. Count the number of different features used for labeling and grouping by each age level. Compare the groups with each other.

2. Add some set and classifying activities to your Activities File.
3. Create a classification game. Try it out with one or more children. Report the results in class. Explain the game and tell about any changes that could be made to improve it. Ask the class for their comments.
4. Visit an early childhood center. Note all the materials available for classification experiences. Write down a description of each classification activity observed. Which were naturalistic, which were informal, and which were structured? Interview the teacher. Find out how she incorporates classification activities into her daily program.
5. Using the evaluation criteria in Activity 8, Unit 9, review any of the software listed below that you have access to:

   Dinosaurs. Advanced Ideas, Inc., Berkeley, Calif.: Apple II's; IBM, Acorn; C64.

   Duck's Playground. Sierra On-Line, Inc. Coarsegold, Calif.:

   Early Games Match Maker. Counterpoint Software, Inc., Minneapolis, Minn.:

   Gertrude's Secrets. The Learning Company, Portola Valley, Calif.:

   Match Up!. Hayden Software Co., Lowell, Mass.: Atari 400, 800, Atari tape 16K, Atari disk 24K; C64.

   Micro Habitats. Reader's Digest Software, Microcomputer Software Division, Pleasantville N.Y.: Apple II's; C64; IBM PC, PCJr.; 48K.

   Soc Pix. American Guidance Service, Circle Pines, Minn.: Apple II family.

REVIEW

A. Define the term *set*.

B. Describe how you would explain the term *set* to young children.

C. Define the term *classification*.

D. Match the features in Column II with the behavior description in Column I.

Column I	Column II
1. The teacher says "Find all the smooth objects."	a. color
2. The mother says "Pour the milk in the glass."	b. shape
3. Mother says "Put on your clothes."	c. size
4. The child says, "All the trees have leaves."	d. material
5. The teacher says "Find all the things you can use to draw."	e. pattern
6. The children make a train with only yellow chairs.	f. texture
7. Mother has Tanya sort out the paper and cloth napkins.	g. function
	h. association
	i. class name
	j. common features
	k. number

8. Tom makes a pile of square blocks.
9. Teacher says, "Find three balloons for me."
10. The child says, "These cookies are bigger than the ones on the table."
11. "Tim," says mother, "you are wearing a checked shirt and striped pants."

UNIT 11 Comparing

OBJECTIVES

After studying this unit, the student should be able to:

- List and define comparison terms
- Identify the concepts learned from comparing
- Do informal measurement and quantity activities with children
- Do structured measurement and quantity comparing activities with children

When comparing, the child finds a relationship between two things or sets of things on the basis of some specific characteristic or attribute. One type of attribute is an informal measurement such as size, length, height, weight, or speed. A second type of attribute is quantity comparison. To compare quantities, the child looks at two sets of objects and decides if they have the same number of items or if one set has more. Comparing is the basis of ordering (Unit 17) and measurement (Units 18 and 19).

Some examples of measurement comparisons are listed:

- John is taller than Mary.
- This snake is long. That worm is short.
- Father bear is bigger than baby bear.

Examples of number comparisons are shown:

- Does everyone have two gloves?
- I have more cookies than you have.
- We each have two dolls—that's the same.

THE BASIC COMPARISONS

To make comparisons and understand them, the child learns the following basic comparisons:

- Informal Measurement

large	small
big	little
long	short
tall	short
fat	skinny
heavy	light
fast	slow
cold	hot
thick	thin
wide	narrow
near	far
later	sooner (earlier)
older	younger (newer)
higher	lower
loud	soft (sound)

- Number

more	less/fewer

The child also finds that sometimes there is no difference when the comparison is made. That is, the items are the same size, same length, same age, and so on. Relative to quantity they discover there is the same amount (or number) of things in two sets that are compared. The concept of one-to-one correspondence and the skills of counting and classifying assist the child in comparing quantities.

ASSESSMENT

During the child's play, the teacher should note any of the child's activities which might show he is comparing. For example, when a bed is needed for a doll and two shoe boxes are available, does he look the boxes over carefully and place the doll in each box in turn in order to get the doll into a box that is the right size? If he has two trucks, one large and one small, does he build a bigger garage for the larger truck? The adult should also note with children old enough to talk if they use the words given in the list of basic comparisons.

In individual interview tasks, the child is asked questions to see if he understands and uses the basic comparison words. The child is presented with some objects or pictures of things which differ or are the same in relation to some attributes or number and is asked to tell if they are the same or different. Tasks (see Appendix) are like these:

SAMPLE ASSESSMENT TASK

Preoperational Ages 4–5

SKILL: The child will be able to point to big (large) and small objects.

MATERIALS: A big block and a small block (a big truck and a small truck, a big shell and a small shell etc.)

PROCEDURE: Present two related objects at a time and say, FIND (POINT TO) THE BIG BLOCK. FIND (POINT TO) THE SMALL BLOCK. Continue with the rest of the object pairs.

EVALUATION: Note if the child is able to identify big and small for each pair.

INSTRUCTIONAL RESOURCE(S):

Charlesworth, R. and Lind, K. (1990) *Math and Science for Young Children.* Albany, N.Y.: Delmar. Unit 11

SAMPLE ASSESSMENT TASK

Preoperational Ages 3–4

SKILL: The child will compare sets and identify which set has more or less (fewer).

MATERIALS: Two dolls (toy animals or cutout figures) and ten cutout posterboard cookies

PROCEDURE: Place the two dolls (toy animals or cutout figures) in front of the child. Say, WATCH, I'M GOING TO GIVE EACH DOLL (OR ____________) SOME COOKIES. Put two cookies in front of one doll and six in front of the other. Say, SHOW ME THE DOLL (____________) THAT HAS MORE COOKIES. Now pick up the cookies and put one cookie in front of one doll and three in front of the other. Say, SHOW ME THE DOLL (____________) THAT HAS FEWER COOKIES. Repeat with different amounts.

EVALUATION: Note whether the child consistently picks the correct amounts. Some children might understand more but not fewer. Some might be able to discriminate if there is a large difference between sets, such as two vs. six, but not small differences, such as four vs. five.

INSTRUCTIONAL RESOURCE(S):
Charlesworth, R. and Lind, K. (1990) *Math and Science for Young Children*. Albany, N.Y.: Delmar. Unit 11

Before giving the number comparison tasks, the teacher should be sure the child has begun to match, count, and classify such as shown in Figures 11–1, 11–2, and 11–3.

NATURALISTIC ACTIVITIES

The young child has many contacts with comparisons in his daily life. At home mother says, "Get up, it's *late*. Mary was up *early*. Eat *fast*. If you eat slowly, we will have to leave before you are finished. Have a *big* bowl for your cereal, that one is too *small*." At school the teacher says, "I'll pick up this *heavy* box, you pick up the *light* one." "Sit on the *small* chair, that one is too *big*." "Let's finish this story." "Remember, the father bear's porridge was too *hot* and the mother bear's porridge was too *cold*."

As the child uses materials, he notices that things are different. The infant finds that some things can be grabbed and held because they are *small* and *light* while others cannot be held be-

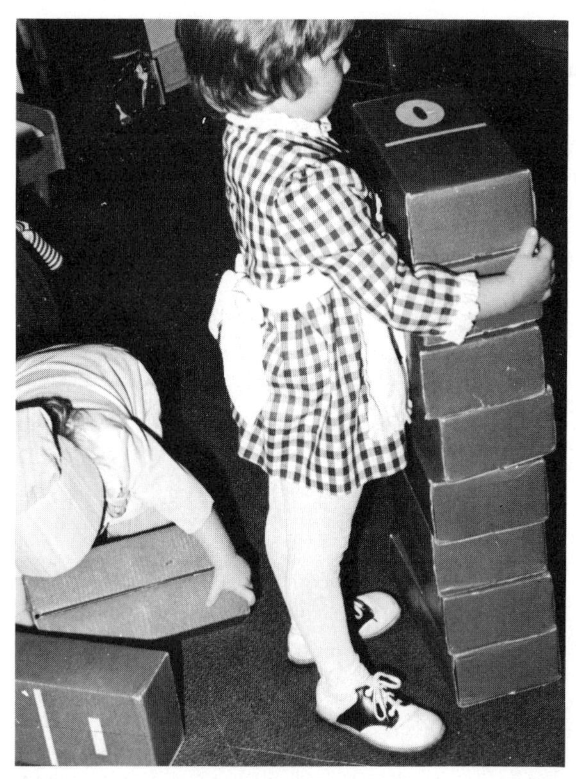

Figure 11–1 The *big* girl builds a *tall* stack, and the *little* girl builds a *short* stack. The *big* girl is *taller* than her stack of blocks.

Figure 11–2 "I can make the blocks taller than me."

Figure 11–3 "I can make it very high."

Figure 11–4 "I'm high!"

cause they are *big* and *heavy*. As he crawls about, he finds he cannot go behind the couch because the space is too *narrow*. He can go behind the chair because there is a *wide* space between the chair and the wall. The young child begins to build with blocks and finds that there are *more small* blocks in his set of blocks than there are *large* ones. He notices that there are people in his environment who are big and people who are small in relation to him. One of the questions most often asked is "Am I a big boy?" or "Am I a big girl?"

INFORMAL ACTIVITIES

Small children are very concerned about size and number, especially in relation to themselves. They want to be bigger, taller, faster, and older (Figures 11–4 and 11–5). They want to be sure they have the same, not less—and if possible more—of things than the other child has. These needs of the young child bring about many situations where the adult can help in an informal way to aid the child in learning the skills and ideas of comparing.

Informal measurements are made in a concrete way. That is the things to be compared are looked at, felt, lifted, listened to, and so on, and the attribute is labeled.

Figure 11–5 "I'm low!"

- Eighteen-month-old Brad tries to lift a large box of toy cars. Mr. Brown squats down next to him, holding out a smaller box of cars, "Here, Brad, that box is too big for your short arms. Take this small box."
- Three-year-olds, Kate and Chris, run up to Mrs. Raymond, "We can run fast. Watch us. We can run faster than you. Watch us." Off they go across the yard while Mrs. Raymond watches and smiles.
- Five-year-olds, Sam and George, stand back to back. "Check us, Mr. Flores. Who is taller?" Mr. Flores says, "Stand by the mirror and check yourselves." The boys stand by the mirror, back to back. "We are the same," they shout. "You are taller than both of us," they tell their teacher.
- It is after a fresh spring rain. The children are on the playground looking at worms. Comments are heard, "This worm is longer than that one." "This worm is fatter." Miss Collins comes up. "Show me your worms. Sounds like they are different sizes." "I think this small, skinny one is the baby worm," says Richard.

Comparative number is also made in a concrete way. When comparing sets of things just a look may be enough if the difference in number is large.

- "Teacher! Juanita has all the spoons and won't give me one!" cries Tanya.

If the difference is small the child will have to use his skill of matching (one-to-one correspondence). He may physically match each item, or he may count—depending on his level of development.

- "Teacher! Juanita has more baby dolls than I do." "Let's check," says Mr. Brown. "I already checked," says Tanya. "She has four and I have three." Mr. Brown notes that each girl has four dolls. Better check again," says Mr. Brown. "Here, let's see. Put each one of your dolls next to one of Juanita's, Tanya." Tanya matches them up. "I was wrong. We have the same."

A child at a higher level of development could have been asked to count.

To promote informal learning, the teacher must put out materials that can be used by the child to learn comparisons on his own. The teacher must also be ready to step in and support the child's discovery by using comparison words and giving needed help with comparison problems which the child meets in his play and other activities.

STRUCTURED ACTIVITIES

Most children learn the idea of comparison through naturalistic and informal activities. For those who do not, more formal experiences can be planned. There are many commercial materials available individually and in kits which are designed to be used to teach comparison skills and words. Also, the environment is full of things that can be used (Figure 11–6). The following are some basic types of activities which can be repeated with different materials.

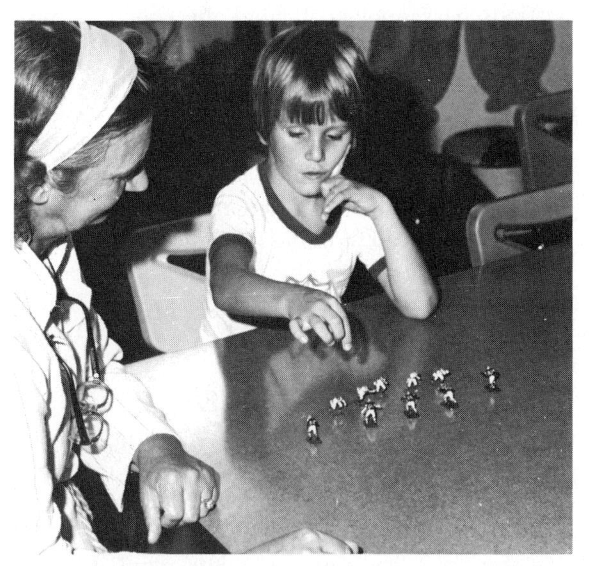

Figure 11–6 "Are there more Saints or more Broncos?"

STRUCTURED ACTIVITIES

COMPARISONS: INFORMAL MEASUREMENTS

OBJECTIVES: To gain skill in observing differences in size, speed, temperature, age, and loudness. To learn the words associated with differences in size, speed, temperature, age, and loudness.

MATERIALS: Use real objects first. Once the child can do the tasks with real things, then introduce pictures and chalkboard drawings.

Comparison	Things to Use
large-small and big-little	buttons, dolls, cups, plates, chairs, books, records, spools, toy animals, trees, boats, cars, houses, jars, boxes, people, pots, and pans
long-short	string, ribbon, pencils, ruler-meter stick, yardstick, snakes, worms, lines, paper strips
tall-short	people, ladders, brooms, blocks, trees, bookcases, flagpoles, buildings
fat-skinny	people, trees, crayons, animals, pencils, books, snowmen
heavy-light	same size but different weight containers (such as shoe boxes or coffee cans taped shut filled with items of different weights)
fast-slow	toy cars or other vehicles for demonstration, the children themselves—their own movements, cars on the street, music, talking
hot-cold	containers of water, food, ice cubes—boiling water, chocolate milk and hot chocolate, weather
thick-thin	paper-cardboard, books, pieces of wood, slices of food (bologna, cucumber, carrot), cookie dough
wide-narrow	streets, ribbons, paper strips, lines (chalk, crayon, paint), doorways, windows
near-far	children and fixed points in the room, places in the neighborhood, map
later-sooner (earlier)	arrival at school or home, two events
older-younger (newer)	people: babies, younger and older children, adults of different ages; any things brought in that have not been in the environment before
higher-lower	swings, slides, jungle gyms, birds in trees, airplanes flying, windows, stairs, elevators, balconies, shelves
loud-soft	voices: singing and talking, claps, piano, drums, records, doors slamming

ACTIVITIES: The basic activity involves the presentation of the two opposites to be compared. They can be objects, cutouts, or pictures—whatever is most appropriate. Then ask the comparison question. For example:

- The teacher places two pieces of paper in front of the children. Each piece is one inch wide. One is six inches long and the other is twelve inches long. WHICH ONE IS LONGER (SHORTER)? If there is no response or an incorrect response, point to the appropriate strip.

LOOK, THIS ONE IS LONGER. THIS ONE IS SHORTER. NOW YOU TELL ME. POINT TO ONE STRIP AND TELL ME IF IT IS THE LONGER OR SHORTER ONE. Try more long and short items using matching and sorting activities.

- The teacher places two identical coffee cans on the table. One is filled with sand; the other is empty. They are both taped closed so the children cannot see inside. PICK UP EACH CAN. TELL ME WHAT IS DIFFERENT ABOUT THEM. If there is no response or an incorrect response, hold each can out in turn to the child. HOLD THIS CAN IN ONE HAND AND THIS ONE IN THE OTHER. (point) THIS CAN IS HEAVY; THIS CAN IS LIGHT. NOW, YOU SHOW ME THE HEAVY CAN, THE LIGHT CAN. If the child has a problem he should do more activities that involve this concept.

The variety of experiences that can be offered with many things that give the child practice exploring comparisons is almost endless.

FOLLOW-UP: On a table, set up two empty containers (so that one is tall and one is short, one is fat and one is thin, or one is big and one is little) and a third container filled with potentially comparable items such as tall and short dolls, large and small balls, fat and thin cats, long and short snakes, big and little pieces of wood, and so on. Have the children sort the objects into the correct empty containers.

COMPARISONS: NUMBER

OBJECTIVES:
- To enable the child to compare sets that are different in number
- To enable the child to use the terms *more*, *less*, *fewer* and *same number*

MATERIALS: Any of the objects and things used for matching, counting, and classifying

ACTIVITIES: The following basic activities can be done using many different kinds of materials.

1. Set up a flannel board with many felt shapes or a magnet board with many magnet shapes. Put up two groups of shapes: ARE THERE AS MANY CIRCLES AS SQUARES? (RED CIRCLES AS BLUE CIRCLES, BUNNIES AS CHICKENS)? WHICH SET HAS MORE? HOW MANY CIRCLES ARE THERE? HOW MANY SQUARES? The children can point, tell with words, and move the pieces around to show that they understand the idea.
2. Have cups, spoons, napkins, or food for snack or lunch: LET'S FIND OUT IF WE HAVE ENOUGH ____________ FOR EVERYONE. Wait for the children to find out. If they have trouble, suggest they match or count.
3. Set up any kind of matching problems where one set has more things than the other: cars and garages, fire fighters and fire trucks, cups and saucers, fathers and sons, hats and heads, cats and kittens, animals and cages, and so on.

FOLLOW-UP: Put out sets of materials which the children can use on their own. Go on to cards with pictures of different numbers of things which the children can sort and match. Watch for chances to present informal experiences:

- Are there more boys or girls here today?
- Do you have more thin crayons or more fat crayons?
- Do we have the same number of cupcakes as we have people?

Mr. Flores introduced his class of four- and five-year-olds to the concept of comparing involving opposites by using books, games, and other materials. To support their understanding of the opposite concepts, he has shown the students how to use the computer software *Stickybear Opposites* (see Activity #5). This program is controlled by two keys, making it a natural for two children working together cooperatively. George and Cindy can be observed making the seesaw go *up* and *down*, watching the plant grow from *short* to *tall*, and comparing the eight (*many*) bouncing balls with the three (*few*) bouncing balls.

EVALUATION

The teacher should note whether the child can use more comparing skills during his play and routine activities (Figure 11–7). Without disrupting his activity, the adult asks questions as the child plays and works:

- Do you have more cows or more chickens in your barn?
- (Child has made two clay snakes) Which snake is longer? Which is fatter?
- (Child is sorting blue chips and red chips into bowls) Do you have more blue chips or more red chips?
- (Child is talking about his family) Who is older, you or your brother? Who is taller?

The assessment tasks in the Appendix may be used for formal evaluation interviews.

SUMMARY

Comparing involves finding the relationship between two things or two groups of things. An informal measurement may be made by comparing two things. Comparing two groups of things incorporates the use of matching, counting, and classifying skills to find out which sets have more, less/fewer, or the same quantities. Naturalistic, informal, and structured experiences support the learning of these concepts.

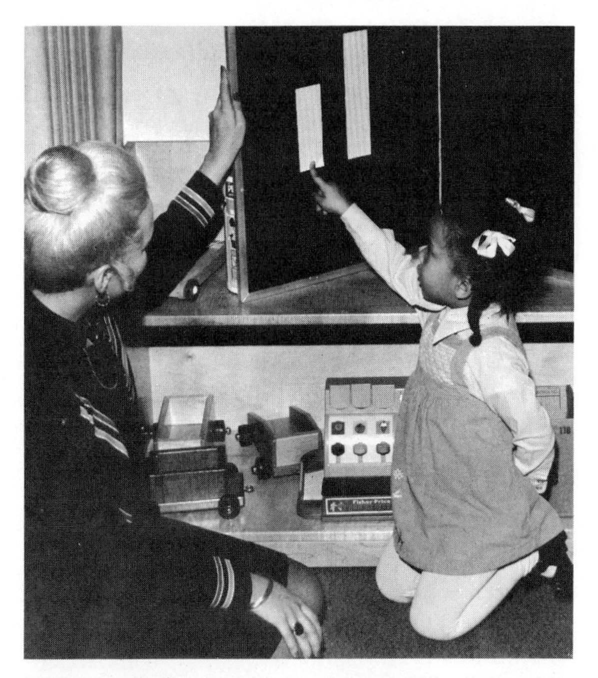

Figure 11–7 "This is the short one. The other one is long."

FURTHER READING AND RESOURCES

Baratta-Lorton, M. 1972. *Workjobs.* Menlo Park, Calif.: Addison-Wesley.

Baratta-Lorton, M. 1976. *Mathematics their way.* Menlo Park, Calif.: Addison-Wesley.

Beaty, J. J., and Tucker, W. H. 1987. *The computer as a paintbrush.* Columbus, Ohio: Merrill.

Richardson, K. 1984. *Developing number concepts using Unifix Cubes.* Menlo Park, Calif.: Addison-Wesley.

SUGGESTED ACTIVITIES

1. Assess one or more children with the sample informal measurement and number tasks in this unit. On the basis of the results, plan activities at each child's level. Prepare the materials and carry out the activities. Note the children's responses. Compare your experiences with those of the other students in the class.
2. Add comparison activities to your Activities File.
3. Observe a group of four- and five-year-olds. Note each instance of the use of comparison words and comparison activities.
4. Plan a comparison activity. Try it out on two classmates. Have them evaluate the presentation and suggest improvements.
5. Using the guidelines in Activity #8, Unit 9, evaluate some of the following software:

   Comparison Kitchen. Developmental Learning Materials, Allen, Tex.: Apple.

   Juggles Rainbow. The Learning Company, Menlo Park, Calif.: Apple IIs; Atari 400, 800; C64.

   Stickybear Opposites. Weekly Reader Family Software, Middletown, Conn.: Apple IIs.

REVIEW

A. Define and give examples of comparing including its two facets: informal measurement and number of quantity comparisons.

B. Match each item in Column II with the correct statement in Column I.

Column I	Column II
1. This stick is longer than yours.	a. Number (quantity) comparison
2. My car goes faster.	b. Speed comparison
3. My baby is bigger than Mary's baby.	c. Weight comparison
4. You gave Pete one more push than you gave me.	d. Length comparison
5. This box is heavier than that one.	e. Height comparison
6. That man is taller than the lady.	f. Size comparison

C. Briefly answer each of the folowing:
1. What are three concepts/skills that serve as the basis of comparing?
2. Explain two comparison assessment tasks.
3. List two examples of naturalistic comparison situations
4. List two examples of informal comparison situations.
5. List two examples of structured comparison activities

D. Describe the role of computer software in developing the comparison concepts and skills.

UNIT 12 Shape

OBJECTIVES

After studying this unit, the student should be able to:

- Describe naturalistic, informal, and structured shape activities for young children
- Assess and evaluate a child's knowledge of shape
- Help children learn shape through haptic, visual, and visual-motor experiences

Each object in the environment has its own shape. Much of the play and activity of the infant during the sensorimotor stage centers on learning about shape. The infant learns through looking and through feeling with hands and mouth. Babies learn that some shapes are easier to hold than others. They learn that things of one type of shape will roll. They learn that some things have the same shape as others. Young children see and feel shape differences long before they can describe these differences in words (Figure 12–1). In the late sensorimotor and early preoperational stages, the child spends a lot of time matching and classifying things. Shape is often used as the basis for these activities.

As the child moves into the middle of the preoperational period, he can learn that there are some basic shapes (called geometric shapes) which have their own names. These are iilustrated in Figure 12-2. First the child learns to label circle, square, and triangle. Then he can learn rectangle, rhombus, and ellipse. Later on, these shape names will be used in geometry, art, and other areas of activity. There are two major purposes for learning about shape:

Figure 12–1 The children experiment with the shape matching toy.

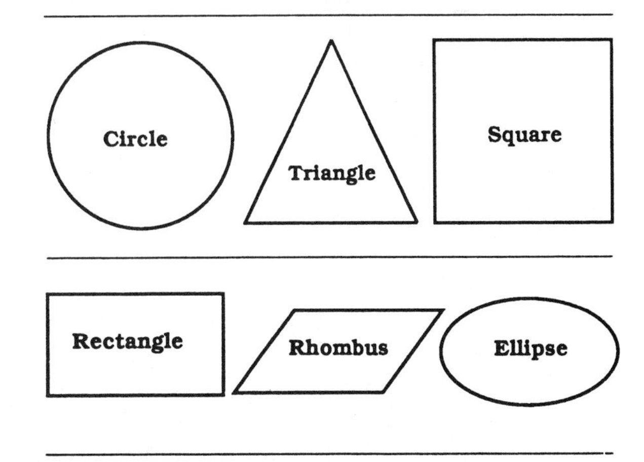

Figure 12–2 Geometric shapes

- It helps children to be more sensitive to similarities and differences in forms in the environment and aids in discriminating one form from another.
- Children learn some labels that they can use when describing things in the environment. ("I put the book on the square table.")

ASSESSMENT

Observational assessment can be done by noticing whether the child uses shape to organize his world. As the child plays with materials, the adult should note whether he groups things together because the shape is the same or similar. For example, a child plays with a set of plastic shape blocks. There are triangles, squares, and circles. Some are red, blue, green, yellow, and orange. Sometimes he groups them by color, sometimes by shape. A child is playing with pop beads of different colors and shapes. Sometimes he makes strings of the same shape; sometimes, of the same color. The child may use some shape names in his everyday conversation (Figure 12–3).

Figure 12–3 The teacher finds out if the child knows the shape names.

The individual interview tasks for shape center on discrimination, labeling, matching, and sorting. Discrimination tasks assess whether the child can see that one form has a different shape from another form. Labeling tasks assess whether the child can find a shape when the name is given and whether he can name a shape when a picture is shown to him. At a higher level, he finds shapes in pictures and in his environment. Matching would require the child to find a shape like one shown to him. A sorting task would be one in which the child must separate a mixed group of shapes into sets (see Unit 10). Two sample tasks follow:

SAMPLE ASSESSMENT TASK

Preoperational Ages 3–4

METHOD: Interview

SKILL: When given the name of a shape the child will be able to point to a drawing of that shape.

MATERIALS: On pieces of white posterboard or on 5½″ × 8″ file cards, draw the following shapes with a black marker (one shape on each card): circle, square and triangle.

PROCEDURE: Place the cards in front of the child. POINT TO THE SQUARE. POINT TO THE CIRCLE. POINT TO THE TRIANGLE.

EVALUATION: Note which, if any, of the shapes the child can identify.

INSTRUCTIONAL RESOURCE(S):
Charlesworth, R. and Linda, K. 1990. *Math and Science for Young Children*. Albany, NY: Delmar. Unit 12.

SAMPLE ASSESSMENT TASK

Preoperational Ages 5–6

METHOD: Interview
SKILL: The child can identify shapes in the environment.
MATERIALS: The natural environment
PROCEDURE: LOOK AROUND THE ROOM. FIND AS MANY SHAPES AS YOU CAN. WHICH THINGS ARE SQUARE SHAPES? CIRCLES? RECTANGLES? TRIANGLES?
EVALUATION: Note how observant the child is. Does he/she note the obvious shapes such as windows, doors, and tables? Does he/she look beyond the obvious? How many shapes and which shapes is he/she able to find?
INSTRUCTIONAL RESOURCE(S):
Charlesworth, R. and Lind, K. 1990. *Math and Science for Young Children*. Albany, NY: Delmar. Unit 12.

NATURALISTIC ACTIVITIES

Naturalistic activities are most important in the learning of shape. The child perceives the idea of shape through sight and touch. The infant needs objects to look at, to grasp, and to touch and taste. The toddler needs different things of many shapes to use as he sorts and matches. He needs many containers (bowls, boxes, coffee cans) and many objects (such as pop beads, ping pong balls, poker chips, and empty thread spools) (Figure 12–4). He needs time to fill containers with these objects of different shapes and to dump the objects out and begin again. As he holds each thing, he examines it with his eyes, hands, and mouth.

The older preoperational child enjoys a junk box filled with things such as buttons, checkers, bottle caps, pegs, small boxes, and plastic bottles which he can explore. The teacher can also put out a box of attribute blocks (wood or plastic blocks in geometric shapes). Geometric shapes and other shapes can also be cut from paper and/or cardboard and placed out for the child to use. Figure 12–5 shows some blob shapes that can be put into a box of shapes to sort.

In dramatic play, the child can put to use his ideas about shape. The preoperational child's play is representational. He uses things to represent something else which he does not have at the time. He finds something that is close to the real thing, and it is used to represent the real thing. Shape is usually one of the elements used when the child picks a representational object:

- A stick or a long piece of wood is used for a gun.
- A piece of rope or old garden hose is used to put out a pretend fire.

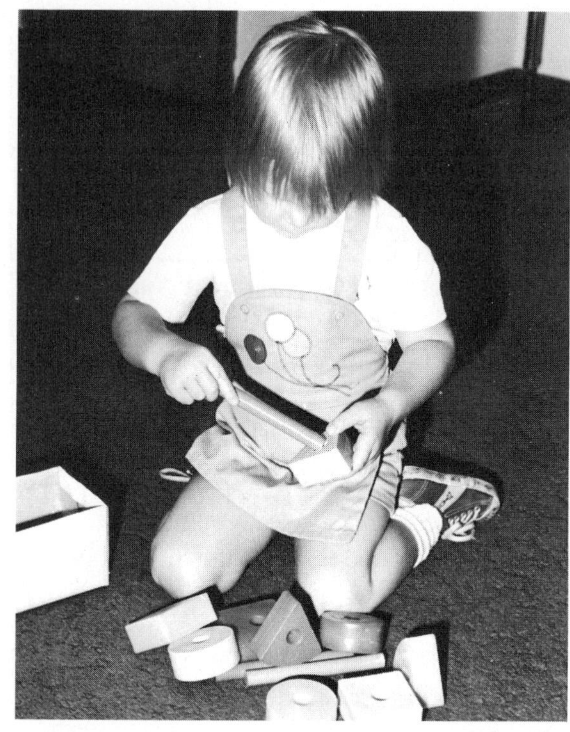

Figure 12–4 The child explores shape.

- The magnet board shapes are pretend candy.
- A square yellow block is a piece of cheese.
- A shoe box is a crib, a bed, or a house—as needed.
- Some rectangular pieces of green paper are dollars, and some round pieces of paper are coins.
- A paper towel roll is a telescope for looking at the moon.

INFORMAL ACTIVITIES

The teacher can let the child know that she notices his use of shape ideas in activities through comments and attention. She can also supply him with ideas and objects which will fit his needs. She can suggest or give the child a box to be used for a bed or a house, some blocks or other small objects for his pretend food, or green rectangles and gray and brown circles for play money.

Figure 12–5 Blob shapes: Make up your own.

Labels can be used during normal activities. The child's knowledge of shape can be used too.

- "The forks have sharp points; the spoons are round and smooth."
- "Put square placemats on the square tables and rectangular placemats on the rectangular tables."
- "We'll have triangle shaped crackers today."
- As a child works on a hard puzzle, the teacher takes his hand and has him feel the empty space with the index finger, "Feel this shape and look at it. Now find the puzzle piece that fits here."
- As the children use clay or play dough, the teacher says, "You are making lots of shapes. Kate has made a ball; Jim, a snake; and Diana, a pancake."

- During cleanup time, the teacher says, "Put the square blocks here and the rectangle blocks here."

The teacher can pay attention and respond when the child calls her attention to shapes in the environment (Figure 12–6). The following examples show that children can generalize; they can use what they know about shape in new situations.

- "Ms. Moore, the door is shaped like a rectangle." Ms. Moore smiles and looks over at George, "Yes, it is."
- "The Plate and the hamburger look round like circles." "They do, don't they," comments Mr. Brown.

Figure 12–6 "I tore a triangle from this paper."

- "Where I put the purple paint, it looks like a butterfly." Mr. Flores looks over and nods.
- "The roof is shaped like a witch's hat." Miss Conn smiles.
- Watching a variety show on TV, the child asks, "What are those things that are shaped like bananas?" (Some curtains over the stage are yellow and do look just like big bananas!) Dad comments laughingly, "That is funny. Those curtains look like bananas."

STRUCTURED ACTIVITIES

Structured shape activities involve two main operations: *discrimination* (seeing or feeling that one shape is the same as or different from another) and *labeling* (giving a name to shapes which are seen and/or felt). Children need both haptic and visual experiences to learn discrimination and labeling (Figure 12–7). These experiences can be described as follows:

- *Haptic activities* use the sense of touch to match and identify shapes. These activities involve experiences where the child cannot see to solve a problem but must use only his sense of touch. The items to be touched are hidden from view. Older children can be blindfolded. The things may be put in a bag or a box or wrapped in cloth or paper. Sometimes a clue is given. The child can feel one thing and then find another that is the same shape. The child can be shown a shape and then asked to find one that is the same. Finally, the child can be given just a name (or label) as a clue.
- *Visual activities* use the sense of sight. The child may be given a visual or a verbal clue and asked to choose from several things the one that is the same shape. Real objects or pictures may be used (Figure 12–8).
- *Visual-motor activities* use the sense of sight and motor coordination at the same time.

Figure 12–7 The teacher observes as the child puts the shapes on the matrix board.

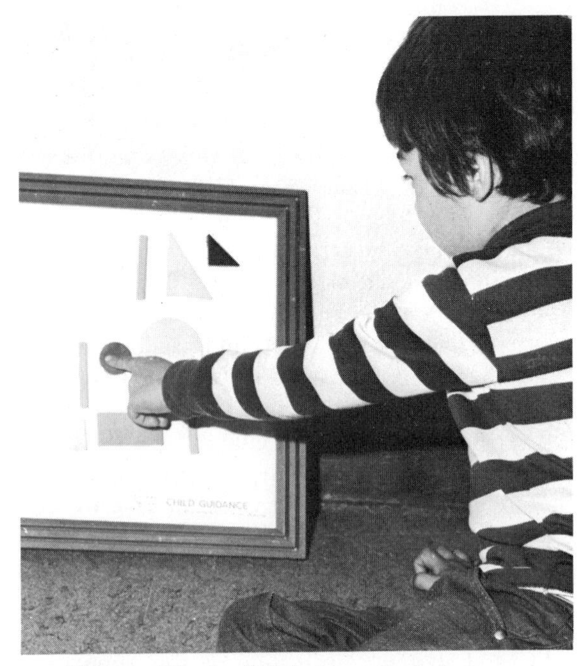

Figure 12–8 "This is a circle."

This type of experience includes the use of puzzles, formboards, attribute blocks, flannel boards, magnet boards, colorforms, and paper cutouts which the child moves about on his own. He may sort the things into

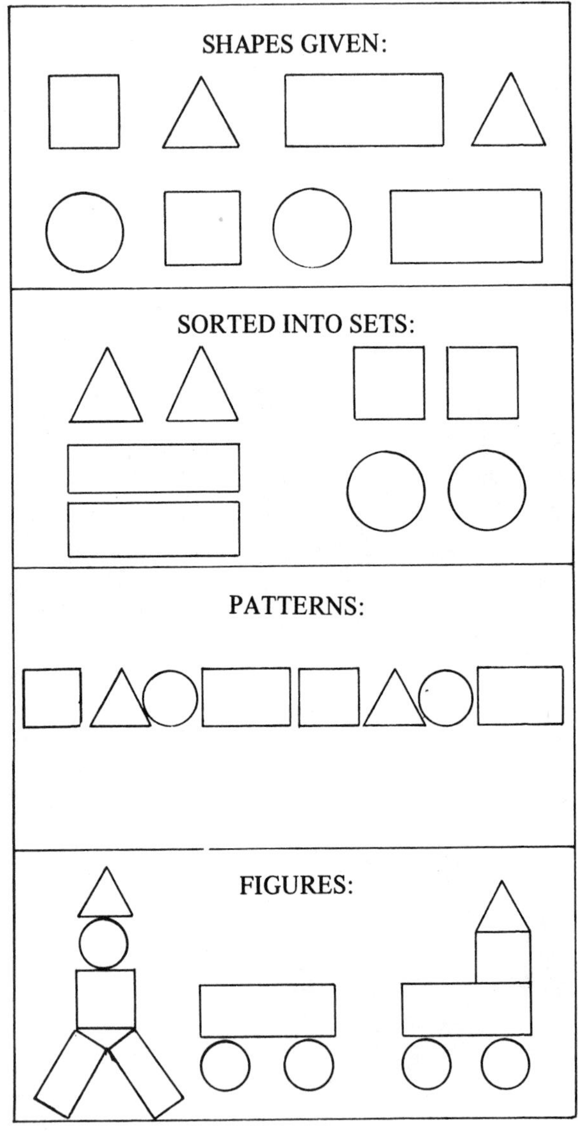

Figure 12–9 Shapes can be sorted into sets, put into a pattern, or made into figures.

sets or arrange them into a pattern or picture. Sorting was described in Unit 7. Examples of making patterns or pictures are shown in Figure 12–9.

As the child does haptic, visual, and visual-motor activities, the teacher can use labels (words such as round, circle, square, triangle, rectangle, shape). The following activities are some examples of basic types of shape experiences for the young child.

STRUCTURED ACTIVITIES

SHAPE: FEELING BOX

OBJECTIVE: To give the child experiences which will enable him to use his sense of touch to label and discriminate shapes

MATERIALS: A medium-sized cardboard box covered with plain Contac® paper with a hole cut in the top big enough for the child to put his hand in but small enough so the child cannot see inside; some familiar objects, such as a toy ear, a small wooden block, a spoon, a small coin purse, a baby shoe, a pencil, and a rock

ACTIVITIES:

1. Show the children each of the objects. Se sure they know the name of each one. Have them pick up each object and name it.
2. Out of his sight, put the objects in the box.
3. The following can then be done:
   - Have another set of identical objects. Hold them up one at a time: PUT YOUR HAND IN THE BOX. FIND ONE LIKE THIS.
   - Have a set of identical objects. Put each one in an individual bag: FEEL WHAT IS IN HERE. FIND ONE JUST LIKE IT IN THE BIG BOX.
   - Use just a verbal clue: PUT YOUR HAND IN THE BOX. FIND THE ROCK (CAR, BLOCK, ETC.)
   - PUT YOUR HAND IN THE BOX. TELL ME THE NAME OF WHAT YOU FEEL. BRING IT OUT AND WE'LL SEE IF YOU GUESSED IT.

FOLLOW-UP: Once the child understands the idea of the "feeling box," a "Mystery box" can be introduced. In this case, familiar objects are placed in the box, but the child does not know what they are. He must then feel them and guess what they are. Children can take turns. Before the child takes the object out, encourage him to describe it (smooth, rough, round, straight, bumpy, it has wheels, and so on). After the child learns about geometric shapes, the box can be filled with cardboard cut-outs or attribute blocks.

SHAPE: DISCRIMINATION OF GEOMETRIC SHAPES

OBJECTIVE: To see that geometric shapes may be the same or different from each other.

MATERIALS: Any or all of the following may be used:

- Magnet board with magnet shapes of various types, sizes, and colors

- Flannel board with felt shapes of various types, shapes, and colors
- Attribute blocks (blocks of various shapes, sizes, and colors)
- Cards with pictures of various geometric shapes in several sizes (they can be all outlines or solids of the same or different colors)

ACTIVITIES: The activities are matching, classifying, and labeling.

- Matching: Put out several different shapes. Show the child one shape, FIND ALL THE SHAPES LIKE THIS ONE.
- Classifying: Put out several different kinds of shapes. PUT ALL THE SHAPES THAT ARE THE SAME KIND TOGETHER.
- Labeling: Put out some shapes—several kinds. Then ask, FIND ALL THE TRIANGLES (SQUARES, CIRCLES, ETC.) or TELL ME THE NAME OF THIS SHAPE. (Point to one at random.)

FOLLOW-UP: Do individual and small group activities. Do the same basic activities with different materials.

SHAPE: DISCRIMINATION AND MATCHING GAME

OBJECTIVE: To practice matching and discrimination skills (for the child who has had experience with the various shapes already)

MATERIALS: Cut out some shapes from cardboard. The game can be made harder by the number of shapes used, the size of the shapes, and the number of colors. Make six Bingo-type cards (each one should be different) and a spinner card which includes all the shapes used:

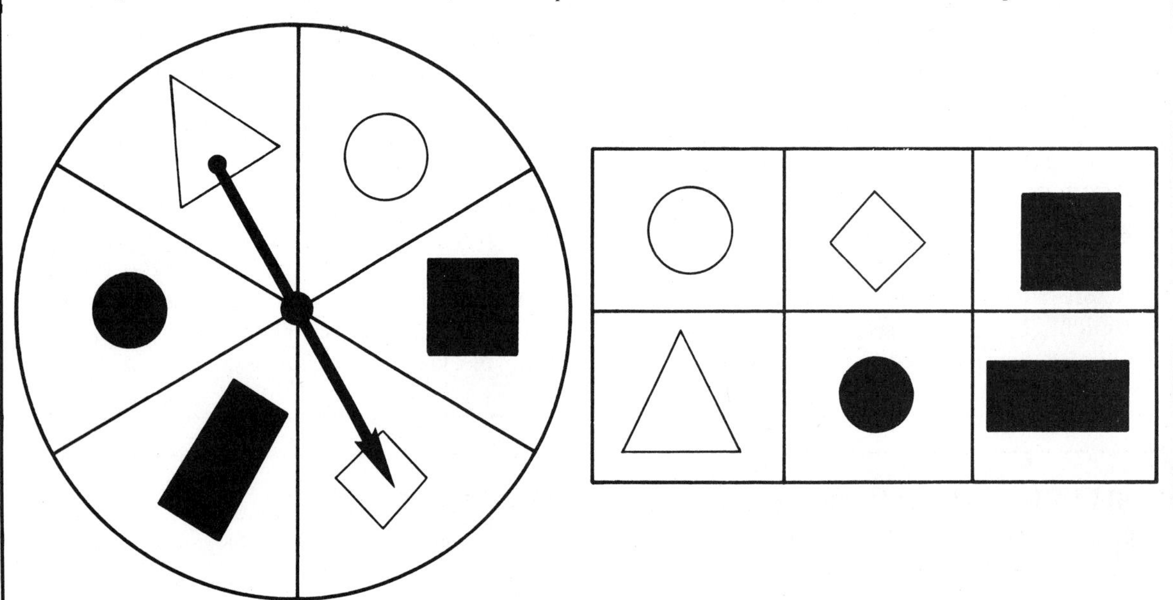

ACTIVITY:
1. Give each child a Bingo card.
2. Have each child in turn spin the spinner. If he has the shape on his card which the spinner points to, he can cover the shape with a paper square or put a marker on it.

FOLLOW-UP: Once the rules of the game are learned the chlldren can play it on their own.

SHAPE: ENVIRONMENTAL GEOMETRY

OBJECTIVE: To see that there are geometric shapes all around in the environment

MATERIALS: The classroom, the school building, the playground, the home, and the neighborhood

ACTIVITIES:
1. Look for shapes on the floor, the ceiling, doors, and windows, materials, clothing, trees, flowers, vehicles, walls, fences, sidewalks, and so on.
2. Make a shape table. Cover the top and divide it into sections. Mark each section with a sample shape.
   Have the children bring things from home and put them on the place on the table that matches the shape of the thing that they bring.

SHAPE: FINDING SHAPES IN PICTURES AND DESIGNS

OBJECTIVE: To sharpen discrimination skills by finding shapes in pictures

MATERIALS: Picture books (see list in Appendix), pictures from workbooks, or pictures drawn by the teacher (Figure 12–10). There are also specific books designed for looking at shapes such as: *Draw me a square, Draw me a triangle, and Draw me a circle* (1970, Nutmeg Press/ Simon and Schuster). Each of these small books follows a format of looking for shapes in the environment as viewed in the illustrations. At the end of each book, the child is asked to "Draw me a square (circle, triangle)."

ACTIVITY: Use the books and pictures with a child or a small group of children. Say, TELL ME WHAT YOU SEE IN THE PICTURE. Note whether any geometric shapes or other shapes which are in the picture are mentioned. Say, FIND THE ________ IN THE PICTURE. HOW MANY ARE THERE?

FOLLOW-UP: Put a large "find the shape" picture on the bulletin board so children can work on it on their own.

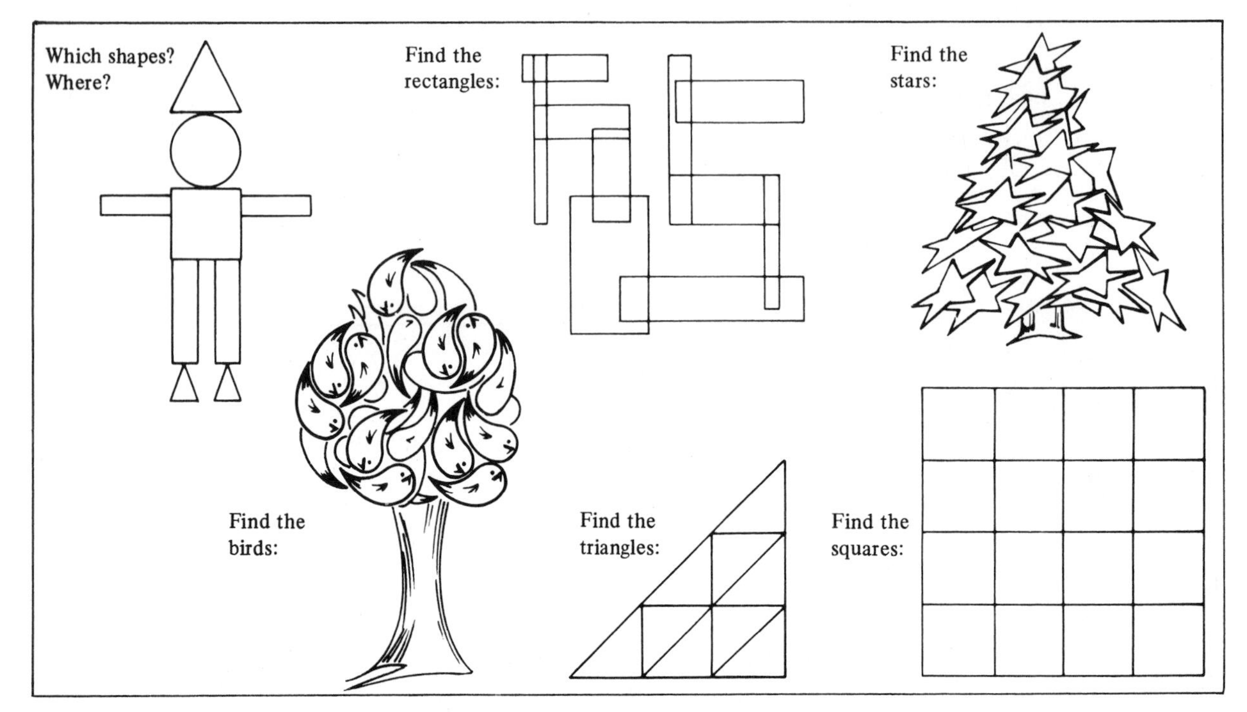

Figure 12–10 Find the shapes.

This week, Miss Collins has introduced her kindergarten class to the concept of shape using many kinds of haptic and visual activities and materials. Now she thinks they are ready to try *Stickybear Shapes* (see Activity #5). Cindy, Kate, Lai, and Richard have already signed up on the computer signup sheet and eagerly await their turns to try out this new software. Cindy is first and after some experimentation catches on to the first game, "Pick It," which involves selecting the correct shape to fill in a missing space in a picture. Kate, who can read, enjoys "Name It," in which she must match a shape name to a shape. All the children enjoy seeing Stickybear dance on the screen whenever they make a correct choice.

EVALUATION

Through observing during free play and during structured experiences, the teacher can see whether the child shows an increase in ideas regarding shape. She observes whether the child uses the word shape and other shape words as he goes about his daily activities. When he sorts and groups materials, the teacher notices whether he sometimes uses shape as the basis for organizing. The adult gives the child informal tasks such as "Put the box on the square table"; "Fold the napkins so they are rectangle shapes"; Find two boxes that are the same shape."

After a period of instruction, the teacher may use interview tasks such as those in the Appendix.

SUMMARY

Each thing the child meets in the environment has shape. The child explores his world and learns in a naturalistic way about the shape of each object in it. Adults help by giving the child things to view, hold, and feel. Adults also teach the child words which describe shapes and the names of geometric shapes such as square, circle, and triangle.

The process of seeing that some shapes are the same and some are different is like the one the child uses later to see that some number and letter symbols are the same and some are different.

FURTHER READING AND RESOURCES

Anselmo, S. 1984. Activities to enhance thinking skills: Matching objects by shape. *Day Care and Early Education*. 12 (1): 43.

Beaty, J. J., and Tucker, W. H. 1987. *The computer as a paintbrush*. Columbus, Ohio: Merrill.

Carson, P., and Dellosa, J. 1977. *All aboard for readiness skills*. Akron, Ohio: Carson-Dellosa.

Dellosa, J., and Carson, P. 1981. *Buzzing into readiness*. Akron, Ohio: Carson-Dellosa.

Dellosa, J., and Carson, P. 1980. *Fluttery readiness*. Akron, Ohio: Carson-Dellosa.

Haas, C. B. 1985. Recipes for fun and learning: Shapes. *Day Care and Early Education*. 13 (1): 41–42.

Scott, L. B., and Garner, J. 1978. *Mathematical experiences for young children*. St. Louis: McGraw-Hill.

SUGGESTED ACTIVITIES

1. Assess a child's shape ideas. Use the sample assessment tasks. Plan an activity at the child's level. Prepare the materials, and use the activity with the child. Evaluate the results.

2. Put together a set of materials for a haptic experience. Try them out on the rest of the class. Have the class members rank the materials as to their value for this sort of activity. Put a list of the best on a 5" x 8" card for the Activities File.

3. Maria Montessori stressed haptic activities. Find one of her books (or a book about her method) in the library. Visit a Montessori classroom. Note how well Montessori's ideas are used. Try some of the materials. Write an evaluation of this aspect of Montessori.

4. Obtain permission from the director to take some haptic, visual, and visual-motor activities to a preschool center. Set those out for the children to explore. If some children seem interested, direct them in the more structured activities. Share with the class what was learned from this experience.

5. Using the guidelines from Unit 9, Activity #8, evaluate the one or more of the following computer programs designed to reinforce shape concepts:

   Shape and Color Rodeo. Developmental Learning Materials, Allen, Tex.: Apple II's.

   Shape-up! Hayden Software Co., Lowell, Mass.: Atari 400, 800, tape 16K, disk 24K; C64.

   Stickers. Springboard Software, Minneapolis, Minn.: Apple II's; IBM PC & PCJr; C64.

   Stickybear Shapes. Weekly Reader Family Software, Middletown, Conn.: Apple II's.

REVIEW

A. Below is a description of four-year-old Tim's activities on a school day. List the thirteen shape activities which are described and indicate whether the activity is naturalistic, informal, or structured.

Tim's mom calls him at 7:30 A.M. and says "It's time to get ready for school, Tim." Tim stretches and squeezes his pillow. He looks at his pillow and snuggles his head into its softness. Mom calls again. She comes into his room. "Come on, Tim. Get up and go wash your face." They walk into the bathroom. Mother gets out a clean washcloth and towel. "Now wash good," she says. "I will get breakfast. What do you want, eggs or cereal?" Time says, "I want one of those round waffles." Mom says "O.K."

Tim takes the washcloth and splashes in the water. He tries to make bubbles from the bar of soap. He pushes the soap to the bottom of the sink. He watches it "pop" back up. He pushes it again. He becomes frustrated at the way it slides through his fingers. He grabs it with both hands and throws it back into the water. Splash! He delights in the splashing. Mother comes back. "Oh, Tim, what a mess. Come on, I told you to wash. It's not time to play with the soap and water. We have to hurry this morning. I have an early meeting." They go into the kitchen. Mom says, "Here is your round waffle. Look, I put it on a square plate. Here's some milk. Now eat, then we'll get dressed."

After Mom helps Tim get dressed, they get in the car and head toward the day care center. Tim looks out the window. He sees apartment buildings, signs, other cars, and buses.

At the day care center, Tim is welcomed by the director. "Good morning Tim." "Good morning Mrs. Adams. How are you today? It sure is getting cold."

Tim's day begins. He looks around the room and runs over by some other boys. They are playing with the unit blocks. Tim starts to build also. He uses long blocks, rectangular blocks, square blocks, and curved blocks. He makes a square enclosure. Miss Collins comes over. She says, "What are you going to put in your square building, Tim?" Tim says, "My truck."

Later Tim goes over to the art activity center. He takes some pipe cleaners and bends and twists them. He makes circles and glues them on some paper. "Look, Teacher, I made circle people."

For snack the children have cubes of cheese and round crackers. Then Tim plays with the wooden puzzles. The puzzles contain many different shapes. Miss Collins says, "Children, come on, it's story time. Today we're going to learn about magic squares, circles, triangles, and rectangles." As Miss Collins reads the story, she stops to ask questions about the shapes. After the story she puts on a record, and the children play a musical shape game. After this game, the children are told to go play with some of the new materials placed in the discovery center. Tim takes out the attribute blocks. He stacks all the same sized circles in a pile. Next, he stacks the squares. "Look John, I made a pile of circles and a pile of squares."

B. Answer each of the following questions.

1. What is the best way to assess a child's use of shape?
2. Describe one shape assessment task.
3. Describe one structured shape activity.

4. Define each: discrimination, labeling, matching, and sorting.

C. Match the shape tasks in Column II with the correct shape activity in Column I.

1. Teacher shows child a circle and says, "What is this shape called?"
2. Teacher says, "Point to the circle."
3. Teacher has a feely box of shapes.
4. Child is shown a circle and told to find one other object just like it.
5. Child plays with shape form fitting box.
6. Children play shape Bingo.

a. Shape discrimination task
b. Shape labeling task
c. Shape matching task
d. Shape sorting task

UNIT 13 Space

OBJECTIVES

After studying this unit, the student should be able to:

- Define the five space ideas and tell how each answers specific questions
- Assess and evaluate a child's ideas about space
- Do informal and structured space activities with young children

Space is a part of geometry just as is shape (Unit 12). There are relationships in space and there is the use of space. The relationship ideas are position, direction, and distance. Use of space includes organization and patterns and construction. Each space concept helps the child answer his own questions:

Space Idea	Question	Answers
Position	Where (am I, are you, is he)?	on-off; on top of-over-under; in-out; into-out of; top-bottom; above-below; in front of-in back of-behind; beside-by-next to; between
Direction	Which way?	up-down; forward-backward; around-through; to-from; toward-away from; sideways; across
Distance	What is the relative distance?	near-far; close to-far from
Organization and pattern	How can things be arranged so they fit in a space?	arrange things in the space until they fit or until they please the eye
Construction	How is space made? How do things fit into the space?	Arrange things in the space until they fit; change the size and shape of the space to fit what is needed for the things

ASSESSMENT

A great deal about the child's concept of space can be learned through observation. The adult notes the child's use of space words:

Does he respond with an appropriate act when he is told the following?

- Put the book **on** the table.
- Please, **take off** your hat.
- You'll find the soap **under** the sink.
- Stand **behind** the gate.
- Sit **between** Kate and Chris.
- Move **away from** the hot stove.
- It's on the table **near** the window. (Figure 13–1)

Figure 13–1 "Room for me too!" The toddler learns about space.

Figure 13–2 John is between two friends.

Does he answer space questions using space words?

- Where is the cat? **On** the bed.
- Where is the cake? **In** the oven.
- Which way did John go? He went **up** the ladder.
- Where is your house? **Near** the corner.

The adult should note the child's use of organization and pattern arrangement during his play activities:

- When he does artwork, such as a collage, does he take time to place the materials on the paper in a careful way? Does he seem to have a design in mind?
- Does the child's drawing and painting show balance? Does he seem to get everything into the space that he wants to have it in, or does he run out of space?
- As he plays with objects, does he place them in straight rows, circle shapes, square shapes, and so on?

The teacher should note the child's use of construction materials such as blocks and containers:

- Does the child make structures with small blocks that toys such as cars and animals can be put into?
- Does he use the large blocks to make buildings that large toys and children can get into?
- Can he usually find the right size container to hold things (such as a shoe box that makes the right size bed for his toy bear)?

The teacher should note the child's use of his own body in space:

- When he needs a cozy place in which to play, does he choose one that fits his size, or does he often get stuck in tight spots?
- Does he manage to move his body without too many bumps and falls?

The individual interview tasks for space center on relationships and use of space. The following are examples of interview tasks:

SAMPLE ASSESSMENT TASK

Preoperational Ages 2–3

METHOD: Interview

SKILL: Given a spatial relationship word the child is able to place objects relative to other objects on the basis of that word.

MATERIALS: A small container such as a box, cup, or bowl and an object such as a coin, checker, or chip.

PROCEDURE: PUT THE (______) IN THE BOX (or CUP or BOWL). Repeat using other space words: ON, OFF OF, OUT OF, IN FRONT OF, NEXT TO, UNDER, OVER.

EVALUATION: Note if the child is able to follow the instructions and place the object correctly relative to the space word used.

INSTRUCTIONAL RESOURCE(S):

Charlesworth, R. and Lind, K. (1990) *Math and Science for Young Children.* Albany, N.Y.: Delmar. Unit 13.

SAMPLE ASSESSMENT TASK

Preoperational Ages 3–4

METHOD: Interview

SKILL: Child will be able to use appropriate spatial relationship words to describe positions in space.

MATERIALS: Several small containers and several small objects. For example, four small plastic glasses and four small toy figures such as a fish, dog, cat, and mouse.

PROCEDURE: Ask the child to name each of the objects so you can use his name for them if it is different from your names. Line up the glasses in a row. Place the animals so that one is *in*, one *on*, one *under*, and one *between* the glasses. Then say, TELL ME WHERE THE FISH IS. Then, TELL ME WHERE THE DOG IS. Then, TELL ME WHERE THE CAT IS. Finally, TELL ME WHERE THE MOUSE IS. Frequently, children will insist on pointing. Say, DO IT WITHOUT POINTING. TELL ME WITH WORDS.

EVALUATION: Note whether the child responds with position words and whether or not the words used are correct.

INSTRUCTIONAL RESOURCE(S):

Charlesworth, R. and Lind, K. (1990) *Math and Science for Young Children.* Albany, N.Y.: Delmar. Unit 13.

Figure 13–3 "Put the people on the cups."

NATURALISTIC ACTIVITIES

It is through everyday motor activities that the child first learns about space. As he moves his body in space, he learns position, direction, and distance relationships and about the use of the space. Children in the sensorimotor and preoperational stages need equipment which lets them place their own bodies on, off, under, over, in, out, through, above, below, and so on. They need places to go up and down, around and through, and sideways and across. They need things which they can put in, on, and under other things. They need things that they can place near and far from other things. They need containers of many sizes to fill, blocks with which to build, and paint, collage, wood, clay and such which can be made into patterns and organized in space. Thus, when the child is matching, classifying, and comparing, he is learning about space at the same time.

The child who crawls and creeps often goes under furniture. At first he sometimes gets stuck when he has not judged correctly the size space under which he will fit. As he begins to pull himself up, he tries to climb on things. This activity is important not only for his motor development but for his spatial learning. However, many pieces of furniture are not safe or are too high. An empty beverage bottle box with the dividers still in it may be taped closed and covered with some colorful Contac® paper. This makes a safe and inexpensive place to climb. The adults can make several, and the child will have a set of large construction blocks. It has been stated in other units that the child needs safe objects to aid in developing basic concepts. Each time the child handles an object, he may learn more than one skill or idea. For instance, Juanita builds a house with some small blocks. The blocks are different colors and shapes. First Juanita picks out all the square, blue blocks and piles them three high in a row. Next she picks all the red rectangles and piles them in another direction in a row. Next she piles orange rectangles to make a third side to her structure. Finally, she lines up some yellow cylinders to make a fourth side. She places two pigs, a cow, and a horse in the enclosure. Juanita has sorted by color and shape. She has made a structure with space for her farm animals (a class) and has put the animals *in* the enclosure.

With the availability of information on outer space flight in movies and on television, children might demonstrate a concept of outer space during their dramatic play activities. For example, they might build a space vehicle with large blocks and fly off to a distant planet or become astronauts on a trip to the moon.

Children begin to integrate position, direction, distance, organization, and pattern and construction through mapping activities. Early mapping activities involve developing more complex spaces such as building houses and laying

Figure 13–4 "Where can I find a parking space for this truck?"

out roads in the sand or laying out roads and buildings with unit blocks. Or the activity could be playing with commercial toys that include a village printed on plastic along with houses, people, animals, and vehicles that can be placed on the village. Provided with these materials, children naturally make these types of constructions, which are the concrete beginnings of understanding maps.

INFORMAL ACTIVITIES

Space is an area where there are many words to be learned and attached to actions. The teacher should use space words (as listed earlier in the unit) as they fit into the daily activities. She should give space directions, ask space questions, and make space comments. Examples of directions and questions are in the assessment section of this unit. Space comments would be such as the following:

- Bob is at the *top* of the ladder.
- Cindy is *close* to the door.
- You have the dog *behind* the mother in the car.
- Tanya can move the handle *backward* and *forward*.
- You children made a house big enough for all of you. (construction)
- You pasted all the square shapes on your paper. (organization and pattern)

Jungle gyms, packing crates, ladders, ramps and other equipment designed for large muscle activity give the child experiences with space. They climb *up*, *down*, and *across*. They climb *up* a ladder and crawl or slide *down* a ramp. On the jungle gym they go *in* and *out*, *up* and *down*, *through*, *above* and *below* and *around*. They get *in* and *out* of packing crates. On swings they go *up* and *down* and *backward* and *forward* and see the world *down below*.

With the large blocks, boxes, and boards they make structures that they can get *in* themselves. Chairs and tables may be added to make a house, train, airplane, bus, or ship. Props such as a steering wheel, fire fighter or police hats, ropes, hoses, discarded radios, and so on inspire children to build structures on which to play. With small blocks, children make houses, airports, farms, grocery stores, castles, trains, and so on. They then use their toy animals, people, and other objects in spatial arrangements, patterns, and positions. They learn to fit their structures into space available: on the floor, on a large table, or on a small table. They might also build space vehicles and develop concrete mapping representations in the sand or with blocks. The teacher can ask questions about their space trip or their geographic construction: "How far is it to the moon?" or "Show me the roads you would use to drive from Richard's house to Kate's house."

As the child works with art materials, he plans what to choose to glue or paste on his paper. A large selection of collage materials such as scrap paper, cloth, feathers, plastic bits, yarn, wire, gummed paper, cotton balls, bottle caps, and ribbon offer a child a choice of things to organize on the space he has. As he gets past the first stages of experimentation, the child plans

Figure 13–5 Children learn about the positions of their bodies in space.

his painting. He may paint blobs, geometric shapes, stripes, or realistic figures. He enjoys printing with sponges or potatoes. All these experiences add to his ideas about space.

STRUCTURED ACTIVITIES

Structured activities of many kinds can be done to help the child with his ideas about space and his skills in the use of space. Basic activities are described for the three kinds of space relations (position, direction, and distance) and the two ways of using space (organization/pattern and construction).

Figure 13–6 Building a structure from paste and scrap materials.

STRUCTURED ACTIVITIES

SPACE: RELATIONSHIPS, PHYSICAL SELF

OBJECTIVE: To help the child relate his position in space to the positions of other people and things

MATERIALS: The child's own body, other people, and things in the environment

ACTIVITIES:

1. Obstacle Course

   Set up an obstacle course using boxes, boards, ladders, tables, chairs, and like items. Set it up so that by following the course, the children can physically experience position, direction, and distance. This can be done indoors or outdoors. As the child proceeds along the course, use space words to label his movement: "Leroy is going *up* the ladder, *through* the tunnel, *across* the bridge, *down* the slide and *under* the table. Now he is *close to* the end."

2. Find Your Friend
   Place children in different places: sitting or standing on chairs or blocks or boxes, under tables, sitting three in a row on chairs facing different directions, and so on. Have each child take a turn to find a friend:
   FIND A FRIEND WHO IS ON A CHAIR (A BOX, A LADDER).
   FIND A FRIEND WHO IS UNDER A TABLE (ON A TABLE, NEXT TO A TABLE).
   FIND A FRIEND WHO IS BETWEEN TWO FRIENDS (BEHIND A FRIEND, NEXT TO A FRIEND).
   FIND A FRIEND WHO IS SITTING BACKWARDS (FORWARDS, SIDEWAYS).
   FIND A FRIEND WHO IS ABOVE ANOTHER FRIEND (BELOW ANOTHER FRIEND).
   Have the children think of different places they can place themselves. When they know the game let the children take turns saying the FIND statements.
3. Put Yourself Where I Say
   One at a time give the children instructions for placing themselves in a position:
   CLIMB UP THE LADDER.
   WALK BETWEEN THE CHAIRS.
   STAND BEHIND TANYA.
   GET ON TOP OF THE BOX.
   GO CLOSE TO THE DOOR (GO FAR FROM THE DOOR)
   As the children learn the game, they can give the instructions.
4. Where Is Your Friend?
   As in Activity #2, *Find Your Friend*, place the children in different places. This time ask WHERE questions. The child must answer in words. Ask WHERE IS (Child's Name)? Child answers, "Tim is under the table," or "Mary is on top of the playhouse."

FOLLOW-UP: Set up obstacle courses for the children to use during playtime both indoors and outdoors.

SPACE: RELATIONSHIPS, OBJECTS

OBJECTIVE: To be able to relate the position of objects in space to other objects

MATERIALS: Have several identical containers (cups, glasses, boxes) and some small objects such as blocks, pegs, buttons, sticks, toy animals, people

ACTIVITIES:

1. Point To
   Place objects in various spatial relationships such as shown below:

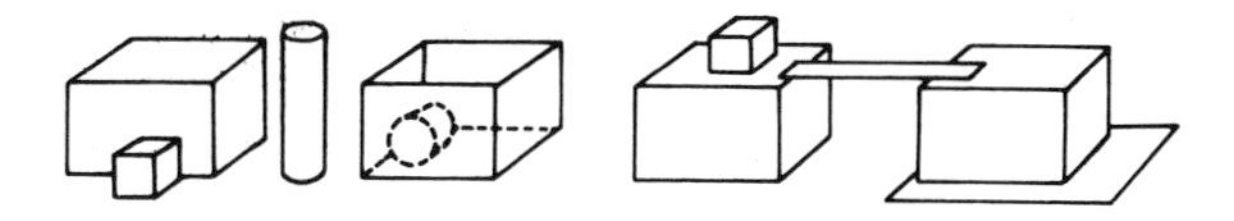

POINT TO (OR SHOW ME) THE THING THAT IS (IN, ON, UNDER, BETWEEN, BEHIND, etc.) A BOX.

2. Put The
   Set some containers out. Place some objects to the side. Tell the child PUT THE object name (IN, ON, THROUGH, ACROSS, UNDER, NEAR) THE CONTAINER.
3. Where Is?
   Place objects as in "1" above and/or around the room. Ask, WHERE IS object? TELL ME WHERE THE object IS. Child should reply using a space word.

FOLLOW-UP: Repeat the activity using different objects and containers. Leave the materials out for the children to use during center time.

SPACE: USE, ORGANIZATION/PATTERN

OBJECTIVE: To organize materials in space in a pattern

MATERIALS: Many kinds of materials are available which will give the child experiences in making patterns in space. Some of these are listed:

1. Geoboards are square boards with attached pegs. Rubber bands of different colors can be stretched between the pegs to form patterns and shapes.
2. Parquetry and pattern blocks are blocks of various shapes and colors which can be organized into patterns.
3. Pegboards are boards with holes evenly spaced. Individual pegs can be placed in the holes to form patterns.
4. Color inch cubes are cubes with one-inch sides. They come in sets with red, yellow, blue, green, orange, and purple cubes.

ACTIVITIES:

1. The children can experiment freely with the materials and create their own patterns.
2. Patterns can be purchased or made for the children to copy.

FOLLOW-UP: After the children have been shown how they can be used, the materials can be left out for use during center time.

SPACE: USE, ORGANIZATION/PATTERN

OBJECTIVE: To organize materials in space in a pattern

MATERIALS: Construction paper, scissors, and glue

ACTIVITY: Give the child one 8 1/2" x 11" piece of construction paper. Give him an assortment of precut construction paper shapes (squares, rectangles, triangles). Show him how many kinds of patterns can be made and glued on the big piece of paper. Then tell him to make his own pattern.
FOLLOW-UP: Offer the activity several times. Use different colors for the shapes, use different sizes, and change the choice of shapes.

SPACE: USE, CONSTRUCTION
OBJECTIVE: To organize materials in space in three dimensions through construction
MATERIALS: wood chips, polythene, cardboard, wire, bottle caps, and other scrap materials, Elmer's glue, heavy cardboard, or scraps of plywood
ACTIVITY: Give the child a bottle of glue and a piece of cardboard or plywood for a base. Let him choose from the scrap materials things to use to build a structure on the base. Encourage him to take his time, plan, and choose carefully which things to use and where to put them.
FOLLOW-UP: Keep plenty of scrap materials on hand so that children can make structures when they are in the mood.

Figure 13–7 The geoboard is used to organize patterns in space.

Figure 13–8 Parquetry blocks can be organized into patterns.

SPACE: USE, CONSTRUCTION

OBJECTIVE: To organize materials in space in three dimensions through construction

MATERIALS: Many kinds of construction materials can be purchased which help the child to understand space and also improve eye-hand coordination and small muscle skills. Some of these are listed:

1. Lego: Jumbo for the younger child, regular for the older child or one with good motor skills
2. Tinker Toys
3. Bolt-it
4. Snap-N-Play blocks
5. Rig-A-Jig
6. Octons, Play Squares (and other things with parts that fit together)

ACTIVITIES: Once the child understands the ways that the toys can be used, he can be left alone with the materials and his imagination.

SPACE, MAPPING

OBJECTIVE: To integrate basic space concepts through simple mapping activities

MATERIALS: Make a simple treasure map on a large piece of posterboard. Draw a floor plan of the classroom, indicating major landmarks (such as the learning centers, doors, windows, etc.) with simple drawings. Draw in some paths going from place to place. Make a brightly colored treasure chest from a shoe box. Make a matching two-dimensional movable treasure chest that can be placed anywhere on the floor plan.

ACTIVITY: Hide the treasure chest somewhere in the room. Place the small treasure chest on the floor plan. Have the children discuss the best route to get from where they are to the treasure using only the paths on the floor plan. Have them try out their routes to see if they can discover the treasure.

FOLLOW-UP: Let class members hide the treasure and see if they are able to place the treasure chest correctly on the floor plan. Let them use trial and error when hiding the treasure, marking the spot on the plan, and finding the treasure. Make up some other games using the same basic format. Try this activity outdoors. Supply some adult maps for dramatic play and observe how children use them.

Mrs. Red Fox's first graders enjoy *Stickybear Townbuilder* (see Activity #6), a computer program that provides them with the raw material for actually building their own town. Buildings and pieces of road are included. Towns can even be saved on a disk and revisited. A cursor car can be used to take a trip around the town. This experience gives practice with basic spatial relationships and an introduction to maps.

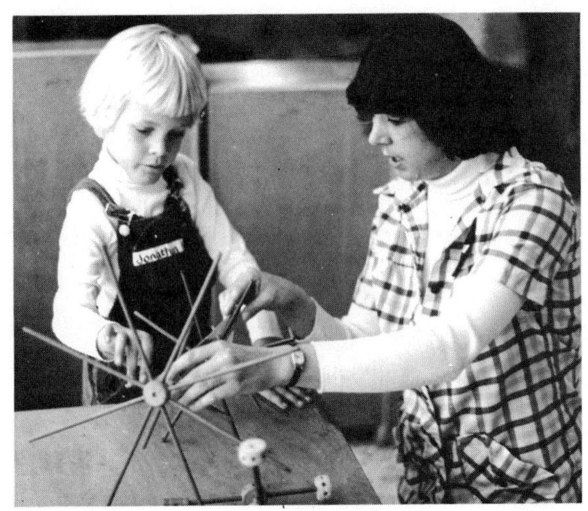

Figure 13–9 Tinker Toys® are made into a three-dimensional structure.

EVALUATION

Informal evaluation can be done through observation. The teacher should note the following as the children proceed through the day:

- Does the child respond to space words in a way that shows understanding?
- Does he answer space questions and use the correct space words?
- Does his artwork and block building show an increase in pattern and organization?
- Does the child handle his body well in space?
- Does his use of geoboards, parquetry blocks, color inch cubes, and/or pegboards show an increase in organization and pattern?

After children have done several space activities, the teacher can assess their progress using the interview tasks in Appendix A.

SUMMARY

Space is an important part of geometry. The child needs to understand the spatial relationship between his body and other things. He must also understand the spatial relationship between things around him. Things are related through position, direction, and distance.

The child also needs to be able to use space in a logical way. He learns to fit things into the space available and to make constructions in space.

FURTHER READING AND RESOURCES

Boals, B., Nalley, J. A., and Vance, M. 1982. IDEAS! Using boxes. *Dimensions.* 10 (3): 79–82.

Eltgroth, M. B. 1986. Before you turn it on. *First Teacher*, 7 (10): 4.

Hillstrom-Svercek, S. 1985. Space: A learning center. *Day Care and Early Education.* 12 (4): 31–36.

Hirsch, E. S., Ed. 1984. *The Block Book, Rev. Ed.* Washington, D.C.: National Association for the Education of Young Children.

Hoot, J. L. 1986. *Computers in early childhood education.* Englewood Cliffs, N.J.: Prentice-Hall.

Provenzo, E. F., and Brett, A. 1984. Creative block play. *Day Care and Early Education.* 11 (3): 6–8.

Reifel, S. 1984. Block construction: Children's developmental landmarks in representation of space. *Young Children.* 40 (1): 61–67.

Scott, L. B., and Garner, J. 1978. *Mathematical experiences for young children.* St. Louis: McGraw-Hill.

SUGGESTED ACTIVITIES

1. Observe some children as they play in an early childhood center. Note instances of behaviors which reflect a child's feelings, ideas, and use of space as listed in the assessment section. Rank the children from those with well developed ideas about space to those whose ideas are not so well developed.

2. Design a jungle gym/climbing structure that would offer all types of spatial experiences. Explain what kind of spatial activity can be experienced on each part.

3. Look through some early childhood materials catalogues. Pick out and list the pattern and construction toys that could be purchased with $100 to spend.

4. Add five space activities to your Activities File.

5. Use four space assessment tasks to interview a three-year-old, a four-year-old, and a five-year-old. Compare the responses of the three children. Discuss the results with other students who try out the assessment tasks. Did they get similar results?

6. Review one or both of the following computer programs using the guidelines suggested in Activity #8, Unit 9.

   Arrow Graphics. Edu-Fun, Milliken Publishing Company, 1100 Research Blvd., St. Louis, MO 63132.

   Stickybear Town Builder. Weekly Reader Family Software, Xerox Educational Publishing Company, 245 Long Hill Rd., Middletown, CT 06457

REVIEW

A. Answer each of the following questions.

   1. How can a teacher assess a child's concept of space through observation?

   2. What are three informal space activities?

   3. What are three structured space activities?

B. Select an item in Column II which applies to each of the phrases in Column I.

	Column I	Column II
1.	Several children have built a block town. "Is Kate's house closer to Richard's or to Sam's?"	a. Position
2.	Lincoln logs	b. Direction
3.	How can things be arranged as they fit in space?	c. Distance
4.	When a child arranges things in space until they fit	d. Organization
5.	Space relations	e. Construction
6.	Near-far	f. Mapping
7.	Parquetry blocks	
8.	"Here We Go 'Round the Mulberry Bush."	
9.	Answers the question, "Where am I?"	
10.	Answers the question, "Which way?"	
11.	Geoboards	
12.	Use of space	

13. Over-under
14. George works with the unit blocks building a garage for his truck.
15. Using toy road graders and shovels, the children build tunnels and roads in the sand table.
16. Up-down

UNIT 14 Parts and Wholes

OBJECTIVES

After studying this unit, the student should be able to:

- Describe the three types of part/whole relationships
- Assess and evaluate a child's knowledge of parts and wholes
- Do informal and structured part/whole activities with young children

The young child learns that wholes have parts. Later the older child learns that parts are *fractions* of the whole. The child must learn the idea of parts and wholes before he can understand fractions. He learns that some things are made up of special (unique) parts, that sets of things can be divided into parts, and that whole things can be divided into smaller parts.

He learns about special parts:

- Bodies have parts (arms, legs, head).
- A car has parts (engine, doors, steering wheel, seats).
- A house has parts (kitchen, bathroom, bedroom, livingroom).
- A chair has parts (seat, legs, back).

He learns that sets of things can be divided:

- He passes out cookies for snack.
- He deals cards for a game of picture rummy.
- He gives each friend one of his toys with which to play.
- He divides his blocks so each child may build a house.

He learns that whole things can be divided into parts:

- One cookie is broken in half.
- An orange is divided into several segments.
- A carrot or banana is sliced into parts.
- The contents of a bottle of soda pop is put into two or more cups.
- A large piece of paper is cut into small pieces.

The young child centers on the number of things he sees. Two-year-old Pete breaks up his graham cracker into small pieces. "I have more than you," he says to Tim who has one whole graham cracker also. Pete does not see that although he has more pieces of cracker he does not have more crackers. Ms. Moore shows Chris a whole apple. "How many apples do I have?" "One," says Chris. "Now watch," says Ms. Moore as she cuts the apple into two pieces. "How many apples do I have now?" "Two!" answers Chris. As the child enters concrete operations, he will see that a single apple is always a single apple even though it may be cut into parts.

Gradually the child is able to see that a whole is made up of parts. He also begins to see that parts may be the same (equal) in size and amount or different (unequal) in size and amount. He compares number and size (Unit 11) and develops the concepts of more, less, and the same. These concepts are prerequisite to the understanding of fractions that are introduced in the primary grades. An understanding of more, less, and the same underlies learning that objects

Figure 14–1 Whole sandwiches are cut into parts.

Figure 14–2 The banana is cut into two parts.

and sets can be divided into two or more equal parts and still maintain the same amount (Figure 14–1).

ASSESSMENT

The teacher should observe as the child works and plays whether he uses the words *part* and *whole* (Figure 14–2). She should note if he uses them correctly. She should note his actions:

- Does he try to divide items to be shared equally among his friends?
- Will he think of cutting or breaking something in smaller parts if there is not enough for everyone (Figure 14–3)?
- Does he realize when a part of something is missing (such as the wheel of a toy truck, the arm of a doll, the handle of a cup)?

Interview questions would be like the following:

Figure 14–3 Now the banana is cut two more times to make four parts.

SAMPLE ASSESSMENT TASK

Preoperational Ages 2–3

METHOD: Interview
SKILL: Child is able to tell which part(s) of objects and/or pictures of objects are missing.
MATERIALS: Several objects and/or pictures of objects and/or people with parts missing. Some examples are:

Objects: A doll with a leg or arm missing
A car with a wheel missing
A cup with a handle broken off
A chair with a leg gone
A face with only one eye
A house with no door

Pictures: Mount pictures of common things on poster board. parts can be cut off before mounting.

PROCEDURE: Show the child each object or picture. LOOK CAREFULLY. WHICH PART IS MISSING FROM THIS?
EVALUATION: Note if the child is able to tell which parts are missing in both objects and pictures. Does he have the language label for each part? Can he perceive what is missing?
INSTRUCTIONAL RESOURCES:

Charlesworth, R. and Lind, K. (1990) *Math and Science for Young Children*. Albany, N.Y.: Delmar. Unit 14.

SAMPLE ASSESSMENT TASK

Preoperational Ages 4–5

METHOD: Interview
SKILL: The child can recognize that a whole divided into parts is still the same amount.
MATERIALS: Apple and knife
PROCEDURE: Show the child the apple. HOW MANY APPLES DO I HAVE? After you are certain the child understands that there is one apple, cut the apple into two equal halves. HOW MANY APPLES DO I HAVE NOW? HOW DO YOU KNOW? If the child says "Two," press the halves together and ask, HOW MANY APPLES DO I HAVE NOW? Then cut the apple into fourths and eighths, following the same procedure.
EVALUATION: If the child can tell you that there is still one apple when it is cut into parts, he is able to mentally reverse the cutting process and may be leaving the preoperational period.
INSTRUCTIONAL RESOURCES:

Charlesworth, R. and Lind, K. (1990) *Math and Science for Young Children*. Albany, N.Y.: Delmar. Unit 14.

SAMPLE ASSESSMENT TASK

Preoperational Ages 5–6

METHOD: Interview

SKILL: The child can divide a set of objects into smaller groups.

MATERIALS: Three small dolls (real or paper cutouts) and a box of pennies or other small objects

PROCEDURE: Have the three dolls arranged in a row. I WANT TO GIVE EACH DOLL SOME PENNIES. SHOW ME HOW TO DO IT SO EACH DOLL WILL HAVE THE SAME AMOUNT.

EVALUATION: Note how the child approaches the problem. Does he give each doll one penny at a time in sequence? Does he count out pennies until there are three groups with the same amount? Does he divide the pennies in a random fashion? Does he have a method for finding out if each has the same amount?

INSTRUCTIONAL RESOURCES:

Charlesworth, R. and Lind, K. (1990) *Math and Science for Young Children*. Albany, N.Y.: Delmar. Unit 14.

NATURALISTIC ACTIVITIES

The newborn infant is not aware that all his body parts are part of him. His early explorations lead him to find out that his hand is connected via an arm to his shoulder and those toes he sees at a distance are hooked to his legs. As he explores objects, he learns that they have different parts also. As he begins to sort and move objects about, he learns about parts and wholes of sets.

The following are some examples of the young child's use of the part/whole idea (Figure 14–4):

- Two-year-old Pete has a hot dog on his plate. The hot dog is cut in six pieces. He gives two pieces to his father, two to his mother, and keeps two for himself.
- Three-year-old Jim is playing with some toy milk bottles. He says to Ms. Brown, "You take two like me."
- Three-year-old Kate is sitting on a stool in the kitchen. She sees three eggs boiling in a pan on the stove. She points as she looks at her mother, "One for you, one for me, and one for Dad."
- Tanya is slicing a carrot. "Look I have a whole bunch of carrots now."

Figure 14–4 A big piece of playdough is divided into smaller parts.

- Juanita is lying on her cot at the beginning of nap time. She holds up her leg. "Mrs. Raymond, is this part of a woman?"
- Bob runs up to Mr. Brown. "Look I have a whole tangerine."

INFORMAL ACTIVITIES

Many times during the day the teacher can help children develop their understanding of parts and wholes. The teacher can use the words *part*, *whole*, *divide* and *half*.

- Today everyone gets *half* of an apple and *half* of a sandwich.
- Too bad, *part* of this game is missing.
- Take this basket of crackers and *divide* them up so everyone gets some.
- No, we won't cut the carrots up. Each child gets a *whole* carrot.
- Give John *half* the blocks so he can build too.

Figure 14–5 Now there are two parts made into balls.

Figure 14–6 Part/whole match-ups are a challenge to the young child.

- We only have one apple left. Let's *divide* it up.
- Point to the *part* of the body when I say the name.

The child can be given tasks which require him to learn about parts and wholes. When the child is asked to pass something, to cut up vegetables or fruit, or to share materials, he learns about parts and wholes (Figures 14–5 and 14–6).

STRUCTURED ACTIVITIES

The child can be given structured experiences in all three types of part/whole relationships. Activities can be done which help the child become aware of special parts of people, animals, and things. Other groups of activities involve dividing sets into smaller sets. The third type of activity gives the child experiences in dividing wholes into parts.

STRUCTURED ACTIVITIES

PARTS AND WHOLES: PARTS OF THINGS

OBJECTIVE: To learn the meaning of the term *part* as it refers to parts of objects, people, and animals

MATERIALS: Objects or pictures of objects with parts missing

ACTIVITIES:

1. The Broken Toys
   Show the child some broken toys or pictures of broken toys. WHAT'S MISSING FROM THESE TOYS? After the child tells what is missing from each toy bring out the missing parts (or pictures of missing parts): FIND THE MISSING PART THAT COMES WITH EACH TOY.
2. Who (or What) Is Hiding?
   The basic game is to hide someone or something behind a screen so that only a part is showing. The child then guesses who or what is hidden. The following are some variations:
   a. Two or more children hide behind a divider screen, a door, or a chair. Only a foot or a hand of one is shown. The other children guess whose body part can be seen.
   b. The children are shown several objects. The objects are then placed behind a screen. A part of one is shown. The chlld must guess which thing the part belongs to. (To make the task harder, the parts can be shown without the children knowing first what the choices will be).
   c. Do the same type of activity using pictures:
      - Cut out magazine pictures and mount on cardboard (or draw your own). Cut a piece of construction paper to use to screen most of the picture. Say, LOOK AT THE PART THAT IS SHOWING. WHAT IS HIDDEN BEHIND THE PAPER?
      - Mount magazine pictures on construction paper. Cut out a hole in another piece of construction paper of the same size. Staple the piece of paper onto the one with the picture on it so that a part of the picture can be seen. Say, LOOK THROUGH THE HOLE. GUESS WHAT IS IN THE PICTURE UNDER THE COVER.

FOLLOW-UP: Play *What's Missing Lotto* game. (Childcraft) or *What's Missing? Parts & Wholes, Young Learners Puzzles* (Teaching Resources)

PARTS AND WHOLES: DIVIDING SETS

OBJECTIVE: To give the child practice in dividing sets into parts (smaller sets)

MATERIALS: Two or more small containers and some small objects such as pennies, dry beans, or buttons

ACTIVITY: Set out the containers (start with two and increase the number as the child is able to handle more). Put the pennies or other objects in a bowl next to the containers. DIVIDE THESE UP SO EACH CONTAINER HAS SOME. Note whether the child goes about the task in an organized way and if he tries to put the same number in each container. Encourage the children to talk about what they have done. DO THEY ALL HAVE THE SAME AMOUNT? DOES ONE HAVE MORE? HOW DO YOU KNOW?

FOLLOW-UP: Increase the number of smaller sets to be made. Use different types of containers and different objects.

PARTS AND WHOLES

OBJECTIVE: To divide whole things into two or more parts

MATERIALS: Real things or pictures of things that can be divided into parts by cutting, tearing, breaking, or pouring

ACTIVITIES:

1. Have the children cut up fruits and vegetables for snack or lunch. Be sure the children are shown how to cut so as not to hurt themselves. Be sure also that they have a sharp knife so the job is not frustrating. Children with poor coordination can tear lettuce, break off orange slices, and cut the easier things such as string beans.
2. Give the child a piece of paper. Have him cut it or tear it. Then have him fit the pieces back together. Have him count how many parts he made.
3. Give the child a piece of play dough or clay. Have him cut it with a dull knife or tear it into pieces. How many parts did he make?
4. Use a set of plastic measuring cups and a larger container of water. Have the children guess how many of each of the smaller cups full of water will fill the one cup measure. Let each child try the one-fourth, one-third, and one-half cups and count how many of each of these cups will fill the one cup.

FOLLOW-UP: Purchase or make some more structured part/whole materials such as:

1. Fraction Pies: circular shapes available in rubber and magnetic versions.
2. Materials which picture two halves of a whole.
   Halves to Wholes (DLM)
   Match-ups (Childcraft and Playskool)
3. Puzzles which have a sequence of difficulty with the same picture cut into two, three, and more parts:
   Basic Cut Puzzlles (DLM)
   Fruit and Animal Puzzles (Teaching Resources)
4. Dowley Doos take-apart transportation toys (Lauri)

Diana, age five, is delighted with *Facemaker* (Spinnaker, see Activity #6), a computer program that enables her to select from a variety of head parts and clothing the combination of her choice. When she has finished creating her cartoon figure, she signals the printer and her creation is printed out for her to keep. This experience gives Diana practice with creating a whole from parts in an imaginative way.

EVALUATION

The adult should observe and note if the child shows increased use of part/whole words and more skills in his daily activities:

- Can he divide groups of things into smaller groups?
- Can he divide wholes into parts?
- Does he realize that objects, people, and animals have parts that are unique to each?

SUMMARY

The young child learns about parts and wholes. The idea of parts and wholes is basic to what the child will learn later about fractions.

The child learns that things, people, and animals have parts. He leams that sets can be divided into parts (sets with smaller numbers of things.) He learns that whole things can be divided into smaller parts or pieces.

Experiences in working with parts and wholes help the young child move from preoperational centering to the concrete view and to understanding that the whole is no more than the sum of all its parts.

FURTHER READING AND RESOURCES

Beaty, J. J., and Tucker, W. H. 1987. *The computer as a paintbrush*. Columbus, Ohio: Merrill.

Burton, G. M. 1985. *Towards a good beginning*. Menlo Park, Calif.: Addison-Wesley.

Scott, L. B. 1978. *Mathematical experiences for young children*. St. Louis: McGraw-Hill.

SUGGESTED ACTIVITIES

1. Visit a Montessori classroom. Examine the materials which teach the ideas of parts and wholes. List these materials and describe how they are used. Report your experience to the class.
2. Assess a young child's level of development in part/whole relations. Prepare an instructional plan based on the results. Prepare the materials and teach the child. Evaluate the results. Did the child learn from the instruction?
3. Check through the Activities File. Be sure it is up to date with activities for each unit topic.
4. Try out some of the structured activities described in this unit with children of different ages.
5. If the school budget allocated forty dollars for part/whole materials, decide on what would be top priority purchases.
6. Using guidelines from Unit 9, Activity #8, evaluate one or more of the following computer programs that are designed to reinforce the concept of parts and wholes:

   Dr. Seuss Fix Up the Mixedup Puzzle. CBS Software, Greenwich, Conn.

   Facemaker. Spinnaker Software, Cambridge, Mass.: Apple II's, Atari, C64, IBM PCJr.

   Mr. and Mrs. Potato Head. Random House, New York: Apple, C64.

   Stickers. Springboard Software, Minneapolis: Apple II's, IBM PC and PCJr, C64.

   Tonk in the Land of Buddy-Bots. Mindscape, Northbrook, Ill.: Apple II's, Atari, C64, IBM PC & PCJr.

REVIEW

A. Select the items in Column II which apply to the items in Column I.

Column I

1. Tina shares her cupcake.
2. Pieces of a monkey puzzle
3. Tom gives Joe some of his crayons.
4. Jim tears a piece of paper.
5. Sue takes half the lump of dough.
6. Chris puts the red blocks in one pile and the blue blocks in another.
7. A doll's arm
8. Stephanie takes the screwdriver from the tool kit.
9. Orange section
10. Piano keys

Column II

a. Things, people, and animals have parts
b. Sets can be divided into parts
c. Whole things can be divided into smaller parts or pieces

B. Answer each of the following.

1. Briefly describe how a child first learns that wholes have parts.
2. What are the three types of part/whole relationships?
3. Describe two informal part/whole relationship activities.
4. Describe two structured part/whole relationship activities.
5. Describe one part/whole assessment task.
6. Describe how to evaluate a child's knowledge of parts and wholes through observation.

UNIT 15 The Language of Concept Formation

OBJECTIVES

After studying this unit, the student should be able to:

- Explain two ways to describe a child's understanding of concept words
- List the concept words used in Units 8 through 14
- List the concept words used for ordering and measuring
- Use books to support the development of math and science language

What the child does and what the child says tells the teacher what the child knows about math and science. The older the child gets, the more important concept words become. The child's language system is usually well developed by age four. That is, by this age, children's sentences are much the same as an adult's. Children are at a point where their vocabulary is growing very rapidly (Figures 15–1 and 15–2).

Figure 15–1 *"One, two, three, four*—I can dial *numbers."*

The adult observes what the child does from infancy through age two and looks for the first understanding and the use of words. Between two and four the child starts to put more words together into longer sentences. He also learns more words and what they mean.

In assessing the young child's concept development, questions are used. Which is the big ball? Which is the circle? The child's understanding of words is checked by having him respond with the right action:

- Point to the big ball.
- Find two chips.

Figure 15–2 "Are there more *short* pencils or more *long* pencils?"

- Show me the picture in which the boy is on the chair.

The above tasks do not require the child to say any words. He needs only point, touch, or pick up something. Once the child demonstrates his understanding of math words by using gestures or other nonverbal answers, he can move on to questions he must answer with one or more words. The child can be asked the same questions as above in a way that requires a verbal response:

- (The child is shown two balls, one big and one small.) "What is different about these balls?"
- (The child is shown a set of objects.) "How many are there in this set?"
- (The child is shown a picture of a boy sitting on a chair.) "Where is the boy?"

The child learns many concept words as he goes about his daily activities. It has been found that by the time a child starts kindergarten, he uses many concept words he has learned in a naturalistic way. Examples have been included in each of the previous units (8 through 14). The child uses both comments and questions. Comments would be like the following:

- Mom, I want *two* pieces of cheese.
- I have a *bunch* of bird seed.
- Mr. Brown, this chair is *small.*
- *Yesterday* we went to the zoo.
- The string is *long.*
- This is the *same* as this.
- The foot fits *in* the shoe.
- This cracker is a *square* shape.
- Look, some of the worms are *long* and some are *short*, some are *fat* and some are *thin*.
- The *first* bean seed I planted is *taller* than the *second* one.
- Outer *space* is *far away*.

Questions would be like these:

- How *old* is he?
- *When* is Christmas?
- *When* will I grow as *big* as you?
- *How many* are coming for dinner?
- Who has *more?*
- What *time* is my TV program?
- Is this a school *day*, or is it *Saturday?*
- What makes the bubbles when the water gets *hot?*
- Why does this roller always go *down* its ramp *faster* than that roller goes *down* its ramp?
- Why are the leaves turning *brown* and *red* and *gold* and falling *down on* the ground?

The answers that the child gets to these questions can help increase the number of concept words he knows and can use (Figure 15–3).

The teacher needs to be aware of using concept words during center time, lunch, and other times when a structured concept lesson is

Figure 15–3 "Hold up *two* fingers."

not being done. She should also note which words the child uses during free times.

The teacher should encourage the child to use concept words even though he may not use them in an accurate, adult way. for example:

- I can count—one, two, three, five, ten.
- Aunt Helen is coming after my last nap. (Indicates future time)
- I will measure my paper. (Holds ruler against the edge of the paper.)
- Last night Grandpa was here. (Actually several days ago)
- I'm six years old. (Really two-years-old)
- I have a million dollars. (Has a handful of play money)

Adults should accept the child's use of the words and use the words correctly themselves. Soon the child will develop a higher level use of words as he is able to grasp higher level ideas. For the two- or three-year-old, any group of things more than two or three may be called a *bunch*. Instead of using *big* and *little* the child may use family words: this is the mommy block and this is the baby block. Time (Unit 19) is one concept that takes a long time to grasp. A young child may use the same word to mean different time periods. The following examples were said by a three-year-old:

- *Last night* we went to the beach (meaning last summer).
- *Last night* I played with Chris (meaning yesterday).
- *Last night* I went to Kenny's house (meaning three weeks ago).

For this child, *last night* means any time in the past. One by one he will learn that there are words that refer to times past such as last summer, yesterday, and three weeks ago.

Computer activities can also add to vocabulary. The teacher uses concept words when explaining how to use the programs. Children enjoy working at the computer with friends and will use the concept words in communication with each other as they work cooperatively to solve the problems presented by the computer (Figure 15–4).

Books also afford a multitude of opportunities for building concept vocabulary. For example, a teacher was observed reading a story called *Julius* by Syd Hoff. The following concept words were used:

- Show me the *front of* the book.
- Show me the *end* of the story.
- The giraffe has a very *long* neck.
- Carry *on top of* their heads.
- *Time* to rest.
- Who's *in* there?
- *On* the ship.
- I'm coming *in* for a landing.
- He landed *on* Julius's nose.
- If someone is *slim*, he is *thin* or *skinny*.
- She is *fat*—the *fattest* lady I've ever seen.
- Mr. and Mrs. *Tiny* are *small*.

Figure 15–4 Working at the computer, children can communicate using concept language.

Figure 15–5 Books contain a rich variety of concept language.

- Mr. *Giant* is *in* the tree.
- We're *bigger* than you.
- Show me the *first* word, the *last* word.

Examples of other books with concepts are included in other units. Appendix B lists many books with concept ideas (Figure 15–5).

Many concept words have already been introduced and more will appear in the units to come. The prekindergarten child continually learns words. The following presents those concept words which most children can use and understand by the time they complete kindergarten.

CONCEPT WORDS

The following words have appeared in Units 8–14:

- **One-to-One Correspondence:** one, pair, more, each, some, group, bunch, set, amount
- **Number and Counting:** zero, one, two, three, four, five, six, seven, eight, nine, ten, how many, count, group, one more than, next, number, computer
- **Sets and Classifying:** sets; descriptive words for color, shape, size, materials, pattern, texture, function, association, class names, and common features; belong with; goes with; is used with; put with; the same
- **Comparing:** more, less, big, small, large, little, long, short, fat, skinny, heavy, light, fast, slow, cold, hot, thick, thin, wide, narrow, near, far, later, sooner, earlier, older, younger, newer, higher, lower, loud, soft (sound)
- **Shape:** circle, square, triangle, rectangle, ellipse, rhombus, shape, round
- **Space:** *where* (on, off, on top of, over, under, in, out, into, out of, top, bottom, above, below, in front of, in back of, behind, beside, by, next to, between); *which way* (up, down, forward, backward, around, through, to, from, toward, away from, sideways, across); *distance* (near, far, close to, far from); map, floor plan (Figure 15–6)
- **Parts and wholes:** part, whole, divide, share, pieces, some, half

Words which will be introduced later are:

- **Ordering:** first, second, third; big, bigger, biggest; few, fewer, fewest; large, larger, largest; little, littler, littlest; many, more, most; thick, thicker, thickest; thin, thinner, thinnest; last, next, then
- **Measurement of volume, length, weight, and temperature:** little, big, medium, tiny, large, size, tall, short, long, far, farther, closer, near, high, higher, thin, wide, deep, cup, pint, quart, gallon, ounces, milliliter, kiloliter, liter, foot, inch, kilometer, mile, meter, centimeter, narrow, measure, hot, cold, warm, cool, thermometer, temperature, pounds, grams, kilograms, milligrams

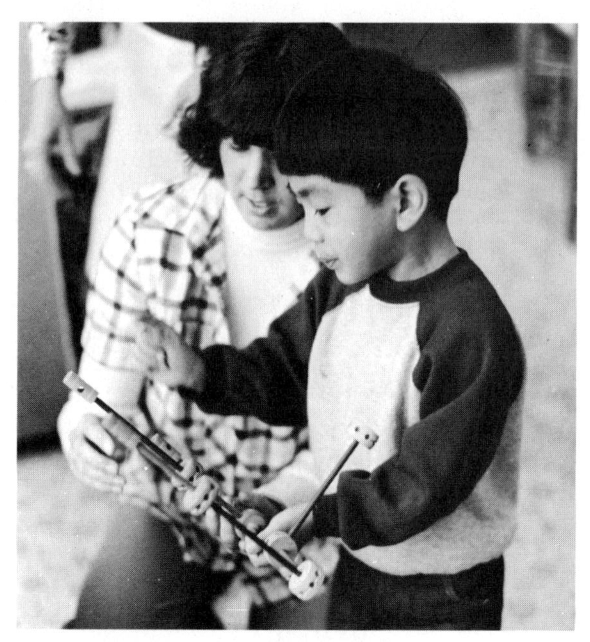

Figure 15–6 "Your airplane can go *high above* the houses."

- **Measurement of time and sequence:** morning, afternoon, evening, night, day, soon, week, tomorrow, yesterday, early, late, a long time ago, once upon a time, minute, second, hour, new, old, already, Easter, Christmas, birthday, now, year, weekend, clock, calendar, watch, when, time, date, sometimes, then, before, present, soon, while, never, once, sometime, next, always, fast, slow, speed, Monday and other days of the week, January and other months of the year, winter, spring, summer, fall
- **Practical:** money, cash register, penny, dollar, buy, pay, change, cost, check, free, store, map, recipe, measure, cup, tablespoon, teaspoon, boil, simmer, bake, degrees, time, hours, minutes, freeze, chill, refrigerate, pour, mix, separate, add, combine, ingredients

Words can be used before they are presented in a formal structured activity. The child who speaks can become familiar with words and even say them before he understands the concepts they stand for (Figure 15–7).

- **Primary Level Words:** addition, subtraction, number facts, plus, add, minus, take away, total, sum, equal, difference, amount, altogether, in all, are left, number line, place value, patterns, ones, tens, hundreds, digit, multiplication, division, equation, times, divide, product, even, odd, fractions, halves, fourths, thirds, wholes, numerator, denominator, measure, inches, feet, yards, miles, centimeter, meter, kilometer

SUMMARY

As children learn math and science concepts and skills, they also add many words to their vocabularies. Math and science have a language that is basic to their content and activities. Language is learned through naturalistic, informal, and structured activities. Computer activities are excellent for promoting communication between and among children. Books are a rich source of concept words. Language and concept development work together to enrich the children's skills and knowledge.

Figure 15–7 *Today* it is sunny outside.

FURTHER READING AND RESOURCES

Ballenger, M., Benham, N., and Hosticka, A. 1984. Children's counting books: Mathematical concept development. *Childhood Education*. September/October: 30–35.

Bayle, L. 1987. *Picture books for preschool nutrition education: A selected annotated bibliography*. Lexington, Mass.

Radencich, M. C. and Bohning, G. 1988. Pop up, pull down, push in, slide out: Natural science action books. *Childhood Education*. 64 (3): 157–161.

Schoenfeld, M. 1987. Rainy weather: Books for children. *Day Care and Early Education*. 15 (2): 44–45.

Smith, D. G. 1986. Organizing libraries for nonreaders. *Day Care and Early Education*, 12 (1): 11–14.

Tischler, R. W. 1988. Mathematics from children's literature. *Arithmetic Teacher*. 35: 42–47.

Journals such as *Childhood Education*, *Dimensions*, *Science and Children* and *Young Children* have book review columns in each issue. Many of the books reviewed apply math and science concepts.

SUGGESTED ACTIVITIES

1. Visit a prekindergarten, a kindergarten, and a primary classroom. Observe for at least thirty minutes in each room. Write down every child and adult math and science word used. Compare the three age groups and teachers for number and variety of words used.
2. Make the concept language assessment materials described in Appendix A. Administer the tasks to four-, five-, and six-year-old children. Compare the number and variety of concept words used. If other students in your class do this activity, compare results.
3. Make a vocabulary list for your Activities File.
4. Observe some young children doing concept development computer activities. Record their conversations and note how many concept words they use.
5. Select one of the concept books suggested in Appendix B. Familiarize yourself with the content. Read the book with one or more young children. Question them to find out how many of the concept words they can use.

REVIEW

A. Briefly answer each of the following.

1. Explain two ways to describe a child's understanding of concept words.
2. List ten concept words used in Units 8–14.
3. List ten concept words used for ordering and measuring.

B. Complete the Concept Word Puzzle.

Down

1. Measurement word
2. Part/whole word
3. Comparing word
4. Shape word
5. Measurement of height
6. Number word
7. Matching word used for things in a group

Across

1. Part/whole word
2. Classifying word
3. Comparing word
4. Classifying word
5. Measurement of time word
6. Ordering
7. Space word

UNIT 16 Fundamental Concepts in Science

OBJECTIVES

After studying this unit, the student should be able to:

- Design lessons that integrate fundamental science and math concepts
- Develop naturalistic, informal, and structured activities that utilize science and math concepts

One-to-one correspondence, number and counting, sets and classifying, comparing, shape, space, and parts and wholes are concepts and skills that are fundamental to science as well as to mathematics. Units 8–14 of this book presented these concepts with a mathematical emphasis. This unit shows how these concepts and skills can be utilized in science with young children.

ONE-TO-ONE CORRESPONDENCE

Science topics such as animals and their homes lend themselves well to matching activities. Not only are children emphasizing one-to-one correspondence by matching one animal to a particular home, they are also increasing their awareness of animals and their habitats. The questions that teachers ask will further reinforce the process skills and science content contained in the activities. The following science-related activities work well with young children:

1. Create a bulletin board background of trees (one tree should have a cavity), a cave, a plant containing a spider web, and a hole in the ground. Hang a beehive from a branch and place a nest on another branch. Ask the children to tell you where each of the following animals might live: owl, bird, bee, mouse, spider, and bear. Place the animal by the appropriate home. Then, make backgrounds and animals for each child and let the children paste the animal where it might live (Figure 16–1). Ask, DO YOU THINK A PLANT IS A GOOD PLACE FOR A SPIDER TO SPIN A WEB? WHY DO YOU THINK SO? (Outdoor insects can land in the web.) Have the children describe the advantages of living in the different homes.
2. Another popular animal matching activity involves the preparation of six animals and six homes. The children must cut out a home and paste it in the box next to the animal that lives there. For example, a bee lives in a beehive, a horse in a barn, an ant in an anthill, a bird in a nest, a fish in an aquarium, and a child in a human home (Figure 16–2). Ask, WHY WOULD AN ANT LIVE IN THE GROUND? Emphasize that animals try to live in places that offer them the best protection or source of food. Discuss how each is suited for its home.
3. Counters are effective in reinforcing the fundamental concept of one-to-one correspondence. For example, when the children study about bears, make and use bear counters that can be put into little bear caves. As the children match the bear to its cave, they understand that there are the

Figure 16–1 Find a home for the bear.

Figure 16–2 Cut out the pictures of animals and homes. Mount them on cardboard for the children to match.

same number of bears as there are caves. Lead the children in speculating why bears might take shelter in caves (convenient, will not be disturbed, hard to find, good place for baby bears to be born, and so on).

4. Children enjoy working with felt shapes. Cut out different sized felt ducks, bears, ponds, and caves for children to match on a flannel board. If children have the materials available, they will be likely to play animal matching games on their own, such as, "Match the baby and adult animals." or "Line up all of the ducks and find a pond for each one." Have children compare the baby ducks to the adult ducks by asking, IN WHAT WAYS DO THE BABIES LOOK LIKE THEIR PARENTS? HOW ARE THEY ALIKE? CAN YOU FIND ANY DIFFERENCES? Emphasize camouflage as the primary reason that baby animals usually blend in with their surroundings. The number of animals and homes can be increased as the children progress in ability.
5. After telling the story of Goldilocks and the Three Bears, draw three bears of different sizes without noses. Cut out noses that will match the blank spaces and have the children match the nose to the bear (Figure 16–3). You might want to do this first as a group and then have the materials available for the children to work with individually. Point out that larger bears have bigger noses.

One-to-one correspondence and the other skills presented in this unit cannot be developed in isolation of content. Emphasize the science concepts and process skills as you utilize animals as a means of developing fundamental skills. Further animal matching and sorting activities can be created by putting pictures into categories of living and nonliving things, vertebrates and invertebrates, reptiles, amphibians, birds, and fish.

When comparing major groups of animals—reptiles, for example—try to include as many representatives of the group as possible. Children can study pictures of turtles (land turtles are called *tortoises* and some freshwater turtles are called *terrapins*), snakes, lizards, alligators, and crocodiles, and match them to their respective homes. The characteristics of each reptile can be compared and contrasted. The possibilities for matching, counting, comparing, and classifying animals are limitless and are a natural math and science integration.

NUMBER AND COUNTING

Emphasizing science content while learning number concepts enables children to relate these subjects to their everyday lives and familiar concrete examples. The following example shows a kindergarten class using concrete experiences to reinforce counting and the concept that apples grow on trees.

When Mrs. Jones teaches an apple unit in the fall, she sorts apples of different sizes and colors. Then she has the children count the apples before they cook them to make applesauce and apple pies.

For additional number concept extensions, she uses a felt board game that requires the children to pick apples from trees. For this activity, she creates two rows of apple trees with four felt trees in each row. The trees are filled with one to nine apples and the numerals are at the bottom of each row. As the children play with the game, Mrs. Jones asks, "How many apples can you pick from each tree?" As Sam picks apples off a tree, he counts aloud, "one, two, three, four, five—I picked five apples from the apple tree." He smiles and points to the numeral five (Figure 16–4).

When doing science activities, do not miss opportunities to count. The following activities emphasize counting:

1. If bugs have been captured for study, have

Figure 16–3 Which nose belongs to baby bear?

your children count the number of bugs in the bug keeper. Keep in mind that insects can be observed for a couple of days but should be released to the area in which they were originally found.

2. Take advantage of the eight tentacles of an octopus to make an octopus counting board. Enlarge an octopus on a piece of posterboard and attach a curtain hook to each tentacle above a number. Make circles with dots (one to eight) to hang on hooks. Ask children to count the number of tentacles that the octopus has. Explain that all octopuses have eight tentacles. Discuss how

Figure 16–4 How many apples did you pick?

Figure 16–5 Count the dots and stick the square on the correct tentacle.

an octopus uses its armlike tentacles. Then, have children count the dots on the circles and put the circles on the hook over the correct number (Figure 16–5).

3. A fishtank bulletin board display emphasizes counting and prepares children for future activities in addition. Prepare construction-paper fish, a treasure chest, rocks, and plants and attach them to a 12″ × 20″ rectangle made from light blue oaktag. Glue some of the fish, the treasure chest, rocks, and plants to the blue background. Then, using thumbtacks, attach a 13″ × 21″ clear plastic rectangle to the top of the light blue oaktag to create a fish tank bulletin board. Before the students arrive the next day, put more fish in the tank. Then ask, "How many fish are swimming

in the tank?" Each day, add more fish and ask the class, "How many fish are swimming in the tank today?" "How many were swimming in the tank yesterday?" "Are there more fish swimming in the tank today or were there more yesterday?" Use as many fish as are appropriate for your students (Figure 16–6).

Children love to count. They count informally and with direction from the teacher. Counting cannot be removed from a context–you have to count something. As children count, emphasize the science in which they are counting. If children are counting the number of baby ducks following an adult duck, lead the children in speculating where the ducks might be going and what they might be doing.

Sequencing and Ordinal Position

Science offers many opportunities for reinforcing sequencing and ordinal position. Observing the life cycle of a caterpillar is a good way to learn about the life of an insect. Like all insects, caterpillars go through four distinct life stages: 1. egg, 2. larva (the caterpillar that constantly eats),

Figure 16–6 Create a fish tank bulletin board for counting fish.

3. pupa (the caterpillar slowly changes inside of a cocoon) and 4. adult (the butterfly or moth emerges from the cocoon). Pictures of each of these stages can be mounted on index cards, matched with numerals, and put in sequence. Children will enjoy predicting which stage they think will happen next and comparing this to what actually happens (Figure 16–7).

Animal Movement

Children can be encouraged to learn how different animals move as they learn one-to-one correspondence and practice the counting sequence.

Arrange the children in a line. Have them stamp their feet as they count, raising their arms in the air to emphasize the last number in the sequence. A bird might move this way, raising its wings in the air in a threatening manner or in a courtship dance. Have children change directions without losing the beat by counting "one" as they turn. For example,

One, two, three, four (up go the arms)
(turn) one, two, three, four (up go the arms)
(turn) one, two, three, four . . .
and so on.

Now, have the children move to different parts of the room. Arm movements can be left out and the children can tiptoe, lumber, or sway, imitating the ways various animals move. A bear has a side-to-side motion, a monkey might swing its arms and an elephant its trunk. This activity is also a good preparation strategy for the circle game suggested in the next paragraph (Figure 16–8).

A Circle Game

Practice in counting, sequence, one-to-one correspondence, looking for patterns, and the process skills of predicting are involved in the following circle game, which focuses attention on the prediction skills necessary in many science activities. Rather than be the source of the "right" answer, Mrs. Jones allows the results of

Figure 16–7 Stages in the life cycle of a butterfly.

Figure 16–8 Move your arms like a bird.

the game to reveal this. For maximum success, refer to the move like an animal activity in the previous section of this unit.

Mrs. Jones asks six children to stand in a circle with their chairs behind them. The child designated as the "starter," Mary, wears a hat and initiates the counting. Mary counts one and each child counts off in sequence, one, two, three, and so on. The child who says the last number in the sequence (six) sits down. The next child begins with one and again the last sits down. The children go around and around the circle, skipping over those sitting down, until only one child is left standing.

Mrs. Jones repeats the activity exactly, starting with the same child and going in the same direction using the same sequence, neither adding nor removing any children. She asks the children to predict who they think will be the last one standing. Mrs. Jones accepts all guesses equally, asking, "Who has a different idea?" until everyone who wishes has guessed. She repeats the activity so children can check their predictions. Then the activity is repeated, but this time Mrs. Jones says, "Whisper in my ear who you think will be the last child standing." The kindergarten teacher plans to repeat this game until every child can predict correctly who will be the one left standing.

The game can be played with any number

of players and with any length number sequence. Rather than evaluate the children's predictions, Mrs. Jones uses the predictions as a good indication of childrens' development. Try this game; you will be surprised by childrens' unexpected predictions.

SETS AND CLASSIFYING

A set is a group of things with common features that are put together by the process of classifying. Children classify when they separate and group things because they belong together for some reason. The following classification activities develop analytical thinking and encourage clear expression of thought in a variety of settings.

1. One way to encourage informal classifying is to keep a button box. Children will sort buttons into sets on their own. Ask, "How do these go together?" Let the children explain.
2. Kindergarten age children can classify animals into mammals, reptiles, and amphibians by their body coverings. They can use pictures, plastic animals, or materials that simulate the body coverings of the animals.
3. Children can classify colors with color-coordinated snacks. As children bring in snacks, help them classify the foods by color. Some examples might include:

   Green: lettuce, beans, peas
   Yellow: bananas, lemonade, corn, cake
   Orange: carrots, oranges, cheese
   White: milk, cottage cheese, bread
   Red: ketchup, cherry-flavored drinks, apples, jams, tomatoes
   Brown: peanut butter, whole wheat bread, chocolate
4. Display a collection of plant parts such as stems, leaves, flowers, fruit, and nuts. Have the children sort all of the stems into one pile, all of the leaves into another, and so on. Ask a child to explain what was done. Stress that things that are alike in some way are put together to form a group. Then, have the children re-sort the plant parts into groups by properties such as size (small, medium, and large).
5. Children might enjoy playing a sorting game with objects or pictures of objects. One child begins sorting a collection of objects. After he is halfway through, the next child must complete the activity by identifying the properties the first child used and continue sorting with the same system.

Using Charts and Lists in Classification

The following example shows a first grade teacher using classification activities to help make her class aware of the technological world around them.

First, Mrs. Red Fox has the children look around the classroom and name all the machines that they can see. She asks, "Is a chair a machine?" Dean answers, "No, it does not move." Mrs. Red Fox follows up with an open question, "Do you know what is alike about machines?" Mary and Judy have had experience in the Machine Center and say, "Machines have to have a moving part." This seems to jog Dean's memory and he says, "And I think they have to do a job." (A machine has to have at least one moving part and do a task.) Mrs. Red Fox writes down all of the suggestion in a list called Machines in Our Classroom. The children add the pencil sharpener, a door hinge, light switch, water faucet, record player, and aquarium to the list. Mrs. Red Fox then asks the children to describe what the machines do and how they make work easier. After discussing the machines found in the classroom, the students talk about the machines that are in their homes (Figure 16–9).

The next day the children brainstorm and make a list called Machines in Our Home, and hang it next to the classroom machine list. When

Figure 16–9 Can a nail be a machine?

the lists are completed, Mrs. Red Fox asks the children, "Do you see any similarities between the machines found at home and the machines found in our classroom?" Trang Fung answers, "There is a clock in our classroom, and we have one in the kitchen, too." Other children notice that both places have faucets and sinks. When the teacher asks, "What differences do you see between the machines found at home and at school?" Dean notices that there isn't a telephone or bicycle at school and Sara did not find a pencil sharpener in her home.

Mrs. Red Fox extends the activity by giving the children old magazines and directs them to cut out pictures of machines to make a machine book. The class also decides to make a large machine collage for their bulletin board.

COMPARING

When young children compare, they look at similarities and differences with each of their senses. Children are able to detect small points of difference and, because of this, enjoy spotting mistakes and finding differences between objects. Activities such as observing appearances, sizes, graphing, and dressing figures give experience in comparing.

1. Compare appearances by discussing likeness between relatives, dolls, or favorite animals. Then, make a block graph (see Unit 20) to show the different hair colors in the room or any other characteristics that interests your students.
2. Size comparisons seem to fascinate young children. They want to know who is the biggest, oldest, tallest, shortest. Compare the lengths of hand spans, arm spans and lengths of paces and limbs. Have a few facts ready. Children will want to know which animals are the biggest, smallest, run the fastest, and so on. They might even decide to test which paper towel is the strongest or which potato chip is the saltiest.

Figure 16–10 Compare hard and soft by sorting objects.

3. Clothes can be compared and matched to paper dolls according to size, weather conditions, occupations, and activities such as swimming, playing outdoors, or going to a birthday party. This comparison activity could be a regular part of a daily weather chart activity.

Collections

What did you collect as a child? Did you collect baseball cards, leaves, small cereal box prizes, shells, or pictures of famous people? Chances are you collected something. This is because young children are natural scientists; they are doing what comes naturally. They collect and organize their environment in a way that makes sense to them. Even as adults, we still retain the natural tendency to organize and collect.

Consider for a moment: Do you keep your coats, dresses, shoes, slacks, and jewelry in separate places? Do the spoons, knives, forks, plates, and cups have their own space in your house? In all likelihood, you have a preferred spot for each item. Where do you keep your soup? The cans are probably lined up together. Do you have a special place just for junk? Some people call such a place a junk drawer. Take a look at your house or room as evidence that you have a tendency to think like scientists and try to bring the observable world into some sort of structure.

In fact, when you collect, you may be at your most scientific. As you collect, you combine the processes of observation, comparison, classification, and measurement, and you think like a scientist thinks. For example, when you add a leaf to your collection, you observe it and compare it to the other leaves in your collection. Then, you ask yourself, "Does this leaf measure up to my standards?" "Is it too big, small, old, drab?" "Have insects eaten its primary characteristics?" You make a decision based on comparisons between the leaf and the criteria you have set (Figure 16–11).

Figure 16–11 Gathering leaves

Similarities and Differences

To find similarities, concepts, or characteristics that link things together may be even more difficult than identifying differences in objects. For example, dogs, cats, bears, lizards, and mice are all animals. But to gain an understanding of a concept, children need to develop the idea that some similarities are more important than others. Even though the lizard has four legs and is an animal, the dog, cat, bear, and mouse are more alike because they are mammals. Unit 35 in this book shows ways of teaching differences in animals.

In addition to quantitative questions such as "How long?" and "How many?", qualitative questions are necessary to bring about keener observations of similarities and differences. Observe the contrast in Chris's responses to his teacher in the following scenario.

Mrs. Raymond asks Chris, "How many seeds do you have on your desk?" Chris counts the seeds, says "Eight," and goes back to arranging the seeds. Mrs. Raymond then asks Chris, "Tell me how many ways your seeds are alike and how they are different." Chris tells his observations as he closely observes the seeds for additional differences and similarities. "Some seeds are small, some seeds are big, some have cracks, some feel rough." Chris's answers began to include descriptions that included shape, size, color, texture, structure, markings, and so on. He is beginning to find order in his observations.

Everyday Comparisons

Comparisons can help young children become more aware of their environment. For example:

1. Go for a walk and have children observe different trees. Ask them to compare the general shape of the trees and the leaves. Let them feel the bark and describe the differences between rough or smooth, peeling, thin or thick bark. As the children compare the trees, discuss ways that they help people and animals (home, shelter, food, shade, and so on).
2. Activities that use sound to make comparisons utilize an important way that young children learn. Try comparing winter sounds with summer sounds. Go for a walk in the winter or summer and listen for sounds. For example, the students in Mrs. Hebert's class heard their boots crunching in snow, the wind whistling, a little rain splashing, and a few even noticed a lack of sound on their winter walk. Johnny commented that he did not hear any birds. This began a discussion of where the birds might be and what the class could do to help the birds that remained in the area all winter.
3. Insect investigations are natural opportunities for comparison questions and observation. Take advantage of the young childrens' curiosity and let them collect four or five different insects for observation. Mr. Wang's class collected insects on a field trip. Theresa found four different kinds of insects. She put them carefully in a jar with leaves and twigs from the area in which she found them and punched holes in the jar lid. Mr. Wang helped her observe differences such as size, winged or wingless, and similarities such as six legs, three body parts, and two antennae. All insects have these characteristics. If your captive animal does not have these, it is not an insect. It is something else. (Figure 16–12)
4. Snack time is a good time to compare the changes in the color and texture of vegetables before and after they are cooked. Ask, "How do they look different?" When you make butter ask, "What changes do you see in the cream?" Make your own lemonade with fresh lemons and ask, "What happens to the lemons?" Then let the children taste the lemonade before and after sugar is added. Ask, "How does sugar change the taste of the lemonade?"

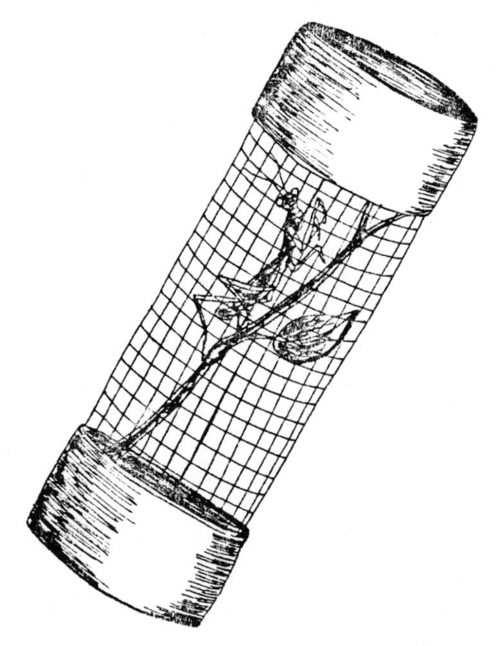

Figure 16–12 A bug collector

SHAPE

Description of the physical environment is essential to the study of any aspect of science. Such descriptions are not possible without an understanding of spatial relationships, which include the study of shapes, symmetry, time, and motion.

Most things have a shape. Children identify and classify objects by their shapes. Basic two-dimensional geometric shapes include the circle, triangle, square, and rectangle. Each of these shapes is constructed from a straight line. The following scenario is a fundamental kindergarten lesson that introduces shapes and applies them to how children learn.

Mrs. Jones arranges children in groups and gives each child a different sponge shape. After the children have had time to explore the outer edges of their sponge shape, she picks up one of the shapes, and asks, "How many sides does this shape have?" Mary and Lai answer, "Three." Mrs. Jones says, "Hold your shape up if it is just like my shape." She makes sure that each child gets a chance to examine each shape. If the shape has sides, the children count the sides of the shape aloud. Then she asks all children who have circles to hold them up and pass them to other members of the group for examination. Yarn is passed out and the children begin to duplicate the shapes on the table top.

The next day, Mrs. Jones asks George to pick up a shape from the table. He selects a circle. Mrs. Jones asks George if he can find a shape in the room that matches the circle (clock, button, table, pan, record). Sam and Mary take turns finding triangle shapes (sailboat sails, diagonally cut sandwiches), and Lai finds square shapes (floor tiles, crackers, cheese slices, napkins; rectangular paper towels, bulletin boards, books, tables). Then, each child stands next to the shape and identifies its name. Mrs. Jones leads the group in shape-matching until everyone finds and reports objects of each shape.

To extend the activity, Mrs. Jones has the children use their sponge shapes to paint pictures. She prepares the paint by placing a wet paper towel in a pie tin and coating the towel with tempera paint. The towel and paint work like a stamp pad in which to dip damp sponge shapes. After the children complete their pictures, she asks, "Which shapes have you used in your picture?" (Figure 16–13).

More Shape Activities

The following activities integrate shape with other subject areas.

1. Construction-paper shapes can be used to create fantasy animal shapes such as the shape from outer space, or the strangest insect-shape ever seen. Have the children design an appropriate habitat and tell the background for their imaginary shape creature.
2. Take students on a shape walk to see how many shapes your children can find in the

Figure 16–13 Vegetables make interesting shapes

objects they see. Keep a record. When you return to your classroom, make a bar graph (see Unit 20) to determine which shape was seen most often.

3. Shape books are an effective way of reinforcing the shape that an animal has. For example, after viewing the polar bears, make a book in the shape of a polar bear. Children can paste pictures in the book, create drawings of what they saw, and dictate or write stories about bears. Shape books can be covered with wallpaper or decorated in the way that the animal appears (Figure 16–14).

Shape and the Sense of Touch

Young children are ardent touchers; they love to finger and stroke different materials. The following activities are a few ways that touch can be used to classify and identify shapes.

1. A classic touch bag gives children a chance to recognize and match shapes with their sense of touch. Some children might not want to put their hands into a bag or box. Be sure and use bright colors and cheerful themes when decorating a touch bag or box. A touching apron with several pockets is an alternative. The teacher wears the apron and students try to identify what is in the pockets.
2. Extend this activity by making pairs of tactile cards with a number of rough- and smooth-textured shapes. Have the children blindfold each other and match the cards with partners. Ask, "Can you find the cards that match?" "How do you know they match?" Or, play Mystery Shapes by cutting different shapes out of sandpaper and mounting them on cards. Ask the children to close their eyes, feel the shape, and guess what it is.
3. Students will enjoy identifying shapes with their feet, so you can make a foot feely box. Students will probably infer that they can feel things better with their fingertips than with their toes. A barefoot trail of fluffy rugs, a beach towel, bare tiles, a pillow, and window screen would be a good warm-up activity for identifying shapes with feet. Students should describe what they feel as they walk over each section of the trail (Figure 16–15).

Figure 16–14 A polar bear book for writing

Figure 16–15 Taking a tactile walk

SPACE

Although young children are not yet ready for the abstract level of formal thinking, they can begin thinking about space and shape relationships. The following science board game being played in a primary classroom is a good example of this relationship.

"Put the starfish on top of the big rock."

"Which rock do you mean?"

"The big gray one next to the sunken treasure chest."

This is the conversation of two children playing a game with identical boards and pieces. One child gives directions and follows them, placing the piece in the correct position on a hidden game board. Another child, after making sure he understands the directions, chooses a piece and places it on the board. Then the first child continues with another set of directions, "Put the clam under that other rock."

The children are having fun, but there's more to their game than might first appear. A great deal about the child's concept of space can be learned through observation once the child responds to directions such as up-down; on-off; on top of; over-under; in front of; behind; etc. Is care taken to place the objects on the paper in a careful way?

Figure 16–16 Katie's cat from a bird's-eye view

Bird's-eye View

Children sometimes have difficulty with concepts involved in reading and making maps. One reason for this is that maps demand that they look at things from an unusual perspective, a bird's-eye view of spatial relations. To help children gain experience with this perspective, try the following strategy:

Ask children to imagine what it's like to look down on something from high above. Say, "When a bird looks down as it flies over a house, what do you think it sees?" You will need to ask some directed questions for young children to appreciate that objects will look different. "What would a picnic table look like to a bird?" "Does your house look the same to the bird in the air as it does to you when you're standing on the street?"

Then tell the children to stand and look down at their shoes. Ask them to describe what they can see from this perspective; then ask them to draw their bird's-eye view. Some children will be able to draw a faithful representation of their shoes; others may only be able to scribble what they see. Give them time to experiment with perspective, and if they are having difficulty getting started, have them trace the outlines of their shoes on a piece of drawing paper. Ask, "Do you need to fill in shoelaces or fasteners?" "What else do you see?" Put the children's drawings on their desks for open house. Have parents locate their child's desk by identifying the correct shoes.

This lesson lends itself to a simple exercise that reinforces a new way of looking and gives parents a chance to share what is going on at school. Ask children to look at some familiar objects at home (television, kitchen table, or even the family pet) from their new bird's-eye perspective and draw one or two of their favorites. When they bring their drawings to school, other children can try to identify the objects depicted.

On the Playground

Playground distance can be a good place to start to get children thinking about space. Use a stopwatch to time them as they run from the corner of the building to the fence. Ask, "How long did it take?" Then, have the children walk the same route. Ask, "Did it take the same amount of time?" Discuss the differences in walking and running and moving faster or slower over the same space.

PARTS AND WHOLES

After students are used to identifying shapes, introduce the concept of balanced proportions called symmetry. Bilateral symmetry is frequently found in nature. *Bilateral* means that a line can be drawn through the middle of the shape and divide it into two. Each of the shapes the students have worked with to this point can be divided into two matching halves. Point out these divisions on a square, circle, and equilateral triangle. Provide lots of pictures of animals, objects, and other living things. Have the children show where a line could be drawn through each of them. Ask, "How might your own body be divided like this?" Provide pictures for the children to practice folding and drawing lines.

Food activities commonly found in early childhood classrooms provide many opportunities for demonstrating parts and wholes. For example, when working with apples, divide them into halves or fourths. Ask, "How many apples will we need for everyone to receive a piece?" When you cut open the apples, let the children count the number of seeds inside. If you are making applesauce, be sure the children notice the difference between a whole cup of apples or sugar and a half cup. A pizza would be a natural food to order and talk about. Ask, "Can you eat the whole pizza?" or, "Would you like a part of the pizza?"

Figure 16–17 A bear has symmetry

During snack time, ask each student to notice how his or her sandwich has been cut. Ask, "What shape was your sandwich before it was cut?" "What shape are the two halves?" "Can you think of a different way to cut your sandwich?" Discuss the whole sandwich and parts of a sandwich.

SUMMARY

One-to-one correspondence is the most basic number skill. Children can increase their awareness of this skill and science concepts as they do science-related matching activities such as putting animals with their homes and imitating animal movement.

Animal and plant life cycles offer many opportunities for reinforcing sequencing and ordinal position, and number concepts can be emphasized with familiar concrete examples of counting.

Sets are composed of items put together in a group based on one or more common criteria. When children put objects into groups by sorting out items that share some feature, they are classifying. Comparing requires that children look for similarities and differences with each of their senses. Activities requiring observation, measuring, graphing, and classifying encourage comparisons.

Adults help children learn shape when they give them things to view, hold, and feel. Each thing that the child meets in the environment has a shape. The child also needs to use space in a logical way. Things must fit into the space available. As children make constructions in space, they begin to understand the spatial relationship between themselves and the things around them.

The idea of parts and wholes is basic to objects, people, and animals. Sets and wholes can be divided into smaller parts or pieces. The concept of bilateral symmetry is introduced when children draw a line through the middle of a shape and divide it into two halves.

Naturalistic, informal, and structured experiences support the learning of these basic science and math skills. A variety of evaluation techniques should be used. Informal evaluation can be done simply by observing the progress and choices of the child. Formal tasks are available in the match section of the Appendix.

FURTHER READING AND RESOURCES

Abruscato, J. 1982. *Teaching Children Science.* Englewood Cliffs, N.J.: Prentice-Hall.

Day, B. 1988. *Early Childhood Education.* New York: MacMillan.

Harlan, W. and Symington, D. 1988. Helping children to observe. *Primary Science: Taking the Plunge.* London: Heinemann Educational Books.

Hunder-Grundin, E. 1979. *Literacy: A Systematic Experience.* New York: Harper & Row.

Kamii, C. 1982. *Number in Preschool and Kindergarten.* Washington, D.C.: National Association for the Education of Young Children.

Koppel, S., and Lind, K. 1985. Follow my Lead: A science board game. *Science and Children.* 23 (1): 48–50.

Lind, K. K. 1984. A Bird's-eye View: Mapping readiness. *Science and Children.* 22 (3), 39–40.

Lind, K. K., and Milburn, M.J. 1988. Mechanized childhood. *Science and Children.* 25 (5): 39–40.

MacMillan Early Skills Program 1982. Number skills. MacMillan Educational Co.

Markle, S. 1988. *Hands-On Science.* Cleveland, Ohio: The Instructor Publications.

Pugh, A. F., and Dukes-Bevans, L. 1987. Planting seeds in young minds. *Science and Children.* 25 (3), 19–21.

Richards, R., Collis, M., and Kincaid, D. 1987. *An Early Start to Science.* London: MacDonald Educational.

Sprung, B., Foschl, M., and Campbell, P.B. 1985. *What Will Happen If . . . ?* New York: Educational Equity Concepts.

Warren, J. 1984. *Science Time.* Palo Alto, Calif.: Monday Morning Books.

SUGGESTED ACTIVITIES

1. How will you evaluate childrens' learning? How will you assess if children are developing science and math concepts and skills?
2. Select a science topic such as air and identify possible ways the science and the math concepts in this unit can be integrated with the curriculum.
3. Teach a lesson that integrates science and math to a group of children. Assess the extent to which you think they understand both the science and the math concept you were emphasizing.
4. Describe your childhood collection to the class. Discuss the basis for your interest and where you kept your collection. Do you continue to add to this collection?
5. Begin a nature collection. Mount and label your collection and discuss ways that children might collect objects in a formal way. Refer to Unit 34 of this book for nature collection suggestions.
6. Observe a group of students doing a science activity. Record the informal math and science discoveries that they make.
7. Design a questioning strategy that will encourage the development of classification and encourage student discussion. Try the strategy with children. Develop questioning strategies for two more of the concepts and skills presented in this unit.

REVIEW

Match each term to an activity that emphasizes the concept.

______	Using the sense of touch to match similar objects	a. One-to-one correspondence
______	Looking down on things from a "bird's-eye view	b. Counting
______	Matching animals to their homes	c. Sets and classifying
______	"How many fish are swimming in the tank?"	d. Comparing
______	Classifying animals into groups of mammals and reptiles	e. Shape
______	Children duplicate yarn triangles and circles	f. Space
______	Christina draws an imaginary line through the middle of a sandwich	g. Parts and whole

SECTION III

Applying Fundamental Concepts, Attitudes, and Skills

UNIT 17 Ordering and Patterning

OBJECTIVES

After studying this unit, the student should be able to:

- Define ordering and patterning
- List and describe four basic types of ordering activities
- Provide for naturalistic ordering and patterning experiences
- Do informal and structured ordering and patterning activities with young children
- Assess and evaluate a child's ability to order and pattern

Ordering is a higher level of comparing (Unit 8). Ordering involves comparing more than two things or more than two sets. It also involves placing things in a sequence from first to last. In Piaget's terms, ordering is called *seriation*. Patterning is related to ordering in that children need a basic understanding of ordering to do patterning. Patterning involves making or discovering auditory, visual, and motor regularities.

Ordering starts to develop in the sensorimotor stage. Before the age of two, the child likes to work with nesting toys (Figure 17–1). Nesting toys are items of the same shape but of varying sizes so that each one fits into the larger ones. If put into each other in order by size, they will all fit in one stack. Ordering involves seeing a pattern that follows continuously in equal increments. Other types of patterns involve repeated sequences that follow a preset rule. Daily routine is an example of a pattern that is learned early. That is, infants become cued into night and day and to the daily sequence of changing, eating, playing, and sleeping. As they experiment with rattles, they might use a regular pattern of movement that involves motor, auditory, and visual sequences repeated over and over. As the sensorimotor period progresses, toddlers can be observed lining up blocks, placing a large one, then a small one, large, small, or red, green, yellow, red, green, yellow.

An early way of ordering is to place a pattern in one-to-one correspondence with a model as in Figure 17–2 (A and B), Page 204. This gives the child the idea of ordering. Next, he learns to place things in ordered rows on the basis of length, width, height, and size. At first

Figure 17–1 The eighteen-month-old girl is challenged by nesting toys.

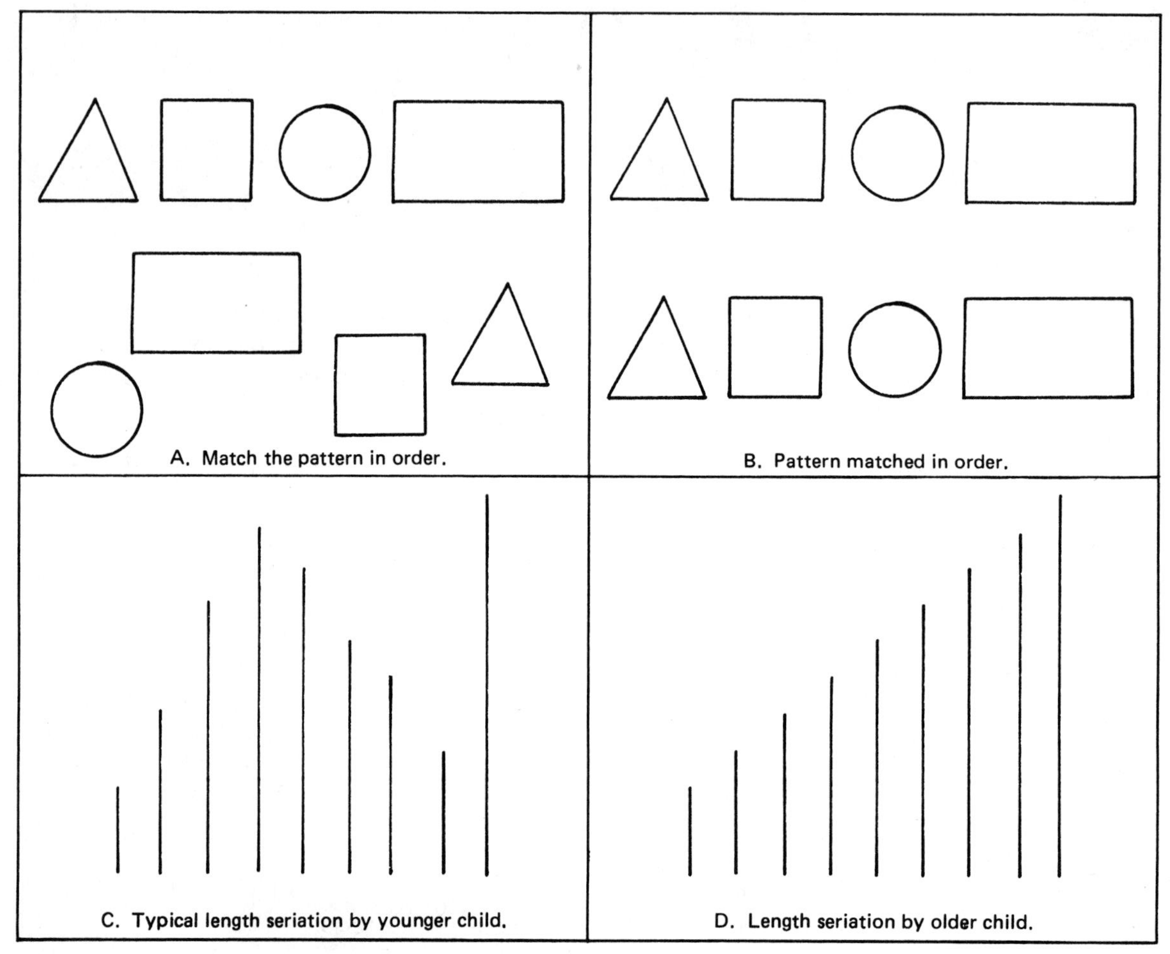

Figure 17–2 Ordering by pattern and size.

the child can think of only two things at one time. When ordering by length, he places sticks in a sequence such as shown in Figure 17–2(C). As he develops and practices, he will be able to use the whole sequence at once and place the sticks as in Figure 17–2(D). Figures 17–3, 17–4, and 17–5 show a child as he goes through this process. As the child develops further, he becomes able to order by other characteristics, such as color shades (dark to light), texture (rough to smooth), and sound (loud to soft).

Once the child can place one set of things in order, he can go on to double seriation. For double seriation he must put two groups of things in order, such as the pictures of girls with umbrellas as can be seen in Figure 17–6. This is a use of matching, or one-to-one correspondence (Unit 8).

Sets of things can also be put in order by the number of things in each set. By ordering sets, each with one more thing than the others, the child learns the concept (idea) of *one more*

Figure 17–3 The child first explores the material.

Figure 17–4 With teacher help, the child begins to learn about ordering.

Figure 17–5 The child uses his senses to check what he has done.

Figure 17–6 Which umbrella belongs to which girl?

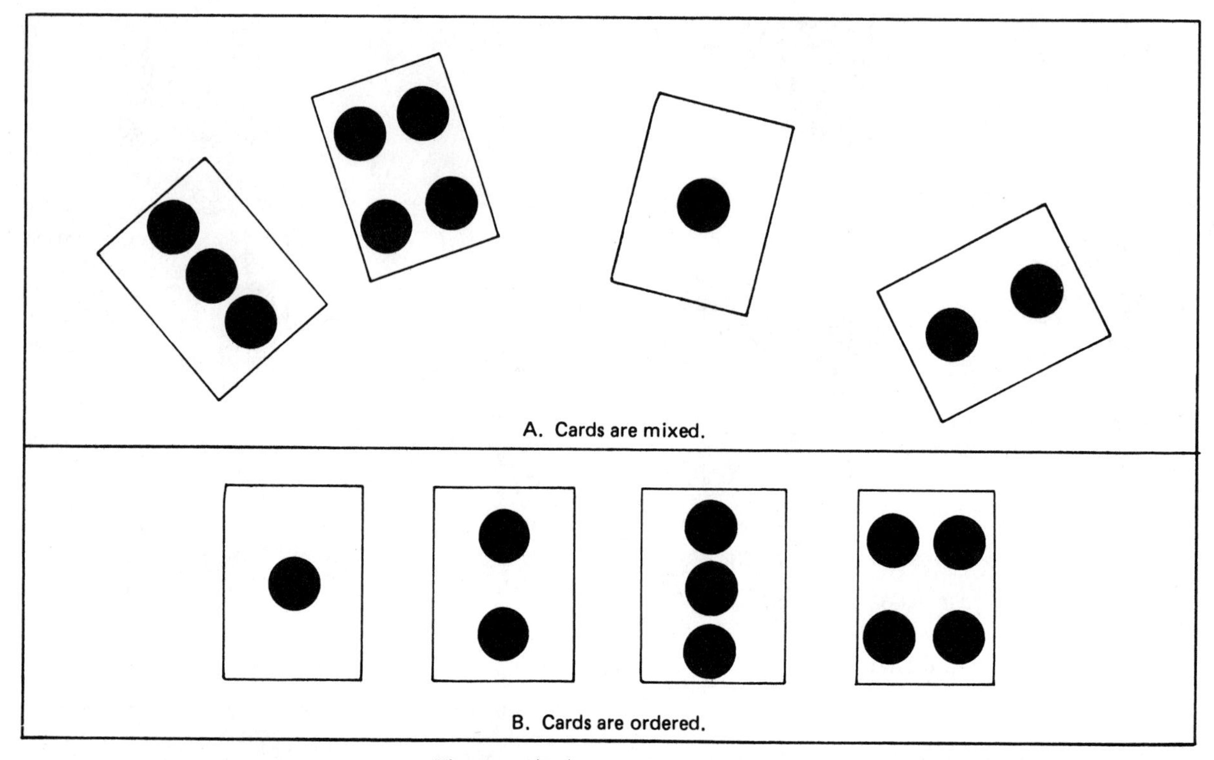

Figure 17–7 Ordering sets

than. In Figure 17–7, some cards with different numbers of dots are shown. In 17–7(A), the cards are mixed. In 17–7(B), the cards have been put in order so that each set has one more dot than the one before.

More complex patterns involve the repetition of a sequence. For example, children might be presented with a pile of shapes such as those depicted in Figure 17–7, shown the pattern, and then asked to select the correct shapes to repeat the pattern in a line (rather than matching underneath, one to one). Patterns can be developed with Unifix Cubes, cube blocks, beads, alphabet letters, numerals, sticks, coins, and many other items. Auditory patterns can be developed by the teacher with sounds such as hand clapping and drum beats or motor activities such as 'jump, jump, sit.' To solve a pattern problem, children have to be able to figure out what comes next in a sequence.

Ordering and patterning words are those such as next, last, biggest, smallest, thinnest, fattest, shortest, tallest, before, and after. Also included are the ordinal numbers: first, second, third, fourth, and so on to the last thing. Ordinal relations as compared with counting are shown in Figure 17–8.

ASSESSMENT

While the child plays, the teacher should note activities which might show the child is learning to order things. Notice how he uses nesting toys. Does he place them in each other so he has only one stack? Does he line them up

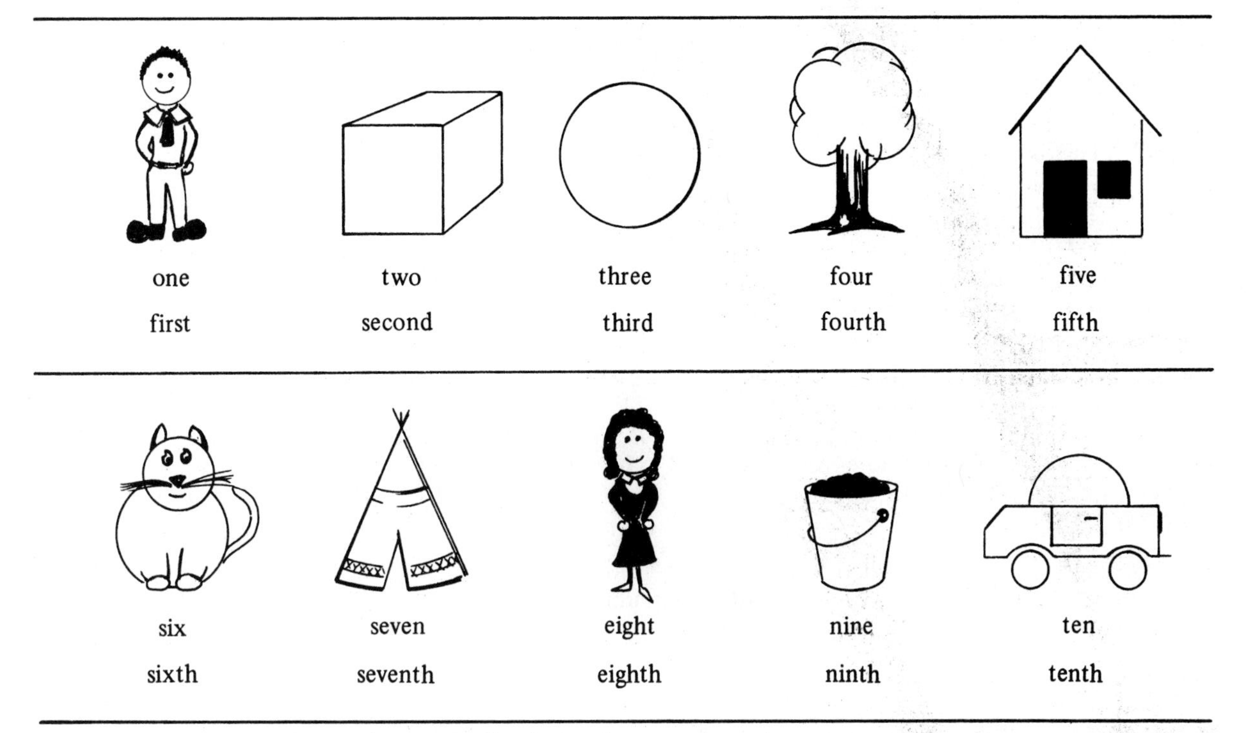

Figure 17–8 Ordinal numbers and counting numbers.

in rows from largest to smallest? Does he use words such as first (I'm first.) and last (He's last.) on his own? In his dramatic play, does he go on train or plane rides where chairs are lined up for seats and each child has a place (first, second, last)?

Also during play, watch for evidence of patterning behavior. Patterns might appear in artwork such as paintings or collages; in motor activity such as movement and dance; in musical activity such as chants and rhymes; in language activities such as acting out patterned stories (e.g. "The Three Billy Goats Gruff" or "Goldilocks and the Three Bears"); or with manipulative materials such as Unifix® Cubes, Teddy Bear Counters, Lego®, building blocks, attribute blocks, beads for stringing, Geoboards, and so on.

Ask the child to order different numbers and kinds of items during individual interview tasks (see Appendix). In Figure 17–9, the child has been asked to place the rectangles in order by width. The following are examples of two assessment tasks:

Figure 17–9 A young child draws his family in order by size.

SAMPLE ASSESSMENT TASK

Preoperational Ages 4–5

METHOD: Interview

SKILL: Child can order up to five objects relative to physical dimensions and identify the ordinal position of each.

MATERIALS: Five objects or cutouts that vary in equal increments of height, width, length, and overall size dimensions.

PROCEDURE: Start with five objects or cutouts. If this proves to be difficult, remove the objects or cutouts, then put out three and ask the same questions.
FIND THE (TALLEST, BIGGEST, FATTEST) or (SHORTEST, SMALLEST, THINNEST). PUT THEM ALL IN A ROW FROM TALLEST TO SHORTEST (BIGGEST TO SMALLEST, FATTEST TO THINNEST). If the child accomplishes the task, ask, WHICH IS FIRST? WHICH IS LAST? WHICH IS SECOND? WHICH IS THIRD? WHICH IS FOURTH?

EVALUATION: Note whether the children find the extremes but mix up the three objects or cutouts that belong in the middle. This is a common approach for preoperational children. Note if children take an organized approach to solving the problem or if they seem to approach it in a disorganized, unplanned way.

INSTRUCTIONAL RESOURCES:
Charlesworth, R. and Lind, K. 1990. *Math and Science for Young Children*. Albany, N.Y.: Delmar.

SAMPLE ASSESSMENT TASK

Transitional Period Ages 5–7

METHOD: Interview

SKILL: Child will place two sets of ten items in double seriation.

MATERIALS: Two sets of ten objects, cutouts, or pictures of objects that vary in one or more dimensions in equal increments such that one item in each set is the correct size to go with an item in the other set. The sets could be children and baseball bats, children and pets, chairs and tables, bowls and spoons, cars and garages, hats and heads, etc.

PROCEDURE: Suppose you have decided to use hats and heads. First, place the heads in front of the child in random order. Instruct the child to line the heads up in order from smallest to largest. Help can be given such as, FIND THE SMALLEST. GOOD, NOW WHICH ONE COMES NEXT? AND NEXT? If the child is able to line up the heads correctly, then put out the hats in a random arrangement. Tell the child, FIND THE HAT THAT FITS EACH HEAD AND PUT IT ON THE HEAD.

EVALUATION: Note how the children approach the problem: whether in an organized or haphazard fashion. Note whether they get the whole thing correct or partially correct. If they get a close approximation, go through the procedure again with seven items or five in order to see if they grasp the concept when fewer items are used. A child going into concrete operations should be able to accomplish the task with two groups of ten. Transitional children may be able to perform the task correctly with fewer items in each group.

INSTRUCTIONAL RESOURCES:
Charlesworth, R. and Lind, K. 1990. *Math and Science for Young Children*. Albany, N.Y.: Delmar.

SAMPLE ASSESSMENT TASK

Transitional Period Ages 5–7

METHOD: Interview

SKILL: Child can copy, extend, and describe patterns made with concrete objects

MATERIALS: Color cubes, Unifix® Cubes, Teddy Bear Counters, attribute blocks, small toys, or other objects that can be placed in a sequence to develop a pattern

PROCEDURE:

1. Copy patterns: One at a time, make patterns of various levels of complexity (each letter stands for one type of item such as one color of a color cube, one shape of an attribute block or one type of toy). For example, A-B-A-B could be red block–green block–red block–green block or big triangle–small triangle–big triangle–small triangle. Using the following series of patterns, tell the child to MAKE A PATTERN JUST LIKE THIS ONE. (If the child hesitates, point to the first item and say, START WITH ONE LIKE THIS):
   a. A-B-A-B
   b. A-A-B-A-A-B
   c. A-B-C-A-B-C
   d. A-A-B-B-C-C-A-A-B-B-C-C
2. Extend patterns: Make patterns as in # 1 but this time say, THIS PATTERN ISN'T FINISHED. MAKE IT LONGER. SHOW ME WHAT COMES NEXT.
3. Describe patterns: Make patterns as in #1 and #2. Say, TELL ME ABOUT THESE PATTERNS (WHAT COMES FIRST? NEXT? NEXT?) IF YOU WANTED TO CONTINUE THE PATTERN WHAT WOULD COME NEXT? NEXT?
4. If the above tasks are easily accomplished, then try some more difficult patterns such as:
   a. A-B-A-C-A-D-A-B-A-C-A-D
   b. A-B-B-C-D-A-B-B-C-D
   c. A-A-B-A-A-C-A-A-D

EVALUATION: Note which types of patterns are easiest for the children. Are they more successful with the easier patterns? With copying? Extending? Describing?

INSTRUCTIONAL RESOURCES:
Charlesworth, R. and Lind, K. 1990. *Math and Science for Young Children*. Albany, N.Y.: Delmar.

NATURALISTIC ACTIVITIES

Just as children's natural development guides them to sort things, it also guides them to put things in order and to place things in patterns. As children sort they often put the items in rows or arrange them in patterns. For example, Kate picks out blocks that are all of one size, shape, and color and lines them up in a row. She then adds to the row by lining up another group of blocks of the same size, shape, and color. She picks out blue blocks and yellow blocks and lines them up, alternating colors. Pete is observed examining his mother's measuring cups and spoons. He lines them up from largest to smallest separately. Then he makes a pattern: cup-spoon-cup-spoon-cup-spoon-cup-spoon.

As speech increases, the child uses order words. "I want to be *first*." "This is the *last* one." "Daddy Bear has the *biggest* bowl." "I'll sit in the *middle*." As he starts to draw pictures, he often draws Moms, Dads, and children and places them in a row from smallest to largest (Figure 17–9).

Figure 17–10 Which rectangle is the widest?

INFORMAL ACTIVITIES

Informal teaching can go on quite often during the child's daily play and routine activities (Figure 17–10). The following are some examples:

- Eighteen-month-old Brad has a set of mixing bowls and measuring cups to play with on the kitchen floor. He puts the biggest bowl on his head. His mother smiles and says, "The *biggest* bowl fits on your head." He tries the smaller bowls, but they do not fit. Mom says, "The *middle sized* bowl and the *smallest* bowl don't fit, do they?" She sits down with him and picks up a measuring cup. "Look, here is the cup that is the *biggest*. These are *smaller*." She lines them up by size. "Can you find the *smallest* cup?" Brad proceeds to put the cups one in the other until they are in one stack. His mother smiles, "You have them all in *order*."
- Five-year-old George, four-year-old Richard, and three-year-old Jim come running across the yard to Mr. Brown. "You are all fast runners." "I was *first*," shouts George. "I was *second*," says Richard. "I was *third*," says Jim. George shouts, "You were *last*, Jim, 'cause you are the *littlest*." Jim looks mad. Mr. Brown says, "Jim was both *third* and *last*. It is true Jim is the *littlest* and the *youngest*. George is the *oldest*. Is he the *biggest?*" "No!" says Jim, "He's *middle size*."
- Mary has some small candies she is sharing with some friends and her teacher. "Mr. Brown, you and I get five because we are the *biggest*. Diana gets four because she's the *next* size. Pete gets three. Leroy gets two and Brad gets one. Michael doesn't get any 'cause he is a baby." "I see," says Mr.

Brown, "You are dividing them so the *smallest* people get the least and the *biggest* get the most."

- Mrs. Red Fox tells her first graders that she would like them to line up boy-boy-girl-girl.
- Second Grader Liu Pei decides to draw a picture each day of her bean sprout that is the same height as it is that day. Soon she has a long row of bean sprout pictures, each a little taller than the one before. Mr. Wang comments on how nice it is to have a record of the bean sprout's growth from the day it sprouted.
- Miss Collins tells the children, "You have to take turns on the swing. Tanya is *first* today."

These examples show how comments can be made which help the child see his own use of order words and activities. Many times in the course of the day, opportunities come up where children must take turns. These times can be used to the fullest for teaching order and ordinal number. Many kinds of materials can be put out for children which help them practice ordering. Some of these things are self-correcting as shown in Figure 17–11.

Figure 17–11 This child works with self-correcting material.

STRUCTURED ACTIVITIES

Structured experiences with ordering and patterning can be done with many kinds of materials. These materials can be purchased, or they can be made by the teacher. Things of different sizes are easy to find at home or school. Measuring cups and spoons, mixing bowls, pots and pans, shoes, gloves, and other items of clothing are easy to get in several different sizes. Paper and cardboard can be cut into different sizes and shapes. Paper towel rolls can be made into cylinders of graduated sizes. The artistic teacher can draw pictures of the same item in graduated sizes. Already drawn materials such as an ordered set of birds (Dellosa & Carson, 1981), Birthday Presents and Animal Families (Carson & Dellosa, 1977) can be used for seriation and for patterning. The following are basic activities that can be done with many different kinds of objects, cutouts, and pictures.

STRUCTURED ACTIVITIES

ORDERING AND PATTERNING: THE BASIC CONCEPT

OBJECTIVE: To help the child understand the idea of order and sequence

MATERIALS: Large colored beads with a string for the teacher and each child

ACTIVITY: The beads are in a box or bowl where they can be reached by each child. Say, WATCH ME. I'M GOING TO MAKE A STRING OF BEADS. Start with three beads. Add more as each child learns to do each amount. Lay the string of three beads down where each child can see it: NOW YOU MAKE ONE LIKE MINE. WHICH KIND OF BEAD SHOULD YOU TAKE FIRST? When the first bead is on: WHICH ONE IS NEXT? When two are on: WHICH ONE IS NEXT?
Use patterns of varying degrees of complexity such as:
1. A-B-A-B
2. A-B-C-A-B-C
3. A-A-B-A-A-C-A-A-D
4. Make up your own patterns

FOLLOW-UP:
1. Make a string of beads. Pull it through a paper towel roll so that none of the beads can be seen. Say, I'M GOING TO HIDE THE BEADS IN THE TUNNEL. NOW I'M GOING TO PULL THEM OUT. WHICH ONE WILL COME OUT FIRST? NEXT? NEXT? and so on. Then pull the beads through and have the children check as each bead comes out.
2. Dye some macaroni with food coloring. Set up a pattern for a necklace. The children can string the macaroni to make their own necklaces in the same pattem.

ORDERING: DIFFERENT SIZES, SAME SHAPE

OBJECTIVE: To make comparisons of three or more items of the same shape and different sizes

MATERIALS: Four to ten squares cut with sides one inch, one and one-quarter inch, one and one-half inch, and so on

ACTIVITY: Lay out the shapes: HERE ARE SOME SQUARES. STACK THEM UP SO THE BIGGEST IS ON THE BOTTOM. Mix the squares up again: NOW, PUT THEM IN A ROW STARTING WITH THE SMALLEST.

FOLLOW-UP: Do the same thing with other shapes and materials.

ORDERING: LENGTH

OBJECTIVE: To make comparisons of three or more things of the same width but different lengths

MATERIALS: Sticks, strips of paper, yarn, string, cuisinaire rods, drinking straws, or anything similar cut in different lengths such that each one is the same difference in length from the next one.

ACTIVITY: Put the sticks out in a mixed order: LINE THESE UP FROM SMALLEST TO LARGEST (LARGEST TO SMALLEST). Help if needed: WHICH ONE COMES NEXT? WHICH ONE OF THESE IS BIGGEST? IS THIS THE NEXT ONE?

FOLLOW-UP: Do this activity with many different kinds of materials.

ORDERING: DOUBLE SERIATION

OBJECTIVE: To match one-to-one two or more ordered sets of the same number of items

MATERIALS: Three Bears flannel board figures or cutouts made by hand: a mother bear, father bear, baby bear, Goldilocks, the three bowls, three spoons, three chairs, and three beds

ACTIVITY: Tell the story. Use all the order words/Biggest, middle sized, smallest, next. Follow-up with questions: WHICH IS THE BIGGEST BEAR? FIND THE BIGGEST BEAR'S BOWL (CHAIR, BED, SPOON). Use the same sequence with each character.

FOLLOW-UP: Let the children act out the story with the felt pieces or cutouts. Note if they use the order words, if they change their voices, and if they match each bear to the right bowl, spoon, chair, and bed.

ORDERING: SETS

OBJECTIVE: To order sets of one to five objects

MATERIALS: Glue buttons on cards or draw dots on cards so there are five cards.

ACTIVITY: Lay out the cards. Put the card with one button in front of the chlld: HOW MANY BUTTONS ON THIS CARD? Child answers. Say, YES, THERE IS ONE BUTTON. FIND THE CARD WITH ONE MORE BUTTON. If child picks out the card with two: GOOD, NOW FIND THE CARD WITH ONE MORE BUTTON. Keep on untll all five are in line. Mix the cards up. Give the stack to the child. LINE THEM ALL UP BY YOURSELF. START WITH THE SMALLEST SET.

FOLLOW-UP: Repeat with other materials. Increase the number of sets as each child learns to recognize and count larger sets. Use loose buttons (chips, sticks, or coins) and have the child count out his own sets. Each set can be put in a small container or on a small piece of paper.

ORDER: ORDINAL NUMBERS

OBJECTIVE: To learn the ordinal numbers *first*, *second*, *third*, and *fourth* (The child should be able to count easily to four before he does these activities.)

MATERIALS: four balls or beanbags, four common objects, four chairs

ACTIVITIES:

1. Games requiring that turns be taken can be used. Just keep in mind that young children cannot wait very long. Limit the group to four children and keep the game moving fast. For example, give each of the four children one beanbag or one ball. Say, HOW MANY BAGS ARE THERE? LET'S COUNT. ONE, TWO, THREE, FOUR. CAN I CATCH THEM ALL AT THE SAME TIME? NO, I CAN'T. YOU WILL HAVE TO TAKE TURNS: YOU ARE FIRST, YOU ARE SECOND, YOU ARE THIRD, AND YOU ARE FOURTH. Have each child say his number, "I am (first, second, third, and fourth)." OKAY, FIRST, THROW YOURS. (throw it back) SECOND, THROW YOURS. (throw it back) After each has had his turn, have them all do it again. This time have them tell you their ordinal number name.
2. Line up four objects. Say, THIS ONE IS FIRST, THIS ONE IS SECOND, THIS ONE IS THIRD, THIS ONE IS FOURTH. Ask the children: POINT TO THE (FOURTH, FIRST, THIRD, SECOND).
3. Line up four chairs. WE ARE GOING TO PLAY BUS (PLANE, TRAIN) Name a child, ______ YOU GET IN THE THIRD SEAT. Fill the seats. Go on a pretend trip. NOW WE WILL GET OFF. SECOND SEAT GET OFF. FIRST SEAT GET OFF. FOURTH SEAT GET OFF. THIRD SEAT GET OFF.

FOLLOW-UP: Make up some games which use the same basic ideas. As each child knows first through fourth, add fifth, then sixth, and so on.

PATTERNING: AUDITORY

OBJECTIVE: To copy and extend auditory patterns

MATERIALS: None needed

ACTIVITY: Start a hand-clapping pattern. Ask the children to join you. LISTEN TO ME CLAP. Clap, clap, (pause), clap, (repeat several times). YOU CLAP ALONG WITH ME. Keep on clapping for 60 to 90 seconds so everyone has a chance to join in. Say, LISTEN, WHEN I STOP YOU FINISH THE PATTERN. Do three repetitions, then stop. Say, YOU DO THE NEXT ONE. Try some other patterns such as "Clap, clap, slap the elbow" or "Clap, stamp the foot, slap the leg."

FOLLOW-UP: Help the children develop their own patterns. Have them use rhythm instruments to develop patterns (such as drum beat, bell jingle, drum beat).

PATTERNING: OBJECTS

OBJECTIVE: To copy and extend object patterns

MATERIALS: Several small plastic toys such as vehicles, animals, peg people; manipulatives (such as Unifix® Cubes, Inch Cubes, or attribute blocks); or any other small objects such as coins, bottle caps, eating utensils, or cups

ACTIVITY: Have the children explore ways to make patterns with the objects. Have them see how many different kinds of patterns they can make.

FOLLOW-UP: Find additional ideas for pattern activities in the books listed in Further Reading and Resources for this unit.

George, a kindergartner, is enjoying using *Match*, a computer program that provides experience in sequencing designs. When he finishes with *Match*, he asks if can change to *Dr. Seuss Fix Up the Mixed-up Puzzle*, which also includes some sequencing activities.

EVALUATION

Note whether the children use more ordering and patterning words and do more ordering and patterning activities during play and routine activities. Without disrupting the children's activities, ask questions or make comments and suggestions.

- Who is the biggest? (the smallest?)
- (As the children put their shoes on after their nap) Who has the longest shoes? (the shortest shoes?)
- Who came in the door first today?
- Run fast. See who can get to the other side of the gym first.
- (The children are playing train) Well, who is in the last seat? He must be the caboose. Who is in the first seat? He must be the engineer.
- Everyone can't get a drink at the same time. Line up with the shortest person first.
- Great, you found a new pattern to make with the Unifix® Cubes!
- Sam made some patterns with the ink pad and stamps.

The assessment tasks in the Appendix can be used for individual evaluation interviews.

SUMMARY

When more than two things are compared, the process is called ordering, or seriation. There are four basic types of ordering activities. The first is to put things in sequence by size. The second is to make a one-to-one match between two sets of related things. The third is to place sets of different numbers of things in order from the least to the most. The last is ordinal number. Ordinal numbers are first, second, third, and so on.

Patterning is related to ordering and includes auditory, visual, and physical motor sequences that are repeated. Patterns may be copied, extended, or verbally described.

FURTHER READING AND RESOURCES

Baratta-Lorton, M. 1976. *Math their way*. Menlo Park, Calif.: Addison-Wesley.

Barson, A., and Barson, L. 1988. Ideas: What comes next? *Arithmetic Teacher*. 35: 26, 29.

Burton, G. M. 1982. Patterning: Powerful play. *School Science and Mathematics.* 82 (1): 39–43.

Burton, G. M. 1985. *Towards a good beginning.* Menlo Park, Calif.: Addison-Wesley.

Carson, P. and Dellosa, J. 1977. *All aboard for readiness skills.* Akron, Ohio: Carson-Dellosa Publishing.

Dellosa, J., and Carson, P. 1981. *Buzzing into readiness.* Akron, Ohio: Carson-Dellosa Publishing.

Richardson, K. 1984. *Developing number concepts using Unifix Cubes.* Menlo Park, Calif.: Addison-Wesley.

SUGGESTED ACTIVITIES

1. Observe children at school. Note those activities that show the children are learning order and patterning. Share your observations with the class.

2. Add ordering and patterning activities to your Activities File.

3. Assemble the materials needed to do the ordering and patterning assessment tasks in this unit. Try out the tasks with several young children. What did the children do? Share the results with the class.

4. While working with children, do at least one structured ordering and/or patterning activity. Share what happens with the class.

5. Look through two or three educational materials catalogs. Pick out ten materials you think would be best to use for ordering and/or patterning activities.

6. Make a set of ordering and a set of patterning materials. Try them out with some young children. If possible, use different age groups.

7. Using the evaluation system from Activity 8 in Unit 9, evaluate one or more of the following computer programs in terms of each program's value for learning about ordering and/or patterning:

   Dr. Seuss Fix Up the Mixed-up Puzzle. CBS Software, Greenwich, Conn.

   Match (In *Kindercomp*). Spinnaker Software, Cambridge, Mass.: Apple, IBM PC & PCJr, Atari and C64.

   Soc Order. American Guidance Service, Circle Pines, Minn.: Apple Family.

REVIEW

A. Indicate the statements that apply to ordering and patterning.

   1. Ordering is the same as counting.
   2. Ordering is called seriation.
   3. Ordering begins in the concrete operational stage of development.
   4. Ordering involves comparing attributes of things.
   5. Ordering involves placing things in sequence from first to last.
   6. Ordering involves comparing more than two things.
   7. Ordering may involve one-to-one correspondence.
   8. Ordering begins with putting two groups of things in order.

9. Ordering is a higher level of comparing.
10. Children need a basic understanding of ordering to do patterning.
11. Patterning usually refers to ordering that involves repeated sequencing.
12. Patterning is not usually observed in child behavior before age four.

B. Answer each of the following.
 1. Define ordering.
 2. What is another name for ordering?
 3. What are the four basic types of ordering activities?
 4. Define patterning.

C. Match each item in Column II with the correct activity in Column I.

Column I

1. Child lines up cards of buttons, one button on the first card to five on the last.
2. Child says, "I'm next."
3. Child lines up straws of various sizes.
4. Child stacks nesting cups.
5. Child chants "Ho, ho, ho. Ha, ha, ha. Ho, ho, ho. Ha, ha, ha."
6. Child strings beads of different sizes according to pattern.
7. Child puts one poker chip with one poker chip.
8. Child puts the different sized flowers in the different pots of the right size.
9. Teacher says "Line up with the tallest child first."
10. Child parks cars: yellow car, blue car, yellow car, blue car.
11. Child says, "You're first."
12. Child builds a tower with a large block at the bottom, the next block is smaller, and the smallest on top.

Column II

a. Size sequence
b. One-to-one comparison
c. Ordering sets
d. Ordinal words or numbers
e. The basic idea of order
f. Double seriation
g. Patterning

UNIT 18

Measurement: Volume, Weight, Length, and Temperature

OBJECTIVES

After studying this unit, the student should be able to:

- Explain how measurement develops in five stages
- Assess and evaluate the measurement skills of a young child
- Do informal and structured measurement with young children
- Provide for naturalistic measurement experiences

Measurement is one of the most useful math skills. *Measurement* involves assigning a number to things so they can be compared on the same attributes. Numbers can be assigned to attributes such as volume, weight, length, and temperature. For example, the child drinks *one cup* of milk. Numbers can also be given to time measurement. However, time is not an attribute of things and so is presented separately (Unit 19). Standard units such as pints, quarts, liters, yards, meters, pounds, grams, and degrees tell us exactly how much (*volume*); how heavy (*weight*); how long, wide, or deep (*length*); and how hot or cold (*temperature*). A number is put with a standard unit to let a comparison be made. Two quarts contain more than one quart, two pounds weigh less than three pounds, one meter is shorter than four meters, and 30 degrees is colder than 80 degrees.

STAGES OF DEVELOPMENT

The concept of measurement develops through five stages as outlined in Figure 18–1. The first stage is a play stage. The child imitates older children and adults. He plays at measuring with rulers, measuring cups, measuring spoons, and scales as he sees others do. He pours sand, water, rice, beans, and peas from one container to another as he explores the properties of volume. He lifts and moves things as he learns about weight. He notes that those who are bigger than he is can do many more activities and has his first concept of length (height). He finds that his short arms cannot always reach what he wants them to reach (length). He finds that he has a preference for cold or hot food and cold or hot bath water. He begins to learn about temperature. This first stage begins at birth and continues through the sensorimotor period into the preoperational period (Figure 18–2).

The second stage in the development of the concept of measurement is the one of making comparisons (Unit 11). This is well under way by the preoperational stage. The child is always comparing: bigger-smaller, heavier-lighter, longer-shorter, and hotter-colder.

The third stage which comes at the end of the preoperational period and at the beginning of

Piagetian Stage	Age	Measurement Stage
Sensorimotor and Preoperational	Birth to Age Seven Years	1. Plays and imitates 2. Makes comparisons
Transition: Preoperational to Concrete Operations	Five to Seven Years	3. Uses arbitrary units
Concrete Operations	Six Years Or Older	4. Sees need for standard units 5. Uses standard units

Figure 18–1 Stages in the development of the concept of measurement

concrete operations is one in which the child learns to use what are called arbitrary units. That is, anything the child has can be used as a unit of measure. He will try to find out how many coffee cups of sand will fill a quart milk carton. The volume of the coffee cup is the arbitrary unit. He will find out how many toothpicks long his foot is. The length of the toothpick is the arbitrary unit. As he goes through the stage of using arbitrary units, he learns concepts he will need to understand standard units (Figure 18–3).

Figure 18–2 Through play, the child learns to measure.

When the child enters the period of concrete operations, he can begin to see the need for standard units. He can see that to communicate with someone else in a way the other person will understand, he must use the same units the other person uses. For example, the child says that his paper is nine thumbs wide. Another person cannot find another piece of the same width unless

Figure 18–3 Arbitrary units are used to measure.

the child and the thumb are there to measure it. But, if he says his paper is eight and one-half inches wide, another person will know exactly the width of the paper. In this case, the thumb is an arbitrary unit, and the inch is a standard unit. The same is true for other units. Standard measuring cups and spoons must be used when cooking in order for the recipe to turn out correctly. If any coffee- or teacup and any spoon is used when following a recipe, the measurement will be arbitrary and inexact, and the chances of a successful outcome will be poor. The same can be said of building a house. If nonstandard measuring tools are used, the house will not come out as it appears in the plans, and one carpenter will not be communicating clearly with another.

The last stage in the development of the concept of measurement begins in the concrete operations period. In this last stage, the child begins to use and understand the standard units of measurement such as inches, meters, pints, liters, grams, and degrees.

Obviously, prekindergartners and most kindergartners are still exploring the concept of measurement. Prekindergartners are usually in stages one (play and imitation) and two (making comparisons). The kindergartners begin in stage two and move into stage three (arbitrary units). During the primary grades, students begin to see the need for standard units (stage four) and move into using standard units (stage five).

HOW THE YOUNG CHILD THINKS ABOUT MEASUREMENT

To find out why standard units are not understood by young children in the sensorimotor and preoperational stages, Piaget must be reviewed. Remember from the first unit that the young child is fooled by appearances. He believes what he sees before him. He does not keep old pictures in mind as he will do later. He is not yet able to conserve (or save) the first way something looks when its appearance is changed. When the ball of clay is made into a snake, he thinks the volume (the amount of clay) has changed because it looks smaller to him. When the water is poured into a different shaped container, he thinks there is more or less—depending on the height of the glass. Because he can focus on only one attribute at a time, the most obvious dimension determines his response.

Two more examples are shown in Figure 18–4. In the first task, the child is fooled when a crooked road is compared with a straight road. The straight road looks longer (conservation of length). In the second task, size is dominant over material and the child guesses that the table tennis ball weighs more than the hard rubber ball. He thinks that since the table tennis ball is larger than the hard rubber ball, it must be heavier.

The young child becomes familiar with the words of measurement and learns which attributes can be measured. He learns mainly through observing older children and adults as they measure. He does not need to be taught the standard units of measurement in a formal way. The young child needs to gain a feeling that things differ on the basis of "more" and "less" of some attributes. He gains this feeling mostly through his own observations and firsthand experimental experiences.

ASSESSMENT

To assess measurement skills in the young child, the teacher observes. She notes whether the child uses the term *measure* in the adult way. She notes if he uses adult measuring tools in his play as he sees adults use them. She looks for the following kinds of incidents:

- Mary is playing in the sandbox. She pours sand from an old bent measuring cup into a bucket and stirs it with a sand shovel. "I'm measuring the flour for my cake. I need three cups of flour and two cups of sugar."
- Juanita is seated on a small chair. Kate kneels in front of her. Juanita has her right

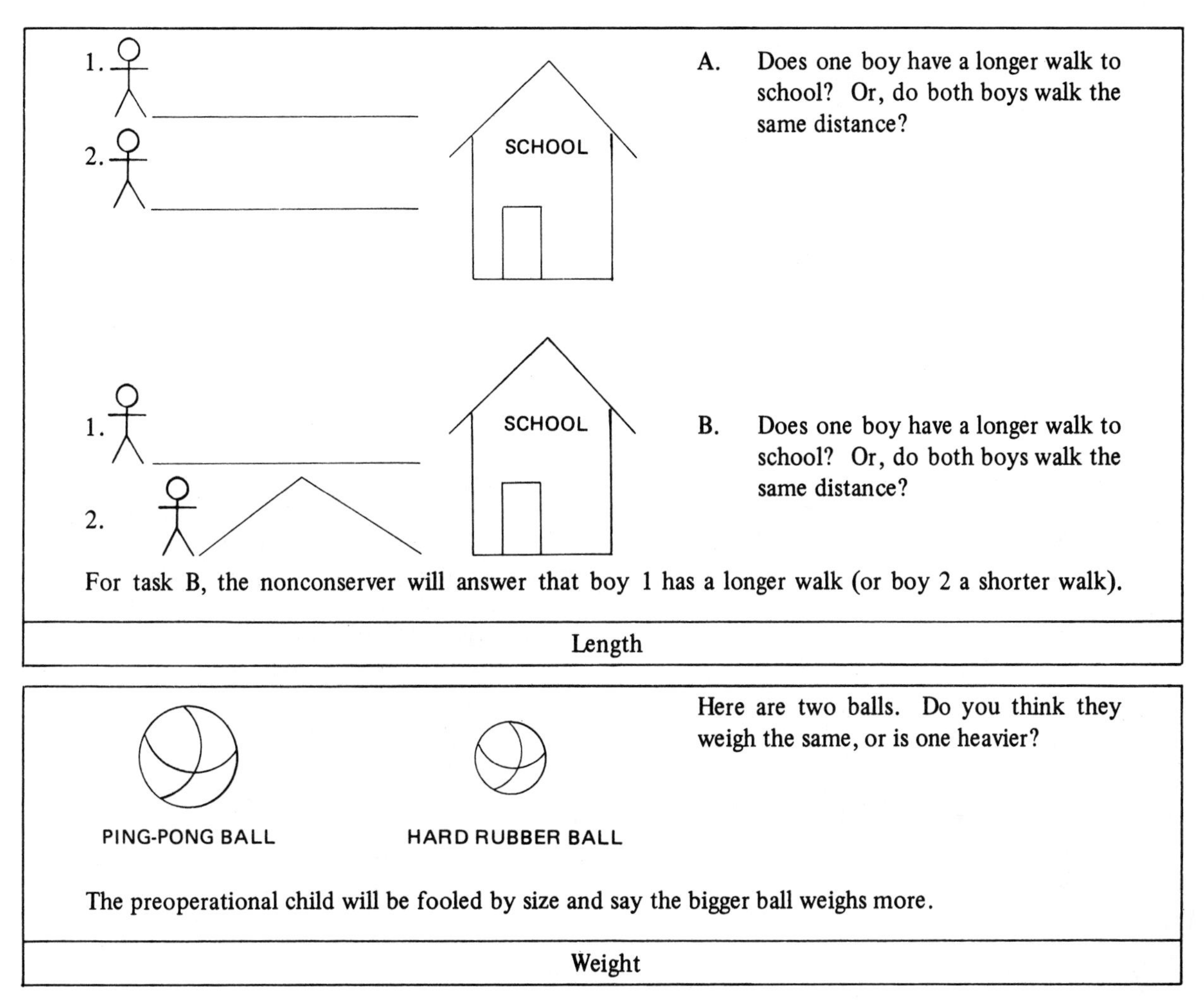

Figure 18–4 Conservation of length and weight

shoe off. Kate puts Juanita's foot on a ruler. "I am measuring your foot for your new shoes."

- The children have a play grocery store. George puts some plastic fruit on the toy scale. "Ten pounds here."
- Tim is the doctor and Bob is his patient. Tim takes an imaginary thermometer from Bob's mouth. "You have a hot fever."

Individual interviews for the preoperational child may be found in Unit 11. For the child who is near concrete operations (past five years of age), the conservation tasks in the Appendix and in Unit 1 may be used to determine if children are conservers and probably ready to use standard units of measurement.

NATURALISTIC ACTIVITIES

Young childrens' concepts of measurement come, for the most part, from their natural, everyday experiences exploring the environment, discovering its properties, and thus constructing their own knowledge. The examples in the assessment section of this unit demonstrate how children's play activities reflect their concepts of measurement. Mary has seen and may have helped someone make a cake. Kate has been to the shoe store and knows the clerk must measure the feet before he brings out a pair of shoes to try on the customer. George has seen the grocer weigh fruit. Tim knows that a thermometer tells how "hot" a fever is. The observant young child picks up these ideas on his own without being told specifically that they are important (Figure 18–5).

Figure 18–5 These girls learn about volume.

The child uses his play activities to practice what he has seen adults do. He also uses play materials to learn ideas through trial and error and experimentation. Water, sand, dirt, mud, rice, and beans teach the child about volume. As he pours these substances from one container to another, he learns about *how much*, or amount. The child can use containers of many sizes and shapes: buckets, cups, plastic bottles, dishes, bowls, and coffee cans. Shovels, spoons, strainers, and funnels can also be used with these materials. When playing with water, the child can also learn about weight if he has some small objects like sponges, rocks, corks, small pieces of wood, and marbles which may float or sink. Any time a child tries to put something in a box, envelope, glass, or any other container, he learns something about volume (Figure 18–6).

The child can begin to learn the idea of linear measure (length, width, height) and area in his play. The unit blocks which are usually found in the early childhood classroom help the child learn the idea of units. He will soon learn that each block is a unit of another block. Two, four, or eight of the small blocks are the same length when placed end to end as one of the longest blocks. As he builds enclosures (houses, garages, farmyards, etc.), he is forced to pick his blocks so that each side is the same length as the one across from it.

The child learns about weight and balance on the teeter-totter. He soon learns that it takes two to go up and down. He also learns that it works best when the two are near the same weight and are the same distance from the middle (Figure 18–7).

The child makes many contacts with temperature. He learns that his soup is hot, warm, and then as it sits out, turns cold. He likes cold milk and hot cocoa. He learns that the air may be hot or cold. If the air is hot, he may wear just shorts or a bathing suit. If the air is cold, he will need a coat, hat, and mittens.

Figure 18–6 Exploring water with assorted containers helps the child learn about volume.

Figure 18–7 Grandmother checks toddler's weight.

INFORMAL ACTIVITIES

The young child learns about measurement through the kinds of experiences just described. During these activities there are many opportunities for informal teaching. One job for the adult as the child plays is to help him by pointing out properties of materials which the child may not be able to find on his own. For instance, if a child says he must have all the long blocks to make his house large enough, the teacher can show him how several small blocks can do the same job. She can show the child how to measure how much string will fit around a box before he cuts off a piece to use.

The teacher can also take these opportunities to use measurement words such as the names of units of measurement and the words listed in Unit 14. She can also pose problems for the child:

- How can we find out if we have enough apple juice for everyone?
- How can we find out how many paper cups of milk can be poured from a gallon container?
- How can we find out if someone has a high fever?

- How can we find out without going outside if we need to wear a sweater or coat?
- How can we find out who is the tallest boy in the class? The heaviest child?
- How many of these placemats will fit around the table?
- Who lives the longest distance from school?

It is the teacher's responsibility to provide environmental opportunities for the exploration and discovery of measurement concepts.

STRUCTURED ACTIVITIES

The young child learns most of his basic measurement ideas through his play and home activities that come through the natural routines of the day. He gains a feeling for the need for measurement and leams the language of measurement. Structured activities must be chosen with care. They should make use of the child's senses. They should be related to what is familiar to the child and expand what he already knows. They should pose problems which will show him the need for measurement. They should give the child a chance to use measurement words to explain his solution to the problem. The following are examples of these kinds of experiences.

STRUCTURED ACTIVITIES

MEASUREMENT: VOLUME

OBJECTIVES:

- To learn the characteristics of volume
- To see that volume can be measured
- To learn measurement words used to tell about volume (more, less, too big, too little, the same)

MATERIALS:

- Sandbox (indoors and/or out), water table (or sink or plastic dishpans)
- Many containers of different sizes: bottles, cups, bowls, milk cartons, cans (with smooth edges), boxes (for dry materials)
- Spoons, scoops, funnels, strainers, beaters
- Water, sand, rice, beans, peas, and seeds, or anything else that can be poured

ACTIVITIES:

1. Allow plenty of time for experimentation with these kinds of materials during playtime.
2. Have several containers of different kinds and sizes. Fill one with water (or sand or rice or peas or beans). Pick out another container. Ask the children: IF I POUR THIS WATER FROM THIS BOTTLE INTO THIS OTHER BOTTLE, WILL THE SECOND BOTTLE HOLD ALL THE WATER? After each child has made his prediction, pour the water into the second container. Ask a child to tell what he saw happen. Continue with several containers. Have the children line them up from the one that holds the most to the one that holds the least.
3. Pick out one standard container (coffee cup, paper cup, measuring cup, tin can). Have one or more larger containers. Say, IF I WANT TO FILL THE BIG BOWL WITH SAND AND

USE THIS PAPER CUP, HOW MANY TIMES WILL I HAVE TO FILL THE PAPER CUP AND POUR SAND INTO THE BOWL? Write down the children's predictions. Let each child have a turn to fill the cup and pour sand into the bowl. Record by making slash marks how many cups of sand are poured. Have the children count the number of marks when the bowl is full. Compare this amount with what the children thought the amount would be. This can be done with many different sets of containers.

FOLLOW-UP: Do the same types of activities using different sizes of containers and common objects. For example, have a doll and three different size boxes. Have the children decide which box the doll will fit into.

MEASUREMENT: WEIGHT

OBJECTIVES:

- To learn firsthand the characteristics of weight
- To learn that weight and size are different attributes (big things may have less weight than small things)
- To learn that light and heavy are relative ideas

MATERIALS:

- Things in the classroom or brought from home, e.g., manipulatives, paper clips, buttons, crayons, pencils, small toys
- A teeter-totter, a board and a block, a simple pan balance
- Sand, sugar, salt, flour, sawdust, peas, beans, rice
- A ball collection with balls of different sizes and materials: ball bearings, table tennis, golf, solid rubber, foam rubber, styrofoam, balsa wood, cotton, balloons

ACTIVITIES:

1. Have the child name things in the room that he can lift and things he cannot lift. Which things can he not lift because of size? Which because of weight? Compare things such as a stapler and a large paper bag (small and heavy and large and light). Have the children line up things from heaviest to lightest.
2. Have the children experiment with the teeter-totter. How many children does it take to balance the teacher? Make a balance with a block and a board. Have the child experiment with different things to see which will make the board balance.
3. A fixed position pan balance can be used for firsthand experiences with all types of things:
   a. The child can try balancing small objects such as paper clips, hair clips, bobby pins, coins, toothpicks, cotton balls, and so on in the pans.
   b. Take the collection of balls. Pick out a pair. Have the child predict which is heavier (lighter). Let him put one in each pan to check his prediction.
   c. Put one substance such as salt in one pan. Have the child fill the other pan with flour until the pans balance. IS THE AMOUNT (VOLUME) OF FLOUR AND SALT THE SAME?

d. Have equal amounts of two different substances such as sand and sawdust in the balance pans. DO THE PANS BALANCE?

FOLLOW-UP: Make some play dough with the children. Have them measure out one part flour and one part salt. Mix in some powder tempera. Add water until the mixture is pliable but not too sticky. See Unit 22 for cooking ideas.

MEASUREMENT: LENGTH AND HEIGHT

OBJECTIVES:

- To learn firsthand the concepts of length and height
- To help the child learn the use of arbitrary units

MATERIALS:

- The children themselves
- Things in the room that can be measured, e.g., tables, chairs, doors, windows, shelves, books
- Balls of string and yarn, scissors, construction paper, magic marker, beans (chips, pennies, or other small counters), pencils, toothpicks, popsicle sticks, unit blocks

ACTIVITIES:

1. Present the child with problems where he must pick out something of a certain length. For example, a dog must be tied to a post. Have a picture of the dog and the post. Have several lengths of string. Have the child find out which string is the right length. Say: WHICH ROPE WILL REACH FROM THE RING TO THE DOG'S COLLAR?
2. LOOK AROUND THE ROOM. WHICH THINGS ARE CLOSE? WHICH THINGS ARE FAR AWAY?
3. Have several children line up. Have a child point out which is the tallest, the shortest. Have the children line up from tallest to shortest. The child can draw pictures of friends and family in a row from shortest to tallest.
4. Draw lines on construction paper. HOW MANY BEANS (CHIPS, TOOTHPICKS OR OTHER SMALL THINGS) WILL FIT ON EACH LINE? WHICH LINE HAS MORE BEANS? WHICH LINE IS LONGEST? Gradually use paper with more than two lines.
5. Put a piece of construction paper on the wall from the floor up to about five feet. Have each child stand next to the paper. Mark his height, write his name by his height. Check each child's height each month. Note how much each child grows over the year.
6. Have an arbitrary unit such as a pencil, a toothpick, a stick, a long block, or a piece of yarn or string. Have the child measure things in the room to see how many units long, wide, or tall the things are.

FOLLOW-UP: Keep the height chart out so the children can look at it and talk about their heights.

MEASUREMENT: TEMPERATURE

OBJECTIVES:

- To give the child firsthand experiences which will help him learn that temperature is the relative measure of heat
- To learn that the thermometer is used to measure temperature
- To experience hot, warm, and cold as related to things, to weather, and to the seasons of the year

MATERIALS: Ice cubes, hot plate, teakettle or pan, pictures of the four seasons, posterboard, magic markers, scissors, glue, construction paper, old magazines with pictures, real thermometers (body, inside, and outside)

ACTIVITIES:

1. Have the children decide whether selected things in the environment are hot, cold, or warm: ice and boiling water, the hot and cold water taps, the radiators, the glass in the windows, their skin, for example.
2. Show pictures of summer, fall, winter and spring. Discuss the usual temperatures in each season. What is the usual weather? What kinds of clothes are worn? Make a cardboard thermometer. At the top put a child in heavy winter clothes, underneath put a child in a light coat or jacket, then a child in a sweater, then one in short sleeves, then one in a bathing suit. Each day discuss the outside temperature relative to what was worn to school.
3. Give the children scissors and old magazines. Have them find and cut out pictures of hot things and cold things. Have them glue the hot things on one piece of poster board and the cold things on another.
4. Show the children three thermometers: one for body temperature, one for room temperature, and one for outdoor use. Discuss when and where each is used.

FOLLOW-UP: Each day the outside temperature can be dicussed and recorded in some way (such as in the second listed activity or on a graph as discussed in Unit 16).

Young children will also enjoy working with the computer program *How to Weigh an Elephant*, which presents problems in weight, volume, and mass designed for children ages four to seven.

EVALUATION

The adult should note the children's responses to the activities given them. She should observe them as they try out the materials and note their comments. She must also observe whether they are able to solve everyday problems that come up by using informal measurements such as comparisons. Use the individual interviews in the Appendix.

SUMMARY

The concept of measurement develops through five stages. Preoperational children are in the early stages: play, imitation, and comparing. They learn about measurement mainly through naturalistic and informal experiences that encourage them to explore and discover. Transitional children move into the stage of experimenting with arbitrary units. During the concrete operations period, children learn to use standard units of measurement.

FURTHER READING AND RESOURCES

Baratta-Lorton, M. 1976. *Mathematics their way.* Menlo Park, Calif.: Addison-Wesley.

Burton, G. M. 1985. *Towards a good beginning.* Menlo Park, Calif.: Addison-Wesley.

Scott, L. B., and Garner, J. 1978. *Mathematical experiences for young children.* St. Louis: McGraw-Hill.

Water is a challenging and enjoyable medium for exploring volume and numerous other concepts. The following articles focus on water play:

Adams, P. K., and Taylor, M. K. 1981. Water play in preschool programs. *Dimensions.* 10 (1): 10–14.

Dukes, L. 1983. IDEAS! Using water. *Dimensions.* 11 (2): 15–18.

Johnson, E. 1981. Water: Wet and wonderful but not necessarily wild. *Day Care and Early Education.* 8 (3): 12–14.

Leigh, C., and Emerson, P. 1985. The miracle of water. *Dimensions.* 13 (2): 4–6.

SUGGESTED ACTIVITIES

1. Observe young children during group play. Record any examples of measurement activities that occur. Which stage of measurement did each activity represent?
2. With a small group of classmates, discuss some ways a home or school environment can encourage children to use measurement skills.
3. Plan some measurement activities. Use one of the activities with a small group of children. Share the results with the class.
4. Look through a toy catalog for measurement materials that can be purchased for young children.
5. Make a balance that could be used with young children.
6. Add at least five structured measurement activities to the Math Activities file.
7. Use the evaluation scheme in Activity 8, Unit 9, to evaluate the computer program *How to Weigh an Elephant*, Panda/Learning Technologies. Dallas, Tex.: Apple, C64/128. Find out if there is any other measurement software for young children.

REVIEW

A. Put the following in order beginning with the first stage of measurement.
   a. Sees need for standard units
   b. Arbitrary units used
   c. Play and imitation
   d. Uses standard units
   e. Makes comparisons

B. List and describe the measurement stages.

C. Five levels of measurement have been discussed. After looking at the following statements, identify the measurement level which best fits each situation.
   1. Tommy says, "My truck is bigger than yours."

2. Mary checks the outdoor thermometer and says, "It's 72 degrees out today."
3. Sara, Vera, and Joe pour sand in and out of various containers.
4. "I weigh fifty pounds. How much do you weigh?"
5. "Dad, two of my shoes are the same length as one of yours."

D. Discuss how you could assess and evaluate a child's measurement skills.

UNIT 19 Measurement: Time

OBJECTIVES

After studying this unit, the student should be able to:

- Describe what is meant by time sequence
- Describe what is meant by time duration
- Explain the three kinds of time
- Do informal and structured time measurement activities with young children

There are two sides to the concept of time. There is sequence and there is duration. Sequence of time has to do with the order of events. It is related to the ideas about ordering presented in Unit 17. While the child learns to sequence things in patterns, he also learns to sequence events. He learns small, middle sized, and large beads go in order for a pattern sequence (Figure 19–1). He gets up, washes his face, brushes his teeth, dresses, and eats breakfast for a time sequence. Duration of time has to do with how long an event takes (seconds, minutes, hours, days, a short time, a long time).

Figure 19–1 Charlene works with beads and learns pattern sequence.

KINDS OF TIME

There are three kinds of time a child has to learn. Time is a hard measure to learn. The child cannot see it and feel it as he can weight, volume, length, and temperature. There are fewer clues to help the child. The young child relates time to three things: personal experience, social activity, and culture.

In his *personal experience*, the child has his own past, present, and future. The past is often referred to as "When I was a baby." "Last night" may mean any time before right now. The future may be "After my night nap" or "When I am big." The young child has difficulty with the idea that there was a time when mother and dad were little and he was not yet born.

Time in terms of *social activity* is a little easier to learn and makes more sense to the young child. The young child tends to be a slave to order and routine. A change of schedule can be very upsetting. This is because time for him is a sequence of predictable events. He can count on his morning activities being the same each day when he wakes up. Once he gets to school, he learns that there is order there too: first he takes off his coat and hangs it up, next he is greeted by his teacher, then he goes to the big playroom to play, and so on through the day.

A third kind of time is *cultural time*. It is the time that is fixed by clocks and calendars. Everyone learns this kind of time. It is a kind of time that the child probably does not really understand until he is in the concrete operations period. He can, however, learn the language (seconds, minutes, days, months, etc.) and the names of the timekeepers (clock, watch, calendar). He can also learn to recognize a timekeeper when he sees one.

LANGUAGE OF TIME

To learn time is as dependent on language as any part of math. Time and sequence words are listed in Unit 15. They are listed again in this unit for easy reference:

- **General words:** time, age
- **Specific words:** morning, afternoon, evening, night, day, noon
- **Relational words:** soon, tomorrow, yesterday, early, late, a long time ago, once upon a time, new, old, now, when, sometimes, then, before, present, soon, while, never, once, next, always, fast, slow, speed, first, second, third, and so on.
- **Specifc duration words:** Clock and watch (minutes, seconds, hours) Calendar (date, days of the week names, names of the month, names of seasons, year)
- **Special days:** birthday, Easter, Christmas, Thanksgiving, vacation, holiday, school day, weekend.

ASSESSMENT

The teacher should observe the child's use of time language. She should note if he makes an attempt to place himself and events in time. Does he remember the sequence of activities at school and at home? Is he able to wait for one thing to finish before going on to the next? Is he able to order things (Unit 17) in a sequence?

The following are examples of the kinds of interview tasks which are included in the Appendix.

SAMPLE ASSESSMENT TASK

Preoperational Ages 4–5

METHOD: Interview

SKILL: Shown pictures of daily events, the child can use time words to describe the action in each picture and place the pictures in a logical time sequence.

MATERIALS: Pictures of daily activities such as meals, nap, bath, playtime, bedtime

PROCEDURE: Show the child each picture. Say, TELL ME ABOUT THIS PICTURE. WHAT'S HAPPENING? After the child has described each picture, place all the pictures in front of him and tell the child, PICK OUT (SHOW ME) THE PICTURE OF WHAT HAPPENS FIRST EACH DAY. After a picture is selected, ask WHAT HAPPENS NEXT? Continue until all the pictures are lined up.

EVALUATION: When describing the pictures, note whether the child uses time words such as breakfast time, lunchtime, play time, morning, night, etc. Note whether a logical sequence is used in placing the pictures in order.

INSTRUCTIONAL RESOURCE(S):

Charlesworth, R., and Lind, K. 1990. *Math and Science for Young Children.* Albany, N.Y.: Delmar.

SAMPLE ASSESSMENT TASK

Preoperational Ages 3–6

METHOD: Interview

SKILL: The child can identify a clock and/or watch and describe its function.

MATERIALS: One or more of the following timepieces: conventional clock and watch, digital clock and watch. Preferably at least one conventional and one digital should be included. If real timepieces are not available, use pictures.

PROCEDURE: Show the child the timepieces or pictures of timepieces. Ask, WHAT IS THIS? WHAT DOES IT TELL US? WHAT IS IT FOR? WHAT ARE THE PARTS AND WHAT ARE THEY FOR?

EVALUATION: Note whether the child can label watch(es) and clock(s), how much he is able to describe about the functions of the parts (long and short hands, second hands, alarms set, time changer, numerals). Note also if the child tries to tell time. Compare knowledge of conventional and digital timepieces.

INSTRUCTIONAL RESOURCE(S):

Charlesworth, R., and Lind, K. 1990. *Math and Science for Young Children*. Albany, N.Y.: Delmar.

NATURALISTIC ACTIVITIES

From birth on, children are capable of learning time and sequence. In an organized, nurturing environment, infants learn quickly that when they wake up from sleep, they are held and comforted, their diapers are changed, and then they are fed. The first sense of time duration comes from how long it takes for each of these events. Infants soon have a sense of how long they will be held and comforted, how long it takes for a diaper change, and how long it takes to eat. Time for the infant is a sense of sequence and of duration of events.

The toddler shows his understanding of time words through his actions. When he is told "It's lunchtime," he runs to his highchair. When he is told it is time for a nap, he may run the other way. He will notice cues which mean it is time to do something new: toys are being picked up, the table is set, or Dad appears at the door. He begins to look for these events which tell him that one piece of time ends and a new piece of time is about to start (Figure 19–2).

As spoken language develops, the child will use time words. He will make an effort to place events and himself in time. It is important for adults to listen and respond to what he has to say. The following are some examples:

- Eighteen-month-old Brad tugs at Mr. Flores' pants leg, "Cookie, cookie." "Not yet Brad. We'll have lunch first. Cookies are after lunch."
- Linda (twenty months of age) finishes her lunch and gets up. "No nap today. Play with dollies." Ms. Moore picks her up, "Nap first. You can play with the dolls later."
- "Time to put the toys away, Kate." Kate (thirty months old) answers, "Not now. I'll do it a big later on."
- Chris (three years old) sits with Mrs. Raymond. Chris says, "Last night we stayed at the beach house." "Oh yes," answers Mrs. Raymond, "You were at the beach last

Figure 19–2 Dad kisses Kate goodbye when it is time to leave for work each day.

summer, weren't you?"
(For Chris anything in the past happened "last night").

- Mr. Flores is showing the group a book with pictures of the zoo. Richard (four years old) comments, "I want to go there yesterday." Mr. Flores says, "We'll be going to the zoo on Friday."
- Cindy (five years old), says, "One time, when I was real small, like three or something,. . ." Her teacher listens as Cindy relates her experience.

It is very important for the young child to have a predictable and regular routine. It is through this routine that the child gains his sense of time duration and time sequence. It is also important for him to hear time words and to be listened to when he tries to use his time ideas. It is especially important that his own time words be accepted. For instance Kate's "a big later on" and Chris's "last night" should be accepted. Kate shows an understanding of the future and Chris of the past even though they are not as precise as an adult would be.

Figure 19–3 This father reads a book about time with his daughter.

INFORMAL ACTIVITIES

The adult needs to capitalize on the child's efforts to gain a sense of time and time sequence. Reread the situations given as examples in the section before this (*Naturalistic Activities*). In each, the adults do some informal instruction. Mr. Flores reminds Brad of the coming sequence. So does Ms. Moore, so does the adult with Kate. Mrs. Raymond accepts what Chris says while at the same time she uses the correct time words "last summer." It is important that adults listen and expand on what children say.

The adult serves as a model for time related behavior. The teacher checks the clock and the calendar for times and dates. The teacher uses the time words in the *Language of Time* section. She makes statements and asks questions:

- "*Good morning*, Tom."
- "*Goodnight*, Mary. See you *tomorrow*."
- "What did you do over the *weekend*?"
- "Who will be our guest for lunch *tomorrow*?"
- "*Next week* on *Tuesday* we will go to the park for a picnic."
- "Let me check the *time*. No wonder you are hungry. It's almost *noon*."

• "You are the *first* one here *today.*"

Children will observe and imitate what the teacher says and does before they really understand the ideas completely.

An excellent tool for informal classroom time instruction is a daily picture/word schedule placed in a prominent place. Figure 19–4 is an example of such a schedule. Children frequently ask, "When do we . . .?", "What happens after this?", and so on. Teachers can take them to the pictorial schedule and help them find the answer for themselves, "What are we doing now?" "Find (activity) on the schedule." "What comes next?" Eventually, children will just have to be reminded to "Look at the schedule," and they will answer their questions by themselves.

STRUCTURED ACTIVITIES

Structured time and sequence activities include sequence patterns with beads, blocks and other objects, sequence stories, work centering around the calendar, and work centering around clocks. Experiences with pattern sequence and story sequence can begin at an early age. The infant enjoys looking at picture books, and the toddler can listen to short stories and begin to use beads and blocks to make his own sequences. The more structured pattern, story, calendar, clock, and other time activities described next are for children older than four and one-half.

STRUCTURED ACTIVITIES

TIME: SEQUENCE PATTERNS

OBJECTIVE: To be able to understand and use the sequence idea of *next*

MATERIALS: Any real things that can be easily sequenced by category, color, shape, size, etc. Some examples are listed.

- Wooden beads and strings
- Plastic eating utensils
- Poker chips or buttons or coins
- Shapes cut from cardboard
- Small toy animals or people

ACTIVITY: In this case, plastic eating utensils are used as the example. There are knives (K), forks (F) and spoons (S) in three colors (C_1, C_2, and C_3). The teacher sets up a pattern to present to the child. Many kinds of patterns can be presented. Any of the following may be used:

- Color: C_1-C_2-C_1-C_2. . . .
  C_2-C_3-C_3-C_2-C_3-C_3. . .
- Identity: K-F-S-K-F-S-. . . ..
  K-S-S-K-S-S. . . ...

Say to the child: THIS PATTERN IS KNIFE, FORK, SPOON (or whatever pattern is set up). WHAT COMES NEXT? When the child has the idea of pattern then set up the pattern and say, THIS IS A PATTERN. LOOK IT OVER. WHAT COMES NEXT?

FOLLOW-UP: Do the same activity with some of the other materials suggested. Also try it with the magnet board, flannel board, and chalkboard.

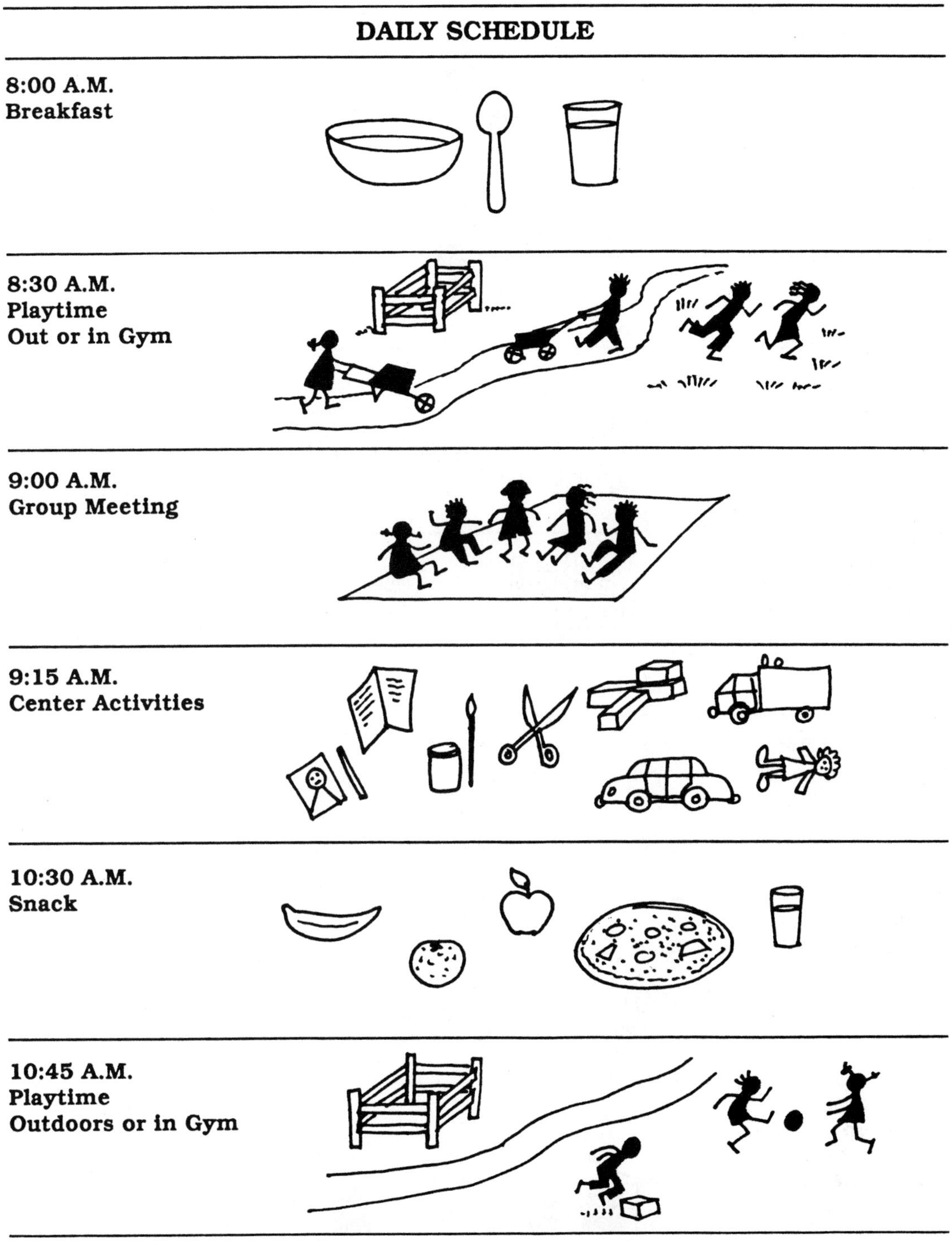

DAILY SCHEDULE

8:00 A.M.
Breakfast

8:30 A.M.
Playtime
Out or in Gym

9:00 A.M.
Group Meeting

9:15 A.M.
Center Activities

10:30 A.M.
Snack

10:45 A.M.
Playtime
Outdoors or in Gym

Figure 19–4 A picture/word daily schedule supports the development of the concept of time sequence (continued on next page).

11:15 - 11:45 A.M.
Story and language development group activities

11:45 - 12:00 A.M.
Wash hands, go to lunch

12:00 - 12:30
Lunch

12:30 - 1:00 P.M.
Playtime (Outdoors or Gym)

1:00 - 2:00 P.M.
Rest

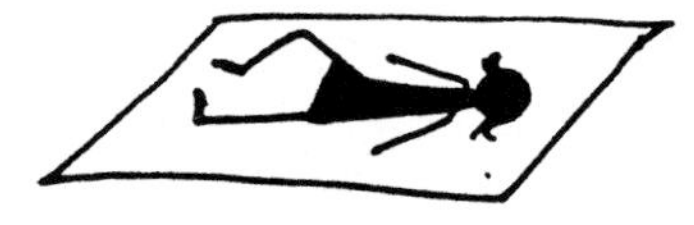

2:00 - 3:00 P.M.
Art, Music, Writing, Reading

3:00 - 3:30 P.M.
Clean-up
Prepare to leave or go to extended day

TIME: SEQUENCE STORIES

OBJECTIVE: To learn sequences of events through stories

MATERIALS: Picture story books which have clear and repetitive sequences of events, such as:

- *The Gingerbread Man*
- *The Three Little Pigs*
- *The Three Billy Goats Gruff*
- *Henny Penny*
- *Caps for Sale*

ACTIVITIES: Read the stories several times until the children are famliar with them. Begin by asking, WHAT HAPPENS NEXT? before going on to the next event. Have the children say some of the repeated phrases such as: "LITTLE PIGS, LITTLE PIGS, LET ME COME IN." "NOT BY THE HAIR ON MY CHINNY-CHIN-CHIN." "THEN I'LL HUFF AND I'LL PUFF AND I'LL BLOW YOUR HOUSE IN." Have the children try to repeat the list of those who chase the Gingerbread Man. Have them recall the whole story sequence.

FOLLOW-UP: Obtain some sequence story cards, such as DLM Sequential Picture Cards or Lakeshore's Logical Sequence Tiles and Classroom Sequencing Card Library. Encourage children to reenact the stories that are read to them. Encourage them to pretend to read familiar storybooks. This kind of activity helps with comprehending the stories and the sequences of events in the them.

TIME: SEQUENCE ACTIVITY, GROWING SEEDS

OBJECTIVE: To experience the sequence of the planting of a seed and the growth of a plant

MATERIALS: Radish or lima bean seeds, styrofoam cups, sharp pencil, six-inch paper plate, some rich soil, a tablespoon

ACTIVITY:

1. Give the child a styrofoam cup. Have him make a drainage hole in the bottom with the sharp pencil.
2. Set the cup on the paper plate.
3. Have the child put dirt in the cup up to about an inch from the top.
4. Have the child poke three holes in the dirt with his pointer finger.
5. Have him put one seed in each hole and cover the seeds with dirt.
6. Have the child put in one tablespoon of water.
7. Place the pots in a sunny place and watch their sequence of growth.
8. Have the children water the plants each day. Have them record how many days go by before the first plant pops through the soil.

Figure 19–5 What comes next?

Figure 19–6 "What day is today?"

FOLLOW-UP: Plant other types of seeds. Make a chart or obtain a chart that shows the sequence of growth of a seed. Discuss which steps take place before the plant breaks through the ground.

TIME: THE FIRST CALENDAR

OBJECTIVE: To learn what a calendar is and how it can be used to keep track of time

MATERIALS: A one-week calendar is cut from posterboard with sections for each of the seven days identified by name. In each section, tabs are cut with a razor blade to hold signs made to be slipped under the tabs to indicate special times and events or the daily weather. These signs may have pictures of birthday cakes, items seen on field trips, umbrellas to show rainy days, the sun to show fair days, and so on.

ACTIVITIES: Each day the calendar can be discussed. Key questions may include:
- WHAT IS THE NAME OF TODAY?
- WHAT IS THE NAME OF YESTERDAY?

- WHAT IS THE NAME OF TOMORROW?
- WHAT DAY COMES AFTER ______?
- WHAT DID WE DO YESTERDAY?
- DO WE GO TO SCHOOL ON SATURDAY AND SUNDAY?
- HOW MANY DAYS UNTIL ______?
- HOW MANY DAYS OF THE WEEK DO WE GO TO SCHOOL?
- WHAT DAY OF THE WEEK IS THE FIRST DAY OF SCHOOL?
- WHAT DAY OF THE WEEK IS THE LAST DAY OF SCHOOL?

FOLLOW-UP: Eventually a monthly calendar can be introduced and each month's pages can be attached together and saved so that the previous month's events can be reviewed.

TIME: THE USE OF THE CLOCK

OBJECTIVE: To find out how we use the clock to tell us when it is time to change activity

MATERIALS: School wall clock and a handmade or a purchased large clockface such as that made by the Judy Company

ACTIVITY: Point out the wall clock to the children. Show them the clockface. Let them move the hands around. Explain how the clockface is made just like the real clockface. Show them how you can set the hands on the clockface so that they are the same as the ones on the real clock. Each day set the clockface for important times (such as clean-up, lunch, time to get up from the nap, etc.). Explain that when the real clock and the clockface have their hands in the same place that it will be time to (do whatever the next activity is).

FOLLOW-UP: Do this every day. Soon each child will begin to catch on and check the clocks. Instead of asking "When do we get up from our nap?," they will be able to check for themselves.

TIME: BEAT THE CLOCK GAME

OBJECTIVE: To learn how time limits the amount of activity that can be done

MATERIALS: Minute Minder or similar timer

ACTIVITIES: Have the child see how much of some activity can be done in a set number of minutes, e.g., three to five:

1. How many pennies can be put in a penny bank one at a time?
2. How many times can he bounce a ball?
3. How many paper clips can he pick up one at a time with a magnet?
4. How many times can he move across the room: walking, crawling, running, going back-

wards, sideways, etc.? Set the timer for three to five minutes. When the bell rings the child must stop. Then count to find out how much was accomplished.

FOLLOW-UP: Try many different kinds of activities and different lengths of time. Have several children do the tasks at the same time. Who does the most in the time given?

TIME: DISCUSSION TOPICS FOR LANGUAGE

OBJECTIVES: To develop time word use through discussion

MATERIALS: Pictures collected or purchased. The following book may be used for this purpose:
Rutland, J. *Time.* New York: Grosset and Dunlap, 1976. ($ 1.95)
Pictures could show:
- Day and night
- Activities which take a long time and a short time
- Picture sequences which illustrate times of day, yesterday, today, and tomorrow
- Pictures which illustrate the seasons of the year
- Pictures which show early and late

ACTIVITIES: Discuss the pictures using the key time words.

FOLLOW-UP: Put pictures on the bulletin board which the children can look at and talk about during their free playtime.

EVALUATION

The teacher should note whether the child's use of time words increases. She should also note whether his sense of time and sequence develops to a more mature level: Does he remember the order of events? Can he wait until one thing is finished before he starts another? Does he talk about future and past events? How does he use the calendar? The clock? The sequence stories? The teacher may use the individual interview tasks in the Appendix.

SUMMARY

The young child can begin to learn that time has duration and that time is related to sequences of events. The child first relates time to his personal experience and to his daily sequence of activities. It is not until the child enters the concrete operations period that he can use units of time in the ways that adults use them.

The young child learns his concept of time through naturalistic and informal experiences for the most part. When he is around the age of four and one-half or five, he can do structured activities also.

FURTHER READING AND RESOURCES

Baratta-Lorton, M. 1976. *Math their way*. Menlo Park, Calif.: Addison-Wesley.

Burton, G. 1985. *Towards a good beginning*. Menlo Park, Calif.: Addison-Wesley.

Kurshan, L. 1984. Telling time. *Day Care and Early Education*. 11 (3): 22–25.

Learning about time. 1982. *First Teacher*. 3 (3).

Scott, L. B., and Garner, J. 1978. *Mathematical experiences for young children*. St. Louis: McGraw-Hill.

Seefeldt, C., and Tinnie, S. 1985. Dinosaurs: The past is present. *Young Children*, 40 (4): 20–24.

Timberlake, P. 1986. Time concepts in the classroom: What is today? *Dimensions*. 15 (1): 5–7.

SUGGESTED ACTIVITIES

1. Observe some young children engaged in group play. Record any examples of time measurement that take place. Which stages of measurement did each incident represent?
2. With a small group of classmates, discuss some ways a home or school environment can encourage children to use time measurement skills.
3. Plan some structured time sequence and time duration measurement activities. Use the activities with a small group of young children. Share the results with the class.
4. Add time sequence and duration measurement activities to your Resource File.

REVIEW

A. Describe what is meant by the sequence of time.

B. Describe what is meant by the duration of time.

C. Listed below are comments made by children. Indicate whether the statement represents (a) sequence of time, (b) duration of time, or (c) none of these.

1. Child with ball says, "It went up."
2. Child says, "Mama read me a story at bedtime."
3. Child says, "I stayed at grandma's for two nights."
4. Child says, "Daddy plays with me for hours and hours."
5. Child says, "It's snack time."
6. Child says, "This ball weighs two tons."
7. Child says, "Teacher, this picture took me a long time to paint."

D. List the three kinds of time and give examples of each.

E. Match each item in Column II with the correct item in Column I.

I	II
1. General time word	a. Tomorrow
2. Specific time word	b. Two hours
3. Relational word	c. This morning
4. Specific duration word	d. Christmas
5. Special days	e. Once
	f. Three years old
	g. Birthday
	h. One minute
	i. Yesterday

UNIT 20 Graphs

OBJECTIVES

After studying this unit, the student should be able to:

- Explain the use of graphs
- Describe the three stages that young children go through in making graphs
- List materials to use for making graphs

Ms. Moore hears George and Sam talking in loud voices. She goes near them and hears the following discussion.

George: "More kids like red than blue."

Sam: "No! No! More like blue!"

George: "You are all wrong."

Sam: "I am not. You are are wrong."

Ms. Moore goes over to the boys and asks, "What's the trouble, boys?" George replies, "We have to get paint to paint the house Mr. Brown helped us build. I say it should be red. Sam says it should be blue."

Sam insists, "More kids like blue than red."

Ms. Moore suggests, "Maybe there is some way we can find out." She takes the boys to a table.

"On this table let's put a piece of blue paper and a piece of red paper and a bowl of red and blue cube blocks."

George's eyes light up, "I see, and then each child can vote, right?"

Sam and George go around the room. They explain the problem to each child. Each child comes over to the table. They each choose one block of the color they like better and stack it on the paper of the same color. When they finish, there are two stacks of blocks as shown in Figure 20–1.

Ms. Moore asks the boys what the vote shows. Sam says, "The red stack is higher. More children like the idea of painting the house red."

"Good," answers Ms. Moore, "would you like me to write that down for you?" Sam and George chorus, "Yes!"

"I have an idea," says George, "Let's make a picture of this for the bulletin board so everyone will know. Will you help us Ms. Moore?"

Ms. Moore shows them how to cut out squares of red and blue paper to match each of

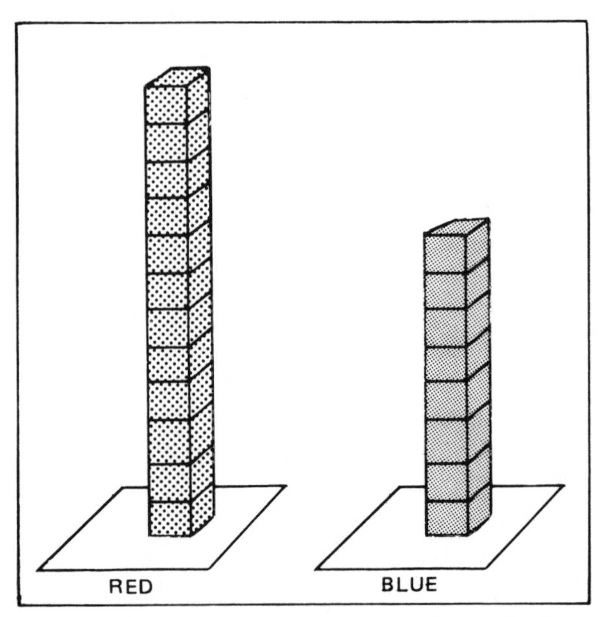

Figure 20–1 A three-dimensional graph that compares children's preferences for red or blue.

the blocks used. The boys write red and blue on a piece of white paper and then paste the red squares next to the word red and the blue squares next to the word blue. Ms. Moore shows them how to write the title: Choose the Color for the Playhouse. Then they glue the description of the results at the bottom. The results can be seen in Figure 20–2.

In the preceding example, the teacher helped the children solve their problem by making two kinds of graphs. Graphs are used to show visually two or more comparisons in a clear way. When a child makes a graph, he uses basic skills of classifying, comparing, counting, and measuring to make a picture of some information. A child who has learned the basics of math will find this to be an interesting and challenging activity.

STAGES OF DEVELOPMENT FOR MAKING AND UNDERSTANDING GRAPHS

The types of graphs that young children can construct progress through five stages of development. The first three stages are described in this unit. The fourth and fifth are included in

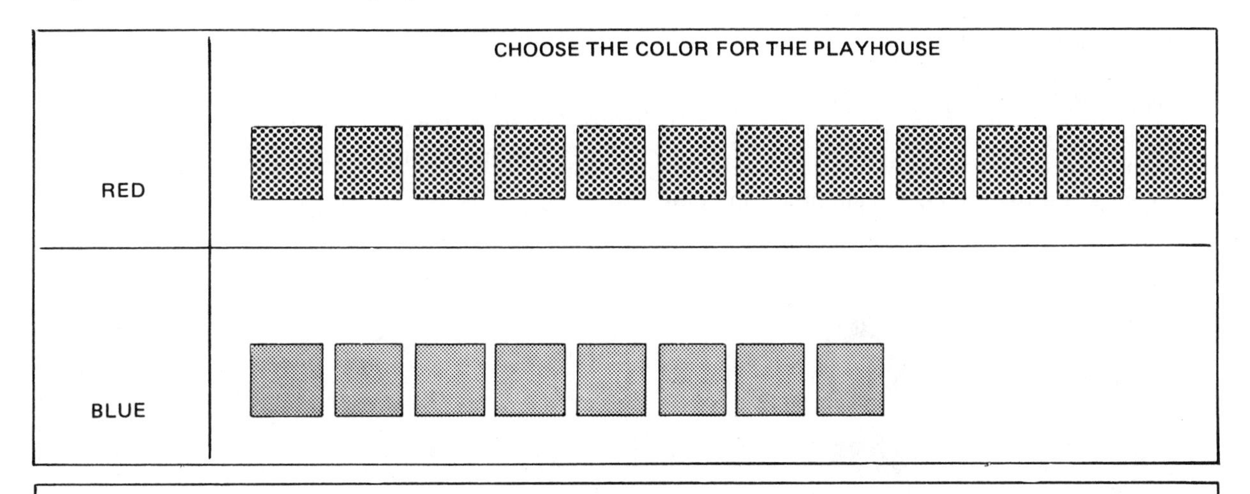

The red row is longer than the blue row. More children like red than like blue. We will buy red paint for our playhouse. 12 like red. 8 like blue.

by George and Sam

Figure 20–2 The color preference graph is copied using squares of red and blue paper.

Unit 25. In stage one, the child uses real objects to make his graph. Sam and George used inch-cube blocks. At this stage only two things are compared. The main basis for comparison is one-to-one correspondence (one block for each child).

In the second stage, more than two items are compared. In addition, a more permanent record is made—such as when Sam and George in an earlier example glued squares of paper on a piece of paper for the bulletin board. An example of this type of graph is shown in Figure 20–3. The teacher has lined off twelve columns on poster board (or large construction paper). Each column stands for one month of the year. Each child is given a paper circle. Crayons, water markers, glue, and yarn scraps are available so each child can draw his own head and place it on the month for his birthday. When each child has put his head on the graph, the children can compare the months to see which month has the most birthdays.

In the third stage, the children progress through the use of more pictures to block charts. They no longer need to use real objects but can start right off with cut out squares of paper. Figure 20–4 shows this type of graph. In this stage, the children work more independently.

DISCUSSION OF A GRAPH

As the children talk about their graphs and dictate descriptions for them, they use concept words. They use words such as:

less than	the same as
more than	none
fewer than	all
longer, longest	some
shorter, shortest	a lot of
the most	higher
the least	taller

Figure 20–3 "When is your birthday?"

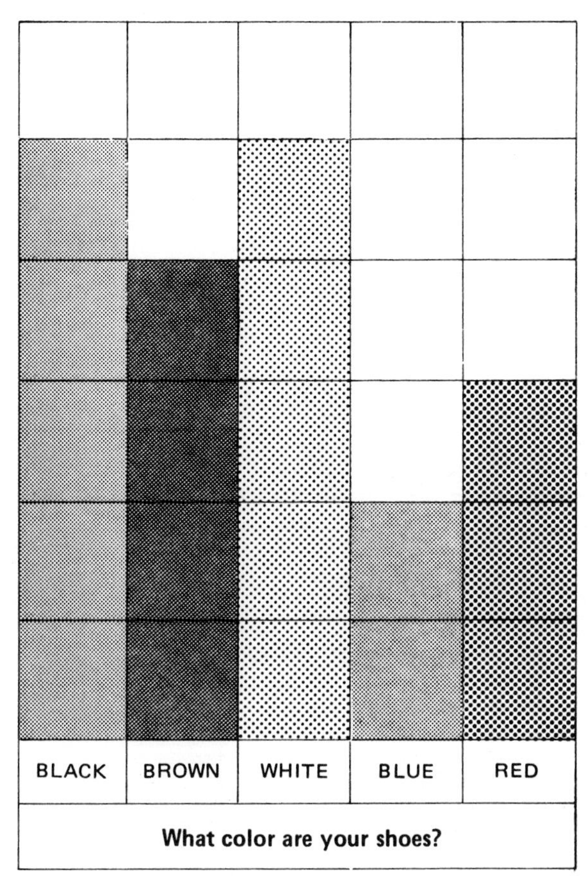

Figure 20–4 A block graph made with paper squares

MATERIALS FOR MAKING GRAPHS

There are many kinds of materials that can be used for the first stage graphs. An example has been shown in which cube blocks were used. Other materials can be used just as well.

At first it is best to use materials that can be kept in position without being knocked down or pushed apart by young children. Stands can be made from dowel rods. A washer or curtain ring is then placed on the dowel to represent each thing or person, Figure 20–5(A). Strings and beads can be used. The strings can be hung from hooks or a rod; the lengths are then compared, Figure 20–5(B). Unifix interlocking cubes, Figure 20–5(C), or pop beads, Figure 20–5(D), can also be used.

Once the children have worked with the more stable materials, they can use the cube blocks and any other things which can be lined up. Poker chips, bottle caps, coins, spools, corks, and beans are good for this type of graph work (Figure 20–6).

At the second stage, graphs can be made with these same materials but with more comparisons made. Then the child can go on to more permanent recording by gluing down cutout pictures or markers of some kind (Figure 20–7).

At the third stage, the children can use paper squares (Figure 20–8). This prepares the way for the use of squared paper. (This will be included in Unit 25.)

TOPICS FOR GRAPHS

Once children start making graphs, they often think of problems to solve on their own. The following are some comparisons that might be of interest:

- number of brothers and sisters
- hair color, eye color, clothing colors
- kinds of pets children have
- heights of children in the class
- number of children in class each day
- sizes of shoes
- favorite TV programs (or characters)
- favorite foods
- favorite colors
- favorite storybooks
- type of weather each day for a month
- number of cups of water or sand which will fill different containers
- time in seconds, to run across the playground
- number of baby hamsters class members predict that their female hamster will bear
- number of days class members predict that it will take for their bean seeds to sprout

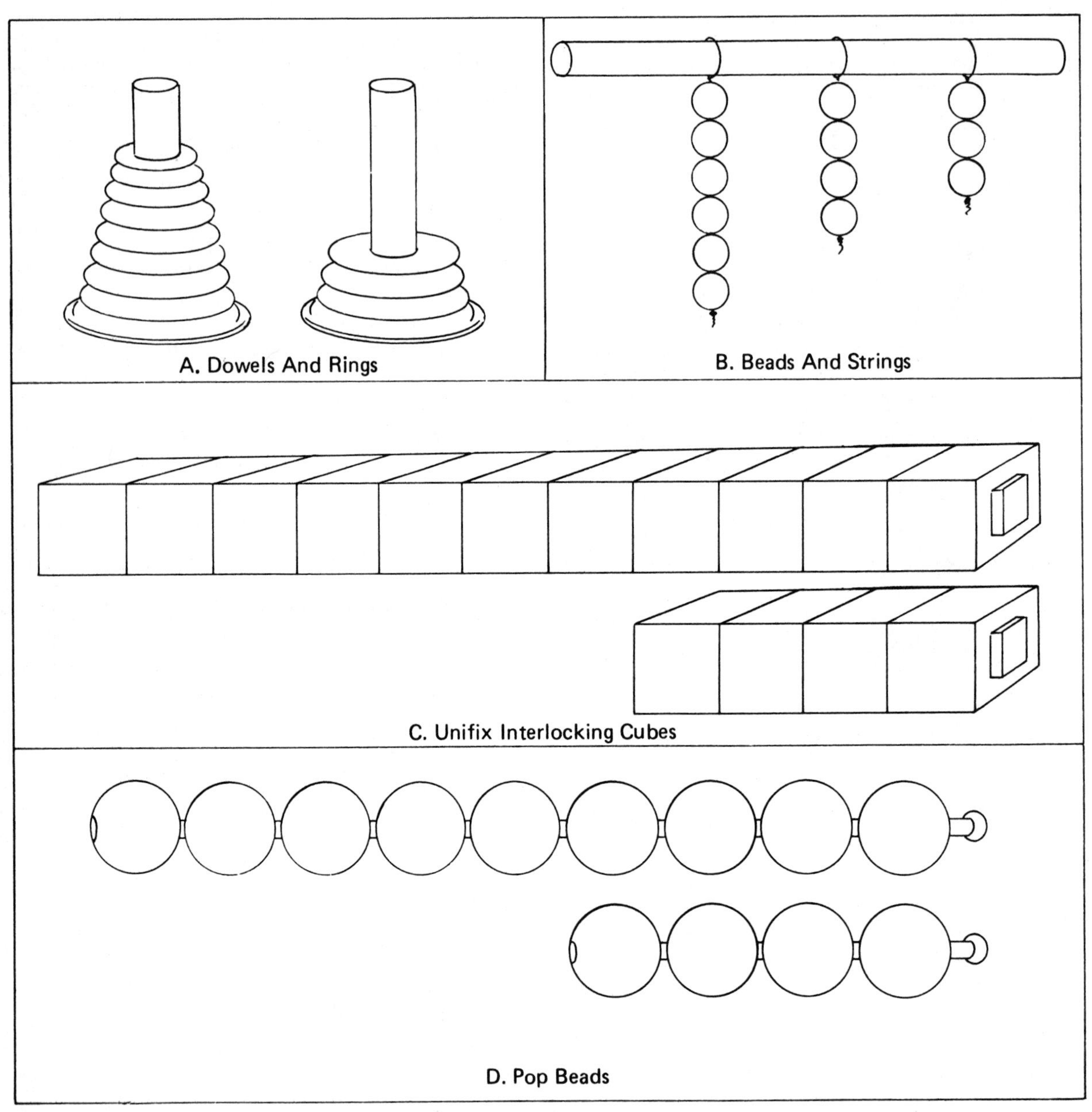

Figure 20–5 Four examples of three-dimensional graph materials that can be made into a stable graph.

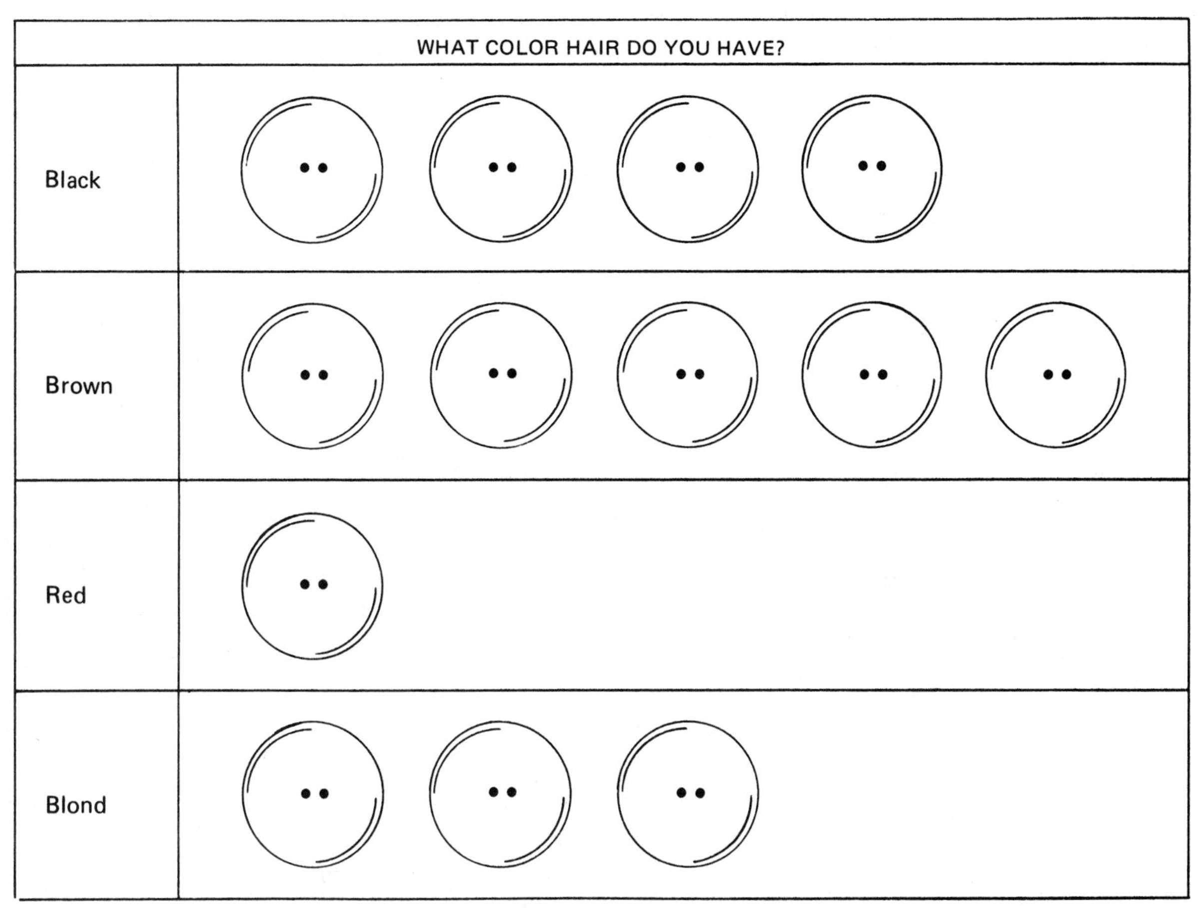

Figure 20–6 Graph made with buttons glued to cardboard.

- data obtained regarding sinking and floating objects (Unit 13)
- comparison of the number of seeds found in an apple, an orange, a lemon, and grapefruit
- students' predictions regarding items that will be attracted and not attracted by magnets
- frequency with which different types of insects are found on the playground
- distance that rollers will roll when ramps of different degrees of steepness are used
- comparison of the number of different items that are placed in a balance pan to weigh the same as a standard weight

SUMMARY

Making graphs provides a means for use of some of the basic math skills in a creative way. Children can put into a picture form the results of classifying, comparing, counting, and measuring activities.

The first graphs are three dimensional

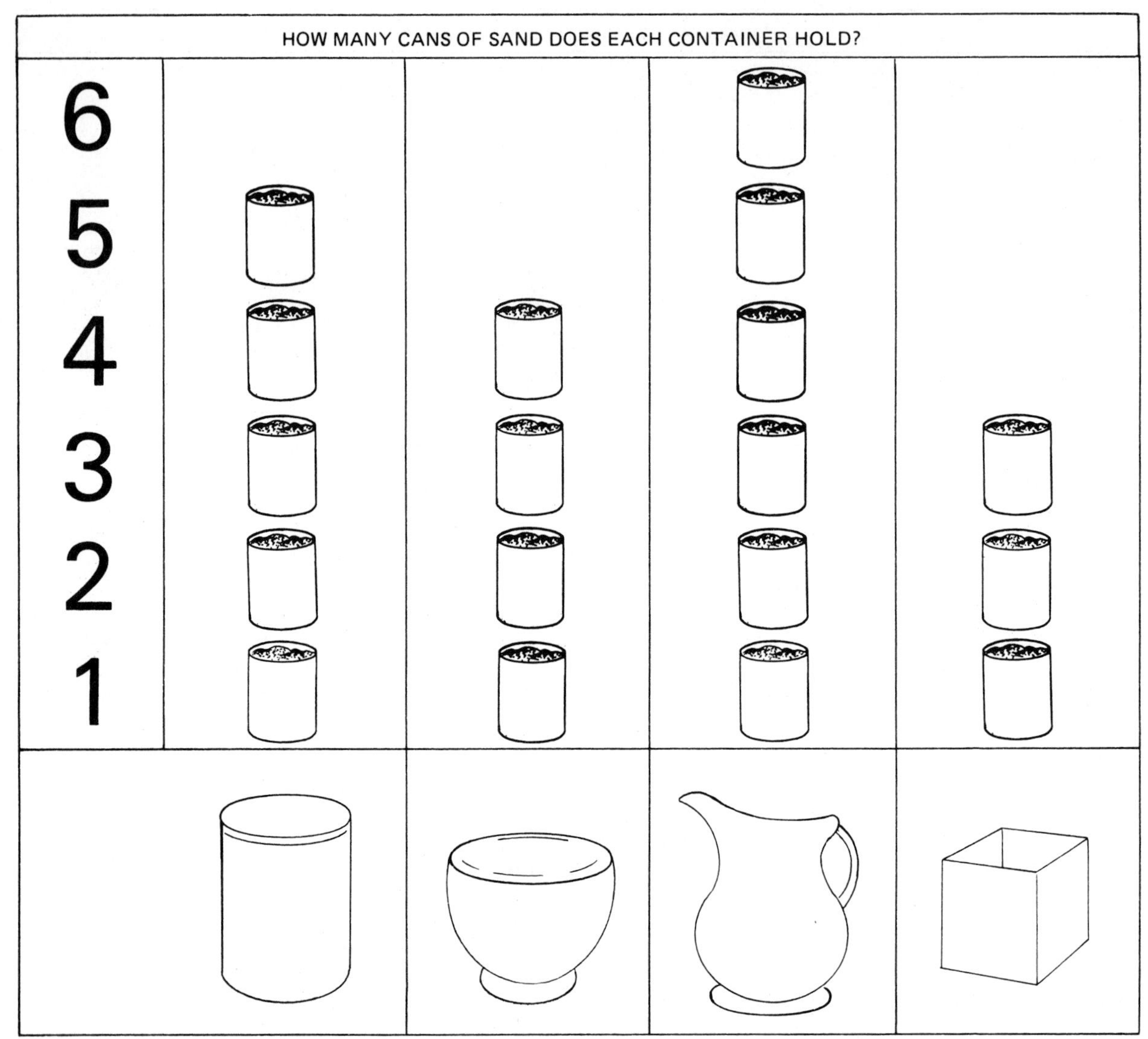

Figure 20–7 Graph made with paper cutouts.

made with real objects. The next are made with pictures and the next with paper squares. Children can discuss the results of their graph projects and dictate a description of the meaning of the graph to be put on the bulletin board with the graph.

Figure 20–8 These girls use inch cubes to make a three-dimensional graph.

FURTHER READING AND RESOURCES

Baratta-Lorton, M. 1976. *Mathematics their way*. Menlo Park, Calif.: Addison-Wesley.

Burton, G. 1985. *Towards a good beginning*. Menlo Park, Calif.: Addison-Wesley.

Church, E. B. 1986. Fun graphs. *Day Care and Early Education*. 14 (2): 20–23.

Pictorial representation. 1967. New York: John Wiley.

SUGGESTED ACTIVITIES

1. Discuss with some teachers how they use graphs in their schools. Share with the class what was learned.
2. Collect materials from which to make graphs.
3. If possible, supply a group of children with graph materials and topics, and observe what kinds of graphs they make.
4. Add graph activities to the Activities File.

REVIEW

A. Explain the use of graphs.

B. List and describe the three stages of development for making and understanding graphs.

C. Place the following types of graphs in the order in which young children are able to make and understand them.

a. picture graphs
b. paper square graphs
c. real object graphs

D. List five materials to use for making graphs.

UNIT 21 Applications of Fundamental Concepts in Preprimary Science

OBJECTIVES

After studying this unit, the student should be able to:

- Apply the concepts and skills related to ordering and patterning, measuring, and graphing to science lessons and units at the preprimary level
- Design science lessons that include naturalistic, informal, and structured activities

This unit presents activities, lessons, and scenarios that focus on the fundamental concepts, attitudes, and skills of ordering and patterning, measuring, and graphing. Units 17 through 20 of this book present these concepts with a mathematical emphasis and should be referred to for assessment strategies and additional information.

ORDERING AND PATTERNING

Ordering and patterning build on the skills of comparing. If children have not had prior experience in comparing, they will not be ready to order and find patterns. When we order we compare more than two things and place things in a sequence. This is called *seriation*. Children must have some understanding of seriation to do the more advanced skill of *patterning*. The following activities involve using a science emphasis of shape, animals, color, and sound while making or discovering visual, auditory, and motor regularities:

1. *Sun, moon, and stars.* Use your flannel board to help children recognize pattern in flannel figures of the moon, sun, and star shapes. Start the activity by discussing the shapes in the night sky. Then, place the moon, sun, and star shapes in a pattern on the flannel board. Make a game of placing a figure in the pattern and having the children decide which figure comes next. As the children go through the process of making a pattern with the flannel figures, they are also reinforcing the concept of shapes existing in the night sky (Figure 21–1).
2. *Animal patterns.* Make patterns for shapes that go together in some way such as zebra, tiger, leopard (wild animals), or pig, duck, and cow (farm animals). After introducing children to the pattern game described above, have them manipulate tagboard animal shapes in the same way.
3. *Changing colors.* Create original patterns in necklaces by dyeing macaroni with food color. Children will be fascinated by the necklace and by the change in the macaroni. Have children help prepare the macaroni: mix one tablespoon of alcohol, three

Figure 21–1 Cindy creates a pattern.

drops of food coloring, and a cup of macaroni in a glass jar. Screw the lid on tight and take turns shaking the jar. Then, lay the macaroni out on a piece of newspaper to dry overnight. Do this for each color you select.

4. *Stringing macaroni.* To help children string the macaroni, tie one end through a piece of tagboard. Then, have students practice stringing the macaroni on the string. When they are comfortable, let each invent a pattern of colors. The pattern is then repeated over and over until the necklace is completed. Have the children say the patterns they are stringing aloud, for example, "Red, blue, yellow; red, blue, yellow," until they complete the necklace. In this way, the auditory pattern is reinforced as well as the visual pattern. To extend this activity, have children guess each other's patterns, tell patterns, and try to duplicate patterns. Children will enjoy wearing their creations when they dress up for a Thanksgiving feast (Figure 21–2).

Figure 21–2 Macaroni necklaces make a pattern.

Sound Patterns

In the following scenario, Mrs. Jones introduces sound patterns to her kindergarten children.

Mrs. Jones brings out plastic bells. She selects three widely varied tones, rings them for the children, and asks, "Can you tell me what you hear?" Jane responds, "Bells, I hear bells!" The teacher asks, "Do any of the bells sound different?" "Yes," the children say. "In what way do they sound different?" Mrs. Jones asks. "Some sound like Tinkerbell and some do not," George responds. After giving the children a chance to ring the bells, Mrs. Jones puts out a red, a yellow, and a blue construction paper square on the table. She asks, "Will someone put the bell with the highest sound on the red card?" (Figure 21–3). After Lai completes the task, Mrs. Jones asks the children to find a bell that has a low sound, and finally, one with the lowest sound. Each bell has its own color square.

Patterns can be found in the sounds that rubber bands make. Construct a rubber band banjo by placing three different widths of rubber bands around a cigar box, the open end of a coffee can, milk carton, or margarine tub. Have children experiment with the rubber band "strings" of varying length and see if they notice differences in sound. Playing around with strings and homemade instruments helps prepare children for future concept development in sound.

Rhythmic patterns can be found in classic poems and songs such as "There Was a Little Turtle." Have children add actions as they sing the song.

There was a little turtle
He lived in a box
He swam in a puddle
He climbed on the rocks
He snapped at a mosquito
He snapped at a flea
He snapped at a minnow
and he snapped at me.

Figure 21–3 Which bell has the highest sound?

He caught the mosquito
He caught the flea
He caught the minnow
But he didn't catch me.

MEASUREMENT: VOLUME, WEIGHT, LENGTH, TEMPERATURE

Measurement is basically a spatial activity that must include the manipulation of objects to be understood. If children are not actively involved with materials as they measure, they simply will not understand measurement. Water and sand are highly sensory science resources for young children. Both substances elicit a variety of responses to their physical properties, and they can be used in measurement activities. Keep in mind that for sand and water to be effective as learning tools, long periods of "messing around" with the substances should be provided. For further information on the logistics of handling water and sand and a discussion of basic equipment needs, see Unit 40 of this book. The following activities and lessons use sand and water as a means of introducing measurement.

1. *Fill it up.* Put out containers and funnels of different sizes and shapes and invite children to pour liquid from one to the other. Aspects of pouring can be investigated. Ask, "Is it easier to pour from a wide- or narrow-mouthed container?" "Do you see anything that could help you pour liquid into a container?" "Does the funnel take longer? Why?" Add plastic tubing to the water center. Ask, "Can a tube be used to fill up a container?" "Which takes longer?" (Figure 21–4).
2. *Squeeze and blow.* Plastic squeeze bottles or turkey basters will encourage children to find another way to fill the containers. After the children have manipulated the baster and observed air bubbling out of it, set up the water table for a race. You will need table tennis balls, tape, and basters. The object is to move the ball across the water by squeezing the turkey baster. Children should practice squeezing the baster and moving the balls across the water. Place a piece of tape down the center of the table lengthwise and begin. Use a piece of string to measure how far the ball is moved. Compare the distances with the length of string. Ask, "How many squeezes does it take to cover the distance?"
3. *A cup is a cup.* Volume refers to how much space a solid, liquid, or gas takes up or occupies. For example, when a one-cup measure is full of water, the volume of water is one cup. Volume can be compared by having children fill small boxes or jars with

Figure 21–4 Can a tube be used to fill a container?

sand or beans. Use sand in similar containers and compare the way the sand feels. Compare the feel of a full container with an empty container. Use a balance scale to dramatize the difference. If you do not have a commercial balance, provide a homemade balance and hanging cups. Directions for constructing balances can be found in Unit 40. Children will begin to weigh different levels of sand in the cups. Statements such as, "I wonder if sand weighs more if I fill the container even higher?" can be overheard. Let children check predictions and discuss what they have found (Figure 21–5).

4. *Sink or float*. Comparisons between big-little and heavy-light are made when children discover that objects of different sizes sink and float. Make a chart to record ob-

Figure 21–5 Children explore volume by weighing sand.

Figure 21–6 Will it sink or float?

servations. After discussing what the terms *sink* and *float* mean, give the children a variety of floating and nonfloating objects to manipulate. Ask, "What do you think will float? Big corks, small marbles?" As children sort objects by size and shape and test whether they sink or float, they are discovering science concepts and learning about measurement (Figure 21–6).

Ms. Moore has her class predict, "Will it sink or float?" Her materials include a leaf, nail, balloon, cork, wooden spoon, and metal spoon. She makes a chart with columns labeled *sink* and *float* and has the children place the leaf, and so on, where they think it will go.

Cindy excitedly floats the leaf. "It floats, it floats!" she says. Because Cindy correctly predicts that the leaf would float, the leaf stays in the *float* column on the chart. Diana predicts that the wooden spoon will sink. "Oh, it's not floating. The wooden spoon is sinking." Diana moves the wooden spoon from the *float* column to the *sink* column on the chart. In this way, children can predict and compare objects as they measure weight and buoyancy (Figure 21–7).

5. *Little snakes.* Many children in preschool and kindergarten are not yet conserving length. For example, they still think that a stick may vary in length if it is changed in some way. Therefore, use care in assessing your students' developmental stages. The child must be able to conserve for measurement activities to make sense. In the following scenario, a teacher of young children assesses the conservation of length while relating length to animals.

   Mrs. Raymond reinforces the shape/

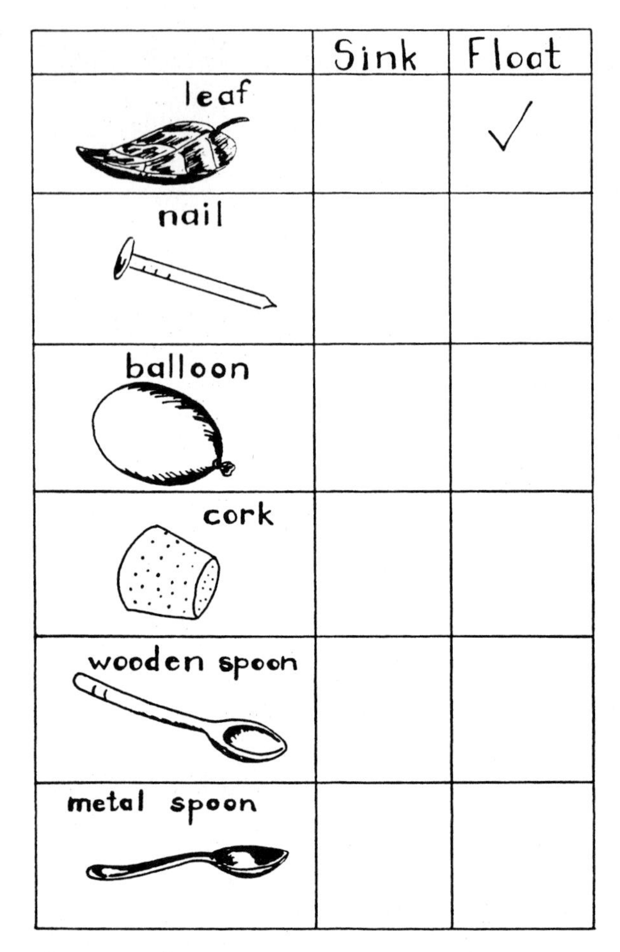

Figure 21–7 Do you think the leaf will sink or float?

space concept by giving each child in her class two pipe cleaners. She instructs them to place the pipe cleaners side by side and asks, "Which pipe cleaner is longer?" Then she tells the children to bend one of the pipe cleaners and asks, "Which is longer now?" She repeats this several times, having the children create different shapes.

Then the children make the pipe cleaners into little snakes by gluing small construction paper heads on one end of each pipe cleaner. Mrs. Raymond shows the children pictures of a snake that is coiled and one that is moving on the ground (try to find pictures of the same kind of snake). She asks, "Is the coiled snake shorter?" "Yes," the children say. "Let's see if we can make our pipe cleaners into coiled snakes," instructs Mrs. Raymond. "Oh," says Jimmie, "This really looks like a coiled snake."

The teacher asks the students to lay out the other pipe cleaners to resemble a moving snake and asks, "Which is longer?" There are a variety of responses as the children manipulate the pipe cleaner snakes to help them understand that the coiled and moving snakes are the same lengths. Then, have the children glue a small black construction paper end on the end of the pipe cleaners. Let them practice coiling and coiling the "snakes" and discuss how snakes move. Ask, "Can you tell how long a snake is when it is coiled up?" "Was your coiled-up snake longer than you thought?" (Figure 21–8).

6. *A big and little hunt.* Height, size, and length can be compared in many ways. For example, Mrs. Jones takes her children on a big and little hunt.

   After taking the children outside, Mrs. Jones has them sit in a circle and discuss the biggest and smallest things that they can see. Then she asks them, "Of all the things we have talked about, which is the smallest?" George and Mary are certain that the leaves are the smallest, but Sam and Lai do not agree. "The blades of grass are smaller," Sam observes. "Are they?" Mrs. Jones asks. "How can we tell which is smaller?" "Let's put them beside each other," Mary suggests. The children compare the length of each item and discuss which is longer and wider.

Figure 21–8 Which snake is longer?

7. *Body part measures.* Children enjoy measuring with body parts, such as a hand or foot length. Ask, "How many feet is it to the door?," and let the children count the number of foot lengths they must take. Or ask, "How many hands high are you?"

   Give children pieces of string and have them measure head sizes. Then take the string and measure other parts of their bodies. Compare their findings.
8. *How long will it take to sink?* Let children play for five minutes with dishpans of water, a kitchen scale, a bucket of water, and an assortment of objects. Then, give the children plastic tubes to explore and measure. As one child drops a marble into a plastic tube filled with water, count in measured beats the amount of time that the marble takes to sink (Figure 21–9). Ask, "Will it sink faster in cold water or hot water?"

Figure 21–9 How long will it take for a marble to sink?

9. *Hot or not.* The following activities and questions correlate with science and, in many cases, integrate into the curriculum of a preschool or kindergarten.
   - Take an outdoor walk to see what influences the temperature. "Is it warmer or colder outside?"
   - Visit a greenhouse. Predict what the temperature will be inside the greenhouse; discuss the reasons for these predictions.
   - Notice differences in temperature in different areas of the supermarket. Ask, "Are some areas colder?" "Which is the warmest area of the supermarket?"
   - Melt crayons for dripping on bottles and painting pictures. Notice the effect of heat on the crayons. Ask, "Have you ever felt like a melted crayon?" "Act out the changes in a melting crayon with your bodies."
   - Put some finger paints in the refrigerator. Have children compare how the refrigerated paints and room-temperature paints feel on their hands.

- Ask, "What happens to whipped cream as we finger paint with it?" Have children observe the whipped cream as it cools off and melts.
- Ask, "What happens to an ice cream cone as you eat it?" "Why do you think this happens?"

GRAPHS

When children make graphs, they use the basic process skills of observing, classifying, comparing, and measuring visually communicated information. In fact, the act of making a graph is in itself the process of communication. A graph is a way of displaying information so that predictions, inferences, and conclusions can be made. Refer to Unit 20 for helpful hints on introducing graphs to children and problem-solving topics that result in graph making.

When using graphs, it is essential that the meaning of the completed graph is clear to the children. With questions and discussion, it is possible for a graph to answer a question, solve a problem, or show data so that something can be understood. In the following scenario, Mrs. Jones uses common kitchen sponges to provide children with an opportunity to manipulate and construct two types of graphs.

Mrs. Jones distributes sponges of different colors to her kindergarten class. After allowing children time to examine the sponges, she suggests that they stack the sponges by color. As the sponges are stacked, a bar graph is naturally created. Mrs. Jones lines up each stack of sponges and asks, "Which color sponge do we have the most of?" Lai says, "The blue stack is the tallest. There are more blue sponges." "Good," answers Mrs. Jones. "Can anyone tell me which stack of sponges is the smallest?" "The yellow ones," says Mary. The children count the number of sponges in each stack and duplicate the three-dimensional sponge graph by making a graph of colored oaktag sponges. The children place pink, blue, yellow, and green oaktag sponges that Mrs. Jones has prepared in columns of color on a tagboard background. In this way, children learn in a concrete way the one-to-one correspondence that a bar graph represents (Figure 21–10).

Mrs. Jones follows up the graphing activity with other sponge activities. Sam says, "Sponges in my house are usually wet." "Yes," chorus the children. George says, "Our sponges feel heavier, too." The children's comments provide an opportunity to submerge a sponge in water and weigh it. The class weighs the wet and dry sponges and an experience chart of observations is recorded. The next day, Sam brings in a sponge with bird seed on it. He says, "I am going to keep it damp and see what happens."

Calendar Graph

Calendars are another way to reinforce graphing, the message of time, and one-to-one correspondence. Mrs. Carter laminates small birthday cakes, and then writes each child's name and birthdate on the cake with a permanent marker. At the beginning of the year, she shows the children the chart and points out

Figure 21–10 Making a sponge graph.

which cake belongs to them. At the beginning of each month, the children with birthdays move their cakes to the calendar. Various animals, trees, and birds can be substituted for cakes (Figure 21–11).

Pets

Graphs can be used to compare sets and give early literacy experiences. Make a list of the pets that your children have. Divide a bristol board into columns to represent every pet named. Cover the board with clear laminate film. Make labels with pet categories such as dogs, cats, snakes, fish, lizards, turtles, and so on. Glue the labels to clothespins. Then, place a labeled clothespin at the top of each column. Have the children write their names in the appropriate column (Figure 21–12).

Graphing Attractions

Simple magnets can be fun to explore. A real object graph can be made with ruled bristol board. Rule the board into columns to equal the number of magnets. Trace the shape of a magnet at the top of each column. Attach a drapery hook below each picture and proceed like Mrs. Carter.

After letting the children explore metal objects such as paper clips, Mrs. Carter places a pile of paper clips on the table along with different types of magnets. She invites Cindy to choose a magnet and see how many paper clips it

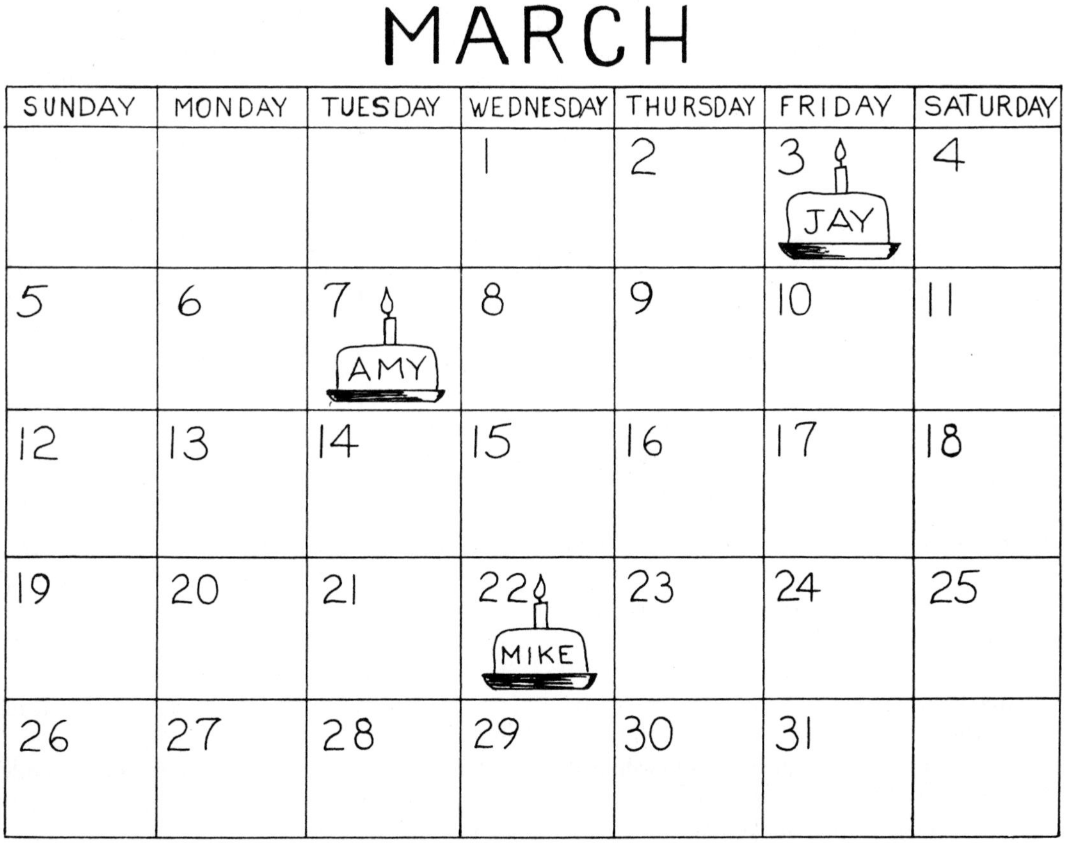

MARCH

SUNDAY	MONDAY	TUESDAY	WEDNESDAY	THURSDAY	FRIDAY	SATURDAY
			1	2	3 JAY	4
5	6	7 AMY	8	9	10	11
12	13	14	15	16	17	18
19	20	21	22 MIKE	23	24	25
26	27	28	29	30	31	

Figure 21–11 Calendars reinforce graphing, time, and one-to-one correspondence.

Figure 21–12 Graphing our pets

will attract. Cindy selects a horseshoe-shaped magnet and hangs paper clips from the end, one at a time, until the magnet no longer attracts any clips. "Good," says Mrs. Carter. "Now, make a chain of the paper clips that your magnet attracts." After Cindy makes the paper clip chain, Mrs. Carter directs her to hang it on the drapery hook under the drawing of the horseshoe-shaped magnet (Figure 21–13).

"I bet the magnet that looks like a bar will attract more paper clips." predicts Richard. He selects the rod magnet and begins to attract as many paper clips as he can. Then, he makes a chain and hangs it on the drapery hook under the rod magnet. The process is repeated until each child in the group has worked with a magnet. Mrs. Carter makes no attempt to teach the higher level concepts involved in magnetic force; rather, her purpose is to increase the children's awareness and give them some idea of what magnets do.

In this scenario, children observe that mag-

Figure 21–13 How many paper clips will the magnet attract?

nets can be different shapes and sizes (horseshoe, rod, disc, the letter U, ring, and so on) and attract a different amount of paper clips. They also have a visual graphic reminder of what they have accomplished. When snack time arrives, Mrs. Carter prepares a plate of metallic-paper-wrapped chocolate candy balls and a jar of peanut butter. She gives each child two large pretzel sticks. The children dip the pretzels into the peanut butter and then "attract" the metallic candy, imitating the way that the magnets attracted the paper clips.

SUMMARY

When more than two things are ordered and placed in sequence, the process is called *seriation*. If children understand seriation, they will be able to do the more advanced skills of patterning. Patterning involves repeating auditory, visual, or physical motor sequences.

Children begin to understand measurement by actively manipulating measurement materials. Water and sand are highly sensory and are effective resources for young children to develop concepts of volume, temperature, length, and weight. Time measurement is related to sequencing events. The use of measurement units will formalize and develop further when children enter the concrete operations period.

Graph making gives children an opportunity to classify, compare, count, measure, and visually communicate information. A graph is a way of displaying information so that predictions, inferences, and conclusions can be made.

FURTHER READING AND RESOURCES

Baratta-Lorton, M. 1976. Mathematics their way. Menlo Park, Calif.: Addison-Wesley.

Burk, D., Snider, A., and Symonds, P. 1988. *Box it or bag it mathematics*. Salem, Ore.: Math Learning Center.

Council for Elementary Science International Sourcebook IV. 1986. *Science experiences for preschoolers*. Columbus, Ohio: SMEAC Information Reference Center.

Forman, G. E., and Kuschner, D. 1983. *Child's construction of knowledge*. Washington, D.C.: National Association for the Education of Young Children.

Fowlkes, M. A. 1985. Funnels and tunnels. *Science and Children*. 22(6): 28–29.

James, J. C., and Granovetter, R. F. 1987. *Waterworks*. Lewisville, N.C.: Kaplan Press.

Hill, D. M. 1977. *Mud, sand and water.* Washington, D. C.: National Association for the Education of Young Children.

McIntyre, M. 1984. *Early childhood and science.* Washington, D. C.: National Science Teachers Association.

Phillips, D. G. 1982. Measurement or mimicry? *Science and Children*. 20 (3): 32–34.

Rockwell, R. E., Sherwood, E. A., and Williams, R. A. 1986. *Hug a tree.* Mt. Rainier, Md.: Gryphon House, Inc.

Shaw, E. L. 1987. Students and sponges—soaking up science. *Science and Children*. 25 (1): 21.

Warren, J. 1984. *Science time.* Palo Alto, Calif.: Monday Morning Books.

SUGGESTED ACTIVITIES

1. What kind of problems do you anticipate in teaching ordering and patterning, measurement, and graphing to young children? Discuss problems you have encountered and relate how they were overcome. How could the problem have been handled in a different way? What were the children like? Videotape or tape record yourself in a teaching situation; this can help to clearly illustrate a problem.

2. To measure is to match things. Observe and record every comparison that one child makes in a day. Record all forms of measuring that you observe this child performing. Teachers will probably tell you that children are not very interested in measuring for its own sake, so find the measuring in the child's everyday school life.

3. Reflect on measurement in your own life. What did you measure this morning? In groups, brainstorm all of the measuring that you did on your way to class. Predict the types of measuring that you will do tonight. Then reflect on ordering, patterning, and graphing in the same way.

4. Design a lesson that emphasizes one of the concepts and skills discussed in this chapter. Teach the lesson to a class and note the reactions of the children to your teaching. Discuss these reactions in class. "Did you teach what you intended to teach?" "How do you know your lesson was effective?" "If not, why not, and how could the lesson be improved?"

REVIEW

1. What is seriation? Why must children possess this skill before they are able to pattern objects?
2. Which activities might be used to evaluate a child's ability to compare and order?
3. How will you teach measurement to young chidren?
4. Explain how graphing assists in learning science concepts.

UNIT 22 Practical Activities and Science Investigations: Math, Science, and Social Studies Work Together

OBJECTIVES

After studying this unit, the student should be able to:

- Describe how children apply and extend concepts through dramatic play
- Describe how children apply and extend concepts through food experiences
- Encourage dramatic role playing that promotes concept acquisition
- Recognize how dramatic role playing and food experiences promote interdisciplinary instruction and learning
- Use dramatic play and food experiences as settings for science investigations, mathematical problem solving, and social learning

The purpose of this unit is to demonstrate how dramatic play and food experiences can enrich and enhance children's acquisition of concepts and knowledge, not only in science and math, but also in social studies. Furthermore, these areas offer rich settings for social learning, science investigations, and mathematical problem solving. In Unit 1 the commonalities between math and science were described. Unit 7 provided a basic unit plan and examples of science units that incorporate math, social studies, language arts, fine arts, motor development, and dramatic play. Units 34 through 38 include integrated units and activities for the primary level. This unit emphasizes the natural play of young children as the basis for developing units of study and uses food experiences to highlight the interdisciplinary nature of practical activities.

Concepts and skills are only valuable to children if they can be used in everyday life. Young children spend most of their waking hours involved in play. Play can be used as a vehicle for the application of concepts. Young children like to feel big and do "big person" things. They like to pretend they are grown up and want to do as many grown-up things as they can. Role playing can be used as a means for children to apply what they know as they take on a multitude of grown-up roles.

Experiences with food also include a variety of opportunities for applying concepts and carrying out scientific investigations and mathematical problem solving. Food experiences are motivating because they provide opportunities for children to do adult activities. They can grow food and shop for groceries; plan and prepare meals, snacks, and parties; serve food and enjoy sharing and eating the results of their efforts. Both dramatic play and food experiences offer opportunites for social education as children learn more about adult tasks, the community, and other cultures. Teachers can use these experiences to assess and evaluate through observation.

DRAMATIC ROLE PLAYING

When children are engaged in dramatic role playing, they practice what it is like to be an adult. They begin with simple imitation of what they have observed. Their first roles reflect what they have seen at home. They bathe, feed, and rock babies. They cook meals, set the table, and eat. One of their first outside experiences is to go shopping. This experience is soon reflected in dramatic play. They begin by carrying things in bags, purses, and other large containers. At first, they carry around anything that they can stuff in their containers. Gradually, they move into using more realistic props such as play money and empty food containers. Next, they might build a store with big blocks and planks. Eventually, they learn to play cooperatively with other children. One child might be the mother, another the father, another the child, and another the store clerk. As the children move toward this stage, teachers can provide more props and background experiences that will expand the raw material children have for developing their role playing. Problem-solving skills are refined as children figure out who will take which role, where the store and home will be located, and develop the rules for the activity (Figure 22–1).

Background experiences include trips, discussions, stories, records, and films. Visits can be made to stores, to restaurants, to the post office, to the bank and other service centers. Stories can be read, trips discussed, records played, and films shown which tell more and add to what the child has experienced. Basic props for shopping include play money, a cash register, purses, wallets, and bags and baskets for carrying the things which are bought. Each type of store or business can then be set up with props which fit the items or services to be bought.

Figure 22–1 Going shopping is a popular dramatic play activity.

Some stores and other places/activities that inspire the application of basic concepts and social knowledge are listed below.

- Toy Store: Children can bring old toys from home or use toys already in the school to buy and sell.

- Grocery Store: Children and teachers can bring empty food containers from home to stock the store. They can make pretend food from play dough and papier mache. Plastic food can also be used.
- Clothing Store: Old clothes can be brought. Departments can be set up for baby clothes, children's clothes, ladies' clothes, and men's clothes. A shoe section can be included.
- Jewelry Store: Old jewelry can be brought from home. Jewelry can be made by stringing paper or macaroni. Watches and clocks can be made from paper and cardboard.
- Service Centers: Post office, bank, gasoline station, auto repair shop, beauty shop, barbershop, doctor's office, restaurant, and fast food business.
- Transportation: Bus, plane, train, boat, and space shuttle trips. Children can buy tickets and discuss times for departure, arrival, and length of the trip.
- Animals: Zoo, veterinarian, circus, farm, or pet shop. Have children bring stuffed animals from home to live in the zoo, visit the vet, act in the circus, live on the farm, or be sold in the pet shop. Classify the animals as to which belong in each setting. Children can predict which animals eat the most, are dangerous to humans, are the smartest, and so on. Provide play money to pay for goods and services.
- Health/Medical: Provide props for medical play. Tie in with discussions of good nutrition and other health practices. They can pay the bill for the services.
- Space Science: Provide props for space travel (e.g., a big refrigerator carton that can be made into a spaceship, paper bag space helmets, and so on). Provide materials for making mission control and for designing other planetary settings.
- Water: Provide toy boats, people, rocks for islands, and the like. Discuss floating and sinking. Outdoors, use water for firefighter play and for watering the garden. Have a container (bucket or large dishpan) that can be a fishing hole, waterproof fish with a safety pin or other metal object attached so they can be caught with a magnet fish bait. Investigate why the magnet/metal combination makes a good combination for pretend fishing. Count how many fish each child catches.
- Simple Machines: Vehicles, a packing box elevator, a milk carton elevator on a pulley, a plank on rollers, and so on, make interesting dramatic play props, and their construction and functioning provide challenging problems for investigation.

Concepts are applied in a multitude of play activities such as those described above. The following are some examples:

- One-to-One Correspondence: Exchanging money for goods or services. Children give one or more pieces of play money for purchases such as a gallon of gasoline, a box of cereal, a haircut, postage stamp, and so on.
- Sets and Classifying: Things brought home must be put in the right place in the right room. To set up a store requires sorting. For example, the grocery store must have places for meat, for dairy products, for frozen food, for fresh food, for cans, for boxes, for soaps, and so on.
- Counting: Things which are bought are counted. He buys one coat, two bottles of milk, three oranges, ten gallons of gasoline, three hamburgers, and so on. The money paid must also be counted.
- Comparing and Measuring: Clothing can be tried on for fit. Real tape measures and rulers can be used to "measure" feet, shoulders, arms, and other body parts for fit. Meat, fruits, and vegetables can be weighed. The doctor and nurse can check the sick person's temperature. Things can

be available in containers of different sizes so that large, small, or middle sized can be bought. Milk cartons can be labeled as quarts, half gallons, and gallons. Customers may ask for a little bit or a lot.

- Spatial Relations and Volume: Things bought must be fit into bags, boxes, and baskets that will hold them.
- Number Symbols: For the child who can name the number symbols, price tags can be made. The child who can write number symbols can make price tags and write bills of sale. They can be shown how to make dollar ($) and cents (¢) symbols for tags. They can pretend to write checks, make bank deposits, and use pretend charge cards. Number symbols can also be put on play coins and bills.

Money is a basic part of most of these activities. For advanced preprimary and primary children who have a beginning understanding of earning and spending money, the computer program *Duck's Playground* can offer entertainment and practice in working and earning money. This program is available from Sierra On-Line, Inc., Coarsegold, Calif.

Pocket calculators are excellent props for dramatic play. Children can pretend to add up their expenses, costs, and earnings. As they explore calculators, they will learn how to use them for basic mathematical operations. Methods for introducing calculators are described in Section IV.

FOOD EXPERIENCES

Experiences with food involve many science, math, and social studies concepts. Food is a familiar thing to young children. They have many naturalistic experiences buying, preparing, serving, and eating it. They might have seen food grown or have even grown some themselves if they have a garden or live in a rural area. As scientists, children observe the growth of food, the physical changes that take place when food is prepared, and the effects of food on growth of humans and animals. They also compare the tastes of different foods and categorize them into those they like and those they dislike, and into sweet and sour, liquid and solid, "junk" and healthful, and groups such as meat/dairy products, cereals/breads, and fruits/vegetables (Figure 22–2).

As mathematicians, children pour and measure, count, cut wholes into parts, and divide full pan or full bowl into equal servings. They count the strokes when mixing a cake, make sure the oven is on the correct temperature setting, and set the clock for the required baking time. At the store, they exchange money for food and fruits and weigh vegetables. They count the days until their beans sprout or the fruit ripens.

Through food experiences, children learn much about society and culture. They can make foods from different cultures. They learn where food is grown, how it is marketed, and how it must be purchased with money at the grocery store. They cooperate with each other and take turns when preparing food. Then they share what they make with others.

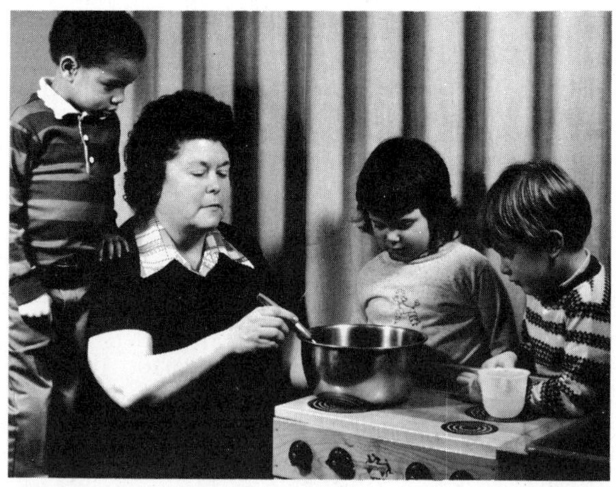

Figure 22–2 The children measure when they make sugar cookies.

Food and Dramatic Play

In the home living center at school, children purchase, cook, serve, and eat food as part of their role playing. It was suggested in Unit 18 that a simple measuring activity could be to make flour and salt dough. This dough can be made into pretend cookies, cakes, and other foods and used as props for dramatic role playing.

Food and Math

The young child can make real things to eat. He can learn that certain amounts of each item to be used in the recipe must be carefully measured in order to come out with something that looks and tastes good to eat. This is an excellent way to introduce the child to the concept of standard units in a natural and informal way. As young as two or three, the child can understand that the recipe says that only so much is to be used. He can be shown that we have special cups and spoons which are used to measure food. In this country, we measure food by volume (cups, teaspoons, and tablespoons). In some other countries, food is measured out by weight. For instance, compare the following recipes, one from England and one from the United States.

From England: Fairy Cakes[1]

8 oz flour

4 oz fat

4 oz sugar

1 or 2 eggs

From the U.S.A.: Play Dough Biscuits[2]

2 cups sifted unbleached white flour

3 3/4 teaspoons baking powder

1 teaspoon salt

1/3 cup oil

3/4 cup milk

Figures 22-3, 22-4, and 22-5 illustrate three popular recipes that young children enjoy making. In addition to measuring, the children count (number of cups, teaspoons, ounces). They also learn to follow a sequence of steps such as measure, mix, bake, cut, and serve. For things that are baked, they learn that a certain temperature must be used and that the baking takes a special length of time. They learn that some foods need a high heat and some a low heat and that some must be chilled and others frozen. Some things are cut into two or more pieces when they are prepared. The finished food must be divided into portions to be served.

The child who lives where food grows has even more opportunities for the use of math skills. The eggs laid by the hens can be counted each day. A graph can be made showing the comparison of the number of eggs collected each day. The eggs can be sold, and the children can record the amount of money received. The number of days from the time each vegetable in the garden is planted until it is picked can be recorded. The number of carrots, beans, squash, and so on that are picked each day can be recorded also.

When the child sets the table in the play area or for a real meal, he uses math skills. He counts and matches the number of people to be served, the number of spaces at the table, the napkins, dishes, spoons, forks, knives, and cups. As he cleans up and puts things away, he sorts things into sets. A silverware holder with a place for each eating utensil offers a structured sorting experience. The shelves for each type of dish and

[1] *Beginnings* (New York: John Wiley and Sons Inc., 1968).

[2] Doreen Croft and Robert Hess, *An Activities Handbook for Teachers of Young Children*, Second edition (Boston: Houghton Mifflin Company, 1975).

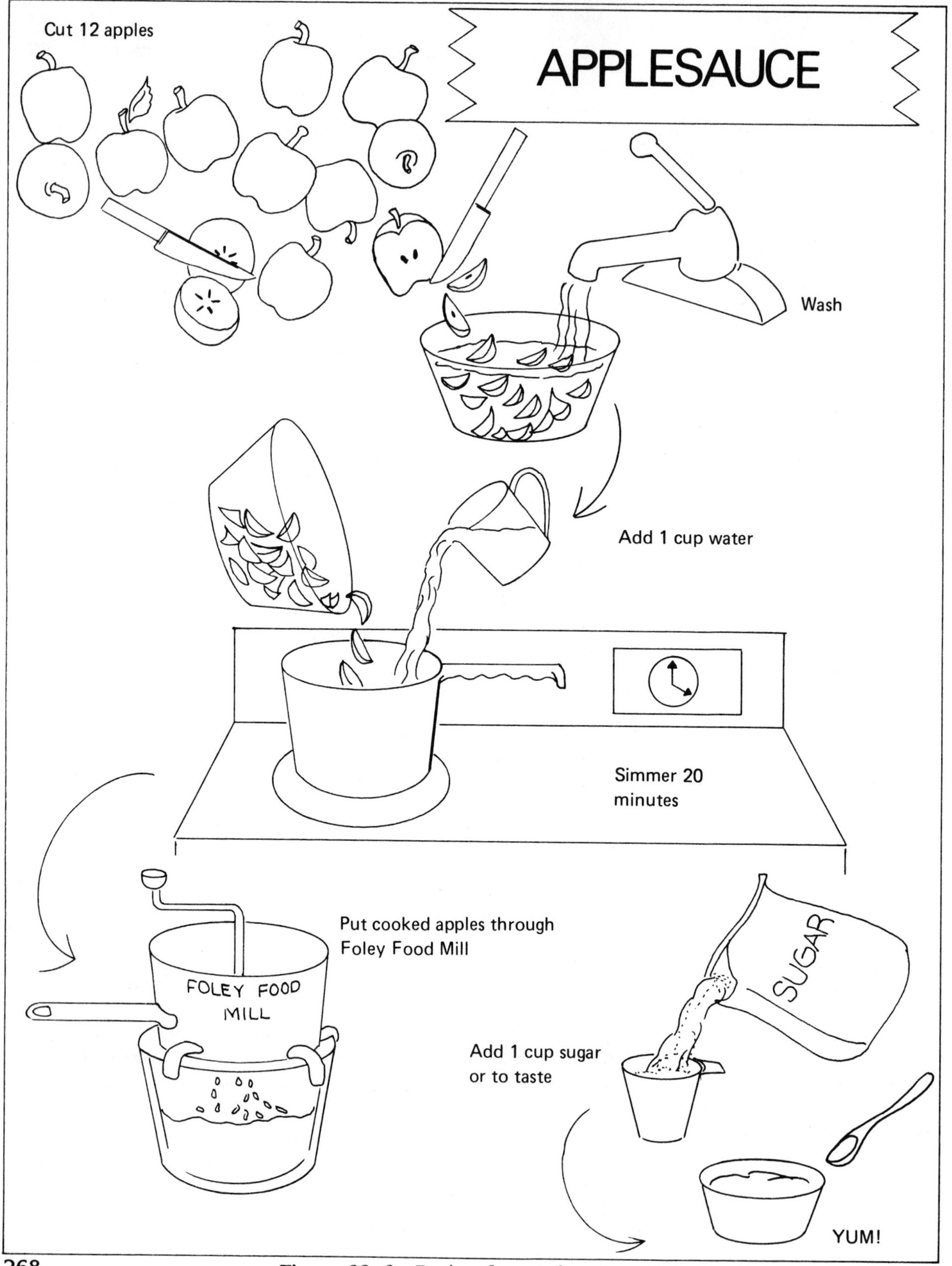

Figure 22–3 Recipe for applesauce

FIRST DAY

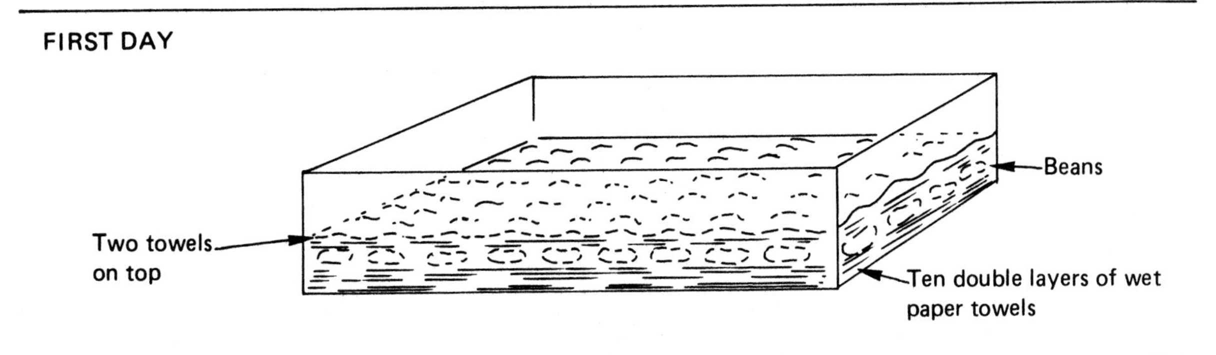

EIGHT DAYS LATER

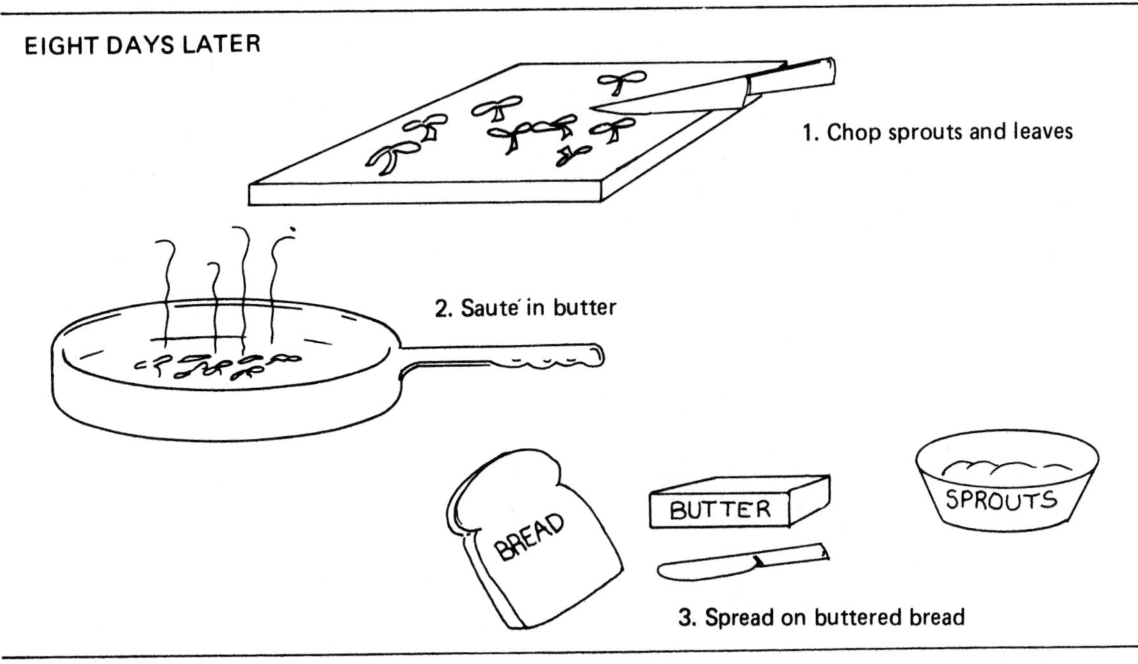

Sprout beans. Place navy beans on ten thicknesses of 2-ply paper towels, covered by two thicknesses of paper towels. Keep beans moist and warm for about eight days or until leaves appear on 1 1/2 inch sprouts. Chop beans, sprouts, and leaves until fine. Saute in butter and salt. Spread on buttered bread.

Figure 22–4 Bean sprout sandwich

bowl and cup can be labeled with pictures to help the children match and sort them into the right places.

Food and Science

Each of the activities described under Food and Math also involves science. Children can be asked to predict what will happen when the wet ingredients (oil and milk) are mixed with the dry ingredients (flour, baking powder, and salt) when making the playdough biscuits. They can then make the mixture and observe and describe its texture, color, and density and compare their results with their predictions. Next, they can

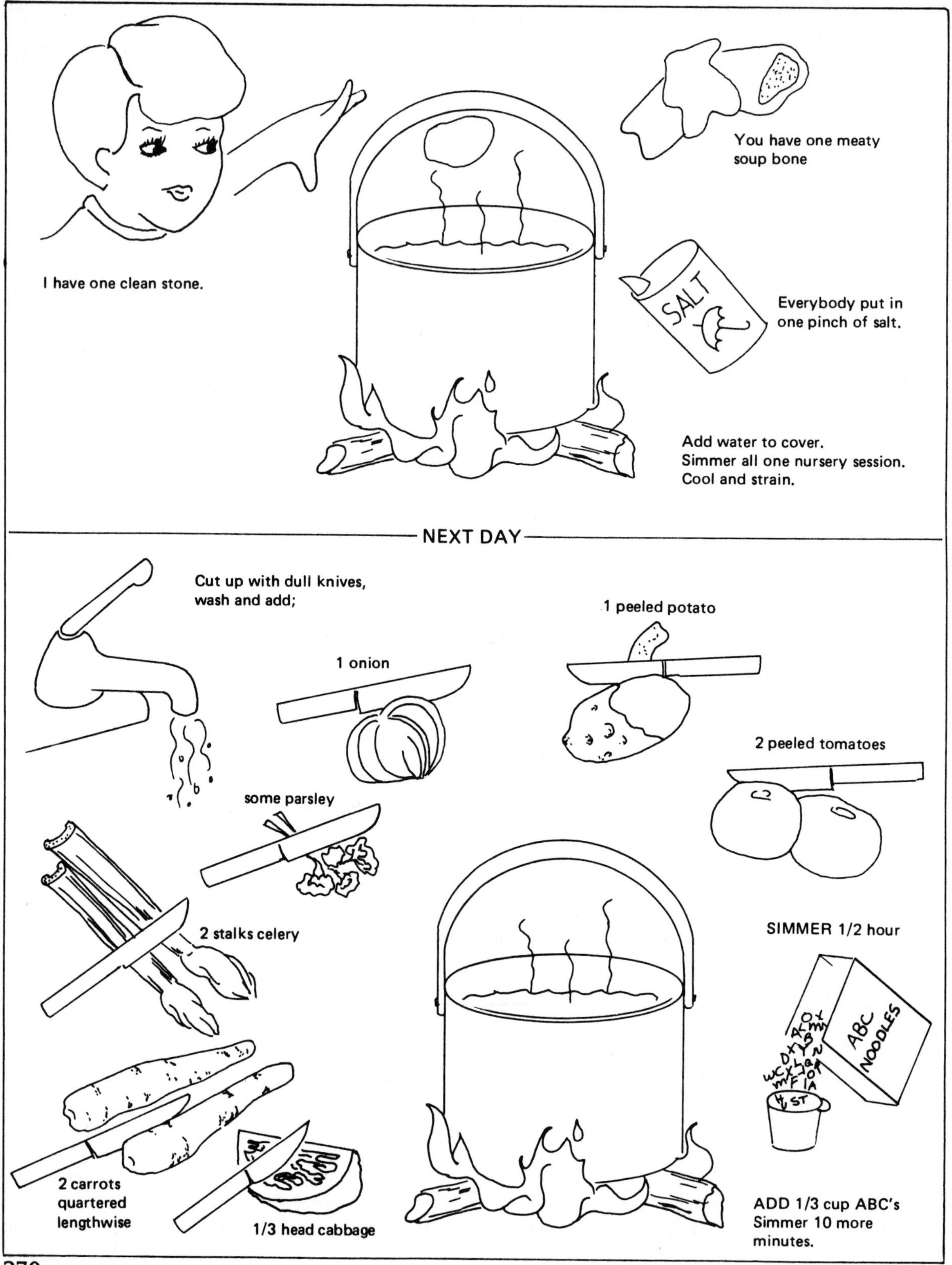

Figure 22–5 Recipe for stone soup

predict what will happen when the dough is baked in the oven. When it is taken out, they can observe and describe the differences that take place during the baking process. They can be asked what would happen if the oven is too hot or if the biscuits are left in too long.

If the children have the opportunity to see eggs produced, this can lead to a discussion of where the eggs come from and what would happen if the eggs were fertilized. They could observe the growth from seed to edible food as vegetables are planted, cultivated, watered, and picked. Applesauce exemplifies several physical changes: from whole to parts, from solid chunks to soft lumps, and to smooth and thick as cutting, heating, and grinding each have an effect (Figure 22–3). Children can also note the change in taste before and after the sugar is added. The bean sprout sandwich is an experience in which even city children can take food from seed to eating (Figure 22–4). Stone soup offers an opportunity for discussion of the significance of the stone (Figure 22–5). What does the stone add to the soup? Does a stone have nutrients? What really makes the soup taste good and makes it nutritious?

Food and Social Studies

Each of the activities described also involves social studies. City children might take a trip to the farm. For example, a trip to an orchard to get apples for applesauce is an enriching and enjoyable experience. They might also take a trip to the grocery store to purchase the ingredients needed in their recipes. Then they can take turns measuring, cutting, adding ingredients, or whatever else is required as the cooking process proceeds (Figure 22–6). Stone soup is an excellent group activity because everyone in the class can add an ingredient. Invite people from different cultures to bring foods to class and/or help the children make their special foods. Children can note similarities and differences across cultures.

Figure 22–6 Scott spreads peanut butter on one piece of bread and jelly on the other. Two pieces put together make a sandwich.

SUMMARY

Dramatic play and food experiences provide math, science, and social studies experiences that afford children an opportunity to apply concepts and skills. They can predict, observe, and investigate as they explore these areas. As children play home, store, and service roles, they match, count, classify, compare, measure, and use spatial relations concepts and number symbols. They also practice the exchange of money for goods and services. Through dramatic play they try out grown-up roles and activities.

Through food experiences children learn the sequence from planting and growing to picking, buying, preparing, serving, and eating. They can also count, measure, and match to serve a practical purpose.

For teachers, these activities offer valuable opportunities for naturalistic and informal instruction as well as time to observe children and assess their ability to use concepts in everyday situations.

FURTHER READING AND RESOURCES

UNITS AND ACTIVITIES

Ayers, J. B., and Ayers, M. N. 1984. Popcorn, Piaget and science for preschoolers. *Dimensions*. 12 (2): 4–6.

Charlesworth, R. 1988. Integrating math, science, and social studies: A unit example. *Day Care and Early Education*. 15 (4) 28–31.

Dalton, D. 1982. Resource list: Non-cooked food preparation. *Dimensions*. 10 (1): 26–29.

Dukes, L. 1986. An apple for the learner. *Day Care and Early Education*. 14 (1): 15–17.

Food and cooking. 1983. *First Teacher*. 4 (11).

Hack, K., and Flynn, V. 1985. Green beans: Gardening with two's. *Day Care and Early Education*. 13 (1): 14–17.

Harlan, J. 1988. *Science experiences for the early childhood years*, 4th ed., Columbus, Ohio: Merrill.

Homestyle recipes for classroom cooks. 1986. *Instructor*. November/December: pp. 66, 73.

Jobe, P., and Washington, R. E. 1981. The kitchen: A unique learning center. *Dimensions*. 9 (2): 110–113.

Martin, S. 1987. IDEAS! Nutrition: Avenue for discovery learning. *Dimensions*. 15 (2): 15–18.

McCree, N. L. 1981. A down-under approach to parent and child food fun. *Childhood Education*. 57 (4): 216–222.

Warner, L. 1981. Some basic concepts in the grocery store center. *Day Care and Early Education*. 8 (4): 14–19.

COOKBOOKS

Christenberry, M. A., and Stevens, B. 1984. *Can Piaget cook?* Atlanta, Ga.: Humanics.

Faggella, K., and Dixler, D. 1985. *Concept cookery*. Bridgeport, Conn.: First Teacher Press.

McClenahan, P., and Jaqua, I. 1976. *Cool cooking for kids*. Belmont, Calif.: Fearon-Pitman.

Teddlie, A. T., and Turner, I. M. 1984. *Lots of action cooking with us*. Ruston, La.: Louisiana Association On Children Under Six.

Wanamaker, N., Hearn, K., and Richarz, S. 1979. *More than graham crackers*. Washington, D.C.: National Association for the Education of Young Children.

SUGGESTED ACTIVITIES

1. Observe young children at play in school. Note if any concept experiences take place during dramatic role playing or while doing food activities. Share what you observe with your class.
2. Add five dramatic play concept activities to your file.
3. Review several cookbooks (the ones suggested in this unit or some others) and examine recipes recommended for young children. Explain to the class which recipe books you think are most appropriate for use with young children.
4. Add to your Activities File five food experiences that enrich children's concepts and promote application of concepts.

5. If possible, view the film "Jenny Is A Good Thing." Information about obtaining the film on a free loan basis is available from Sponsor Service Desk, Modern Talking Picture Service, Inc., 1212 Avenue of the Americas, New York, New York 10036.

REVIEW

A. Briefly answer each of the following:
   1. Why should dramatic role playing be included as a math, science, and social studies concept experience for young children?
   2. Why should food activities be included as math, science, and social studies concept experiences for young children?
   3. How can teachers encourage dramatic role playing that includes concept experiences?
   4. Describe four food activities that can be used in the early childhood math, science, and social studies program.

B. Indicate which item (a) or (b) in Column II matches each activity in Column I.

I	II
1. Planting bean seeds	a. Dramatic role play experience
2. Buying bananas	b. Food experience
3. Using a cash register	
4. Making and selling lemonade	
5. Having an auto repair shop	
6. Setting the snack table	

SECTION IV

Symbols and Higher-Level Activities

UNIT 23 Symbols

OBJECTIVES

After studying this unit, the student should be able to:

- List the six number symbol skills
- Describe four basic types of self-correcting number symbol materials
- Set up an environment that supports naturalistic and informal number symbol experiences
- Do structured number symbol activities with children

Number symbols are called numerals. Each numeral represents an amount and acts as a shorthand for recording *how many*. The young child sees numerals all around. (Figure 23–1). He has some idea of what they are before he can understand and use them. He sees that there are

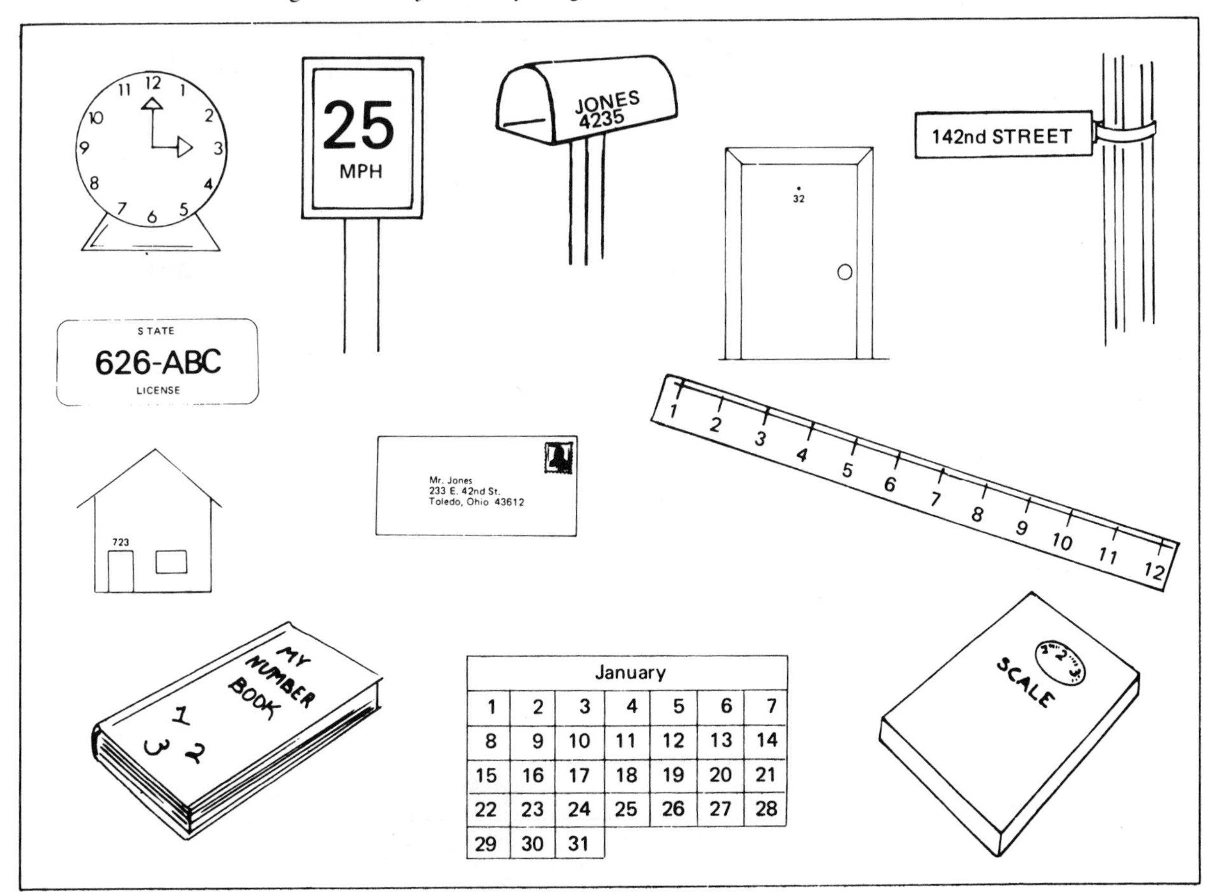

Figure 23–1 Numerals are seen everywhere in the environment.

numerals on his house, the phone, the clock, and the car license plate. He may have one or more counting books. He may watch a children's TV program where numeral recognition is taught. Sometime between the age of two and the age of five a child learns to name the numerals from zero to ten. However, the child is usually four or more when he begins to understand that each numeral stands for a set of things of a certain amount that is always the same. He may be able to tell the name of the symbol "3" (three) and count three objects, but he may not realize that the "3" can stand for the three objects. This is illustrated in Figure 23–2.

It can be confusing to the child to spend time on drill with numerals until he has had many concrete experiences with basic math concepts. Most experiences with numerals should be naturalistic and informal.

THE NUMBER SYMBOL SKILLS

There are six number symbol skills that young children acquire during the preoperational period:

- He learns to recognize and say the name of each numeral (Figure 23–3).
- He learns to place the numerals in order: 0-1-2-3-4-5-6-7-8-9-10.
- He learns to associate numerals with sets: "1" goes with one thing.
- He learns that each numeral in order stands for one more than the numeral that comes before it. (That is, two is one more than one, three is one more than two, and so on.)
- He learns to match each numeral to any set of the size that the numeral stands for and to make sets that match numerals.

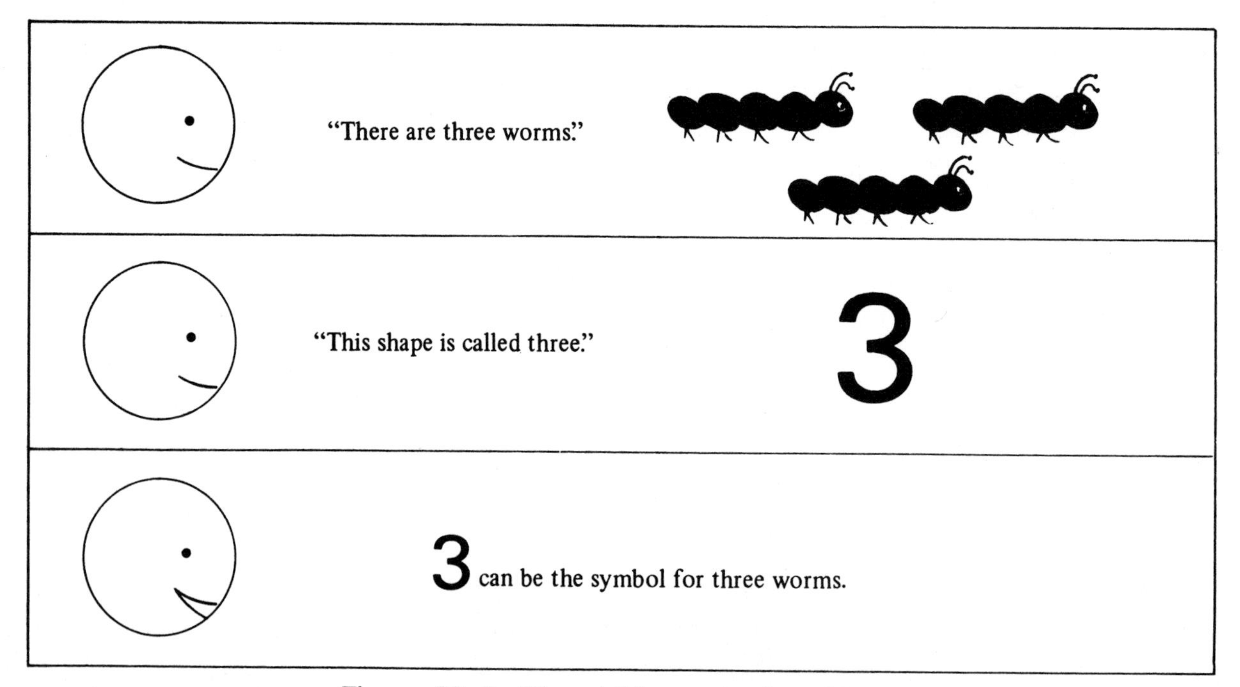

Figure 23–2 The child counts the objects, learns the symbol, and realizes that the symbol can stand for the set.

Figure 23–3 "It's a one!"

- He learns to reproduce (write) numerals.

The first four skills are included in this unit. The last two are the topics for Unit 24.

Assessment

The teacher should observe whether the child shows an interest in numerals. Does he repeat the names he hears on television? Does he use self-correcting materials that are described in the part of this unit on informal learning and teaching? What does he do when he uses these materials? Individual interviews would include the types of tasks which follow.

SAMPLE ASSESSMENT TASK

Preoperational Ages 3–6

METHOD: Interview

SKILL: Child is able to recognize numerals zero to ten presented in sequence.

MATERIALS: 5″ × 8″ cards with one numeral from zero to ten written on each

PROCEDURE: Starting with zero, show the child each card in numerical order from zero to ten. WHAT IS THIS? TELL ME THE NAME OF THIS.

EVALUATION: Note if the child uses numeral names (correct or not), indicating he knows the kinds of words associated with the symbols. Note which numerals he can label correctly.

INSTRUCTIONAL RESOURCE(S):

Charlesworth, R., and Lind, K. 1990. *Math and Science for Young Children.* Albany, N.Y.: Delmar.

SAMPLE ASSESSMENT TASK

Preoperational Ages 4–6

METHOD: Interview

SKILL: Child is able to sequence numerals from zero to ten.

MATERIALS: 5″ × 8″ cards with one numeral from zero to ten written on each

PROCEDURE: Place all the cards in front of the child in random order. PUT THESE IN ORDER. WHICH COMES FIRST? NEXT? NEXT?

EVALUATION: Note whether the child seems to understand that numerals belong in a fixed sequence. Note how many are placed in the correct order and which, if any, are labeled.

INSTRUCTIONAL RESOURCE(S):

Charlesworth, R., and Lind, K. 1990. *Math and Science for Young Children*. Albany, N.Y.: Delmar.

SAMPLE ASSESSMENT TASK

Preoperational Ages 5 and Older

METHOD: Interview

SKILL: Child is able to identify numerals that are "one more than."

MATERIALS: 5″ × 8″ cards with one numeral from zero to ten written on each
PROCEDURE: Place the numeral cards in front of the child in order from zero to ten. TELL ME WHICH NUMERAL MEANS ONE MORE THAN TWO. WHICH NUMERAL MEANS ONE MORE THAN SEVEN? WHICH NUMERAL MEANS ONE MORE THAN FOUR? (If the child answers these, then try LESS THAN.)

EVALUATION: Note whether the child is able to answer correctly.

INSTRUCTIONAL RESOURCE(S):

Charlesworth, R., and Lind, K. 1990. *Math and Science for Young Children*. Albany, N.Y.: Delmar.

NATURALISTIC ACTIVITIES

As the young child observes his environment, he sees numerals around him. He sees them on clocks, phones, houses, books, food containers, television programs, money, calendars, thermometers, rulers, measuring cups, license plates, and on many other objects in many places. He hears people say:

- My phone number is 622-7732.
- My house number is 1423.
- My age is six.
- I have a five-dollar bill.
- The temperature is 78 degrees.
- Get a five-pound bag of rabbit food.
- We had three inches of rain today.
- This pitcher holds eight cups of juice.

Usually children start using the names of the number symbols before they actually match them with the symbols:

- Tanya and Juanita are ready to take off in their space ship. Juanita does the countdown, "Ten, nine, eight, three, one, blast off!"

- Tim asks Ms. Moore to write the number seven on a sign for his race car.
- Sam notices that the thermometer has numbers written on it.
- Diana is playing house. She takes the toy phone and begins dialing, "One-six-two. Hello dear, will you please stop at the store and buy a loaf of bread."
- "How old are you, Pete?" "I'm six," answers two-year-old Pete.
- "One, two, three, I have three dolls."
- Tanya is playing house. She looks up at the clock. "Eight o'clock and time for bed," she tells her doll. (The clock really says 9:30 A.M.).

Children begin to learn number symbols as they look and listen and then imitate in their play what they have seen and heard.

Figure 23–4 Putting in and taking out tactile numerals requires the use of touch and sight.

INFORMAL ACTIVITIES

During the preoperational period most school activities with numerals should be informal. Experimentation and practice in perception with sight and touch are most important (Figure 23–4). These experiences are made available by means of activities with self-correcting manipulative materials. Self-correcting materials are those which the child can use by trial and error to solve a problem as he works by himself (Figure 23–5). The material is made in such a way that it can be used with success with very little help. Manipulative materials are things which have parts and pieces which can be picked up and moved by the child to solve the problem presented by the materials. The teacher observes the child as he works. She notes whether the child works in an organized way and whether he sticks with the material until he has the task finished.

Figure 23–5 Self-correcting materials can be used by one child.

There are four basic types of self-correcting manipulative math materials that can be used for informal activities. These materials can be bought or they can be made. The four basic groups of materials are those which teach dis-

Figure 23–6 Self-correcting materials can be used cooperatively by more than one child.

crimination and matching, those which teach sequence (or order), those which give practice in association of symbols with sets, and those which combine association of symbols and sets with sequence. Examples of each type are illustrated in Figures 23–7 through 23–10.

The child can learn to discriminate one numeral from the other by sorting packs of numeral cards. He can also learn which numerals are the same as he matches. Another type of material which serves this purpose is a lotto-type game. The child has a large card divided equally into four or more parts. He must match individual numeral cards to each numeral on the big card. These materials are shown in Figure 23–7 (A and B). He can also experiment with felt, plastic, magnet, wooden, rubber, and cardboard numerals.

There are many materials which teach sequence or order. These may be set up so that parts can only be put together in such a way that when the child is done, he sees the numerals are in order in front of him. An example would be the Number Worm® by Childcraft, Figure 23–8(A). Sequence is also taught through the use of a number line or number stepping stones. The Childcraft giant Walk-On Number Line® lets the child walk from one numeral to the next in order, Figure 23–8(B). The teacher can set out numerals on the floor (such as Stepping Stones® from Childcraft) which the child must step on in order, Figure 23–8(C).

The hand calculator lends itself to informal exploration of numerals. First, show the students how to turn the calculator on and off. Tell them to watch the display window and then turn on the calculator. A zero will appear. Explain that when they first turn on their calculators a zero will always appear. Then tell them to turn on their calculators and tell you what they see in the window. Have them practice turning their calculators on and off until you are sure they all understand this operation. Next, tell them to press [1]. Ask them what they see in the window. Note if they tell you they see the same number. Next, have them press [2]. A two will appear in the window next to the one. Show them that they just need to press the [C] key to erase. Then let them explore and discover on their own. Help them by answering their questions and posing questions to them, such as, "What will happen if . . . ?"

Many materials can be purchased which help the child associate each numeral with the set that goes with it. Large cards which can be placed on the bulletin board (such as Childcraft Poster Cards®) give a visual association, Figure 23–9(A). Numerals can be seen and touched on textured cards which can be bought (such as Childcraft Beaded Number Cards®). Numeral cards can be made using sandpaper for the sets of dots and for the numerals, Figure 23–9(B). Other materials require the child to use visual and motor coordination. He may have to match

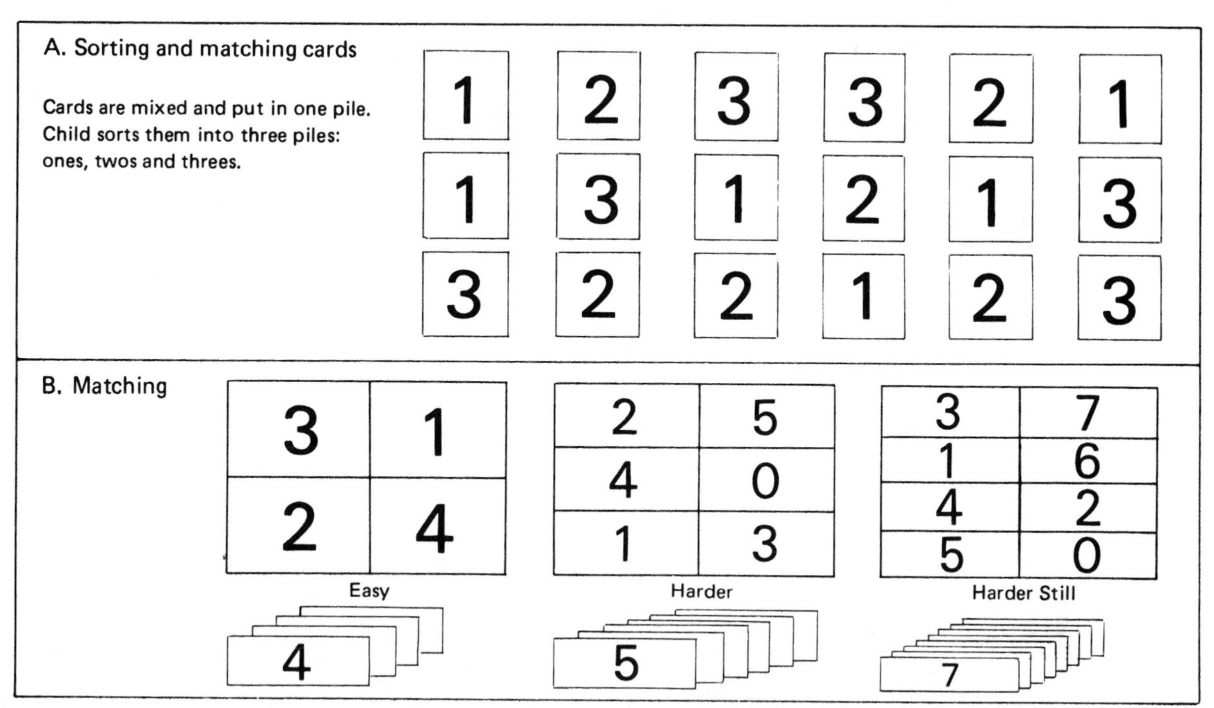

Figure 23–7 Sorting and matching

Figure 23–8 Materials that help the child learn numeral sequence

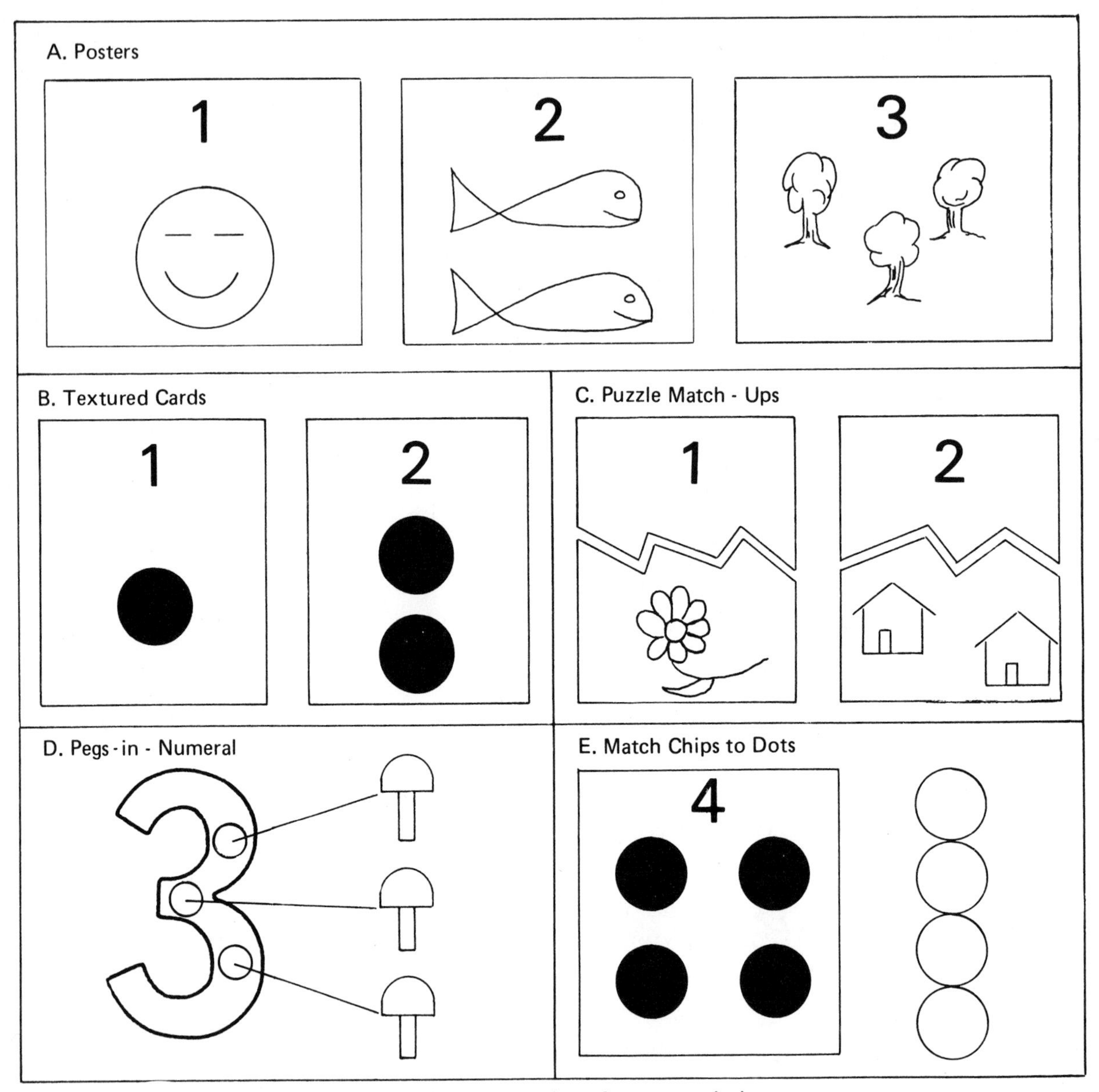

Figure 23–9 Numeral and set association

puzzle-like pieces (such as Math Plaques® from Childcraft), Figure 23–9(C). He may put pegs in holes (such as Peg Numerals® from Childcraft), Figure 23–9(D). Unifix® inset pattern boards require the same type of activity. The teacher can make cards which have numerals and dots the size of buttons or other counters. The child could place a counter on each dot, Figure 23–9(E).

Materials which give the child experience

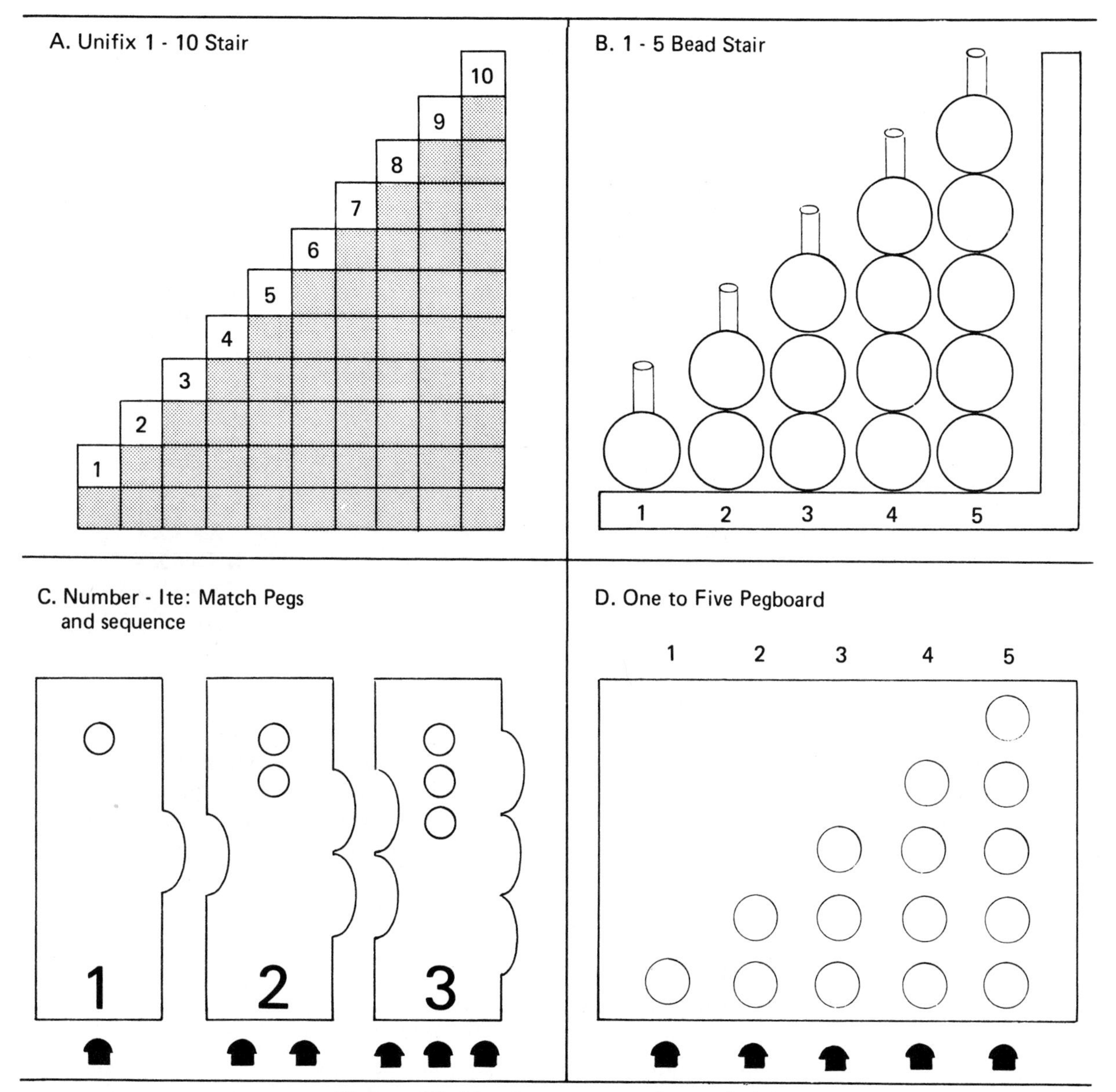

Figure 23–10 Sequence and association

with sequence and association at the same time are also available. These are shown in Figure 23–10. It can be seen that the basis of the materials is that the numerals are in a fixed order and the child adds some sort of counter which can only be placed in the right amount. Unifix stairs are like the inset patterns but are stuck together, Figure 23–10(A). Other materials illustrated are counters on rods, Figure 23–10(B), pegs in holes (Number-Ite® from Childcraft),

Figure 23–10(C), or 1-5 pegboard, Figure 23–10(D).

The teacher's role with these materials is to show the child how they can be used and then step back and watch. After the child has learned to use the materials independently, the teacher can make comments and ask questions:

- How many pegs on this one?
- Can you tell me the name of each numeral?
- You put in the four pegs that go with that numeral four.
- How many beads are there here? (point to stack of one) How many here? (stack of two) (and so on.)
- Good, you separated all the numerals into piles. Here are all "ones" and here are all "twos."

Most children will learn through this informal use of materials to recognize and say the name of each numeral, to place the numerals in order, to see that each numeral stands for one more than the one before it, and to associate numerals with amounts. However, some children will need the structured activities described next.

STRUCTURED ACTIVITIES

By the time young children finish kindergarten, they should be able to:

- Recognize the numerals from zero to ten
- Place the numerals from zero to ten in order
- Know that each numeral represents a set one larger than the numeral before (and one less than the one that comes next)
- Know that each numeral represents a set of things

Numeral	Amount in Set
0	
1	X
2	XX
3	XXX
4	XXXX
5	XXXXX
6	XXXXXX
7	XXXXXXX
8	XXXXXXXX
9	XXXXXXXXX
10	XXXXXXXXXXX

He may not always match the right numeral to the correct amount, but he will know that there is such a relationship. The five-year-old child who cannot do one or more of the tasks listed needs some structured help.

NUMERALS: RECOGNITION

OBJECTIVE: To learn the names of the number symbols

MATERIALS: Write the numerals from zero to ten on cards.

ACTIVITY: This is an activity that a child who can name all the numbers can do with a child who needs help. Show the numerals one at a time in order, THIS NUMERAL IS CALLED ____. LET'S SAY IT TOGETHER, ____. Do this for each numeral. After ten, say, I'LL HOLD THE CARDS UP ONE AT A TIME. YOU NAME THE NUMERAL. Go through once. Five minutes at a time should be enough.

FOLLOW-UP: Give the child a set of cards to review on his own.

NUMERALS: SEQUENCE AND ONE MORE THAN

OBJECTIVE: To learn the sequence of numerals from zero to ten

MATERIALS: Flannel or magnet board, felt or magnet numerals, felt or magnet shapes (such as felt primary cutouts, or magnet geometric shapes)

ACTIVITY: Put the "zero" up first at the upper lefthand corner of the board. WHAT IS THIS NUMERAL CALLED? If the child cannot tell you, say: THIS IS CALLED "ZERO." Put the "one" numeral up next to the right. WHAT IS THIS NUMERAL CALLED? If the child cannot tell you, say: THIS IS "ONE." SAY IT WITH ME, "ONE." Continue to go across until the child does not know two numerals in a row. Then, go back to the beginning of the row. TELL ME THE NAME OF THIS NUMERAL. YES, "ZERO." WHAT IS THE NAME OF THE NEXT ONE? YES, IT IS "ONE" SO I WILL PUT ONE RABBIT HERE. Put one rabbit under the "one." THE NEXT NUMERAL IS ONE MORE THAN "ONE." WHAT IS IT CALLED? After the child says "two" on his own or with your help, let him pick out two shapes to put on the board under the "two." Keep going across until you have done the same with each numeral he knows, plus two that he does not know.

FOLLOW-UP: Have the child set up the sequence. If he has trouble, ask: WHAT COMES NEXT? WHAT IS ONE MORE THAN ______? Leave the board and the numerals and shapes out during playtime. Encourage the children who know how to do this activity to work with a child who does not.

NUMERALS: RECOGNITION, SEQUENCE, ASSOCIATION WITH SETS, ONE MORE THAN

OBJECTIVE: To help the child to integrate the concepts of association with sets, with one more than, while learning the numeral names and sequence

MATERIALS: Cards with numerals zero to ten and cards with numerals and sets zero to ten

ACTIVITIES:

1. I'M GOING TO PUT DOWN SOME CARDS. EACH ONE HAS A NUMERAL ON IT. THEY GO UP TO TEN. SAY THE NAMES WITH ME IF YOU KNOW THEM.
2. HERE IS ANOTHER SET OF CARDS WITH NUMERALS. Give the cards with numerals and sets to the child. MATCH THESE UP WITH THE OTHER CARDS. LET'S SAY THE NAMES AS YOU MATCH.

FOLLOW-UP: Let the child do this activity on his own. Encourage him to use the self-correcting materials also.

Figure 23–11 Cube blocks can be used informally to match numerals.

Most of the computer programs described in Unit 9 include numeral recognition. *Knowing Numbers* by Learning Well and *Sequence*, part of the *Kindercomp* software from Spinnaker, also provide experiences with numeral recognition. *Sequence* takes more advanced students up into two- and three-digit numerals.

EVALUATION

The teacher may question each child as he works with the self-correcting materials. She should note which numerals he can name and if he names them in sequence. She may interview him individually using the assessment questions in the Appendix.

SUMMARY

Numerals are the symbols used to represent amounts. The name of each numeral must be learned. The sequence, or order, must also be learned. The child needs to understand that each numeral represents a set that is one larger than the one before (and one less than the one that comes next).

Most children learn the properties and purposes of numerals through naturalistic and informal experiences. There are many excellent self-correcting materials which can be bought or made for informal activities. Any structured activities should be short in length.

Figure 23–12 The flannel board numerals can be moved until they are in the correct sequence.

FURTHER READING AND RESOURCES

INFORMATION AND ACTIVITIES

Adler, D. A. 1981. *Calculator fun.* New York: Franklin Watts.

Baratta-Lorton, M. 1972. *Workjobs.* Menlo Park, Calif.: Addison-Wesley.

Burton, G. 1985. *Towards a good beginning.* Menlo Park, Calif.: Addison-Wesley.

Englehardt, J. M., Ashlock, R. B., and Wiebe, J. H. 1984. *Helping children understand and use numerals.* Boston: Allyn & Bacon.

Scott, L. B., and Garner, J. 1978. *Mathematical experiences for young children.* New York, N.Y.: McGraw-Hill.

MATERIALS TO MAKE/GAMES TO PLAY

Carson, P., and Dellosa, J. 1979. *Holiday learning activity ideas.* Akron, Ohio: Carson-Dellosa Publishing.

Carson, P., and Dellosa, J. 1977. *All aboard for readiness skills.* Akron, Ohio: Carson-Dellosa Publishing.

Dellosa, J., and Carson, P. 1981. *Buzzing into readiness.* Akron, Ohio: Carson-Dellosa Publishing.

Dellosa, J., and Carson, P. 1980. *Fluttery readiness.* Akron, Ohio: Carson-Dellosa Publishing.

Schutte, B. 1978. *Readiness gameboards.* Palos Verdes Peninsula, Calif.: Frank Schaffer.

Young, S. 1988. *Math with calculators: Resources for teachers.* Menlo Park, Calif.: Addison-Wesley.

SUGGESTED ACTIVITIES

1. Observe children to see how they use numerals. Share your experience with your class.
2. Talk with preschool and kindergarten teachers about how they teach structured number symbol skills. Share with your class what you learned.
3. Add number symbol skill activities to your Resource File.
4. Make a number symbol skill activity material and share it with the class.
5. Go to a toy store and examine the materials available that are used to teach number symbols.
6. Do a structured number symbol activity, first with one child and, then, with a small group of children.
7. Use the evaluation system described in Activity 8, Unit 9, to review and evaluate the following computer software:

   Knowing Numbers. Learning Well. Roslyn Heights, N.Y.

   Sequence from *Kindercomp.* Spinnaker Software. Cambridge, Mass.: Apple, IBM PC, PCJr, Atari, C64.

REVIEW

A. Answer each of the following.

1. What are numerals?
2. What are the six number symbol skills?
3. What is meant by self-correcting manipulative materials?
4. What are the four basic types of self-correcting manipulative materials used to teach number symbol skills?
5. What is the teacher's role when using self-correcting manipulative materials?

B. Below are listed some numeral materials. Tell what they could help a child learn in the numbers skill area.

1. Magnet board 0-5 numerals, various magnetic animals
2. Matchmates® (match plaques)
3. A set of eleven cards containing the number symbols 0 to 10
4. Lotto® numeral game
5. Number Worm®
6. Pegs-in-Numerals®
7. Unifix 1-10 Stair®
8. Walk on Line®
9. Number-Ite®
10. 1-2-3 Puzzle®
11. Hand calculator
12. Personal computer

UNIT 24 Sets and Symbols

OBJECTIVES

After studying this unit, the student should be able to:

- Describe the three higher-level tasks that children do with sets and symbols
- Set up an environment that provides for naturalistic and informal sets and symbols activities
- Plan and do structured sets and symbols activities with young children
- Assess and evaluate a child's ability to use math sets and symbols

The activities in this unit build on many of the ideas and skills presented in earlier units: matching, number and counting, sets and classifying, comparing, ordering, and symbols.

The experiences in this unit will be most meaningful to the child who can

- Match things one-to-one and match sets of things one-to-one
- Recognize sets of one to four without counting and count sets up to at least ten things without a mistake
- Divide large sets into smaller sets and compare sets of different amounts
- Place sets containing different amounts in order from least to most
- Name each of the numerals from zero to ten
- Recognize each of the numerals from zero to ten
- Be able to place each of the numerals in order from zero to ten
- Understand that each numeral stands for a certain number of things
- Understand that each numeral stands for a set of things one more than the numeral before it and one less than the numeral after it

When the child has reached the objectives in the preceding list, he can then learn

Figure 24–1 As mother watches, Lisa puts the car with two red dots on the hood in the garage with the numeral "2" on the roof.

- To match a symbol to a set. That is, if he is given a set of four items, he can pick out or write the numeral *4* as the one that goes with that set.
- To match a set to a symbol. That is, if he is given the numeral *4*, he can make or pick out a set of four things to go with it.
- To reproduce symbols. That is, he can learn to write the numerals.

Moving from working with sets alone to sets and symbols and finally to symbols alone must be done carefully and sequentially. In *Workjobs II*, Mary Baratta-Lorton describes three levels of increasing abstraction and increasing use of symbols: the concept level, connecting level, and symbolic level. These three levels can be pictured as follows:

Concept Level	0 0 0 0
Connecting Level	0 0 0 0 4
Symbolic Level	4

Units 9, 10, and 11 worked at the concept level. In Unit 23 the connecting level was introduced informally. The major focus of this Unit is the connecting level. In Unit 25 the Symbolic Level will be introduced.

ASSESSMENT

If the children can do the assessment tasks in Units 8 through 14, 17, and 23, then they have the basic skills and knowledge necessary to connect sets and symbols. In fact, they may be observed doing some symbol and set activities on their own if materials are made available for exploration in the math center. The following are some individual interview tasks:

SAMPLE ASSESSMENT TASK

Preoperational/Concrete Ages 5–7

METHOD: Interview

SKILL: Child will be able to match symbols to sets using numerals from zero to ten and sets of amounts zero to ten.

MATERIALS: 5″ × 8″ cards with numerals zero to ten, ten objects (e.g., chips, cube blocks, buttons)

PROCEDURE: Lay out the cards in front of the child in numerical order. One at a time show the child sets of each amount in this order: 2, 5, 3, 1, 4. PICK OUT THE NUMERAL THAT TELLS HOW MANY THINGS ARE IN THIS SET. If the child does these correctly, go on to 7, 9, 6, 10, 8, 0 using the same procedure.

EVALUATION: Note which sets and symbols the child can match. The responses will indicate where instruction can begin.

INSTRUCTIONAL RESOURCE(S):

Charlesworth, R., and Lind, K. 1990. *Math and Science for Young Children*. Albany, N.Y.: Delmar.

SAMPLE ASSESSMENT TASK

Preoperational/Concrete Ages 5–7

METHOD: Interview

SKILL: Child will be able to match sets to symbols using sets of amounts zero to ten and numerals from zero to ten.

MATERIALS: 5″ × 8″ cards with numerals zero to ten, 60 objects (e.g., chips, cube blocks, coins, buttons)

PROCEDURE: Lay out the numeral cards in front of the child in a random arrangement. Place the container of objects within easy reach. MAKE A SET FOR EACH NUMERAL. Let the child decide how to organize the materials.

EVALUATION: Note for which numerals the child is able to make sets. Note how the child goes about the task. For example, does he sequence the numerals from zero to ten? Does he place the objects in an organized pattern by each numeral? Can he recognize some amounts without counting? When he counts does he do it carefully? His responses will indicate where instruction should begin.

INSTRUCTIONAL RESOURCE(S):

Charlesworth, R., and Lind, K. 1990. *Math and Science for Young Children*. Albany, N.Y.: Delmar.

SAMPLE ASSESSMENT TASK

Preoperational/Concrete Ages 5–7

METHOD: Interview

SKILL: Child can reproduce (write) numerals from zero to 10

MATERIALS: Pencil, pen, black marker, black crayon, white paper, numeral cards, from zero to ten

PROCEDURE: HERE IS A PIECE OF PAPER. PICK OUT ONE OF THESE (point to writing tools) THAT YOU WOULD LIKE TO USE. NOW, WRITE AS MANY NUMBERS AS YOU CAN. If the child is unable to write from memory, show him the numeral cards. COPY ANY OF THESE THAT YOU CAN.

EVALUATION: Note how many numerals the child can write and if they are in sequence. If the child is not able to write the numerals with ease, this indicates that responding to problems by writing is not at this time an appropriate response. Have him do activities in which movable numerals or markers can be placed on the correct answers.

INSTRUCTIONAL RESOURCE(S):

Charlesworth, R., and Lind, K. 1990. *Math and Science for Young Children*. Albany, N.Y.: Delmar.

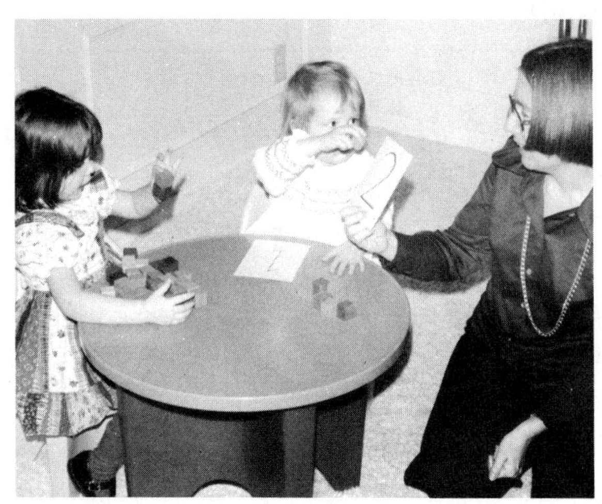

Figure 24–2 "Find this many blocks."

NATURALISTIC ACTIVITIES

As the child learns that sets and symbols go together, this will be reflected in his daily play activities.

- Mary and Dean have set up a grocery store. Dean has made price tags, and Mary has made play money from construction paper. They have written numerals on each price tag and piece of money. Sam comes up and picks out a box of breakfast cereal and a carton of milk. Dean takes the tags, "That will be four dollars." Sam counts out four play dollar bills. Dean takes a piece of paper from a note pad and writes a "receipt." "Here, Sam."
- Brent has drawn a picture of a birthday cake. There are six candles on the cake and a big numeral "6." This is for my next birthday. I will be six."
- The flannel board and a set of primary cutouts have been left out in a quiet corner. George sits deep in thought as he places the numerals in order and counts out a set of cutouts to go with each numeral.

Each child uses what he has already learned in ways that he has seen adults use these skills and concepts.

INFORMAL ACTIVITIES

The child can work with sets and numerals best through informal experiences. Each child needs a different amount of practice. By making available many materials that the child can work with on his own, the teacher can help each child have the amount of practice he needs. Each child can choose to use the set of materials that he finds the most interesting.

Workjobs and *Workjobs II* are excellent resources for sets and symbols activities and materials. The basic activities for matching symbols to sets and sets to symbols involve the use of the following kinds of materials:

1. Materials where the numerals are 'fixed' and counters are available for making

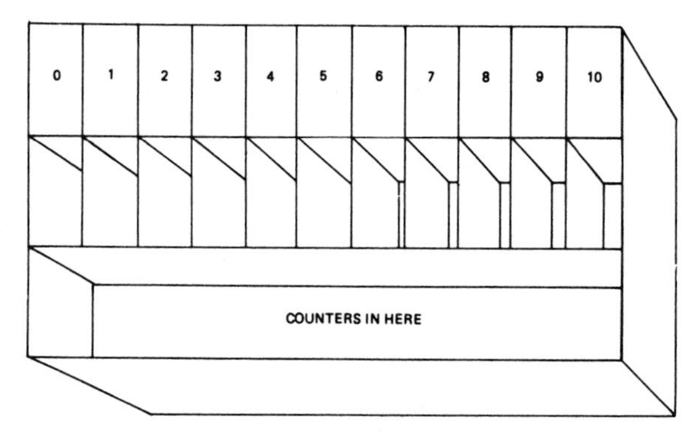

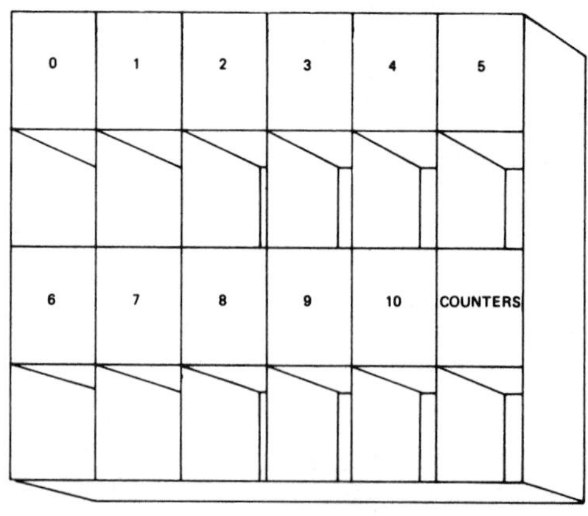

the sets. These are called counting trays and may be made or purchased. They may be set up with the numerals all in one row or in two or more rows.

2. Materials where there are movable containers on which the numerals are written and counters of some kind. There might be pennies and banks, cups and buttons, cans and sticks, and similar items.
3. Individual numeral cards with a space for the child to make a set to match.
4. Sets of real things or pictures of things which must be matched to numerals written on cards.

Each child can be shown each new set of materials. He can then have a turn to work with them. If the teacher finds that a child has a hard time, she can give him some help and make sure he takes part in some structured activities.

Informal experiences in which the child writes numerals come up when the child asks how to write his age, phone number, or address. Some children who are interested in writing may copy numerals that they see in the environment—on the clock and calendar or on the numeral cards used in matching and set-making activities. These children should be encouraged and helped if needed. The teacher can make or buy a set of sandpaper numerals. The child can trace these with his finger to get the feel of the shape and the movement needed to make the numeral. Formal writing lessons should not take place until the child's fine muscle coordination is well developed. For some children this might not be until they are seven or eight years of age.

STRUCTURED ACTIVITIES

Structured activities with symbols and sets for the young child are done in the form of games. One type of game has the child match sets and numerals using a theme such as "Letters to the Post Office" or "Fish in the Fishbowl." A second is the basic "board" game. A third type of game is the Lotto or Bingo type. In each case, the teacher structures the game. However, once the children know the rules, two or more can play on their own. One example of each game is described. With a little imagination, the teacher can think of variations. The last three activities are for the child who can write numerals.

SETS AND SYMBOLS: FISH IN THE FISHBOWL

OBJECTIVE: To match sets and symbols for the numerals zero through ten

MATERIALS: Sketch eleven fishbowls about 7″ × 10″ on separate pieces of cardboard or posterboard. On each bowl write one of the numerals from zero to ten. Cut out eleven fish, one for each bowl. On each fish, put dots—from zero on the first to ten on the last.

ACTIVITY: Play with two or more children. Line up the fishbowls (on a chalk tray is a good place). One at a time, have each child choose a fish, sight unseen. Have him match his fish to its own bowl.

FOLLOW-UP:

1. Make fish with other kinds of sets such as stripes or stars.
2. Line up the fish and have the children match the fishbowls to the right fish.

Figure 24–3 The girls put the fish with the right number of dots or stripes in each fishbowl.

SETS AND SYMBOLS: BASIC BOARD GAMES

OBJECTIVE: To match sets and symbols

MATERIALS: The basic materials can be purchased or can be made by the teacher. Basic materials would include

- A piece of posterboard (18″ × 36″) for the game board
- Clear Contac® or laminating material
- Marking pens
- Spinner cards, plain 3″ × 5″ file cards, or a di
- Place markers (chips, buttons, or other counters)

Materials for three basic games are shown: The game boards can be set up with a theme for interest such as the race car game. Themes might be *Going to School*, *The Road to Happy Land*, or whatever the teacher's imagination can think of.

ACTIVITY: The basic activity is the same for each game. Each child picks a marker and puts it on start. Then each in turn spins the spinner (or chooses a card or rolls the di) and moves to the square that matches.

FOLLOW-UP:

1. The children can learn to play the games on their own.
2. Make new games with new themes. Make games with more moves and using more numerals and larger sets to match.

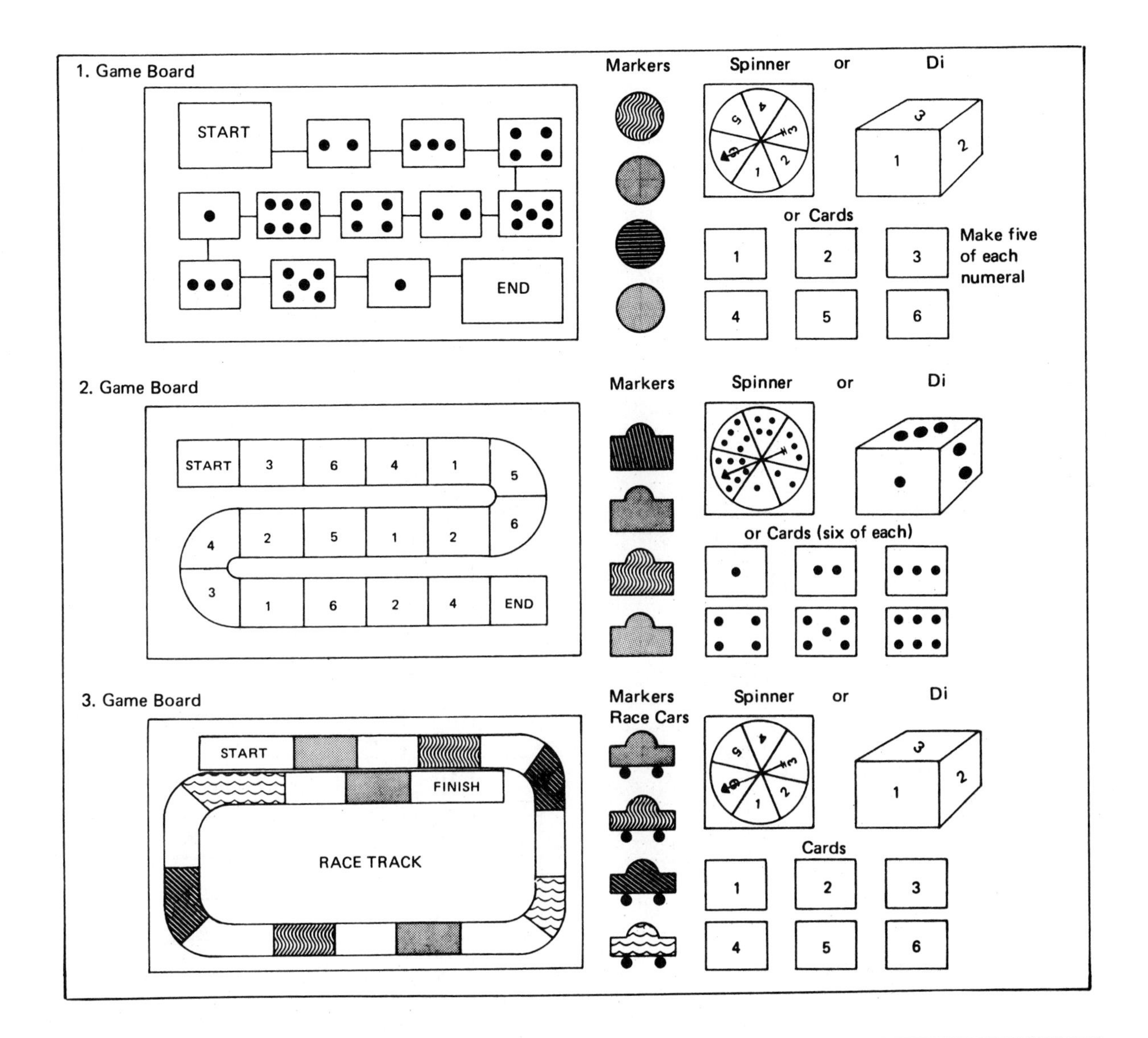

SETS AND SYMBOLS: LOTTO AND BINGO GAMES

OBJECTIVE: To match sets and symbols

MATERIALS: For both games, there should be six basic game cards each with six or more squares (the more squares, the longer and harder the game). For Lotto, there is one card to match each square. For Bingo, there must be markers to put on the squares, also. For Bingo, squares on the basic game cards are repeated; for Lotto, they are not.

ACTIVITIES:

1. Lotto Game
   Each child receives a basic game card. The matching cards are shuffled and held up one at a time. The child must call out if the card has his mark on it (dot, circle, triangle) and then match the numeral to the right set. The game can be played until one person fills his card or until everyone does.
2. Bingo Game
   Each child receives a basic game card. He also receives nine chips. The matching set cards are shuffled. They are held up one at a time. The child puts a chip on the numeral that goes with the set on the card. When someone gets a row full in any direction, the game starts again.

FOLLOW-UP: More games can be made using different picture sets and adding more squares to the basic game cards. Bingo cards must always have the same odd number of squares up and down and across (3×3, 5×5, 7×7).

1. Lotto Game

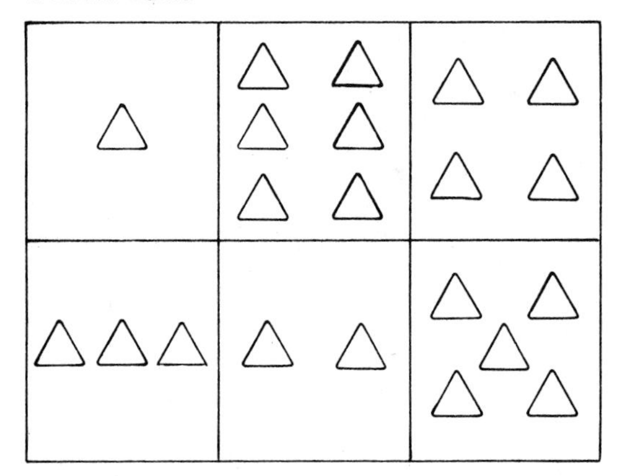

Make six cards for six players

Matching cards:
Make a set for each big card.

1	2	3	4	5	6
△	△	△	△	△	△

2. Bingo Game

1	2	3
8	0	4
7	6	5

2	5	3
7	0	9
4	6	1

6	1	5
9	0	4
7	10	8

10	2	6
3	0	4
8	1	7

10	6	1
9	0	7
8	3	2

7	3	9
2	0	8
1	6	5

4	5	9
8	0	1
10	3	2

10	3	8
1	0	2
6	4	5

Markers: ● ● ● ,54 chips

Matching cards:
11 cards with sets zero to 10:

[ ] [●] and so on.

SETS AND SYMBOLS: MY OWN NUMBER BOOK

OBJECTIVE: To match sets and symbols

MATERIALS: Booklets made with construction paper covers and several pages made from newsprint or other plain paper, hole puncher and yarn or brads to hold book together, crayons, glue, scissors, and more paper or stickers

ACTIVITY: The child writes or asks the teacher to write a numeral on each page of the book. The child then puts a set on each page. Sets can be made using

a. Any kind of stickers with gummed backs which can be wet and stuck on.
b. Cutouts made by the child.
c. Drawings done by the child.

FOLLOW-UP: Have the children show their books to each other and then take the books home. Also read to the children some of the number books listed in Unit 21.

SETS AND SYMBOLS: WRITING NUMERALS TO MATCH SETS

OBJECTIVE: To write the numeral that goes with a set

MATERIALS: Objects and pictures of objects, chalk and chalkboard, crayons, pencils, and paper

ACTIVITY: Show the child some objects or pictures of objects. WRITE THE NUMERAL THAT TELLS HOW MANY ______ THERE ARE. The child then writes the numeral on the chalkboard or on a piece of paper.

FOLLOW-UP: Get some clear acetate. Make some set pictures that can be placed in acetate folders for the child to use on his own. Acetate folders are made by taking a piece of cardboard and taping a piece of acetate of the same size on with some plastic tape. The child can write on the acetate with a crayon and then erase his mark with a Kleenex or a soft cloth.

Counting books are another resource for connecting sets and symbols. In most counting books the numerals are included with each set to be counted. Caution must be taken in selecting counting books. Ballenger, Benham, and Hostricka (1984) suggest the following criteria for selecting counting books:

1. Be sure that the numerals always refer to how many and not to ordinal position or sequence.

Figure 24–4 Dean learns to write numerals.

2. The numeral names should also always refer to how many.
3. The narrative on the page should clearly identify the set of objects that the numeral is associated with.
4. The illustrations of objects to be counted and connected to the numeral on each page should be clear and distinct.
5. When ordinals are being used, the starting position (e.g., first) should be clearly identified.
6. When identifying ordinal positions, the correct terms should be used (e.g., first, second, third).
7. When numerals are used to indicate ordinal position, they should be written as 1st, 2nd, 3rd, and so on.
8. The numerals should be uniform in size (not small numerals for small sets and larger numerals for larger sets).
9. The book should emphasize the concept of one-to-one correspondence.
10. When amounts above ten and their associated numerals are illustrated, the amounts should be depicted as a group of ten plus the additional items.

Books highly recommended by Ballenger et al. are:

> Davis, B. S. 1972. *Forest hotel-A counting story*. Racine, Wis.: Western Publishing.
> Gretz, S. 1969. *Teddy bears 1 to 10*. Chicago: Follett.

Most of the books they reviewed had major faults, so take care when selecting books that connect sets and symbols.

Computers and calculators can also be used for helping children acquire the set and symbol connection. Most of the software listed in Units 9 and 23 as supportive of counting and symbol recognition also connects sets and symbols. These programs should also be evaluated with the same criteria suggested for books. Students could play games with their calculators such as closing their eyes, pressing a key, identifying the numeral, and then selecting or constructing a set that goes with the numeral. There are also many self-correcting computer/calculator-type toys available that children enjoy using. Some examples are *Learning-Window®*, *Videosmarts®*, and *Talkboard®*.

EVALUATION

With young children most of the evaluation can be done by observing their use of the materials for informal activities. The adult can also notice how the children do when they play the structured games.

For children about to enter first grade, an individual interview should be done using the assessment interviews in the unit and in the Appendix.

SUMMARY

When the child works with sets and symbols, he puts together the skills and ideas learned earlier. He must match, count, classify, compare, order, and associate written numerals with sets.

He learns to match sets to symbols and symbols to sets. He also learns to write each number symbol. The child uses mostly materials which can be used informally on his own. He can also learn from more structured game kinds of activities, number books, computer games, and calculator activities.

FURTHER READING AND RESOURCES

Ballenger, M., Benham, N., and Hosticka, A. 1984. Children's counting books *Childhood Education*. September/October: pp. 30–35.

Baratta-Lorton, M. 1972. *Workjobs*. Menlo Park, Calif.: Addison-Wesley.

Baratta-Lorton, M. 1979. *Workjobs II*. Menlo Park, Calif.: Addison-Wesley.

Burton, G. M. 1985. *Towards a good beginning*. Menlo Park, Calif.: Addison-Wesley.

Engelhardt, J. M., Ashlock, R. B., and Wiebe, J. H. 1984. *Helping children understand and use numerals*. Boston: Allyn and Bacon.

Hiebert, J. 1989. The struggle to link written symbols with understandings: An update. *Arithmetic Teacher*. 36 (7): 38–44.

Richardson, K. 1984. *Developing number concepts using Unifix Cubes*. Menlo Park, Calif.: Addison-Wesley.

Scott, L. B., and Garner, J. 1978. *Mathematical experiences for young children*. St. Louis, Mo.: McGraw-Hill.

Also see the games and materials resources suggested in Unit 23.

SUGGESTED ACTIVITIES

1. Observe children during play. Look for how they match symbols to sets, sets to symbols, or reproduce symbols.
2. Add assessment activities in the area of sets and symbols to the Assessment File.
3. Make one instructional material which can be used by children during play to help them match symbols to sets or sets to symbols. Share your material with the class. If possible, donate it to a preschool or kindergarten and check to find out if and how it is used.
4. Design a structured sets and symbols activity for each of the following:

   a. a game which matches sets and numerals

   b. a basic board game

   c. a Lotto or Bingo game.

   Use one of these with a child or a small group of children. Share your experience with the class.
5. Examine catalogs to see what kinds of materials are available that can be used to help children learn sets and symbols.
6. Review one or more computer programs listed in Units 9 and 23 using the procedure suggested in Activity 8, Unit 9 and the criteria suggested for children's counting books in this unit.
7. Go to a bookstore, the children's literature library, and/or your local public library. Identify three or more counting books and evaluate them using the criteria suggested by Ballenger et al.

REVIEW

A. Indicate whether each of the following statements describes a skill the child must have before he goes on to the activities discussed in this unit or a higher level activity.

1. Reproduce numerals
2. Understand that each numeral stands for a certain number of things
3. Match a set to a symbol
4. Match things one-to-one and match sets one-to-one
5. Recognize sets of sizes one to four
6. Count sets of at least ten things without a mistake
7. Recognize each of the numerals from zero to ten
8. Name each of the numerals from zero to ten
9. Understand that each numeral stands for a set of things one more than the numeral before it and one less than the numeral after it
10. Match a symbol to a set
11. Place sets containing different amounts in order from least to most
12. Divide large sets into smaller sets of different amounts

B. Match each activity in Column II with the correct item in Column I.

I	II
1. To match a symbol to a set	a. writes numerals
2. To reproduce symbols	b. picks three rabbits to go with numeral "3"
3. To match a set to a symbol	c. writes numeral "5" to go with five boats

C. Indicate **All** the correct answers to each numbered item.

1. Materials for matching symbols to sets and sets to symbols include
   a. Expensive components
   b. Materials where the numerals are fixed
   c. Materials which can be used to make sets
   d. Materials which include movable containers on which the numerals are written
2. Young children who are interested in writing numerals
   a. Should not be allowed to do so
   b. Should be advanced to first grade
   c. May copy numerals they see in the environment
   d. Can be given a set of sandpaper numerals
3. Structured set and symbol activities include
   a. The game Fish in the Fishbowl
   b. Lotto games
   c. Ten pins
   d. Dominoes
4. Materials to help children reproduce symbols include
   a. Sand
   b. Crayons
   c. Chalk
   d. Paper
5. Informal sets and symbols activities children do include
   a. Lemonade-selling stands
   b. Bean bag toss
   c. Put the car in the right garage

   d. Place the set card next to the numeral card

6. In selecting books and computer software to use in helping children acquire the sets and symbols connection be sure:
   a. The members of each set are clearly depicted and all the same size.
   b. The numerals for each set are uniform in size.
   c. Sets greater than ten are not divided into groups of ten plus the additional items.
   d. The numerals and numeral names clearly refer to how many.

UNIT 25 Higher-Level Activities and Concepts

OBJECTIVES

After studying this unit, the student should be able to:

- List the seven areas in which higher-level concept activities are described in the unit
- Describe the three higher levels of classification
- Plan higher level activities for children who are near the stage of concrete operations

  The experiences in this unit include
  - Further application of skills the child learns through the activities in units
  - Activities for the child who develops at a fast rate and can do the higher level assessment tasks with ease
  - Activities for the older child who needs concrete experiences and variety

  The specific areas of experiences are
  - Classification
  - Shape
  - Spatial relations
  - Concrete whole number operations
  - Graphs
  - Symbolic level activities
  - Quantities above ten

ASSESSMENT

The teacher looks at the child's level in each area. Then she makes a decision as to when to introduce these activities. When introduced to one child, any one activity could capture the interest of another child who might be at a lower developmental level. Therefore, it is not necessary to wait for all the children to be at the highest level to begin. Children at lower levels can participate in these activities as observers and as contributors. The higher level child can serve as a model for the lower level child. The lower level child might be able to do part of the task following the leadership of the higher level child. For example, if a floor plan of the classroom is being made, the more advanced child might design it while everyone draws a picture of a piece of furniture to put on the floor plan. The more advanced child might get help from the less advanced child when he makes a graph. The less advanced child can count and measure; the more advanced child records the results. Children can work in pairs to solve concrete addition, subtraction, multiplication, and division problems. They can move into higher levels of symbol use and work with numerals and quantities greater than ten. They can also work together exploring calculators and computer software.

CLASSIFICATION

The higher levels of classification are called *multiple classification*, *inclusion relations*, and *hierarchical classification*. Multiple classification requires the child to classify things in more than one way and to solve matrix problems. Figures 25–1 and 25–2, page 000, illustrate these two types of multiple classification. In Figure 25–1, the child is shown three shapes, each in three sizes and in three colors. He is asked to put the ones together that belong together. He is then asked to find another way to divide the shapes.

The preoperational child will not be able to do this. He centers on his first sort. Some games are suggested which will help the child move to concrete operations.

Matrix problems are illustrated in Figure 25–2. A simple two-by-two matrix is shown in the A part. In this case, size and number must both be considered in order to complete the matrix. The problem can be made more difficult by making the matrix larger (there are always the same number of squares in each row, both across

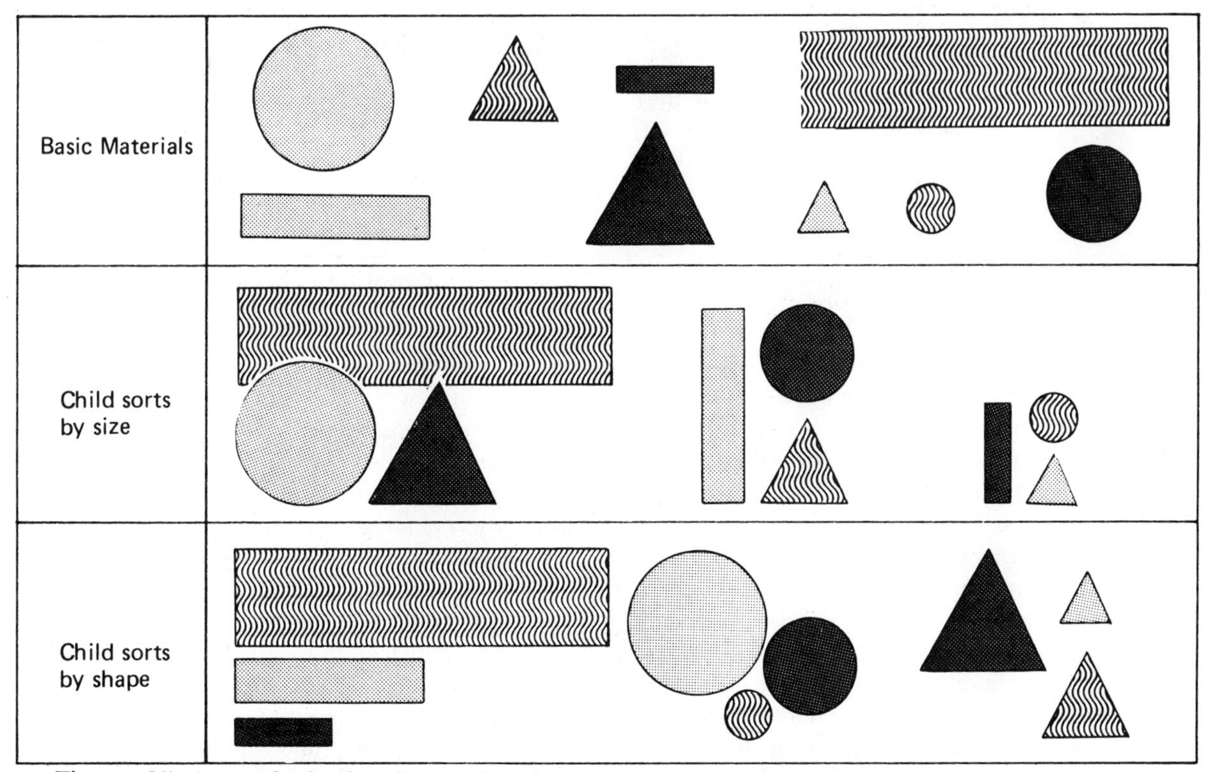

Figure 25–1 *Multiple classification* involves sorting one way and then sorting again using different criteria.

Figure 25–2 The matrix problem is another type of multiple classification.

and up and down.) The B part shows a four by four matrix. The easiest problem is to fill in part of a matrix. The hardest problem is to fill in a whole blank matrix as is illustrated in the C part.

The preoperational child cannot see that one class may be included within another (class inclusion). For example, the child is shown ten flowers: two roses and eight daisies. The child can divide the flowers into two groups: roses and daisies. He knows that they are all flowers. When asked if there are more flowers or more daisies, he will answer "More daisies." He is fooled by what he sees and centers on the greater number of daisies. He is not able to hold in his mind that daisies are also flowers. This problem is shown in Figure 25–3.

Hierarchical classification has to do with classes being within classes. For example, black kittens ⇒ kittens ⇒ house cats ⇒ cats ⇒ mammals. As can be seen in Figure 25–4, this forms a *hierarchy*, or a series of ever-larger classes. Basic level concepts are usually learned first. This level includes categories such as dogs, monkeys, cats, cows, and elephants as illustrated in Figure 25–4. Superordinate-level concepts such as mammals, furniture, vehicles, and so on are learned next. Finally, children learn subordinate categories such as domestic cats and wildcats, or types of chairs such as dining room, living room, rocking, kitchen, folding, and so on.

Another interesting aspect of young children's concept learning is their view of which

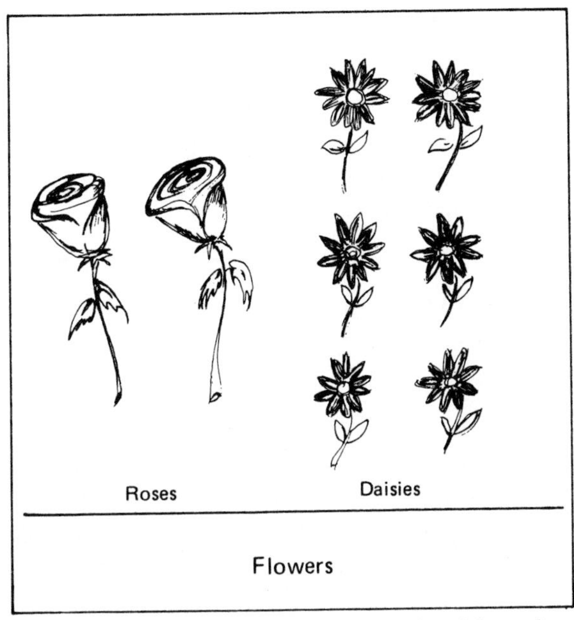

Figure 25–3 *Class inclusion* is the idea that one class can be included in another.

characteristics the members of a class have in common. Although preoperational level children tend to be perceptually bound when they attempt to solve many types of conceptual problems, they are able to classify on category membership when shown things that are perceptually similar. For example, four-year-olds were shown pictures of a blackbird, a bat, which looked much like the blackbird, and a flamingo. They were told that the flamingo gave its baby mashed-up food and the bat gave its baby milk. When asked what the blackbird fed its baby, they responded that it gave its baby mashed-up food. In this case the children looked beyond the most obvious physical attributes.

Another type of characteristic that is interesting to ask young children about is their view of what is inside members of a class. When young children are asked if members of a class all have the same "stuff" inside, preschoolers tend to say that yes, they have. That is, all dogs,

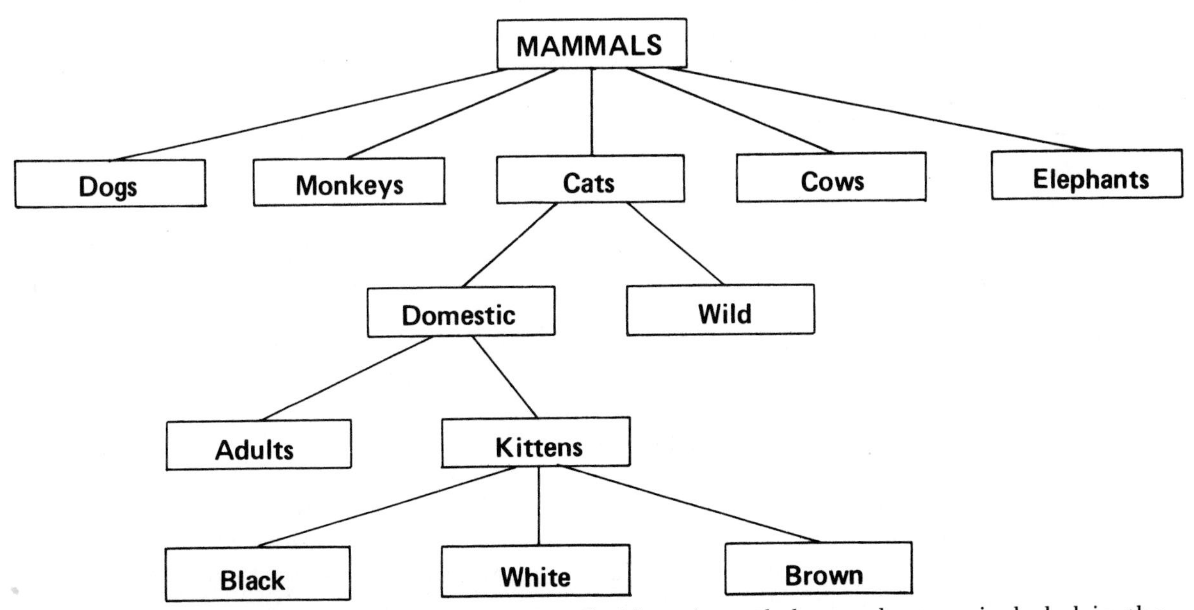

Figure 25–4 In a *hierarchical classification,* all things in each lower class are included in the next higher class.

people, chairs, and dolls are the same inside. Children are aware of more than just observable similarities. By second grade, they can discriminate between natural and synthetic items. That is, they realize that living things such as dogs, people, or apples are, for the most part, the same inside as other dogs, people, or apples, although the insides of different types of chairs, dolls, or other manufactured items are not necessarily the same. For the younger children, category membership overwhelms other factors.

The following activities will help the transitional child (usually age five to seven) to enter concrete operations.

HIGHER-LEVEL CLASSIFICATION: MULTIPLE CLASSIFICATION, RECLASSIFY

OBJECTIVE: To help the transitional child learn that groups of objects or pictures can sometimes be sorted in more than one way

MATERIALS: Any group of objects or pictures of objects which can be classified in more than one way: for example, pictures or cardboard cutouts of dogs of different colors (brown and white), sizes (large and small), and hair lengths (long and short)

ACTIVITY: Place the dogs in front of the child. WHICH DOGS BELONG TOGETHER or ARE THEY THE SAME? Note whether he groups by size, color, or hair length. NOW, WHAT IS ANOTHER WAY TO PUT THEM IN SETS?. . . .CAN THEY BE PUT LIKE (name another way)? Put them in one pile again if the child is puzzled. OKAY, NOW TRY TO SORT THE ____ FROM THE ____. Repeat using different criteria each time.

FOLLOW-UP: Make other sets of materials. Set them up in boxes where the child can get them out and use them during free play time. Make some felt pieces to use on the flannel board.

HIGHER-LEVEL CLASSIFICATION: MULTIPLE CLASSIFICATION, MATRICES

OBJECTIVE: To help the transitional child see that things may be related on more than one criteria

MATERIALS: Purchase or make a matrix game. Start with a two-by-two matrix and gradually increase the size (three-by-three, four-by-four, and so on). Use any of the criteria from Unit 7 such as color, size, shape, material, pattern, texture, function, association, class name, common feature, or number. Make a game board from posterboard or wood. Draw or paint permanent lines. Use a flannel board, make the lines for the matrix with lengths of yarn. An example of a three-by-three board is shown. Start with three-dimensional materials, then cutouts, then cards.

ACTIVITIES: Start with the matrix filled except for one space and ask the child to choose from two items the one that goes in the empty space. WHICH ONE OF THESE GOES

HERE? After the item is placed: WHY DOES IT BELONG THERE? Once the child understands the task, more spaces can be left empty until it is left for the child to fill in the whole matrix.

FOLLOW-UP: Add more games which use different categories and larger matrices.

Items to be placed:

1. Color and size: red, green and yellow apples:

2. Position and class: dog, cat, bird; rightside-up, upside-down and sideways.

3. Number and shape: triangles, squares, circles; rows of one, two, and three.

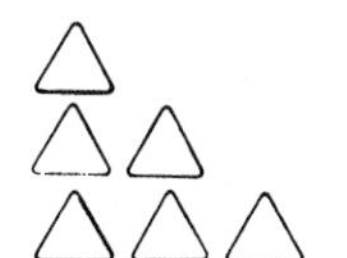
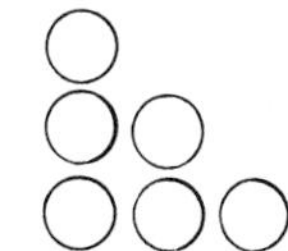
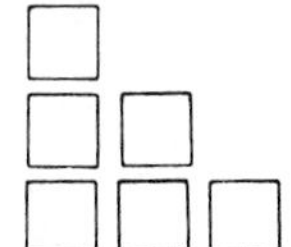

HIGHER-LEVEL CLASSIFICATION: CLASS INCLUSION

OBJECTIVES: To help the transitional child see that a smaller set may be included within a larger set

MATERIALS: Seven animals. Two kinds should be included (such as horses and cows, pigs and chickens, dogs and cats). There should be four of one animal and three of the other. These can be cutouts or toy animals.

ACTIVITY: Place the animals within an enclosure (a yarn circle or a fence made of blocks). WHO IS INSIDE THE FENCE? Children will answer "horses," "cows," "animals." SHOW ME WHICH ONES ARE HORSES (COWS, ANIMALS). ARE THERE MORE HORSES OR MORE ANIMALS? . . . HOW DO YOU KNOW? . . . LET'S CHECK (use one-to-one correspondence).

FOLLOW-UP: Play the same game. Use other categories such as plants, types of material, size, and so on. Increase the size of the sets.

HIGHER-LEVEL CLASSIFICATION: HIERARCHICAL

OBJECTIVE: To help the transitional child see that each thing may be part of a larger category (or set of things)

MATERIALS: Make some sets of sorting cards. Glue pictures from catalogs and/or workbooks onto file cards or posterboard. The following are some that can be used:

1. one black cat, several house cats, cats of other colors, one tiger, one lion, one panther, one bobcat, one dog, one horse, one cow, one squirrel, one bear
2. one duck, three swans, five other birds, five other animals
3. one teaspoon, two soup spoons, a serving spoon, two baby spoons, three forks, two knives

ACTIVITIES: Place the cards where they can all be seen. Give the following instructions:

1. FIND ALL THE ANIMALS. . . .FIND ALL THE CATS. . . .FIND ALL THE HOUSE CATS . . . .FIND ALL THE BLACK CATS. Mix up the cards and lay them out again. PUT THEM IN SETS THE WAY YOU THINK THEY SHOULD BE. When the child is done, WHY DID YOU PUT THEM THAT WAY? Mix them up and lay them out. IF ALL THE ANIMALS WERE HUNGRY, WOULD THE BLACK CAT BE HUNGRY? IF THE BLACK CAT IS HUNGRY ARE ALL THE ANIMALS HUNGRY?
2. FIND ALL THE ANIMALS. . . .FIND ALL THE BIRDS. . . . FIND THE WATER BIRDS . . . . FIND THE DUCK. Mix up the cards and lay them out again. PUT THEM IN SETS THE WAY YOU THINK THEY SHOULD BE. When the child is done, WHY DO THEY BELONG THAT WAY? Mix them up and lay them out again. IF ALL THE BIRDS WERE COLD WOULD THE DUCK BE COLD? IF THE DUCK WAS COLD WOULD ALL THE WATER BIRDS BE COLD? IF ALL THE ANIMALS WERE COLD WOULD THE WATER BIRDS BE COLD?
3. FIND ALL THE THINGS THAT WE EAT WITH. . .FIND ALL THE KNIVES. . . .FIND ALL THE FORKS. . . .FIND ALL THE SPOONS. Mix them up and lay them out again. PUT THEM IN SETS THE WAY YOU THINK THEY BELONG. When the child is done, WHY DO THEY BELONG THAT WAY? Mix them up and lay them out again. IF ALL THE SPOONS ARE DIRTY WOULD THE TEASPOON BE DIRTY?. . .IF ALL THE THINGS WE EAT WITH WERE DIRTY WOULD THE BIG SPOON BE DIRTY?. . .IF THE TEASPOON IS DIRTY, ARE ALL THE OTHER THINGS WE EAT WITH DIRTY TOO?

FOLLOW-UP: Make up other hierarchies. Leave the card sets out for the children to sort during play. Ask them some of the same kinds of questions informally.

HIGHER-LEVEL CLASSIFICATION: WHAT'S INSIDE?

OBJECTIVE: To help the transitional child understand that natural kinds of categories have more internal common characteristics than manufactured items

MATERIALS: The children, some X-ray plates and/or a skeleton, pictures of the insides of the human body, several different types of old dolls

ACTIVITY: Ask, WHAT KIND OF STUFF IS INSIDE OF PEOPLE? DO ALL PEOPLE HAVE THE SAME STUFF INSIDE? Encourage the children to brainstorm. Write their contributions on chart paper. Show them the X-ray plates and/or the skeleton. Ask, WHAT IS THIS? If it's the X-ray plate ask, WHAT DOES IT SHOW? Help them if they don't know. Say, FEEL YOUR BODY. WHERE CAN YOU FEEL BONES? CAN YOU TELL WHERE YOU HAVE SOME VEINS? Compare with picture of the inside of the human body. Have them find on a friend and on themselves where they think different organs are.
Next show them the dolls. Ask, DO ALL THESE DOLLS HAVE THE SAME STUFF INSIDE? WHAT IS INSIDE THESE DOLLS? Write their responses on chart paper. Let them feel the dolls. Write down any further ideas that they have. Look inside the dolls. Let them confirm or disconfirm their guesses.

FOLLOW-UP: Do the same activity with other groups of natural and synthetic objects.

SHAPE

Once the child can match, sort, and name shapes he can also reproduce shapes. This can be done informally. The following are some materials that can be used.

Geoboards can be purchased or made. A *geoboard* is a square board with nails (heads up) or pegs sticking up at equal intervals. The child is given a supply of rubber bands and can experiment in making shapes by stretching the rubber bands between the nails.

A container of pipe cleaners or straws can be put out. The children can be asked to make as many different shapes as they can. These can be glued onto construction paper. Strips of paper, toothpicks, string, and yarn can also be used to make shapes.

SPATIAL RELATIONS

Children can learn more about space after playing the treasure hunt game described in Unit 13 by reproducing the space around them as a floor plan or a map. Start with the classroom for the first map. Then move to the whole building, the neighborhood, and the town or city. Be sure the children have maps among their dramatic play props.

HIGHER-LEVEL ACTIVITIES: SPATIAL RELATIONS, FLOOR PLANS

OBJECTIVE: To relate position in space to symbols of position in space

MATERIALS: Large piece of posterboard or heavy paper; markers, pens, construction paper, glue, crayons, scissors; some simple sample floor plans

ACTIVITY:

1. Show the children some floor plans. WHAT ARE THESE? WHAT ARE THEY FOR?. . .IF WE MAKE A FLOOR PLAN OF OUR ROOM WHAT WOULD WE PUT ON IT? Make a list.

2. Show the children a large piece of posterboard or heavy paper. WE CAN MAKE A PLAN OF OUR ROOM ON HERE. EACH OF YOU CAN MAKE SOMETHING THAT IS IN THE ROOM. JUST LIKE ON OUR LIST. THEN YOU CAN GLUE IT IN THE RIGHT PLACE. I'VE MARKED IN THE DOORS AND WINDOWS FOR YOU. As each child draws and cuts out an item (a table, shelf, sink, chair) have him show you where it belongs on the plan and glue it on.

FOLLOW-UP: After the plan is done, it should be left up on the wall so the children can look at it and talk about it. They can also add more things to the plan. The same procedure can later be used to make a plan of the building. Teacher and children should walk around the whole place. They should talk about which rooms are next to each other, which rooms are across from each other. Sticks or straws can be used to lay out the plan.

HIGHER-LEVEL ACTIVITIES: SPATIAL RELATIONS, MAPS

OBJECTIVE: To relate position in space to symbols of position in space

MATERIALS: Map of the city, large piece of posterboard or heavy paper, marking pens, construction paper, glue, crayons, scissors

ACTIVITY: Show the children the map of the city (or county in a rural area). Explain that this is a picture of where the streets would be if the children were looking down from a plane or a helicopter. Mark each child's home on the map with a small label with his name. Mark where the school is. Talk about who lives closest and who lives farthest away. Each child's address can be printed on a card and reviewed with him each day. The teacher can help mark out the streets and roads. The children can then cut out and glue down strips of black paper for the streets (and/or roads). Each child can draw a picture of his home and glue it on the map. The map can be kept up on the wall for children to look at and talk about. As field trips are taken during the year, each place can be added to the map.

FOLLOW-UP: Encourage the children to look at and talk about the map. Help them add new points of interest. Help children who would like to make their own maps. Bring in maps of the state, the country, and the world. Try to purchase United States and world map puzzles.

GRAPHS

The fourth level of graphs introduces the use of squared paper. The child may graph the same kind of things as discussed in Unit 20. He will now use squared paper with squares that can be colored in. These should be introduced only after the child has had many experiences of the kinds described in Unit 20. The squares should be large. A completed graph might look like the one shown in Figure 25–5.

CONCRETE WHOLE NUMBER OPERATIONS

Once children have a basic understanding of one-to-one correspondence, number and counting, and comparing, they can sharpen their

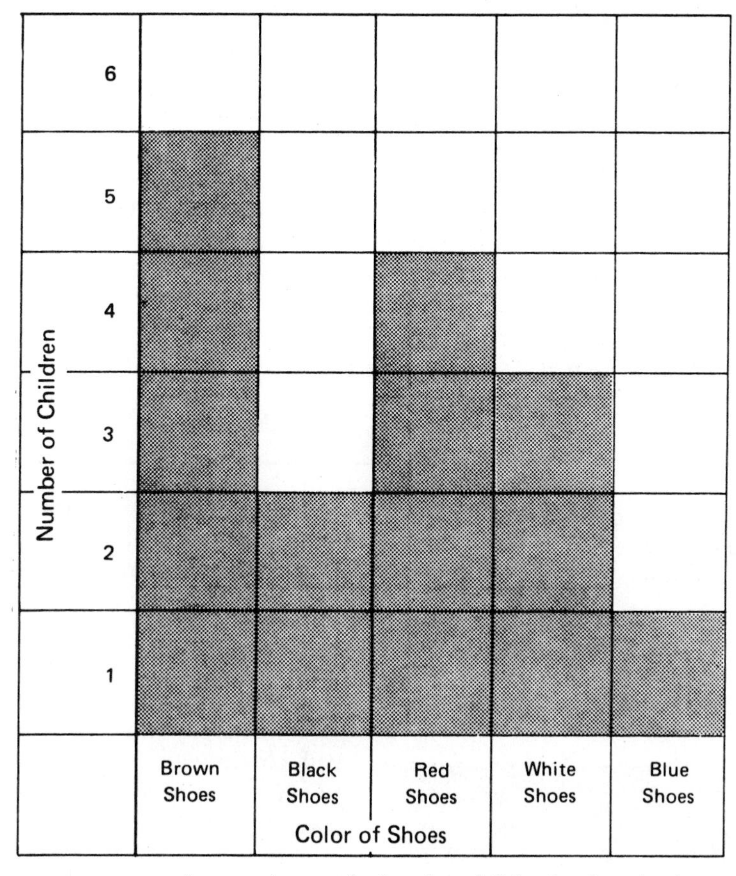

Figure 25–5 Square paper graphs can be made by the child who has had many experiences with simpler graphs.

problem-solving skills with concrete whole number operations. That is, they can solve simple addition, subtraction, division, and multiplication problems using concrete materials. You can make up problems such as those below. Provide the children with ten counters (pennies, chips, Unifix® cubes, and so on) or the real items or replicas of the real items descibed in the problems.

ADDITION:
- If Mary has three pennies and her mother gives her one more penny, how many pennies will Mary have?
- George wants two cookies for himself and two for Richard. How many cookies should he ask for?

SUBTRACTION:
- Mary has six pennies. She gives three pennies to her sister. How many does she have now?

- George has six cookies. He gives Richard three. How many does he have left?

MULTIPLICATION:

- Mary gives two pennies to her sister, two to her brother, and two to her friend Kate. How many pennies did she give away?
- Tanya had three friends visiting. Mother said to give each one three cookies. How many cookies should Tanya get from the cookie jar?

DIVISION:

- Lai has three dolls. Two friends come to play. How many dolls should she give each friend so they will each have the same number?
- Lai's mother gives her a plate with nine cookies. How many cookies should Lai and each of her friends take so they each have the same number?

Problems can be devised with stories that fit unit topics.

As children grow and develop and have more experiences with whole number operations, they learn more strategies for solving problems. They gradually stop using the less efficient strategies and retain the more efficient ones. For example, four- and five-year-olds usually begin addition with the *counting all* strategy. That is, for a problems such as "John has three cars and Kate gives him two more. How many does John have now?", the four- or five-year-old will count all the cars (one-two-three-four-five). Five's will gradually change to counting on, that is, considering that John has three cars, they then count on two more (four-five) (See Figure 25–6). Even older children who are using recall ($3 + 2 = 5$) will check with counting all and counting on. When observing children working with division, note whether or not they make use of their concept of one-to-one correspondence (for example, when giving three people equal numbers of cookies, note whether they passout the cookies one at a time, consecutively to each of the three recipients).

Figure 25–6 Derrick is engaged in concrete addition as he adds two more to his group so he will have five.

THE SYMBOLIC LEVEL

Children who can connect sets and symbols (Unit 24), identify numerals zero to nine and do concrete addition and subtraction problems can move to the next step, which is connecting sets and symbols in addition and subtraction (Figure 25–7).

For addition, the child can be told that the sign + means to add more to the initial group. Write some problems on cards such as $1 + 2$, $2 + 2$, $3 + 1$, and so on. Using one problem card

Figure 25–7 Concrete addition at the connecting level.

at a time, have the children place the correct number of counters next to each number, then ask, "Which Number do we need to tell how many you have now?" Have them pick the numeral from a set of zero to nine numeral cards. Next, have them count out that amount of counters and check with one-to-one correspondence whether or not they have the same amount as they counted out for the problem card. Once they have the idea, they can be given several problems cards to work out on their own. When they are able to do concrete addition with symbols on their own, a similar sequence can be gone through with subtraction.

As the children work with these concrete symbol/set addition and subtraction problems, they will begin to store the basic facts in their memories and retrieve the answers without counting. They can then do problems without objects. Just have the objects on hand in case they are needed to check the answers.

Once children can write numerals easily, they might start writing down answers rather than picking the number from the zero to nine numeral cards. They can also make up their own problems, solve them, then trade with other students. Excellent resources for games and materials are *Workjobs II*, *Mathematics their way*, and *Developing number concepts using Unifix® Cubes*. Unit 27 provides a more detailed description of the procedure for introducing the symbols needed for whole number operations.

The calculator is fun to experiment with when children reach this point. They can identify the +, −, and = signs and explore doing some of their own problems.

QUANTITIES ABOVE TEN

Once children can count ten objects correctly, can identify the numerals one to ten, have a good grasp of one-to-one correspondence, and can accurately count by rote past ten, they are ready to move on to counting quantities above ten. They can acquire an understanding of quantities above ten through an exploration of the relationship of groups of ten with additional amounts. For example, they can count out ten Unifix® Cubes and stick them together. Then they can pick one more Unifix® Cube. Ask if they know which number comes after ten. If they cannot provide an answer, tell them that eleven comes after ten. Have them take another cube. Ask if they know what comes after eleven. If they do not know the answer, tell them twelve. Go as far as they can count by rote accurately. When they get to twenty, have them put their cubes together, lay them next to the first ten, and compare the number of cubes in the two rows. See if they can tell you that there are two tens. Give them numeral cards for eleven to nineteen. See if they can discover that the right-hand numeral matches up with how many more than ten that numeral stands for.

Once the children understand through nineteen, they can move on to twenty, thirty, forty, and so on. By exploring the number of tens and ones represented by each numeral they will discover the common pattern from twenty to ninety-nine. That is, that the two in twenty means two tens, and no ones; the two in twenty-one means two tens and the one means one one, etc., and the same pattern holds true through ninety-nine.

COMPUTER SOFTWARE

Young children will enjoy exploring computer software (see Activity 6) that reinforces higher level concepts and skills. *Alligator Alley* is an arcade-type game where a correct answer to a math problem placed on an apple results in the alligator eating the apple. Practice in the introductory steps of graphing can be obtained by playing *Bumble Games*. The Peanuts characters motivate children as they play a matching memory game, which builds skills in addition and subtraction in *Peanuts Math Matcher*.

SUMMARY

Children in the transitional stage can develop their concepts further through higher level concrete experiences. Experiences are described in the areas of classification, shape, spatial relations, graphs, concrete addition, subtraction, multiplication and division, symbolic activities, and quantities above ten. Calculator and computer activities are also suggested.

FURTHER READING AND RESOURCES

Baratta-Lorton, M. 1979. *Workjobs II*. Menlo Park, Calif.: Addison-Wesley.

Baratta-Lorton, M. 1976. *Mathematics their way*, Menlo Park, Calif.: Addison-Wesley.

Behounek, L. J., Rosenbaum, L. B., and Burcalow, J. V. 1988. Our class has twenty-five teachers. *Arithmetic Teacher*. 36(4): 10–13.

Richardson, K. 1984. *Developing number concepts using Unifix® Cubes*. Menlo Park, Calif.: Addison-Wesley.

Spence, C., and Martin, C. S. 1988. Mathematics + social studies = learning connections. *Arithmetic Teacher*. 36 (4): 2–5.

Stice, C. F., and Alvarez, M. C. 1987. Hierarchical concept mapping in the early grades. *Childhood Education*. 64: 86–96.

Whitin, D. J. 1989. Bring on the buttons. *Arithmetic Teacher*. 36 (5): 4–6.

SUGGESTED ACTIVITIES

1. Review children's magazines. Make a list of the activities suggested that relate to helping children learn math symbols. Share these with your class.
2. Observe kindergarten age children. What evidence was noted that some children were ready for and some were using higher-level concept activities? Share what you observed with your class.
3. Interview a kindergarten teacher. Ask how she plans concept activities for children who are progressing towards the concrete operational stage.
4. Select, prepare, and present two higher-level concept activities for transitional children. Share these with your class.
5. Add higher-level concept activities to your Activities File.
6. Review some computer software that supports higher-level concepts and skills. Use the evaluation format suggested in Activity 8, Unit 9. Some suggested software follows:

Alligator Alley. DLM Teaching Resources. Allen, Tex.: Apple.

Beginning Math Concepts. Orange Cherry Media Software. Bedford Hills, N.Y.

Bumble Games. Learning Co. Portola Valley, Calif.

Hey Taxi! OL—Opportunities for Learning. Chatsworth, Calif.: Apple II, 48K.

Peanuts Math Matcher. OL—Opportunities for Learning. Chatsworth, Calif.: Apple II, 64K.

Soc Sort and Soc Match. American Guidance Service. Circle Pines, Minn.: Apple II+, IIe, IIc.

REVIEW

A. Name the areas that have higher-level concept activities.

B. Describe the three higher levels of classification.

C. Match the terms in Column II with the correct item in Column 1.

I	II
1. Apple, fruit, food	a. Multiple classification
2. Matrices	b. Class inclusion
3. Two balls, five toy cars—are there more cars or more toys?	c. Hierarchical classification
4. Shape, size, and color	
5. Pencils, crayons, chalk, writing implements	
6. Three carrots and two stalks of celery—are there more carrots or more vegetables?	

D. Given the following situations, tell which higher-level activity the child seems interested in and how the teacher should respond.

1. Mary says, "Susie lives across the street from me."
2. Jim takes the box of shape attribute blocks and begins to stack them into different piles.
3. Karen takes some plastic fruit and wooden people and places them on the balance scale.
4. Alissa points to the picture in the story book and says, "See the round apple."
5. Alissa and Jim are playing. Alissa says, "Blue is my favorite color." Jim says he likes yellow best. He says, "More kids like yellow than blue." Alissa says, "No, they don't."
6. "John, give me two more blocks so that I will have six, too."
7. Mrs. Jones notices that Sam is laboriously writing math problems such as 2 + 3, 4 + 1, 1 + 1 and counting out beans to find out the answers.
8. Mrs. Carter notices that five-year-old George can rote count to ninety-nine accurately.
9. Tanya explains that all people have blood, bones, and fat inside, but all dolls are filled with cotton.

UNIT 26 Higher-Level Activities Used in Science Units and Activities

OBJECTIVES

After studying this unit, the student should be able to:

- Plan and do structured lessons in sets and symbols, classification, and measurement
- Apply the concepts, skills, and attitudes found in math to lessons in science
- Design higher level lessons that develop science concepts for children closer to the concrete level of development

Children should be exposed to a greater number of higher level experiences as they near the concrete level of development. In this way, they are able to develop fundamental concepts. The experiences in this unit build on many of the ideas presented in Units 23 through 25 of this book: sets and symbols, classification, shape, spatial relations, measurement, and graphs.

SETS, SYMBOLS, AND CLASSIFICATION

Classifying Vegetables

Although children will be asked to bring vegetables from home for this lesson, you will need to prepare a variety of vegetables for maximum learning. Seed and garden catalogs and magazine pictures can provide illustrations of how the vegetables grow. Three major categories of vegetables that work well with children are: *leaf* and *stem vegetables* (cabbage, lettuce, spinach, mustard, parsley); *root vegetables* (sweet potatoes, carrots, beets, turnips, onions); and *seed vegetables* (cucumbers, peas, beans, corn, soybeans). Mrs. Red Fox conducts the following activity in the fall with her first grade class:

First, she invites the children to join her on a rug to examine the different vegetables. Then she asks her class to describe the vegetables in front of them. They answer, "Long, rough, smooth, peeling, hard, scratchy, lumpy, bumpy, crunchy, white, orange." Mrs. Red Fox holds up a potato and asks, "Can someone describe this potato? "I can," Trang Fung says. "The potato is bumpy and brown." "Good," says Mrs. Red Fox. "Who can find another vegetable that is bumpy?" "A pea is a little bumpy," suggests Sara. Mrs. Red Fox writes the word *bumpy* on a 3 × 4 inch card and groups the potato and pea together. Then she holds up a carrot and asks, "Is there another vegetable that looks like this one?" Dean studies the assortment of vegetables and says, "How about the pumpkin?" "Yes, the pumpkin and the carrot are both orange." Mrs. Red Fox writes the word *orange* on a card and places the carrot and pumpkin together with the card. The children continue to classify by characteristics until each vegetable is in a group. Mrs. Red Fox refrigerates the vegetables overnight and readies them for a series of activities that emphasize a variety of concepts.

1. *Where do they grow?* The class discusses how the different vegetables looked while they were growing. After all, for all the children know, potatoes come from produce sections of supermarkets and corn comes from a can. The children have fun matching the vegetable with pictures of the vegetable growing. Mrs. Red Fox makes a game of matching actual vegetables with how they look as they grow (Figure 26–1).
2. *Digging for potatoes.* After matching actual vegetables with pictures of growing vegetables, children practice digging for potatoes. Mrs. Red Fox fills the sandbox with rows of potatoes and children take turns digging. Carrots, beets, and turnips are planted with their leafy tops above the soil, and children take turns gardening (Figure 26–2).
3. *What is inside*? Bowls of peas are set out to be opened and explored, and the idea of beginning a garden begins to occur to children.

Figure 26–1 George matches a carrot with the picture of a carrot that the teacher is holding.

Figure 26–2 Digging vegetables reinforces where they grow.

4. *Under, on, and above.* Mrs. Red Fox makes a bulletin board backdrop of a garden, and the children match cutout vegetables to where they grow. Onions are placed *under the ground*; peas, beans, and corn are shown *above the ground*; and lettuce, cabbage, and parsley are displayed *on the ground.*
5. *Patterns.* Children place the vegetables in patterns. Trang Fung made a pattern of potatoes, pumpkins, celery, and cucumbers. She said, "Bumpy, bumpy, bumpy, smooth. Look at my pattern. Dean, can you make a pattern?" Dean began his own pattern. "I am not going to tell you my pattern," said Dean. He laid out mustard, carrots, peas, cabbage, potato, and beans. Trang Fung smiled and continued the pattern. "On the ground, under the ground, and above the ground, spinach, onion, corn." she said (Figure 26–3).
6. *Vegetable prints.* As the week continued, some children wondered if the inside of the vegetables were alike and different, too. So, Mrs. Red Fox cut up several vegetables

Figure 26–3 Can you match my pattern?

and let the children explore printing with vegetables. She poured tempera paint on sponges for a stamp pad. Then the children dipped the vegetable in paint and printed on construction paper. Mrs. Red Fox labeled some of the prints with vegetable words and hung them up for decoration. The prints were then classified by color and characteristics. Some children made patterns with their prints and asked others to guess the pattern and which vegetable made the pattern.

Stone Soup

Children can also learn about the vegetable members of the vegetable and fruit group through literature and cooking. After vegetables are identified as members of the vegetable and fruit food group, read the book *Stone Soup* and act out the story on the flannel board.

Print the recipe on poster board (see Unit 22), add picture clues, and have the children wash vegetables for cooking. The children will quickly learn to use vegetable peelers and table knives as they cut the vegetables into small pieces. Then, act out the story once again. This time, really add the stone and vegetables to a crockpot to cook for the day. Bouillon cubes, salt, and pepper will spice up the soup. Do not be surprised if younger children think that the stone made all of the soup.

Prepare a set of cards with pictures of the ingredients and have students put the cards in the correct sequence of the cooking activity. The children will want to make an experience chart of preparing the stone soup. They could even graph their favorite parts of the soup (Figure 26–4).

Animal Sets

Stuffed animals are familiar to the children and can be grouped in many ways. One way is to group stuffed animals on a table by sets, a group of one, two, three, and so on. Ask the children to draw the set of animals at their table on a piece of white paper. After the children have drawn the animals, include early literacy experiences by having them dictate or write a story about the animals. Staple the drawings at each table into a book. The children at each table have made an animal book of ones, twos, and so on. Read each book with the children and make all books available for independent classroom reading.

More First Mapping Experiences

The everyday experience of reading maps is abstract and takes practice to develop. The following mapping activities include the use of symbols and build on the early "Bird's-Eye View" mapping experiences in Unit 21 and the mapping experiences suggested in Units 13 and 25.

1. *Tangible mapping*. Children will not be able to deal with symbols on maps if they have not had experience with tangible mapping. Observe children as they create roads, valleys, and villages in clay or in the sand

Figure 26–4 What did we add to the soup first? Then what did we add?

Figure 26–5 Children will make hills and valleys in the sand.

table (Figure 26–5). Keep in mind the perspective of *looking down* as children place objects on a huge base map. Such a map can be made from oilcloth and rolled up when not in use.

2. *Pictorial mapping*. Take children to the top of a hill or building and ask them to draw what they see. This activity can emphasize spatial relations and relative locations. After you have returned from a field trip, discuss what was seen on the bus ride. Use crayons or paint to construct a mural of the trip.
3. *Semipictorial mapping*. Children will use more conventional symbols when they construct a semipictorial map. As they discover that pictures take up a lot of room, they search for symbols to represent objects. Colors become symbolic and can be used to indicate water and vegetation.
4. *Base map*. The base map contains the barest minimum detail filled in by the teacher, i.e., outline, key streets, and buildings. This is a more abstract type of map, thus the area must be known to the child. Add tangible objects to the abstract base, i.e., toy objects and pictures (Figure 26–6). A flannel base map is a variation.
5. *Caution*. Never use one map to do many things. If your mapping experience tries to do too much, children cannot rethink the actual experience and relationships. Each map-making experience must fulfill some specific purpose. Each map must represent something in particular that children look for and understand.

Exploring Pumpkins: October Science

If you live where pumpkins grow, take a field trip to purchase some; if not, buy some at the grocery. You will need a pumpkin for each child. Plan to organize the children in groups with an adult helper. The following activities use the senses to integrate the skills of measuring, counting, classifying, and graphing in the exploration of pumpkins.

Figure 26–6 Children add tangible objects to their map.

Time to Explore

Give children time to examine their pumpkins and their stems. Ask, "How does the pumpkin feel?" "How does it smell?" "Do all pumpkins have stems?" Later, when the pumpkins are carved, encourage the children to describe the differences in texture between the inside and outside of their pumpkins.

Ask, "Which pumpkin is the heaviest?" "Who has the lightest pumpkin?" After the children decide, bring out a scale and make an accurate measurement. The children probably cannot comprehend what the scale means or read the numbers, but they like to weigh things anyway. Tape the weight (mass) on the bottom of each pumpkin. Then, when jack-o'-lanterns are created, the children can compare the difference in mass.

Have children measure the circumference of a pumpkin with yarn. Ask, "Where shall we put the yarn on the pumpkin?" "Why?" Instruct the children to wrap the yarn around the middle of the pumpkin. Then, help children cut the yarn and label it. Write the child's name on masking tape and attach it to the yarn length. After the children have measured their pumpkins and labeled their yarn, have them thumbtack the yarn length to a bulletin board. Ask, "Which pumpkin is the smallest around the middle?" "Which is the largest?" "How can you tell?" Have the children refer to the yarn graph. Some kindergarten children might be ready to lay the yarn length along a measuring stick and draw horizontal bar graphs instead of yarn ones (Figure 26–7).

Children will want to measure height as well as circumference. The major problem to solve will probably be where to measure from—the top of the stem or the indentation where the stem was. This time, cut a piece of yarn that fits around the height of the pumpkin. Label the yarn and attach it to pieces of masking tape.

Figure 26–7 How big is your pumpkin?

Order the yarn lengths from shortest to tallest. Ask, "Which yarn is the longest?" "Can you tell me who has the tallest pumpkin?"

Observing Pumpkins

Mrs. Jones has kindergarten children count the number of curved lines along the outside of the pumpkin. She asks, "How many lines are on your pumpkin?" She instructs the children to help each other. One child places a finger on the first line of the pumpkin and another child counts the lines. Then Mrs. Jones asks, "Do all the pumpkins have the same number of lines?" "No," answers Mary, "My pumpkin has more lines than Lai's pumpkin." Some children even begin to link the number of lines with the size of the pumpkin as they compare findings.

Mrs. Jones continues asks, "How does your pumpkin feel?" George says his pumpkin is rough, but Sam insists that his is smooth. "Yes," Mrs. Jones observes. "Some pumpkins are rough and some are smooth." "Are they the same color?" This question brings a buzz of activity as the children compare pumpkin color. Mrs. Jones attaches the name of each child to his or her pumpkin and asks the children to bring their pumpkins to the front table. She asks the children to group the pumpkins by color variations of yellow, orange, and so on. Some of the children are beginning to notice other differences in the pumpkins such as brown spots and differences in stem shapes.

The children gather around Mrs. Jones as she carves a class pumpkin. She carefully puts the seeds and pulp on separate plates and asks the children to compare the inside color of the pumpkin with the outside color. She asks, "What colors do you see?" "What colors do you see inside of the pumpkin?" "I see a lighter color," Mary says. "Do all pumpkins have light insides?" George is more interested in the stringy fibers. "This looks like string dipped in pumpkin stuff." "I am glad you noticed, George," Mrs. Jones comments. "The stringy stuff is called fiber. It is part of the pumpkin." Students learn the word *pulp* and also compare the seeds for future activities.

Children will also notice that *pumpkins smell.* Ask them to describe the smell of their pumpkins. Have them turn their pumpkins over and smell all parts. Then ask them to compare the smell of the inside of their pumpkins with the outside rind and with the seeds, fiber, and pulp. "Does the pumpkin smell remind you of anything?" Record dictated descriptions and create an experience chart of pumpkin memories. Ask, "What do you think of when you smell pumpkins?"

As children make jack-o'-lantern faces, discuss the shapes that they are using. Then, using a felt-tip pen, ask them to draw faces on the surfaces of their pumpkins. Closely supervise children as they carve their pumpkins. Instruct them to cut away from their bodies. They will need help in holding the knife, and their participation might be limited to scooping out the pumpkin. Note: Apple corers work well for these experiences and are easier to use.

After the jack-o'-lanterns are cut and have been admired, Mrs. Jones begins a measurement activity. She asks, "How can we tell if the pumpkin is lighter than it was before it was carved?" "Lift it," say the children. "Yes," says Mrs. Jones, "that is one way to tell, but I want to know for sure." George suggests, "Let's use the scale again." But before the pumpkins are weighed, Mrs. Jones asks the children to whisper a prediction in her ear. She asks, "Do you think the pumpkin will weigh more or less after it has been carved and the seeds and pulp removed? Whisper 'More', 'Less', or 'The same'." As the children respond, Mrs. Jones writes their responses on paper pumpkins and begins a more, less, and the same graph of predictions (Figure 26–8).

Last but not least in a series of pumpkin

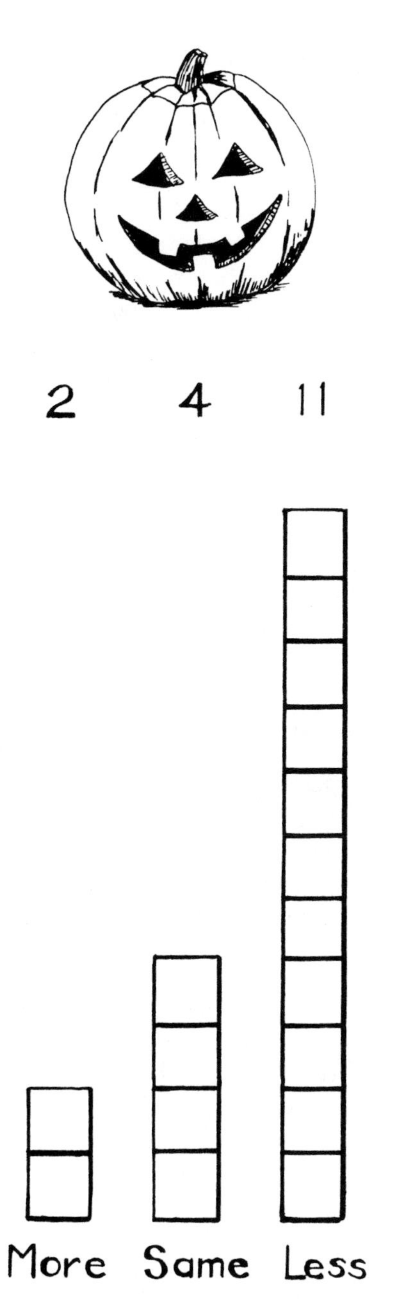

Figure 26–8 Did your pumpkins weigh more, less, or the same after the seeds were removed?

activities is tasting. Pumpkin pulp can be made into pumpkin bread or, for science and Halloween party time, make drop cookies with raisins for eyes, nose, and mouth. Or, for a tasty snack, cut some of the pumpkin pulp into chunks for cooking. Once cooked, dot the pieces with butter and add a dash of nutmeg.

Try roasting some pumpkin seeds. Have the children wash the seeds to remove the pulp and fiber. The seeds can be dried between layers of paper towels, then roasted in a single layer on a cookie sheet. Bake in a 350-degree oven for thirty to forty minutes. The children can tell when the seeds are ready by their pale, brown color. To make a comparison lesson, show them seeds that you have previously roasted and invite them to tell you when the color of the roasting seeds match yours. Cool the seeds, then have a snack.

MEASURING THE WORLD AROUND US

Playgrounds, yards, and sidewalks provide many opportunities for measuring in science. The following activities emphasize the science and measurement in outdoor adventures.

1. *How far will it blow?* After talking about air and how it moves, take the children outside to determine how far the wind will blow dandelion seeds. Draw bull's-eye-like circles on the playground with chalk. Label the circles. Have a child stand in the middle of the circle and hold a mature dandelion up to the wind. Record which direction the wind blows the seeds and the circle in which most of them landed. This would be a good time to discuss wind as a way of dispersing seeds.
2. *Weed watch.* Place a stick next to a growing plant such as a dandelion or similar weed. Have the children mark the height of the weed on the stick. Check the weed each

day for a week and see how tall it gets. (You might have to talk to the groundskeeper before trying this activity.) Children enjoy seeing the weeds grow. Discuss differences and possible factors in weed growth.

3. *How long does it take?* After students have had time to explore the water center, select some of the containers used in the center. Lead the children to an area that has pinecones, acorns, or other natural objects available. Recall the amount of time it took to fill the containers with water. Then, give a container to each group of children and instruct them to fill it with specific objects. Note the time that it takes to fill the containers. Vary the activity by assigning different groups of contrasting objects, i.e., small, big, rough, and so on, with which to fill the containers.
4. *Line them up.* Have children count the objects that they have collected in the containers. Ask each group to line up the objects. Compare the number of objects that it took to fill each container. If the objects collected were different, compare the number of each type of object that was needed to fill the container.
5. *How can we tell?* Children will probably find many ways to compare objects they have found. After they have examined and compared their objects visually, encourage them to weigh several of the objects. Ask, "Which container is the heaviest?" "How can we tell?" A balance provides an objective measure. Balance the content of one container (acorns) against the content of another (walnuts). Have children predict which will be heavier. Some might want to draw a picture of what they think will happen.

Popcorn Time

Children will sequence events as they act out a favorite snack. Have two or three children become popcorn by asking them to crouch down in the middle of a masking tape circle or hula hoop. As you pour imaginary oil on them, have the rest of the class make sizzling noises and wait for them to pop. When ready, each child should jump up burst open like a popcorn kernel, and leap out of the popper circle. Children will want to take turns acting out the popcorn sequence.

Then, place a real popcorn popper on a sheet in the middle of the floor. Seat the children around the popper, remove the popcorn lid, and watch the popcorn fly. Before eating the popcorn, measure how far the popcorn popped with Unifix® cubes or string. (Be careful of the hot popper.) Ask, "Which popcorn flew the farthest?" "Which flew the shortest?"

SPATIAL RELATIONS

Inside and Outside

Young children are curious about their bodies. They are familiar with outside body parts, which they can see and touch, but they are just beginning to notice that things happen inside their bodies. To increase this awareness and reinforce the concepts of inside and outside, Mrs. Jones has children look at, feel, and listen to what is going on inside of their bodies in the following scenario:

Mrs. Jones fills a garbage bag with an assortment of items—a wound-up alarm clock, rubber balls, book, a few sticks, and a bunch of grapes in a small sandwich bag. She places the bag on a chair in the front of the room, allowing it to drape down to show the outlines of some of the objects inside. She asks, "What is on the chair?" George says, "A garbage bag." "Is anything on the chair besides the bag?" asks the teacher. "No," the children respond. Mrs. Jones invites the children to gather around the bag and feel it. She asks, "Do you hear or feel anything?" "Yes," Mary says. "I feel something sharp." "And squishy," adds Sam. George is certain that he hears a clock ticking, and Lai feels a round and firm object that moves (Figure 26–9).

After the children guess what might be in the bag, Mrs. Jones opens it and shows the children what was inside. "Did you guess correctly?" "I did," says George. "I heard something ticking." "Good," answers Mrs. Jones. "How did you know that sticks were in the bag?" Mary says, "I could feel them poking through the plastic." "Yes", Sam adds, "the grapes must have been the squishy stuff."

After the children discuss the contents of the bag, Mrs. Jones takes the bag off the chair, asks the children to return to their places, and invites Mary to sit in the chair. She asks, "Now, what is on the chair?" The class choruses, "Mary." "Yes, Mary is on the chair," says Mrs. Jones. "How is Mary like the bag?" After a few responses, George says, "Mary has something inside of her, too." The children come up to look at Mary, but do not touch her. Mrs. Jones asks, "Do you see what's inside showing through the way it did with the bag?" The children notice bones, knuckles, the funny bone, kneecap, and some veins. Mrs. Jones asks Mary to flex her arm muscles so the children can see muscles moving beneath her skin.

Figure 26–9 Do you feel anything inside the bag?

Mrs. Jones encourages the children to discover other muscles and feel them working. She has them stretch out on mats on the floor, curl up tightly, then slowly uncurl. "What have your muscles done?" Then she has them stretch out like a cat and curl up into a ball. Facial muscles are fun for the children. Mrs. Jones asks them to find and use all the muscles that they can on their faces. They wiggle noses, flutter eyelids, tighten jaws, and raise eyebrows. The teacher asks, "Does your tongue have muscles in it?" "How do you know?" Finally, the children move to music as they focus on what is happening inside their bodies. Mrs. Jones has them dancing like marionettes, stiffly with a few joints moving; then bending and curling the way rag dolls do; and finally, dancing like people, with muscles, joints, and bones controlling their movements.

The children are fascinated with the thought of something important inside of them. They make toy stethoscopes from funnels and rubber tubing and take turns finding and listening to their heartbeats. Some begin to count the beats; others enjoy tapping a finger in time with the sound of their hearts.

Mrs. Jones has the children simulate the way a heart works by folding their hands one over the other and squeezing and releasing them rhythmically. (This motion is somewhat like the way the heart muscles move to expand and contract the heart, pushing blood through.) She has them place their clasped hands close to their ears

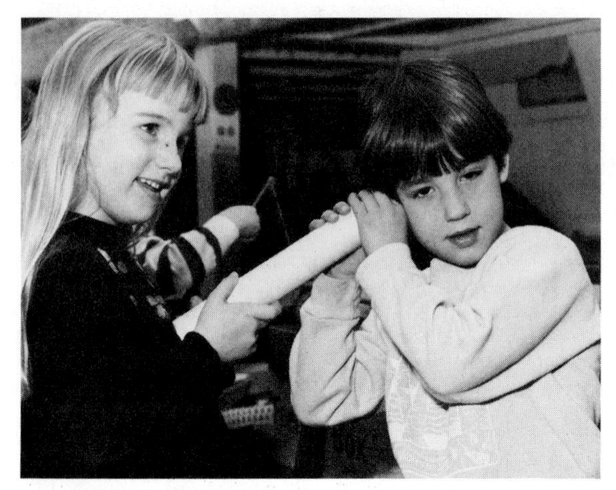

Figure 26–10 What do you hear inside me?

and asks, "What do you hear?" "Is this squeezing sound like the soft thumping you heard through the stethoscope?" "How is it different?" (Figure 26–10).

Turtle Steps

Young computer users can talk to an electronic turtle and tell it where they want it to go in LOGO programs. They are actually talking to the "turtle," the leading edge of the drawing pen in turtle graphics, with a simplified form of Apple LOGO. By issuing commands in KINDER, the less difficult form of LOGO, Mrs. Campbell's students are able to draw pictures on their Apple 11+ and Apple IIe monitors. Then, using a program called, "Triple Dump," described in the LOGO Tool Kit manual, they print their drawings.

Because Mrs. Campbell knows that children learn best by doing, she makes an analogy between the way children's bodies move and how the turtle draws on the screen. Before the children go to the computer terminals, she gets them ready by playing "turtle." Mrs. Campbell invites the children to put on their turtle caps, line up, and pretend that they are a band of marching turtles, who can move only when commanded (a variation of "Mother, May I?"). She directs:

"Turtles, go forward two turtle steps."
"Turtles, go back one step."
"Turtles, turn right three steps."
"Turtles, left."

Oops! She has caught some turtles in an important mistake. For the turtles to move, they must be told how many steps to take. If children have trouble knowing left from right, she labels their hands with washable felt-tip pens and continues to direct them around the room (Figure 26–11).

Another preparatory game is "Turtle in the Pond." Mary is designated Mary Turtle and stands in a masking tape square "pond" on the floor (about three meters to a side). As other children give directions, Mary Turtle moves around the pond. "Turtle, forward three steps," commands George. The children take turns giving commands until the turtle has moved all around the pond.

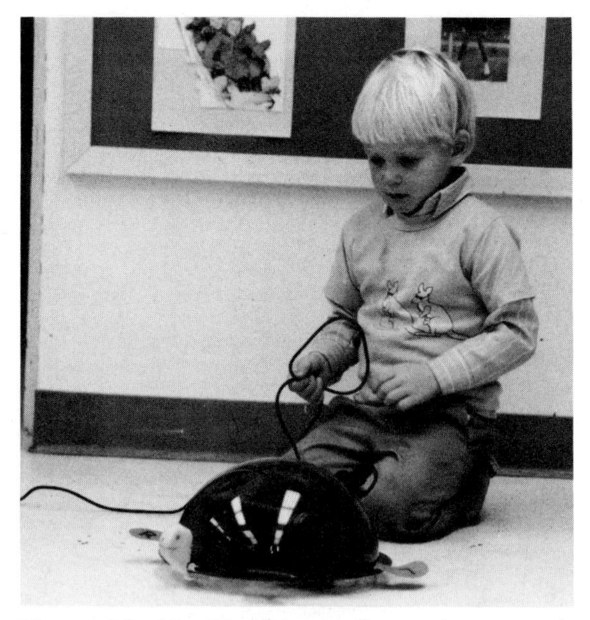

Figure 26–11 Turtles, go forward two turtle steps.

"How do we know where Mary has already been?" asks George. "Good point," says Mrs. Campbell. "Mary Turtle has not left a trail to follow." The teacher gives Mary a ball of string to unwind as she takes her steps. In this way, a record of moves will be left (just like on the computer screen). The children direct Mary Turtle, who trails a string, around the pond.

"It is time for the turtle to hide," announces Mrs. Campbell. Mary jumps out of the pond and leaves the string design behind her. To start over again, the children clean the pond and leave a new trail.

A maze also helps children get ready for computer graphics. After making a masking tape maze on the floor, students take turns directing turtles through it with forward, right, and left directions. These are the basic commands for turtle graphics. A polluted pond can be simulated by having the turtle avoid books, chairs, and even children. Mrs. Campbell asks, "Can you direct the turtle around the debris?"

When students are comfortable with activities such as the ones introduced, Mrs. Campbell takes them to the computer keyboard. Her activities have helped the children get ready to do graphics on the computer. Many versions of simplified turtle graphics are currently available for a variety of machines (Figure 26–12).

Figure 26–12 Children are ready to do graphics on the computer.

Lessons and activities such as Turtle Steps and Inside and Outside make children aware of the fact that there are different ways of looking at things. Taking a different view of things is a good way of reinforcing spatial relations and of beginning to see things like a scientist.

SUMMARY

Children in the transitional stage develop fundamental concepts in sets and symbols, classification, shape, spatial relations, measurement, and graphs as they are exposed to higher level experiences. As children near the concrete operational level of development, they continue the need for hands-on exploration.

Science activities at this level allow children to pull together skills and ideas learned earlier. Most children incorporate these skills through naturalistic and informal experiences. Structured activities are also appropriate, but should be short in length.

FURTHER READING AND RESOURCES

Buckleitner, W. 1989. *Survey of Early Childhood Software.* Ypsilanti, Mich.: High/Scope Press.

Burk, D., Snider, A., and Symonds, P. 1988. *Box it or bag it mathematics.* Salem, Ore.: Math Learning Center.

Campbell, S. 1985. Preschoolers meet a high tech turtle. *Science and Children.* 22 (7), 37-39.

Davidson, J. I. 1989. *Children and Computers Together in the Early Childhood Classroom*. Albany, N.Y.: Delmar.

Council for Elementary Science International Sourcebook IV. 1985. *Science experiences for preschoolers*. Columbus, Ohio: SMEAC Information Reference Center.

Council for Elementary Science International Sourcebook III. *Understanding the healthy body*. 1983. Columbus, Ohio: SMEAC Information Reference Center.

Harlan, J. 1988. *Science Experiences for Early Childhood Years*. Columbus, Ohio: Charles Merrill.

Lind, K. K. 1985. The inside story, *Science and Children*. 22(4). 122–123.

———. 1984. A bird's-eye view: Mapping readiness. *Science and Children*. 22(3). 39–40.

McIntyre, M. 1984. Pumpkin science. *Early childhood and science*. Washington, D.C.: National Science Teachers Association.

Paterson, D. 1981. *Stone soup*. Mahwah, N.J.: Troll Associates.

Rockwell, R. E., Sherwood, E. A., and Williams, R. A. 1986. *Hug a tree*. Mr. Rainier, Md.: Gryphon House.

SUGGESTED COMPUTER PROGRAMS

Greeting Card option of *Print Shop*. Broderbund Software.

Computergarten. Scholastic, Inc. (Pre-K–1)

Micros For Micros: Estimation. Lawrence Hall of Science. (K–2)

SUGGESTED ACTIVITIES

1. Select a computer program for young children. What precomputer activities will you do with children before using the terminal? How will you use the computer program in your teaching? Will you conduct any follow-up activities?
2. Design a graphing experience for transitional children. Select an appropriate science topic, describe procedures, and develop a questioning strategy. Share the lessons in groups and teach the lesson to a class.
3. Accompany a class on a field trip or walk around the block. Make a picture map of the field trip. Record the types of comments children make as they recall their trip. What types of objects made an impression on the children? Are the comments in keeping with what you know about children and the way they think?

REVIEW

1. List areas discussed in this unit that have higher-level concept activities.
2. Name three types of mapping activities that can be used with young children. Why use these strategies with children?
3. Why is it essential that children have time to explore materials when learning a science concept?
4. Should a computer be used in the classroom? Describe possible benefits. What should children do before they go to the keyboard?

SECTION V

Mathematics Concepts and Operations for the Primary Grades

UNIT 27 Operations with Whole Numbers

OBJECTIVES

After studying this unit, the student should be able to:

- Describe the whole number operations of addition, subtraction, multiplication, and division
- Introduce the whole number operations to primary grade children
- Administer whole number operations assessment tasks
- Introduce whole number operations notation and number sentence format following a three-step process
- Develop instructional activities and materials for whole number operations at the primary level

Children naturally engage in the whole number operations of addition, subtraction, multiplication, and division prior to reaching the primary grades. In previous units (see Units 23, 24, and 25), the beginnings of the whole number operations are described as they grow out of naturalistic and informal experiences. Prior to entering first grade, young children also usually have an understanding of number symbols as they represent quantities. During the primary period (grades one through three), children gradually learn the meaning of action symbols such as + (add), − (subtract), × (multiply), ÷ (divide), = (equals), < (less than), and > (greater than).

Teachers are expected to use state and locally developed lists of objectives and selected textbooks to provide a structure for planning instruction. Unfortunately, primary teachers tend to rely too heavily on textbooks, workbooks, and dittoed support materials. Conventionally, students are expected to be able to do paper and pencil arithmetic even though it might be developmentally inappropriate. Opportunities for the use of exploration as a route to the construction of concepts and operations are too seldom observed in the primary classroom. Students usually sit at individual desks with social interaction kept at a minimum (if allowed at all).

Constance Kamii is a major critic of the conventional approach to mathematics instruction in the primary grades. She presents her point of view in her books *Young Children Reinvent Arithmetic* and *Young Children Continue to Reinvent Arithmetic*. As a Piagetian, she believes that just as our ancestors invented it, children reinvent arithmetic through their own actions and needs rather than learning through what someone else tells them. Children need to reinvent arithmetic through naturalistic and informal exploration with naturally occurring problems and through group games. Kamii explains that paper and pencil worksheet approaches remove the child from the logical thinking that is the heart of arithmetic. Kamii's emphasis on group games and social interaction as the basis for the understanding of arithmetic has its roots in Piaget's view that social interaction is essential as a stimulus for construction of knowledge. The following are examples of children solving problems through peer interaction:

- Three nonconservers are going to drink juice. The server pours juice into glass A. Now he must pour equal amounts into glasses B and C, which are different in size and shape from each other and glass A. The children discuss how to do this in a fair manner and arrive at a solution: use glass A as a measuring cup to fill the other two glasses.
- Derrick has written 7 + 4 = 10 and Brent has written 7 + 4 = 12. Their teacher has them explain to each other why they think that their respective answers are correct. Soon they discover that they are both wrong.

A danger in primary math instruction is that students will be pushed too fast before they have developed the cognitive capacity to understand the logical reasoning underlying the operations. Keep in mind that children must be in the concrete operations period before they can successfully meet primary level expectations. Beginning in kindergarten, standardized testing is performed each year, and teachers are pressured to teach the concepts and skills that are included in the tests. This pressure leads teachers to instruct arithmetic as a rote memory activity that has no logical meaning to the children. Children should be allowed to move at their own pace through primary math just as they did prior to entering first grade. They should also be allowed to invent their own procedures for solving problems. Herbert Ginsburg claims that young children naturally invent their own methods and should be able to use and experiment with them. For example, young children usually learn on their own to use counting methods for addition. When adding 2 + 2, the child might say, "One, two, three, four" using fingers or objects. If left on their own, they will eventually stop counting, having internalized 2 + 2 = 4.

In this text, a sequence of concept instruction is described with the caution to the teacher to keep in mind that children move at their own pace. As in previous units, naturalistic, informal, and structured activities are described. Structured activities emphasize the use of concrete materials, with paper and pencil introduced through children's natural interests when they are ready (Figure 27–1).

Figure 27–1 Children often count on their fingers when they are first developing the concept of addition.

ACTION AND RELATIONAL SYMBOLS

At the primary level, children are usually introduced to the action and relational symbols. Action symbols show that some quantities have been or will be acted upon, or changed, in some way (+, −, ×, ÷); relational symbols show that quantities are in some way related (=, <, >). These symbols appear in number sentences that symbolize an operation, such as:

- 2 + 3 = 5 (two things put together in a group with three things is the same amount as five things)
- 5 > 2 (five is more or greater than two)

Kamii and others have found that young children

often learn to deal with these symbols without a genuine understanding of how they relate to real quantities. Children should work with operations mentally through concrete experiences before connecting these operations to symbols and using complete conventional written number sentences such as $1 + 5 = 6$, $5 - 3 = 2$, $6 > 2$, and so on.

Ed Labinowicz, in his book *Learning from Children: New Beginnings for Teaching Numerical Thinking*, reviews the research on young children's understanding of action and relational symbols. He concludes that first grade textbooks introduce the formal symbols of equality sentences much too early. He and Constance Kamii agree that full number sentences (e.g., $4 + 2 = 6$) should not be introduced until grade two. Children should be encouraged to devise their own notation systems and apply them as a bridge to formal notation. Children need experiences in joining and separating quantities and verbalizing about their actions prior to really understanding what symbols represent. Just filling in blanks in a workbook or marking answers in a standardized test booklet does not indicate that children understand the deeper meaning of number sentences.

Labinowicz suggests that the formal number sentence be introduced gradually as follows:

1. Provide extensive experiences in which children solve problems using counting and other strategies that they invent. For example, if a child wants to know how many grapes he will have left if he has seven and gives three to his friends let him record the information in his own way.
2. Encourage students to try out and explore their own systems of informal notation.
3. Use word problems in real-life settings to develop the language of number sentences. Delay the introduction of formal number sentences until students can understand that they are a shorthand representation of words.
4. Just as numerals were introduced gradually in stages, introduce the parts of number sentences with numeral notation first, then operational signs, and finally relational signs. *Workjobs II*, *Math their way*, and *Developing number concepts using Unifix® cubes* provide basic sequences and ideas for materials.

Some of these ideas for introducing formal symbolic notation will be described further as each whole number operation is discussed.

Assessment

Observations and interviews can be used to assess children's progress in constructing operations with whole numbers. Assessment should be done through concrete activities with real-life or pretend situations. Paper and pencil tests are not appropriate for primary students until they can read and comprehend story problems on their

Figure 27–2 Solving a mutiple classification problem signals that the child is ready to move into whole number operations.

own. Assessment examples will be provided as each whole number operation is discussed.

To find out if children are ready to move on to whole number operations, use the assessment interviews in Appendix A, Concrete Operations: Level: 8. These tasks include conservation of number, knowledge of symbols and of sets and symbols, multiple classification, and class inclusion (Figure 27–2).

ADDITION

Constructing the concept of addition involves an understanding that adding is putting together groups of objects to find out how many there are. It also involves learning the application of terms such as *total*, *sum*, and *equal*, as well as the action, signs (+ and =), which represent these terms, and connecting these amounts to symbols. Before children make these connections, they must understand quantity and what happens when quantities are combined.

Assessment

Assessment of children's understanding of addition is more than finding out if they know the so-called number facts. It is important to observe the process each child goes through in dealing with quantities. Observing the process and questioning children regarding what they have done will reveal what they do and do not understand. Their mistakes can be informative and used to help the children develop a more accurate knowledge of arithmetic.

Observations can be made during naturalistic and informal activities. Dean figures out that if he has two dimes and his grandmother gives him four more, he will have six dimes. Liu Pei decides that if there are three children at one table, two at another, and four at a third, she will have to get nine pieces of paper to pass cut. Ann realizes that instead of counting everyone to find out if there are enough pencils, she can record the number at each table, find the total, and compare it with the number of pencils in the box.

Observations can also be made during structured activities. Sara's teacher tells her to take groups of six cube blocks and place them in as many combinations of group sizes as she can. Does Sara realize that, however she arranges six objects (in groups of three and three; one and five; two, two, and two; two and four; or six and zero), there are still six altogether? The children are playing a card game called double war. Each player has two stacks of cards, which are turned over two cards at a time. The player with the higher sum gets the other player's two cards. The teacher can observe whether or not the children figure out the correct totals and whether they help each other.

As with the assessment of other math concepts; addition can also be measured using an interview approach. The following is a sample task.

SAMPLE ASSESSMENT TASK

Concrete Operations Ages 6–8

METHOD: Interview

SKILL: Child is able to combine sets to form new sets up to ten

MATERIALS: Twenty counters (cube blocks, Unfix® cubes, chips): ten of one color and ten of another

PROCEDURE: Have the child select two groups of counters from each color so that the total is ten or less. Say, PUT THREE YELLOW CUBES OVER HERE AND FIVE BLUE CUBES OVER HERE. Child completes task. Say, NOW TELL ME, IF YOU PUT ALL THE CUBES IN ONE BUNCH, HOW MANY CUBES DO YOU HAVE ALTOGETHER? HOW DO YOU KNOW? Do this with combinations that add up to one through ten.

EVALUATION: Note if the child is able to make the requested groups with or without counting. Note the method used by the child to decide on the sum:

1. Does he begin with one and count all the blocks?
2. Does he count on? That is, in the example above, does he put his two small groups together and then say, "Three blocks, four, five, six, seven, eight." "I have eight now."
3. Does he just say, "Eight, because I know that three plus five is eight."

INSTRUCTIONAL RESOURCES

Charlesworth, R., and Lind, K. 1990. *Math and Science for Young Children.* Albany, N.Y.: Delmar.

Instruction

Instruction begins with naturalistic and informal experiences that familiarize children with quantities and how they relate to each other. Students can be guided toward constructing their own concepts if they are provided with games and word or story problems to solve and encouraged to make up their own problems.

In *Young Children reinvent arithmetic*, Constance Kamii and Georgia DeClark describe a number of types of games that can support the development of addition concepts. The following are examples.

ADDITION: DOUBLE WAR

OBJECTIVE: Constructing combinations of addends up to four

MATERIALS: Two decks of cards with different patterns on the back

PLAYING THE GAME: Start using the cards with addends up to four (aces, twos, threes, and fours). Two children play together. To play the game, the children begin with half the cards in each of the two decks, which are stacked facedown next to each other in front of them. Without looking at the cards, they simultaneously turn over the top two cards from each deck. Each finds the total of his two cards. The child with the highest total keeps all four cards.

FOLLOW-UP: Add the cards with the next higher addends as the students become adept with the first four. If the game takes too long, remove some of the smaller addends as the larger are included.

ADDITION: BOARD GAMES

OBJECTIVE: Constructing combinations of addends up to six

MATERIALS: A pair of dice, a marker for each player (four), and a board game. Board games can be purchased or teacher made. Teacher-made games can be designed with themes that fit units in science and social studies. Some basic board game patterns and the materials needed for construction are described in Unit 24. Board games at the primary level can be designed with more spaces than those in Unit 24 since the older students may move farther on each turn and will have longer attention spans.

PLAYING THE GAME: Each player, in turn, rolls the dice, finds the sum of the roll, and moves the marker that many spaces.

FOLLOW-UP: Bring in new games as the students become skilled at playing the old ones. As the students become adept at playing board games, purchase or make some with pitfalls. That is, on some spaces the player might have to move backward or lose a turn (i.e., when a player lands on a red space, the dice are rolled again, and the player moves backward the sum of the dice, or else when a player lands on a certain space, a turn is lost).

There have already been many examples of word or story problems in this text. Placing operations in the context of real-life situations makes them come alive for the students so they can see the practical applications of mathematics. Richardson, in *Developing number concepts using Unifix® cubes*, suggests that the children act out stories using real objects from around the room as props. For example, Derick brings six books from the library center and Theresa brings four. "How many did they bring altogether?" Derick and Theresa actually demonstrate by going to the library center and obtaining the number of books in the problem. Trang Fung joins Dean and Sara. "How many children were there to start with?" "How many are there now?" Again, the children act out the situation. Addition story problems can be devised using the following basic patterns:

1. (Person #1) has (number) (item).
   (Person #2) gives (person #1) (number) (item).
   How many (name of item) does (person #1) have now?
2. (Person #1) has (number) (item).
   (Person #2) has (number) (item).
   How many do they have altogether?
3. (Person #1) has (smaller number) of (item)
   (Person #2) has (number) more than (person #1).
   How many (item) does (#2) have?

As children get more advanced, the numbers can be larger and more addends can be included. More complex, or *nonroutine*, problems also should be used (see Unit 33).

Problems can also be acted out using objects such as small toys or cube blocks, Unifix® cubes, or other objects.

ADDITION: STORY PROBLEMS

OBJECTIVE: To construct the concept of addition by solving story problems

MATERIALS: Twenty small toys that fit a current unit. For example:
- Miniature dinosaurs during a dinosaur unit.
- Miniature dogs, cats, horses, etc., during a pet unit.
- Miniature farm animals during a farm unit.
- Miniature vehicles during a safety unit.

DEVELOPING THE PROBLEMS: Let the students act out the problems as you tell the stories.
- FIND THREE PLANT-EATING DINOSAURS. FIND FOUR MEAT-EATING DINOSAURS. HOW MANY DINOSAURS DO YOU HAVE?
- MARY HAS THREE DOGS. JACK HAS TWO CATS. HOW MANY PETS DO THEY HAVE?
- YOU HAVE SIX COWS ON YOUR FARM. YOU BUY THREE MORE COWS. HOW MANY DO YOU HAVE?
- OFFICER SMITH GIVES TICKETS TO TWO CARS FOR SPEEDING. OFFICER VARGAS GIVES THREE TICKETS FOR CARS GOING TOO SLOWLY ON THE INTERSTATE. HOW MANY TICKETS DID THEY GIVE?
- PRETEND YOUR CUBE BLOCKS ARE HORSES. THE FARMER HAS TWO HORSES. ONE HORSE HAS A FOAL. HOW MANY HORSES DOES THE FARMER HAVE?
- PRETEND YOUR UNIFIX® CUBES ARE TRUCKS. THERE ARE FOUR TRUCKS PARKED IN THE GARAGE. FIVE MORE TRUCKS COME IN DURING THE DAY. HOW MANY TRUCKS ARE IN THE GARAGE?

FOLLOW-UP: Create problems to fit units and other activities and events.

Once the children have had some experiences with teacher-made problems, they can create their own problems.

ADDITION: CREATING PROBLEMS USING DICE OR A FISHBOWL

OBJECTIVE: The children will create their own addition problems.

MATERIALS: A pair of dice or a container (fishbowl) full of numerals written on small pieces of cardboard cut into fish shapes, objects such as cube blocks, chips, or Unifix® cubes

PLAYING THE GAME: Either by rolling the dice or picking two fish, each child obtains two addends. He counts out the amount for each addend and then tells how many objects or fish he has.

FOLLOW-UP: Once students are having an easy time making up problems using the dice or the written numerals as cues, suggest that they write or dictate their favorite problem, draw it, and write or dictate the solution (Figure 27–3). For example, Brent's dog is expecting pups. He writes, "I have one dog. I hope she has five pups." Then he draws his dog and the five pups. He writes, "Then I will have six dogs."

Figure 27–3 Children enjoy making up their own problems, writing or dictating them, and drawing a picture of the problem.

Using number symbols is referred to as *notation*. Gradually, number symbols can be connected to problems as you find that the children understand the process of addition and understand class inclusion. Although most first-graders can fill in the blanks correctly on worksheets, this does not indicate a real understanding of what notation means. To find out if a child really understands notation, present a problem such as the following:

Show the child several (four, five, or six) counters, then show how you add some (two, three, or four) more. Then ask, WRITE ON YOUP PAPER WHAT I DID. Even at the end of first grade you will find very few children who will write the correct notation i.e., $5 + 3 = 8$. It is very common for first-graders to write the first and last numeral (5 8) or to write all three (5 3 8) and omit the action symbols. They may also be unable to tell you what they did and why. It is very important that the use of notation be an integral part of concrete problem-solving activities.

In *Developing number concepts using Unifix® cubes*, Richardson suggests that formal instruction in connecting symbols to the process cf addition begins with modeling of the writing of equations. After acting out a problem such as described previously, write the problem on the chalkboard, explaining that this is another way to record the information. For example, ANOTHER WAY TO WRITE THREE COWS PLUS SIX COWS MAKES NINE COWS IS: $3 + 6 = 9$ (THREE PLUS SIX EQUALS NINE). Help the children learn what the plus sign means by playing games and doing activities that require the use of the plus sign with the equals sign. For example, try the following:

ADDITION: USING NOTATION AT THE CONNECTING LEVEL
OBJECTIVE: Children will connect symbols to problems using numerals and the plus action symbol
MATERIALS: Objects to count and one die
PLAYING THE GAME: Children take turns rolling the die to find out how many to add. For example, a three is rolled. Each child counts out three counters, and the teacher writes 3 + on the board. A five is rolled. The students count out groups of five to put with their groups of three and the teacher writes 3 + 5.
FOLLOW-UP: After working in small groups with the teacher, students can work independently with problems written on cards: 2 + 3, 4 + 6, and so on. As you go by, observe what they do and ask them to read the problems to you.

When the children are comfortable with connecting the symbols to the problems and using the plus symbol, they can begin to write the notation themselves. Start with problems in which you write the notation and have the children copy what you do before they go on to independent work. For example, have everyone pick five groups of five counters each. Then tell them to separate each group of five counters as many ways as they can and you write the results:

$1 + 4 = 5$
$2 + 3 = 5$
$1 + 1 + 3 = 5$
$2 + 2 + 1 = 5$
$5 + 0 = 5$

$$\begin{array}{r}1\\+4\\\hline 5\end{array}\quad\begin{array}{r}2\\+3\\\hline 5\end{array}\quad\begin{array}{r}1\\1\\+3\\\hline 5\end{array}\quad\begin{array}{r}2\\2\\+1\\\hline 5\end{array}\quad\begin{array}{r}5\\+0\\\hline 5\end{array}$$

After you write each equation, have the children write it on a piece of paper and put it next to the counters they have counted out. Follow up by having the students work independently, finding out how many ways they can break up amounts up to six and write the equations. When they are doing well up to six, have them move on to seven and above. After children work with notation this way, they can go on to solving purely numerical problems in workbooks with a real understanding of what they are doing.

SUBTRACTION

To conceptualize subtraction is to develop an understanding that subtracting involves taking objects away to find out how many are left or comparing groups of objects to find out the difference between them. It also involves learning the application of terms such as minus, difference, and equal, as well as the action signs (− and =) which represent these terms. Subtraction also involves thinking about *more than* and *less than* and the symbols (> and <), which stand for these relationships. It also includes connecting to symbols as a shorthand notation for concrete operations. As with addition, before children make the connections to and between symbols, they must understand quantity and what happens when something is taken away from a group or when two groups are compared.

Assessment

As with addition, assessment of children's understanding of subtraction is more than finding out if they know the so-called number facts. It is important to observe the process each child goes through in dealing with quantities. Observing the process and questioning children regarding what they have done will reveal what they do and do not understand. Their mistakes can be informative to the teacher and used to help the children develop a more accurate knowledge of arithmetic.

Observations can be made during naturalistic and informal activities. Chan figures out that if there are ten more minutes until school is out and Ms. Hebert says that in five minutes they will start to get ready to leave, that they will have five minutes to get ready. Jason has ten bean seeds. He decides that he can give Ann four, and the six he will have left will be enough for his seed sprouting experiment. Six children are allowed to work in the science center at one time. Brent notices that there are only four children there now. He suggests to Derick that they hurry over while there is room for two more. Vanessa observes that there is room for eight children in the library center while six at a time may work in the math center. Thus there is room for two more in the library center than in the math center.

Observations can be made during structured activities. Dean is trying to figure out how many different amounts he can take away from five. Mrs. Red Fox notes that Dean is well organized and systematic as he constructs one group of five after another and takes a different amount away until he has the combinations five minus zero, one, two, three, four, and five. Derick and Liu Pei are playing Double War. They have to subtract the amount that is smaller from the amount that is larger. Mr. Wang can note whether or not the children can figure out the correct differences and whether they help each other.

Subtraction can be assessed using an interview approach. The following is a sample task.

SAMPLE ASSESSMENT TASK

Concrete Operations Ages 6–8

METHOD: Interview

SKILL: Child is able to subtract sets to make new sets using groups of ten and smaller

MATERIALS: Twenty counters (cube blocks, Unifix® cubes, chips): ten of one color and ten of another and a small box or other small container

PROCEDURE: Pick out a group of ten or fewer counters. Say, I HAVE SEVEN CUBES. I'M GOING TO HIDE SOME IN THE BOX. (Hide three in the box.) NOW HOW MANY DO I HAVE LEFT? HOW MANY DID I HIDE? If the child cannot answer, give him seven of the other color cubes, ask him to take three away, and tell you how many are left. Do this with amounts of ten and less. For the less mature or younger child, start with five and less.

EVALUATION: Note if the child is able to solve the problem and the process used. Note whether the child has to count or if he just knows without counting.

INSTRUCTIONAL RESOURCES

Charlesworth, R., and Lind, K. 1990. *Math and Science for Young Children*. Albany, N.Y.: Delmar.

Instruction

Just as with addition, instruction begins with naturalistic and informal experiences that familiarize children with quantities and how they relate to each other. Students can be guided toward constructing their own concepts if they are provided with games and word or story problems to solve and encouraged to make up their own problems. Once the students evidence an understanding of addition, subtraction can be introduced. Children can work with both addition and subtraction problems so that they can learn the clues for deciding which operation to use.

The game Double War can be modified and played as a subtraction game by having the player with the largest difference between his pair of cards keeping all four cards (Figure 27–4). As with addition, begin with numbers up to four and gradually include higher numbers as the children get adept at the game. Board games can also be purchased or devised that use subtraction as the operation that indicates which way to move. A board game could be made where all the moves are backward. The theme might be running away from a wild animal or going home from a friend's house. Or dice could be thrown and each move would be the difference between the two.

Figure 27–4 Double War can be adapted for a subtraction game.

Of course, word or story problems are an essential ingredient in the instruction of subtraction just as they are for addition. Problems should be set in real-life contexts and acted out. For example, suppose Derick and Theresa have brought ten books from the library center. The activity could be continued by asking children to take different numbers of books back to the library center and finding out how many are left after each trip. Place children in groups of different sizes. If there are five children in this group and three in this group, which group has more? How many more? How will we find out? There are a number of different basic patterns that can be used for subtraction story problems.

1. (Person #1) had (number) (item).
   (Person #1) gave (number) (item) to (person #2).
   How many (item) does (person #1) have now?
2. (Person #1) has (number) (item).
   (Person #1) gave some to (person #2).
   Now (person #1) has (number) (item).
   How many (item) did he give to (person #2)?
3. (Person #1) has some (item).
   (Person #2) gave (person #1) (number) more.
   Now (person #1) has (number).
   How many (item) did (person #1) have in the beginning?
4. (Person #1) had some (item).
   (Person #1) gave (number) to (person #2).
   Now (person #1) has (number) (items).
   How many (items) did (person #1) have in the beginning?
5. (Person #1) has (number) (item).
   (Person #2) has (larger number) (item).
   How many more (item) would (person #1) need to have as many (item) as (person #2)?

6. (Person #1) has (larger number) of (item).
   (Person #2) has (number) (item).
   How many more (item) would (person #2) need to have as many (item) as (person #1)?
7. (Person #1) and (person #2) have (number) (item) altogether.
   (Person #1) has (number) (item).
   How many (item) does (person #2) have?
8. (Person #1) has (number) (item).
   (Person #2) has (smaller number) (item).
   How many (item) more does (person #1) have than (person #2)?
9. (Person #1) has (number) (item).
   (Person #2) has (smaller number) (items).
   How many (items) less than (person #1) does (person #2) have?
10. (Person #1) has (number) (items).
    (Person #2) has (number) less than (person #1).
    How many (item) does (person #2) have?
11. (Person #1) has (number) (items).
    (Person #1) has (number) more than (person #2).
    How many (item) does (person #2) have?
12. (Person #1) has (number) (item).
    (Person #1) has (number) (item) less than (person #2).
    How many (item) does (person #2) have?

Different persons (or animals or things) can be filled in with different amounts to formulate story problems. For example, the following problem is based on pattern number eight:

> Johnny has five rabbits.
> Mary has three rabbits.
> How many more rabbits does Johnny have than Mary?

As with addition, once the children have experiences with teacher-devised story problems, they can dictate or write their own. The dice/fishbowl game can be modified for subtraction. Children can also dictate or write original problems, draw them, and write or dictate the solutions (see Figure 27–3).

As with addition, subtraction notation can be introduced gradually. Number symbols can be connected to problems as you find the children have an understanding of the process of subtraction. To find out if a child really understands notation, the same type of procedure can be used as for addition. That is, show the child several counters (five, six, or seven). Have him tell you how many you have. Hide one or more of the counters and ask the child to show you on paper what you did. Do not be surprised if very few end-of-the-first-grade students and many second graders will not be able to write the correct equation.

Formal introduction of subtraction can begin with modeling. Act out a problem and then explain that there is another way to record the information. Write the number sentence for the problem on the chalkboard. For example, ANOTHER WAY TO WRITE FIVE RABBITS TAKE AWAY THREE RABBITS IS $5 - 3 = 2$ (FIVE MINUS THREE EQUALS TWO). Help the children learn what the minus sign means by playing games and doing activities that require the use of the minus sign with the equals sign. For example, try the following.

SUBTRACTION: USING NOTATION AT THE CONNECTING LEVEL

OBJECTIVE: The children will connect symbols to problems using numerals and the minus action symbol.

MATERIALS: Objects to count and two dice

PLAYING THE GAME: The children take turns rolling the dice to find out which numbers to subtract. First they identify the larger number and count out that amount of counters. The teacher writes [larger number −]. Then they remove the smaller number of counters and the teacher continues [larger number − smaller number]. Then the children identify how many are left in the original pile and the teacher completes the equation [larger number − smaller number = difference]. For example, they roll six and two. They make a group of six counters. Teacher writes 6 −. Then they remove two counters and the teacher continues 6 − 2. Then they identify the difference (four). The teacher finishes the equation, $6 - 2 = 4$.
FOLLOW-UP: After working in small groups with the teacher, students can work independently with problems written on cards: $5-1$, $3-2$, etc. They can also make up problems using dice or pulling numbers out of a fishbowl. As you observe what the children are doing, stop and ask them to read the problems to you.

As with addition, when the students are comfortable with connecting the symbols to the problems and using the minus symbol, they can begin to write the notation themselves. Start with problems where you write the notation and the children copy you before they go on to independent work. For example, have everyone pick six counters. Have them see what kinds of problems appear as different amounts are taken away. After each problem, take a new group of six so the problems can be compared. You write the results of each take-away on the chalkboard (i.e., $6 - 1 = 5$) and have the children copy it on a piece of paper and put it next to the counters they have counted out. Follow up by having the students work independently, finding out how many subtraction problems they can discover starting with groups of different amounts up through six. When they are doing well up to six, have them move on to seven and above.

The notation for greater (more) than (>) and less than (<) is conventionally introduced in first grade along with subtraction but is usually not really understood until grade three. For the most part, students in early primary grades work with more and less using concrete materials such as described in Unit 11, Comparing. As they begin to understand the concepts, they can apply them to playing games. For example, lotto and bingo boards (see Unit 24) can be used. For bingo the players can roll a die and cover a square on their card containing a number or set that is more than or less than the number rolled. For lotto they would pick a numeral or set card and again cover a card on the board that was either more than or less than the numeral or set on the card. As students become familiar with the action symbols, cards could be used that indicate that they pick an amount or numeral that is > (more or greater than) or < (less than) one of those on the card. They could then move on to using cards that indicate an amount such as [____ > 2] or [____ < 5] and thus require the use of addition or subtraction to arrive at a selection.

MULTIPLICATION

Conceptualizing multiplication requires that the students understand what equal quantities are. Then they can proceed to learn that multiplication is a shorthand way of adding equal quantities. That is, 4 x 3 is the Same as $3 + 3 + 3 + 3$. Multiplication also involves learning the application of terms such as *factors* (the two numbers that are operated on) and *product* (the result of the operation). Students also learn the action terms *times* and *equals* and connect them to the action signs (× and =). Multiplication with concrete objects was introduced prior to the primary level and continues at this level for most primary students. Notation and the more formal aspects may be introduced toward the end of the primary level, but students are not usually proficient at the most fundamental level until fourth grade.

Assessment

Assessment of children's understanding of multiplication, as with the other whole number operations, is more than finding out if children know the number facts. It is important to observe the process each child goes through in dealing with quantities and ask questions that will provide a view of the thought behind their actions.

Observations can be made during naturalistic and informal activities. Use the terms *rows*, *stacks*, *groups*, and *sets* to refer to equal sets that will be added. There is no rush to use the term *times*. Many children learn to recite the times tables by heart without any understanding of what *times* really means. First children must understand that when they multiply they are counting groups of objects, not individual objects. Watch for incidents when children work with equal groups. For example, Dean comments that every child at his table has three carrot sticks. Theresa makes sure that each of the six children working in the science center receives four bean seeds to plant. Chan tells Ms. Hebert that he has purchased three miniature dinosaurs for each of the five friends invited to his birthday party. She asks him if he can figure out how many he bought altogether.

Multiplication can also be assessed by using an interview approach. The following is a sample task.

SAMPLE ASSESSMENT TASK

Concrete Operations Ages 7–8

METHOD: Interview

SKILL: Child is able to demonstrate readiness for multiplication by constructing equal groups of different sizes from groups of the same size

MATERIALS: Twenty counters (cube blocks, Unifix® cubes, chips)

PROCEDURE: Make two groups of six counters each. Ask the child, MAKE THREE GROUPS OF TWO CHIPS (BLOCKS, CUBES) EACH WITH THIS BUNCH OF SIX CHIPS (BLOCKS, CUBES). When the child finishes (right or wrong), point to the other group of counters, NOW MAKE TWO GROUPS OF THREE WITH THESE CHIPS (BLOCKS, CUBES).

EVALUATION: Note if the child is able to make the two different subgroups. Children who are not ready for multiplication will become confused and not see the difference between the two tasks.

INSTRUCTIONAL RESOURCES

Charlesworth, R., and Lind, K. 1990. *Math and Science for Young Children*. Albany, N.Y.: Delmar.

Instruction

Just as with addition and subtraction, instruction in multiplication begins with naturalistic and informal experiences that familiarize children with quantities and how they relate to each other. Students can be guided toward constructing their own concepts if they are provided with games and word or story problems to solve and encouraged to make up their own problems. Richardson, in *Developing number concepts using Unifix® cubes*, suggests that the children should first be asked to look for equal groups in the environment. How many tables have four chairs? How many girls have two barrettes in their hair? How many children have three cookies for dessert? How many parts of the body can they identify that come in groups of two? What parts do cars have that come in groups of four?

Of course, word or story problems are an essential ingredient in the instruction of multiplication just as they are for subtracticn and addition. Problems should be set in real-life contexts and acted out. Dean gives four children two crayons each. How many crayons did he pass out? Chan makes three stacks of books. He puts three books in each stack. How many books does he have? Ann gives each of the five people at her table four pieces of paper. How many pieces of paper did she pass out?

Build models of multiplication problems with the students. Counters can be stacked, put in rows, and placed in groups as illustrated in Figure 27–5. For example, working with inch cubes:

- Make three stacks of four cubes each.
- Make four rows of five cubes each.
- Make six groups of two cubes each.

Notation can be introduced gradually. Number symbols can be connected to problems as you find that the children have an understanding of the process of multiplication. Formal introduction of multiplication can begin with modeling. Act out a problem and then explain

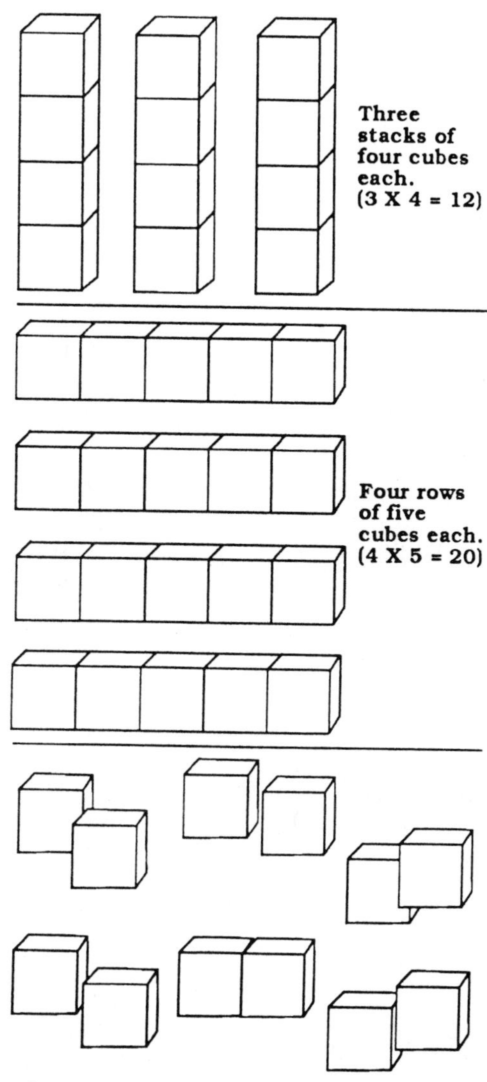

Figure 27–5 Equal rows, stacks, and groups are the basis of multiplication.

that there is another way to record the information. Write the number sentence for the problem on the chalkboard. For example: ANOTHER WAY TO WRITE "DEAN GAVE FOUR CHILDREN TWO CRAYONS EACH" IS [4

GROUPS OF 2 = 8]. ANOTHER WAY TO WRITE "CHAN HAS THREE STACKS OF THREE BOOKS" IS [3 STACKS OF 3 = 9]. Do several problems in this manner. Then introduce the multiplication sign. First model. Explain that there is another shorter way to write problems. Erase "stacks of" or "groups of" and write × in its place. Then write problems on the board and have the children work them out with their counters. Next, have them work out problems and copy you as you write the whole equation, i.e., [4 × 2 = 8]. Finally, have them make models and write the equations on their own.

The following is an example of an independent activity that can be done with multiplication using notation.

MULTIPLICATION: USING NOTATION AT THE CONNECTING LEVEL

OBJECTIVE: The children will connect symbols to problems using numerals and the times action symbol.

MATERIALS: Counters, a die, several small containers, and a sheet for recording the problems

ACTIVITY: Children can work on their own writing equations. First they decide how many containers to use. They can roll the die or just pick a number. They line up the cups and then, starting with zero, one at a time, fill the cups and write out the resulting equation. For example, if they pick four cups, they would first put zero blocks in each cup and write out the equation, then one block in each, two, three, and so on. Their work would look like this:

FOLLOW-UP: Develop some more independent activities using the resources suggested at the end of the unit.

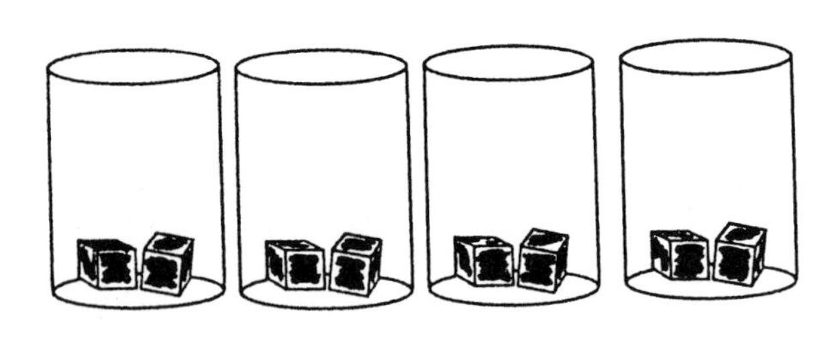

To find out if a child really understands notation, the same type of procedure can be used as was used for addition and subtraction. That is, show the child three or more eqivalent groups. Explain that you are going to put the groups together into one group and ask him to write what you did. For example, if you show three groups of four, he should write [3 × 4 = 12].

DIVISION

Division is an activity that children engage in frequently during their natural everyday activities (see Unit 25). They are encouraged to share equally, and they are often asked to pass out items so that everyone has the same amount

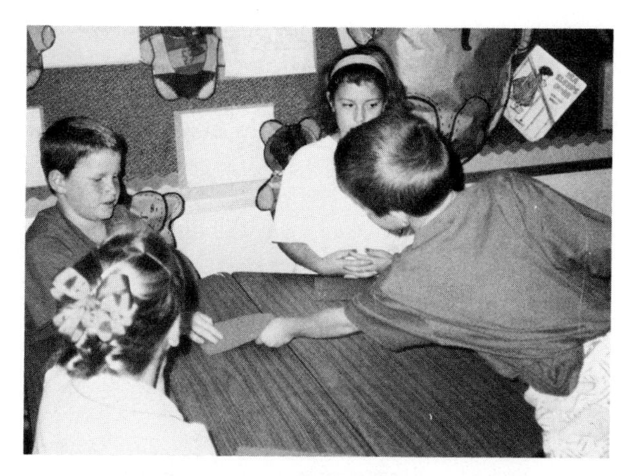

Figure 27–6 Passing out things so that everyone has an equal amount is an initial step in division.

(Figure 27–6). Formal instruction in division is usually introduced toward the end of the primary period during the third grade, but children are not expected to be proficient in doing division problems until the fifth grade. Division is used to solve two types of problems:

- Grouping. The process of grouping is used to find out how many subgroups of a particular size a larger group contains. For example, George has fifteen blocks. He wants to make towers that are five blocks high. How many can he make?
- Sharing. Sharing is the process of dividing a larger group into a particular number of groups to find out how many items will be included in that number of subgroups. Six children will work in the Science Center exploring the reaction of different types of items when they are touched by magnets. There are thirty-two items. How many will each child get? Are there any left over?

The children do not have to distinguish between these types of problems. They serve mainly as a guide for making up problems for them to explore. Eventually they will learn the terminology of division [*dividend* ÷ *divisor* = *quotient*] and if some are left over, *remainder*.

Assessment

Assessment of children's understanding of division focuses on the processes they use to group and to share.

SAMPLE ASSESSMENT TASK

Concrete Operations Ages 7–8

METHOD: Interview

SKILL: Child can demonstrate an understanding that division consists of grouping or sharing objects

MATERIALS: Thirty counters (cube blocks, Unifix® cubes, chips) and five small containers (such as clear plastic glasses)

PROCEDURE: Put out eight chips and four containers. DIVIDE UP THE CHIPS SO THAT EACH CUP HAS THE SAME AMOUNT. When the chips are divided ask, HOW MANY CUBES DO YOU HAVE IN EACH CUP? The child should respond "Two in each cup" rather than "I have two, two, and two." Try the same procedure with more cups and larger amounts to divide. Then try it with uneven amounts. Note if the child becomes confused or can recognize that there are more than are needed. Also do some sharing problems. That is, for example, put out sixteen chips. I WANT TO GIVE THREE FRIENDS THE SAME AMOUNT OF CHIPS. HOW MANY WILL EACH ONE RECEIVE? ARE THERE ANY LEFT OVER?

EVALUATION: Note how the children handle the problem. Do they proceed in an organized fashion? Can they deal with the remainders?

INSTRUCTIONAL RESOURCES

Charlesworth, R., and Lind, K. 1990. *Math and Science for Young Children*. Albany, N.Y.: Delmar.

Instruction

Division also begins with naturalistic and informal experiences. Children can be given many tasks that give them division experiences. Passing out items, putting items into groups to be shared, and finding out if there is enough for everyone are opportunities to develop the division concept (Figure 27–6). Games and story problems can be used as guides in supporting the child's construction of the concept of division as he ventures into more formal activities.

As with the other whole number operations, begin formal instruction by doing concrete problems. As Richardson suggests, tell the children stories and have them act them out. Start with real objects from the classroom. For example:

- Ann has sixteen pieces of paper. Each child in her group needs four pieces. How many children can receive four pieces of paper?
- Jason, Chan, and Vanessa want to feed the guinea pig. The guinea pig gets six pellets of food. How many pellets can each child give it?

Next, have the children act out similar stories, using counters to represent real objects. Have the children make many models by constructing rows and stacks and dividing them into groups.

DIVISION: MAKING MODELS

OBJECTIVE: The children will construct models of division.

MATERIALS: Counters (cube blocks, Unifix® cubes, chips) and several 16-ounce clear plastic cups

ACTIVITY: Using different amounts initially and having the children divide them up into groups of different sizes and into different numbers of groups, have the students do many problems using the following patterns:

1. MAKE A ROW (TRAIN) WITH (number) OF BLOCKS (CUBES, CHIPS). HOW MANY STACKS OF (number) CAN YOU MAKE?
   DIVIDE YOUR ROW (TRAIN) OF (number) INTO (number) OF ROWS. HOW MANY CUBES (BLOCKS, CHIPS) ARE IN EACH ROW?
2. GET (number) CUPS. DIVIDE (number) OF CUBES INTO EACH CUP SO THAT THERE IS THE SAME AMOUNT OF CUBES IN EACH CUP. Continue with different numbers of cups and counters.

FOLLOW-UP: Develop some more independent activities using the resources suggested at the end of the unit.

Division notation can be introduced with modeling. Act out problems just as you did with the other whole number operations. For example:

- JOHN HAS TWELVE CRACKERS. Write [12] on the board. HE HAS THREE FRIENDS. HE WANTS TO GIVE HIMSELF AND EACH FRIEND THE SAME NUMBER OF CRACKERS. Write [12 ÷ 4]. EACH CHILD GOT THREE CRACKERS. Write [12 ÷ 4 = 3].
- THE CHILDREN ARE GOING TO EXPLORE HOW PENDULUMS WORK. THERE ARE FOUR PENDULUMS AND EIGHT CHILDREN. HOW MANY CHILDREN WILL HAVE TO SHARE EACH PENDULUM? EIGHT CHILDREN (write 8), DIVIDED BY FOUR PENDULUMS (write [8 ÷ 4]) EQUALS TWO CHILDREN MUST SHARE EACH PENDULUM (write [8 ÷ 4 = 2]).

After you have modeled several problems, the children can go to the next step by acting out the problems and copying what you write. Next, you can give them problems that they can act out with counters and write the equations themselves. When the children have completed the equations, you can write them on the board and they can check theirs. Check each child's model and equation. Note if there are any difficulties and help children figure out how to act out and write the equation correctly. Move on to giving the children written problems (i.e., 10 ÷ 2) and have them act them out using counters. Finally, have them make up their own problems, act them out, draw them, and write them.

To find out if a child really understands notation, use the same procedure as suggested for the other whole number operations. That is, act out division and ask the children to write what you did. For example, count out fifteen counters and divide them into five groups of three. See if the children can write [15 ÷ 5 = 3] and tell you that fifteen divided into five groups makes three in each group.

COMPUTERS AND CALCULATORS

Computers and hand calculators are very useful tools for supporting the exploration of whole number operations and the properties of whole numbers. There are a multitude of computer programs available for working with basic whole number operations. Several are included in the Activities section of this unit. Most of these computer programs are designed to help children remember the basic addition, subtraction, multiplication, and division facts. Many have interesting graphics that catch the children's attention and make drill and practice fun. With the capability of letting the children know right away whether or not the response is correct, the programs give the children immediate feedback and allow them to move along at their own pace.

During preprimary activities, children have explored some of the basic calculator capabilities. During primary activities, the calculator can be used for further exploration, for checking and comparing with manual calculations, and for problem solutions. A very basic activity with the calculator is the exploration of multiples. Young children are fascinated with rhymes such as "Two, four, six, eight, who do we appreciate." This type of counting is called *skip counting*. Skip counting in the example defines the multiples of two, that is, all the numbers that result when multiplied by 2. That is, $2 \times 2 = 4$, $2 \times 3 = 6$, $2 \times 4 = 8$, and so on. Children can explore these properties with the calculator through calculator counting (Figure 27–7).

EVALUATION

Evaluation, just as with assessment, should be done first with concrete tasks, observing both the process and the product. The tasks in Appendix A can be used for evaluation as well as for initial assessment. Standardized achievement tests should not be administered until the stu-

COUNT WITH YOUR CALCULATOR

Push (C)(0)(+)(2)(=)

What do you see? ____________

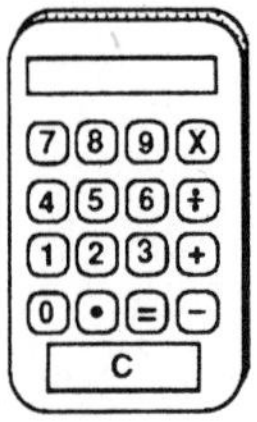

Push (=) again

What do you see? ____________

What will the calculator show if you push (=) again? ____________

Do it. Were you correct? ____________

Push (=)(=)(=).

Guess what the calculator will show each time. ________ ________ ________

What happened? ____

Complete the following. Then use your calculator to see if you are correct.

Push (C)(0)(+)(3)(=)

The calculator will show ____________

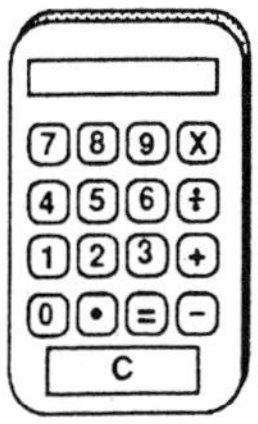

Push

(=) ____________

(=) ____________

(=) ____________

Push (C)(0)(+) (4)(=)

The calculator will show ____________

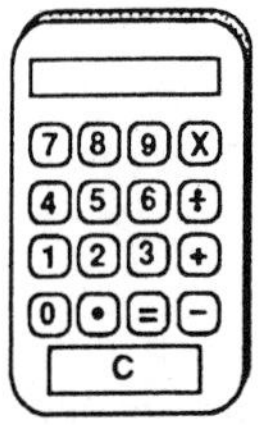

Push

(=) ____________

(=) ____________

(=) ____________

Figure 27–7 The calculator can be used to count by equal multiples.

dents have the concepts internalized with concrete activities. For guidelines for testing young children, see the NAEYC position statement on standardized testing of young children 3 through 8 years of age.

SUMMARY

The whole number operations of addition, subtraction, multiplication, and division have begun to develop through naturalistic and informal experiences during the preprimary years. As children's cognitive development takes them into the concrete operational level, they are ready for the move into learning about action symbols and written notation. Conventionally, children start with addition and subtraction in the beginning of the primary period (grade one). They move on to more complex addition and subtraction and the introduction of multiplication in the second grade. Division is usually introduced in the third grade. Each operation begins with the informal activities and acting out problems with concrete objects and moves gradually into the use of formal notation and written problems. Story or word problems that put the operations into real-life contexts are the core through which the whole number operations are constructed in the young child's mind.

FURTHER READING AND RESOURCES

Baratta-Lorton, M. 1976. *Mathematics their way.* Menlo Park, Calif.: Addison-Wesley.

Baroody, A. J. 1987. *Children's mathematical thinking.* Menlo Park, Calif.: Addison-Wesley.

Burton, G. M. 1985. *Towards a good beginning.* Menlo Park, Calif.: Addison-Wesley.

Howden, H. 1989. Teaching number sense. *Arithmetic Teacher.* 36(6): 6–11.

Kamii, C. K. 1985. *Young children reinvent arithmetic.* New York: Teachers College Press.

Kamii, C. K. 1989. *Young Children Continue to Reinvent Arithmetic.* New York: Teachers College Press.

Labinowicz, E. 1985. *Learning from children: New beginnings for teaching numeral thinking.* Menlo Park, Calif.: Addison-Wesley.

Lewis K. E. 1985. From manipulatives tc computation: Making the mathematical connection. *Childhood Education.* 61: 371–374.

Richardson, K. 1984. *Developing number concepts using Unifix Cubes.* Menlo Park, Calif.: Addison Wesley.

Scott, L. B., and Garner, J. 1978. *Mathematical experiences for young children.* St. Louis: McGraw-Hill.

Starkey, M. A. 1989. Calculating first graders. *Arithmetic Teacher.* 37(2): 6–7.

See the monthly issues of the *Arithmetic Teacher* for activities, materials, and reviews of computer software.

SUGGESTED ACTIVITIES

1. Develop at least two of the independent activities suggested by Kathy Richardson and/or the group games described by Constance Kamii in their respective books. Add the instructions to your Activities File. If possible, try out the activities with one or more primary grade children. Report to the class regarding what you developed, whom you tried it with, what happened, and how well you believe it worked.

2. Do a comparison of concrete and paper and pencil modes of math assessment. Select three whole number operations assessment tasks from Appendix A. Devise a paper and pencil test that includes the same kinds of problems. Use the interview tasks with two primary students. Return a week later and use the conventional paper and pencil test with the same two students. Take careful notes on the process used by the students to solve the problems. Write a report describing the two types of assessment tasks, the results including the students' performance, and your comparison and evaluation of the two methods of assessment.

3. Observe in two or more primary classrooms during times when math activities are occurring. Report to the class, including a description of what you observed, how the instruction compared with the methods described in this unit, and including suggestions of any changes you would make if you were teaching in those classrooms.

4. Review some whole number operations soft-

ware using the format described in Unit 9, Activity 3. You might try one of the following programs if available:

Alligator Alley. DLM Teaching Resources. Allen, Tex.: Apple IIe.

Balancing Bear. Sunburst. Pleasantville, N.Y.: Apple II.

Challenge Math. Sunburst. Pleasantville, N.Y.: C64; Apple II.

Digitosaurus. Sunburst. Pleasantville, N.Y.: 256K IBM PC; 256K Tandy 1000.

Factmaster: Addition and Subtraction and *Factmaster: Multiplication and Division*. Stone and Associates, La Jolla, Calif.: IBM-PC and PC Jr.

Math Blaster. Davidson. Rancho Palos Verdes, Calif.: C64, Apple.

Math Rabbit. American Guidance Service. Circle Pines, Minn.: 64K Apple II, IBM-PC.

REVIEW

A. Describe what is included in whole number operations.

B. Match the action symbols in Column I with the words they stand for in Column II.

Column I	Column II
1. +	a. multiply, times
2. −	b. equals
3. ×	c. more than or greater than
4. ÷	d. add, plus
5. =	e. less than
6. <	f. divide
7. >	g. subtract, minus

C. Select the statements that are correct:

a. Constance Kamii believes that children learn arithmetic by being told about it.

b. Kamii believes that group games are excellent vehicles for supporting the construction of math concepts.

c. Paper and pencil worksheets are excellent for helping children develop the logic of arithmetic.

d. Kamii believes that children should always work alone at their desks when learning math.

D. Explain why standardized testing is a threat to the development of logical thinking.

E. When introducing whole number sentences (equations), a three-stage sequence should be followed. List the steps.

F. Explain briefly the instructional processes for addition, subtraction, multiplication, and division.

UNIT 28 Patterns

OBJECTIVES

After studying this unit, the student should be able to:

- Describe patterning as it applies to primary age mathematics
- Explain why the cognitive developmental level of primary children makes looking for patterns an especially interesting and appropriate activity
- Assess primary grade children's understanding of patterning
- Plan and teach patterning activities for primary children

Patterning and ordering were introduced in Unit 17. Ordering, or putting things into a sequence, is basic to patterning. Patterning is the process of discovering auditory, visual, and motor regularities. There are many regularities in the number system that children must understand. During the primary years, children work with more complex problems with concrete materials, connect concrete patterns to symbols, and learn to recognize some of the patterns and higher level sequences in the number system.

This unit focuses on extending the concept of patterning to more complex patterns and connecting symbols and patterns. It also describes activities for looking at patterns in the real world.

ASSESSMENT

Look back at Unit 17 for a description of naturalistic and informal patterning behaviors that can be observed during children's activities. By the primary grades children should be able to copy and extend patterns with ease. During the primary grades they develop the ability to extend patterns further, make more complex patterns, become more adept at describing patterns with words, build their own patterns, and see patterns in numbers. See Unit 17 for a sample assessment task procedure for pattern copying, extending patterns, describing patterns and more difficult extensions. The following are examples of higher level assessment tasks.

SAMPLE ASSESSMENT TASK

Concrete Operations Ages 6–8

METHOD: Interview
SKILL: The child can extend complex patterns in three dimensions by predicting what will come next
MATERIALS: Inch or centimeter cubes, Unifix® cubes, or other counters that can be stacked
PROCEDURE: Present the child with various patterns made of stacked counters. Ask the child to describe the pattern and to continue it as far as he can. Stack the blocks as follows one pattern at a time:

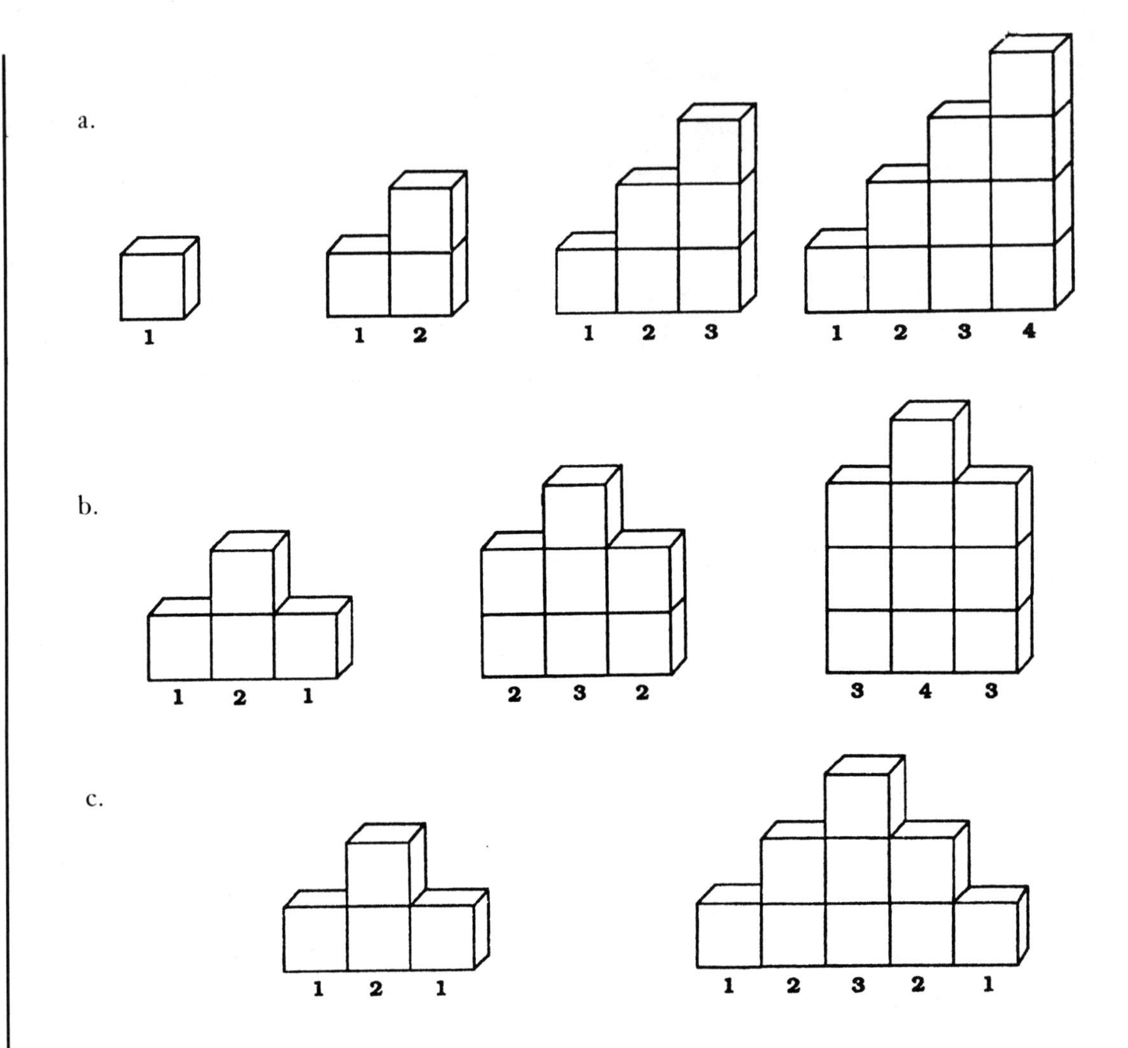

For each pattern ask, TELL ME ABOUT THIS PATTERN. WHAT COMES NEXT? HOW DO YOU KNOW? CONTINUE THE PATTERN FOR ME.

EVALUATION: Note if the child can continue each pattern and state his rationale. Note where the child might need further help and practice.

INSTRUCTIONAL RESOURCES

Charlesworth, R., and Lind, K. 1990. *Math and Science for Young Children*. Albany, N.Y.: Delmar.

SAMPLE ASSESSMENT TASK

Concrete Operations Ages 7–9

METHOD: Interview
SKILL: The child can use a 00 to 99 chart to discover and predict number multiple patterns.
MATERIALS: Inch or centimeter cubes, Unifix© cubes, or other counters, and a 00 to 99 chart (Figure 28–1)
PROCEDURE: Start a pattern using multiples of two blocks. Ask the child, CIRCLE OR MARK THE AMOUNT IN MY GROUP ON THE CHART. If the child has a problem, show him the 02 and circle it for him if necessary. Next to the group of two blocks construct a group of four blocks. Use the same procedure as above. Continue up to ten. Then ask, SHOW ME WHICH NUMBERS YOU WOULD CIRCLE IF I KEPT CONTINUING WITH THIS PATTERN. When children can predict accurately with multiples of two, try threes, fours, fives, and so on.
EVALUATION: Note whether the children can connect the numbers in the pattern to the numerals on the chart and whether they can predict what comes next. If they cannot accomplish these tasks, note where their errors are: Do they need more help with basic pattern construction? With counting? With connecting sets to symbols? With finding numbers on the chart?

INSTRUCTIONAL RESOURCES

Charlesworth, R., and Lind, K. 1990. *Math and Science for Young Children.* Albany, N.Y.: Delmar.

ACTIVITIES

Children who have reached concrete operations are in a stage of cognitive development in which they are naturally seeking out the rules and regularities in the world. Patterning activities fit the natural inclinations and interests of children in this stage. The following examples are adapted from Richardson's *Developing number concepts using Unifix® cubes* and Baratta-Lorton's *Mathematics their way*. Both of these resources contain many patterning activities.

PATTERNING: INCREASING PATTERNS

OBJECTIVE: To copy and extend patterns using objects
MATERIALS: Counters such as chips, cube blocks or Unifix® cubes; paper and pencil
ACTIVITY: The following are examples of patterns that can be developed. In each case, the teacher models the first three elements in the pattern and then the children are asked to predict what comes next and to extend the pattern as far as they can. The children may write down the pattern in numerals under each element and compare the patterns with both the objects and the numeral representations.

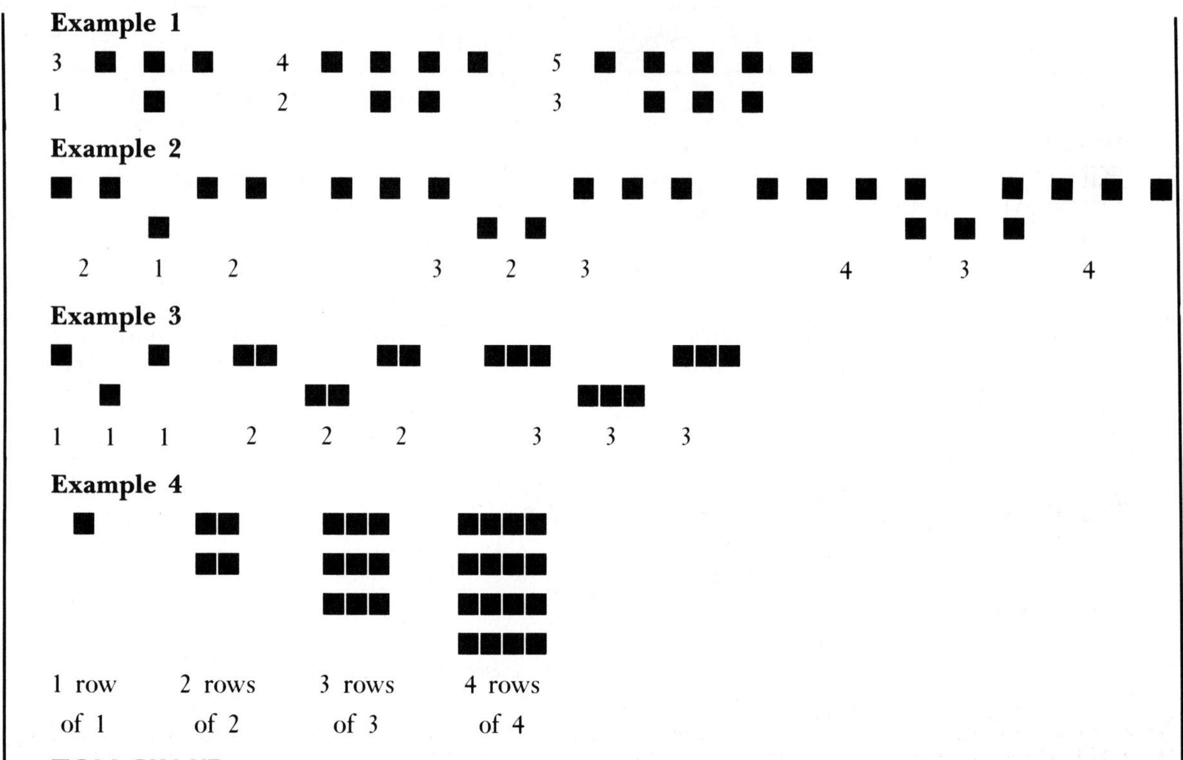

FOLLOW-UP: Have children work with various types of patterns until you believe they have grasped the concept. Then present the higher level pattern activities that follow.

PATTERNING: TASK CARDS

OBJECTIVE: To copy and extend patterns using task cards

MATERIALS: Counters such as chips, inch or centimeter cubes, or Unifix® cubes; task cards with the first three steps in a pattern. For example:

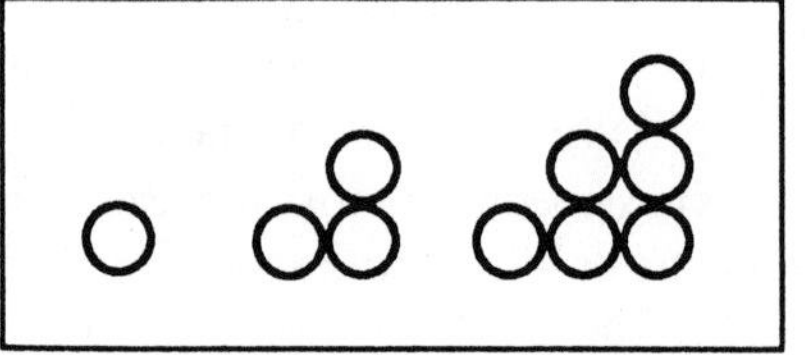

1	2	3	4
2 2	4 4	6 6	8 8

00	01	02	03	04	05	06	07	08	09
10	11	12	13	14	15	16	17	18	19
20	21	22	23	24	25	26	27	28	29
30	31	32	33	34	35	36	37	38	39
40	41	42	43	44	45	46	47	48	49
50	51	52	53	54	55	56	57	58	59
60	61	62	63	64	65	66	67	68	69
70	71	72	73	74	75	76	77	78	79
80	81	82	83	84	85	86	87	88	89
90	91	92	93	94	95	96	97	98	99

Figure 28–1 00 to 99 Chart

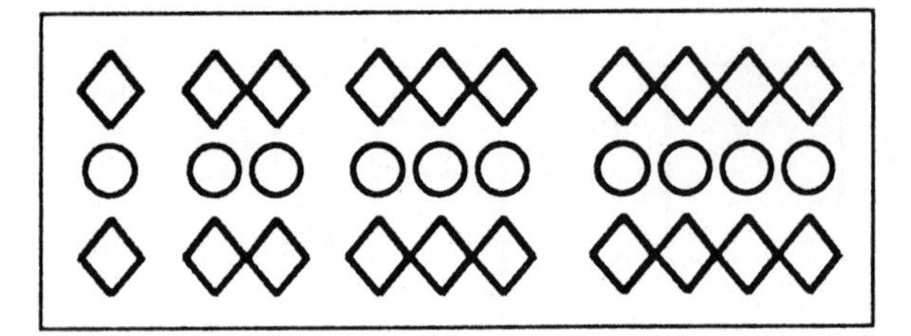

Task cards with the first step in a pattern.

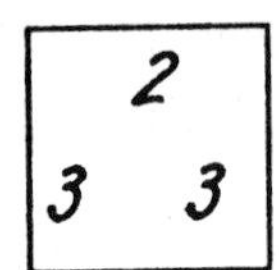

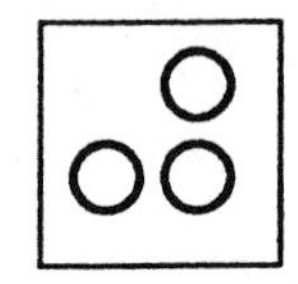

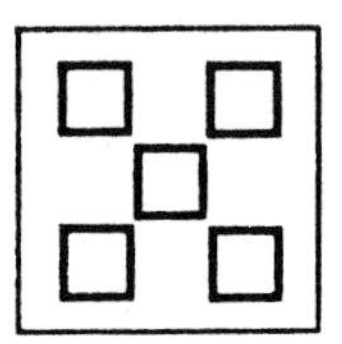

ACTIVITY: Provide the task cards and a good supply of counters. The children copy the models and then proceed to extend the patterns. With the three-step models, the pattern is set. With the one-step models, the children can create their own rules for extending the patterns. Always have them explain their pattern to you.

FOLLOW-UP: When the children have the concept of working from the abstract to the concrete with the task cards, have them create their own patterns, first with objects and then drawing them using the objects as models.

PATTERNING: 00 TO 99 CHART ACTIVITIES

OBJECTIVE: For children who have the concept of place value (see Unit 30) for the one's and ten's place, 00 to 99 chart activities can be used. The objective is to perceive patterns on the chart in pictorial form.

MATERIALS: Copies of the 00 to 99 chart (see Figure 28–1) and counters

ACTIVITY: On the chart the children can color in or mark the amounts in their patterns. For example, they made the following pattern:

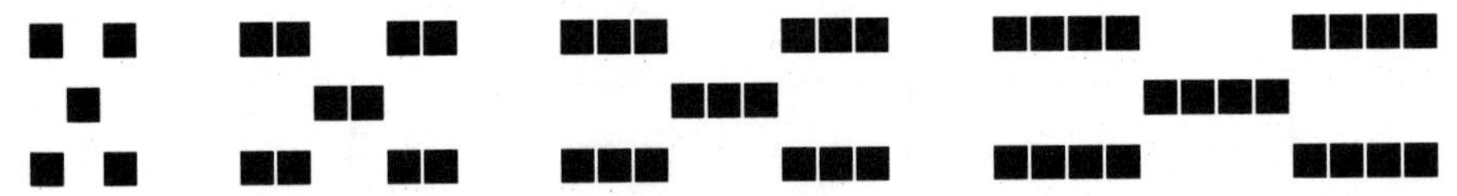

The children then mark off on the chart the amount in each part of the pattern: 5, 10, 15, 20, and so on.

FOLLOW-UP: The children can transfer to the charts from patterns you provide and then move on to patterns that they devise themselves (Figure 28–2).

Figure 28–2 Developing patterns begins with concrete materials such as Unifix Cubes.

PATTERNING: EXPLORING NATURAL MATERIALS

OBJECTIVE: To be able to observe and describe patterns in natural materials

MATERIALS: Fruits and vegetables such as cabbage, onion, orange, lemon, grapefruit, apple, and walnut; magnifying glass, pencil, and paper

ACTIVITY: Let the children explore and examine the whole fruits and vegetables. Encourage them to describe what they see and feel. Suggest that they examinine them with the magnifying glass and draw them if they wish. After a few days, ask them to predict what each item looks like inside. Then cut each one in half. Talk about what they discover. Compare what they see with what they predicted. Suggest that they draw the inside patterns.

FOLLOW-UP: Have each child use his or her pictures to make a "What's inside?" book. The children can write or dictate what they know about each item.

PATTERNING: MULTIPLES GRAPHS

OBJECTIVE: To collect data regarding natural patterns and depict the data on graphs

MATERIALS: 1. Number line templates. These templates are made from heavy tagboard. The numbers are written across. A hole is cut or punched below each one so that the numbers can be copied.

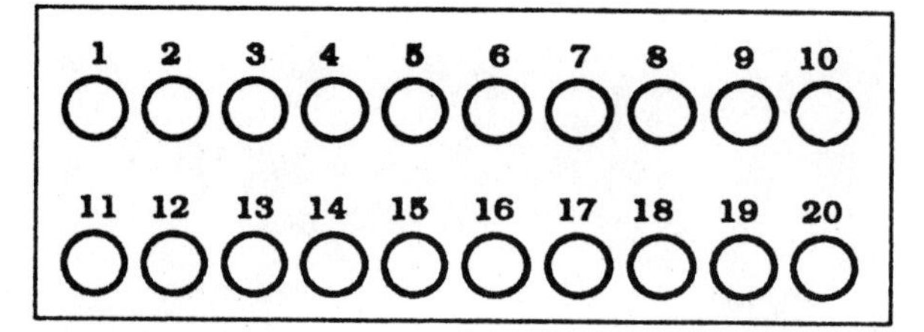

2. Large sheets of manila paper, rulers, crayons, markers, picture magazines

ACTIVITY: With the children discuss a question such as:
- How many eyes do five people have among them?
- How many legs do three chairs have among them?

Have the children draw lines about three to four inches apart on a large piece of manila paper with their rulers. Have them copy their number line at the bottom of the paper using a template. Have them draw or cut out and paste pictures of people's faces or of chairs with legs on their paper as depicted in Figure 28–3. Then have them record the number of eyes (legs) down the right side of the paper and circle the corresponding numerals on the number line. Ask them to examine the number line and describe the pattern they have made.
FOLLOW-UP: Have the children think of other items that they could graph in multiples to create patterns.

PATTERNING: DIVISION
OBJECTIVE: To make division patterns
MATERIALS: Large pieces of paper (11″ × 18″), and smaller pieces of paper (4¼″ × 5½″), scissors, crayons or markers, and glue
ACTIVITY: With the whole group, start with a large piece of paper. Have a child cut it in half and give the half to another child. Have each child cut the paper in half and give one part away. Keep a record of the number of cuts and the number of children until everyone has a piece of paper. Next give each child a large piece of paper and a small piece of paper. Have them glue the smaller piece at the top of the larger sheet (Figure 28–4). Have them take another small piece and cut it in half and glue the two parts on the large paper below the first whole piece (Figure 28–4). Have them take another small piece and cut it in half and then cut each half in half. Glue these four parts on the large sheet. Let them continue as long as they wish. Have them record the number of pieces in each row on the right-hand side of the chart.
FOLLOW-UP: Have the more advanced children cut three parts each time and see what kind of a pattern they make.

Figure 28–3 Multiples Graph

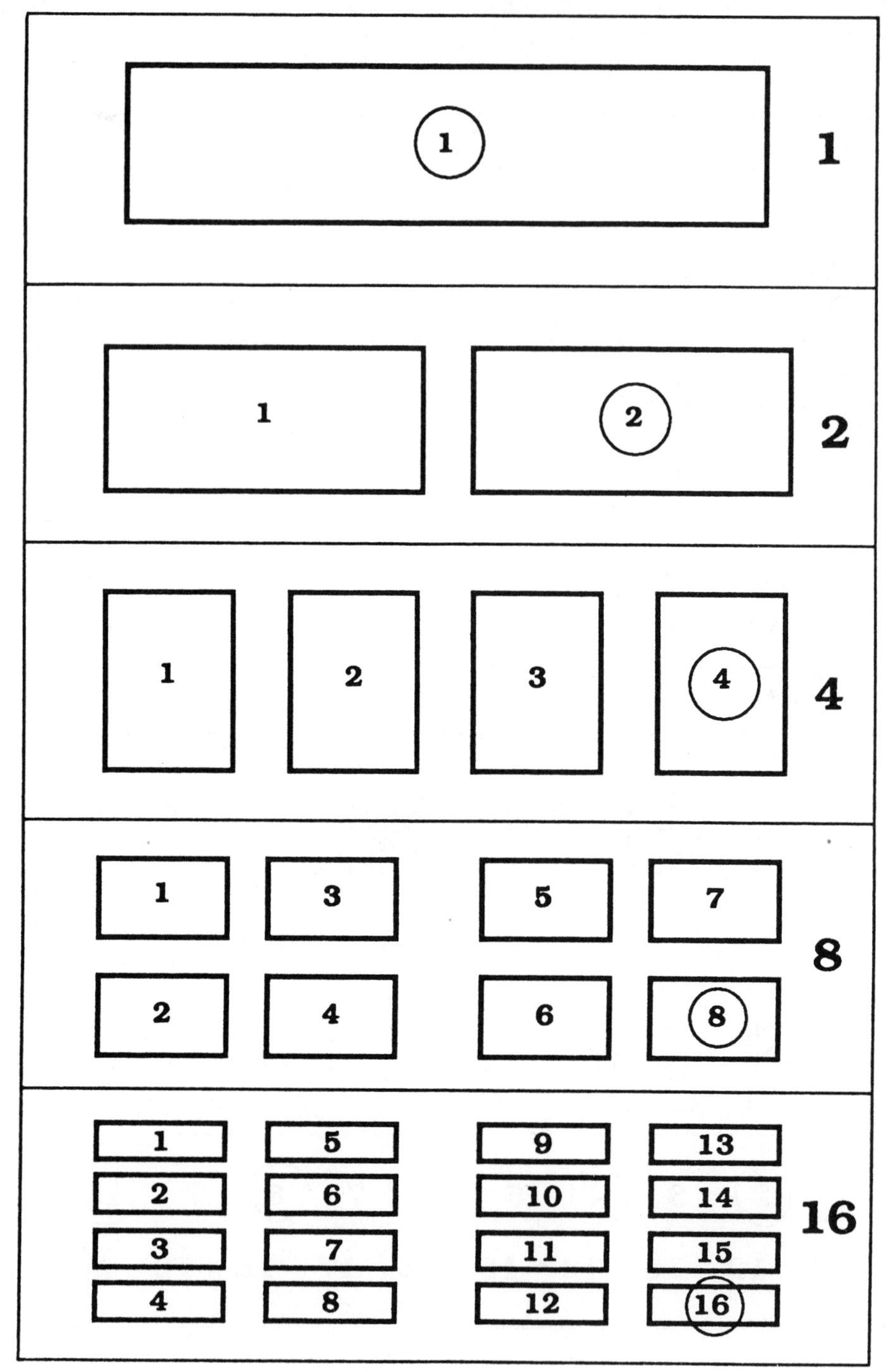

Figure 28–4 Patterning: Division

As already described in Unit 27, calculator activities are interesting ways to look at number patterns. After some practice with patterns as suggested in Unit 27, have the children discover and extend patterns with their calculators (Figure 28–5). For example:

FIND THE RULE AND FINISH THE PATTERN
Use your calculator.

7, 9, 11

PRACTICE PATTERN: 1, 3, 5, ____, ____, ____ RULE: +2

1. Rule: +2
   Pattern: 2, 4, 6, ____, ____, ____
2. Rule: +3
   Pattern: 3, 6, 9, ____, ____, ____
3. Rule: ______
   Pattern: 2, 5, 8, ____, ____, ____
4. Rule: ______
   Pattern: 12, 16, 20, ____, ____, ____
5. Rule: ______
   7, 9, ____, 13, ____, ____, ____
6. Rule: ______
   5, ____, 15, ____, ____, 30, 35, ____, ____

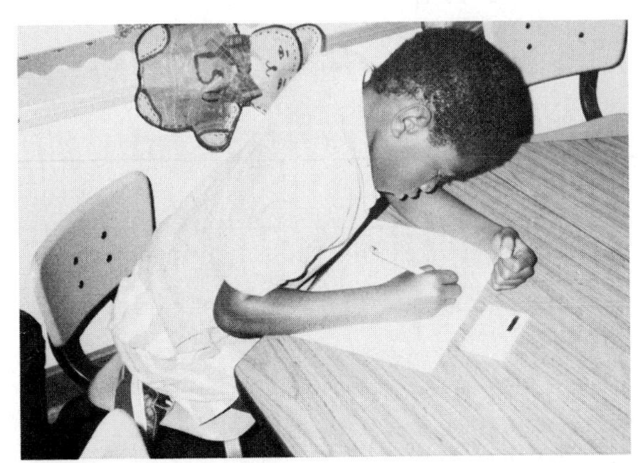

Figure 28–5 The hand calculator can be used to find number patterns.

Make up some more patterns for the children to explore with calculators.

As with other math concepts, pattern computer software is available. See the activities at the end of this unit and at the end of Unit 17 for suggestions.

EVALUATION

Note how the children deal with the suggested pattern activities. Use the assessment tasks in the Appendix for individual evaluation interviews.

SUMMARY

Primary level children extend the work they did with patterning at earlier levels. Now they begin to identify and work with patterns in the number system as they connect numerals to patterns and develop number patterns using counting and calculators. Number patterns are the basis for multiplication and division.

FURTHER READING AND RESOURCES

Adler, D.A. 1981. *Calculator fun.* New York: Franklin Watts.

Baratta-Lorton, M. 1976. *Mathamatics their way.* Menlo Park, Calif.: Addison-Wesley.

Frank, A. R. 1989. Counting skills—a foundation for early mathematics. *Arithmetic Teacher.* 37(1): 14–17.

Richardson, K. 1984. *Developing number concepts using Unifix® Cubes.* Menlo Park, Calif.: Addison-Wesley.

SUGGESTED ACTIVITIES

1. Administer the sample assessment tasks to a first grader, a second grader, and a third grader. Write a report describing what you did, how the children responded, and how they compared with each other. Include suggestions for further activities for each child.
2. Assemble the materials for one or more of the suggested instructional activities. Try out the activities with a small group of primary children. Report to the class on what you did and how the children responded. Evaluate the children's responses and the appropriateness of the activities. Describe any changes you would make when using these activities another time.
3. Using the evaluation system from Activity 8 in Unit 9, evaluate one or more of the following computer programs relative to its value for learning about patterning:

   Bounce. Sunburst. Pleasantville, N.Y.: 64K Apple II Family.

   Tonk in the Land of Buddy-Bots. Mindscape, Inc. Northbrook, Ill.: 64K Apple II Family; Atari; C-64; IBM PC with graphic, color adapter, and enhanced; IBM PCjr, memory required.

   Train set. Harper and Row. Scranton, Pa.: Apple II series; C-64.

REVIEW

A. Describe patterning as it applies to primary age mathematics. Give at least one example.

B. Explain why the cognitive developmental level of primary children makes looking for patterns an especially appropriate and interesting activity.

C. Match the following types of activities to the examples below.

1. Copy and extend patterns using objects.
2. Copy and extend patterns using task cards.
3. Observe and describe patterns in natural materials.
4. Collect data from naturally occurring objects and depict the data on a graph.

5. Make a division pattern.

Examples:

a. Chan carefully draws the pattern that appears in half of an orange.

b. Sara examines three groups of blocks and proceeds to create the group she believes should come next.

c. Theresa draws lines across a large sheet of paper, copies her number line with a template, then looks through a magazine for animals with four legs.

d. Vanessa has a large sheet of paper and several smaller sheets. First she glues one small sheet at the top of the large sheet and writes the numeral 1 to the right.

e. Brent examines some patterns drawn on a card. Then he takes some Unifix® cubes and makes the same designs and adds two more.

D. Describe how a 00 to 99 chart might be used to assist in illustrating number multiple patterns.

UNIT 29 Fractions

OBJECTIVES

After studying this unit, the student should be able to:

- Explain how the concept of fractions is based on an understanding of part/whole relationships
- Explain why primary children should not be rushed into using fraction notation
- Assess primary children's understanding of the concept of fractions
- Plan and teach fraction lessons appropriate for primary children

The fundamental concept of parts and wholes as the basis for the understanding of the concept of fractions was introduced at the beginning levels in Unit 14. Through naturalistic, informal, and structured experiences, preprimary children become familiar with three aspects of the part/whole concept: things have special parts, a whole object can be divided into parts, and sets of things can be divided into smaller sets. They also become familiar with the application of the terms *more*, *less*, and *same*. During the primary level, young children expand on the concrete activities that they engaged in during the preprimary level. It is important not to introduce notation and symbols too soon. Even nine-year-olds have difficulty with fractions at the symbolic level. This would indicate that for most children fraction symbols cannot safely be introduced until the end of the primary period (the latter part of grade three) and may not be fully understood until well into the intermediate level (grade four or higher). Fraction problems cannot be solved by counting as can whole number problems. This factor makes them much more abstract and thus more difficult.

At the presymbolic level, work with fractions should be limited to halves, thirds, and fourths. These are the fractions we deal with most frequently in life, and if children develop an understanding of them, they should be able to transfer this knowledge to fractions in general. Children can learn fraction terminology relative to concrete experiences without being concerned with the corresponding symbols. Terms such as *one-half*, *one-third*, and *one-fourth* can be associated with parts of concrete objects and subgroups of larger groups of concrete objects. During the primary period, children continue to work with fractions as part/whole relationships. They can work with volume, regions, length, and sets. Experiences with foods (such as cutting up a carrot) and cooking (measuring ingredients) involve volume. Regions are concrete and easy to work with. They involve working with shapes such as circles, rectangles, squares, and triangles. Lengths can also be divided into parts. Long, narrow pieces of paper, string, thread, and ribbon are useful for this type of activity. When working with sets, a whole set of objects serves as the unit to be divided into smaller subsets.

ASSESSMENT

In Unit 14, both observational and interview tasks for assessment of part/whole concepts were described. When children understand that a whole can be divided into parts, that when a quantity is divided its whole is conserved, and that the size of each part gets smaller as the number of equal divisions increases, then they are ready to understand fractions. If children evidence some of the behaviors described in Unit 14 as indicating an understanding of the part/whole concept and are able to respond successfully to the Unit 14 assessment tasks, then try the following kinds of higher level assessment tasks.

SAMPLE ASSESSMENT TASK

Concrete Operations Ages 6–8

METHOD: Interview

SKILL: The child can divide a rectangle into smaller equal parts.

MATERIALS: A supply of paper rectangles of equal size (8½″ × 2¾″) in four different colors and a pair of scissors

PROCEDURE: Show the child a paper rectangle. THIS IS A RECTANGLE. Place two more rectangles (color # 2) below the first one. HERE ARE TWO MORE RECTANGLES. ARE ALL THREE THE SAME SIZE? Be sure the child agrees. Let him compare them to be sure. NOW I'M GOING TO FOLD ONE OF THE RECTANGLES (color #2) SO BOTH PARTS ARE THE SAME. Fold the rectangle. NOW YOU FOLD THIS OTHER ONE (also color # 2) JUST LIKE I DID. Offer assistance if necessary. The three rectangles should look like this:

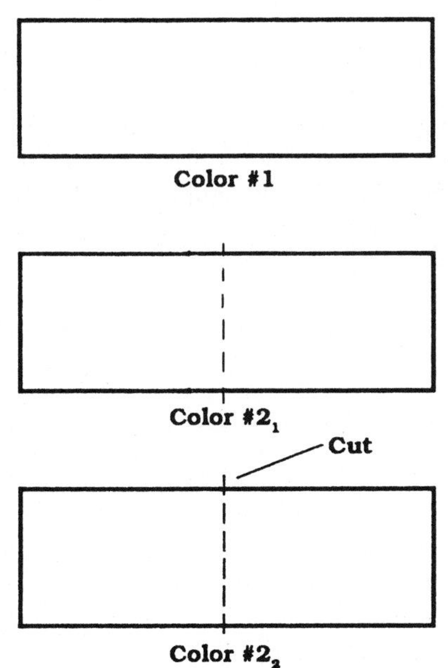

ARE THE PARTS OF (color #2) RECTANGLE THE SAME SIZE AS THE PARTS OF THIS ONE? (also color #2) SHOW ME HOW YOU KNOW. I'M GOING TO CUT THIS ONE (second color #2) ON THE FOLD. HOW MANY PARTS DO I HAVE NOW? IF I

PUT THEM BACK TOGETHER, WILL THEY BE THE SAME SIZE AS THIS WHOLE RECTANGLE? (color #1) AS YOUR RECTANGLE? WHAT IS A SPECIAL NAME FOR THIS AMOUNT OF THE WHOLE RECTANGLE? Point to the half. If the response is one-half, go through the procedure again with one-third and one-fourth, using colors #3 and #4 respectively.

EVALUATION: Note whether the child has to check on the equivalency of the three rectangles. Can he keep in mind that the parts still equal the whole, even when cut into two or more parts? Does he know the terms *one-half*, *one-third*, and/or *one-fourth*?

INSTRUCTIONAL RESOURCE

Charlesworth, R., and Lind, K. 1990. *Math and Science for young Children.* Albany, N.Y.: Delmar.

SAMPLE ASSESSMENT TASK

Concrete Operations Ages 6–8

METHOD: Interview

SKILL: The child can divide a set of objects into smaller groups when given directions using the term *one-half.*

MATERIALS: Ten counters (cube blocks, chips, Unifix® cubes, or other concrete objects)

PROCEDURE: Place the counters in front of the child. I HAVE SOME *(name of counters).* DIVIDE THESE SO THAT WE EACH HAVE ONE-HALF OF THE GROUP. If the child completes this task easily, go on to nine counters and ask him to divide the group into thirds, and eight counters and ask him to divide the group into fourths.

EVALUATION: Note the method used by the child. Does he use counting or does he pass the counters out: "One for you and one for me . . . ."? Does he really seem to understand the terms *one-half*, *one-fourth*, and *one-third*?

INSTRUCTIONAL RESOURCE

Charlesworth, R., and Lind, K. 1990. *Math and Science for young Children.* Albany, N.Y.: Delmar.

ACTIVITIES

In Unit 14, naturalistic and informal activities were emphasized as the foundation for structured experiences. These activities should be encouraged and continued. Primary children need to continue to have time to explore materials and construct their concept of parts and wholes through their own actions on the environment. There are many materials that children can explore independently in developing the foundations for the understanding of fractions. Any of the usual kinds of counting objects can be grouped into a set, which can then be divided (or partitioned) into smaller sets. Some materials for dividing single objects or shapes into parts are illustrated in Figure 29–1. The examples in-

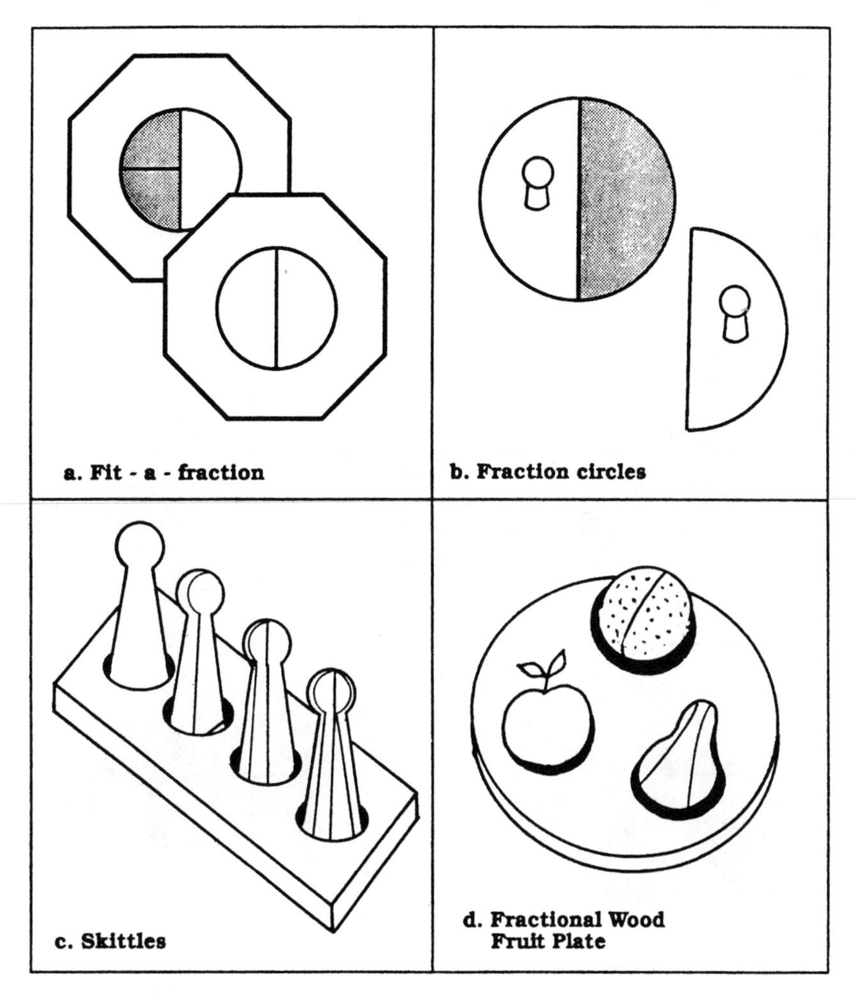

Figure 29–1 Examples of materials that can be used to explore how whole things can be divided into equal parts.

clude Fit-a-Fraction (Lakeshore), Fraction Circles and Skittles (Nienhuis-Montessori), and Fractional Wood Fruit Plate (ETA). Figure 29–2 illustrates materials that promote the construction and comparison of parts and wholes. These materials include unit blocks (Lakeshore), Cuisinaire™ rods (Cuisinaire of America), puzzles (Lakeshore), and Lego™ (Lakeshore).

The following are structured activities that can be used to develop fraction concepts.

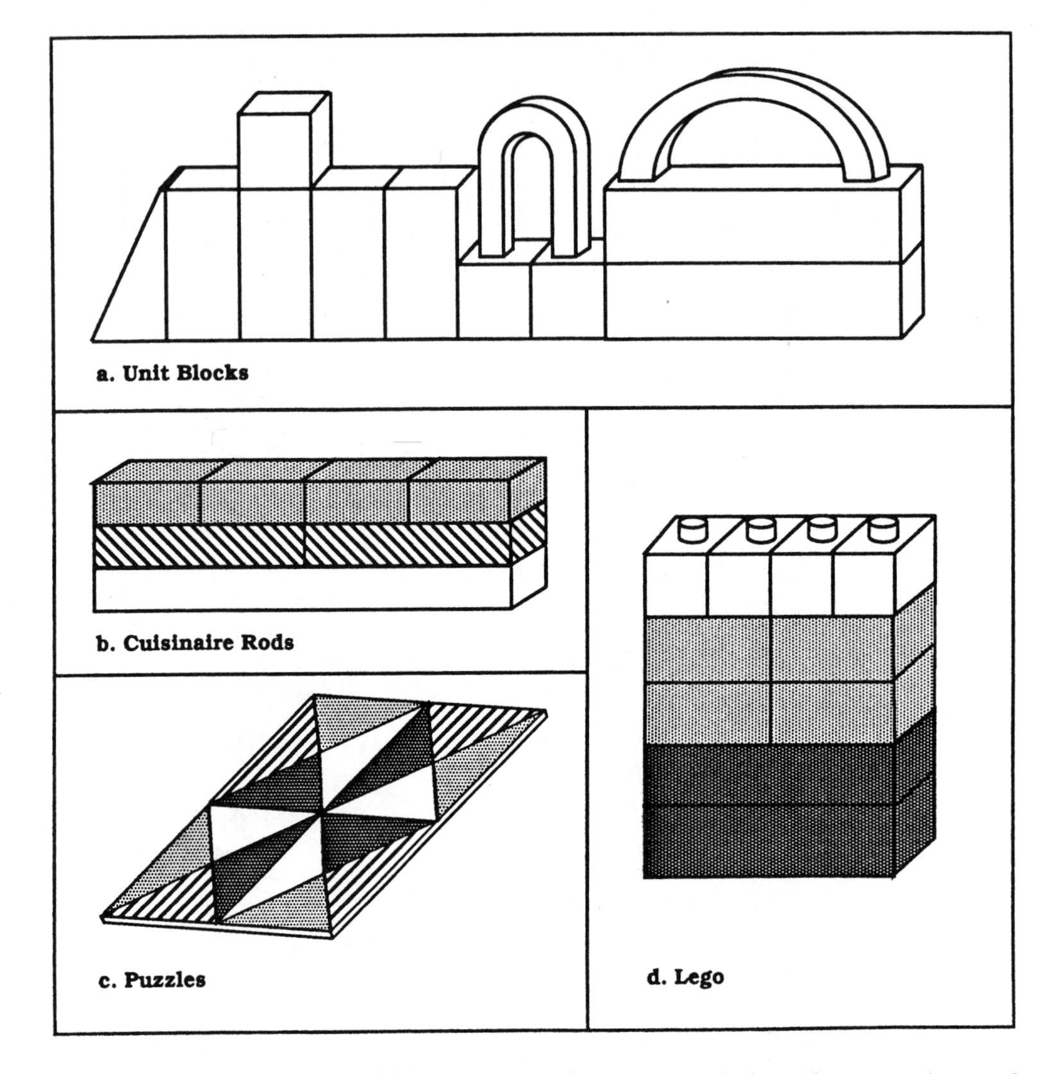

Figure 29–2 Materials which promote the construction and comparison of parts and wholes.

FRACTIONS: CONSTRUCTION PAPER MODELS

OBJECTIVE: To conceptualize fractional parts of wholes using construction paper and/or poster board models

MATERIALS: Make your own models out of construction paper and/or poster board. Use a different color for each fractional part. For example, a blue whole circle, a red circle the same size cut into halves, a yellow circle cut into thirds, and a green circle cut into fourths:

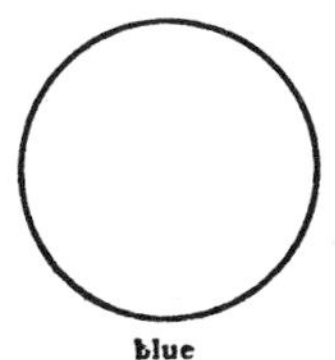
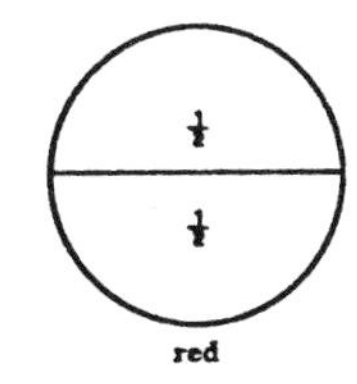
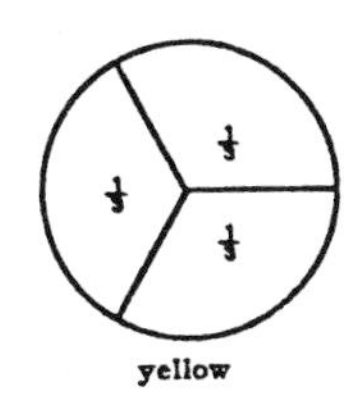
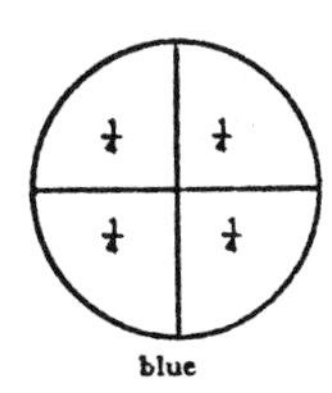

blue red yellow blue

The same colors can be used to make a set of fraction rectangles and a set of fraction squares.

ACTIVITY: Let the children explore the sets of fraction models. After they have worked with them several times, ask them some questions such as:

- What have you learned about those shapes?
- How many red pieces make a circle the same size as the blue circle? How many yellow pieces? How many green pieces?
- Use the same prodedure with the rectangles and the squares.

FOLLOW-UP: Label the parts as halves, thirds, and fourths of the wholes. Provide the students with matching construction paper shapes and have them make their own models by folding. Start with rectangles and squares.

FRACTIONS: COMPARING SIZES

OBJECTIVE: To compare sizes of halves, thirds, and fourths

MATERIALS: Same as in previous activity

ACTIVITY: Put out the models one set at a time. Ask the children, WHICH IS LARGER—A HALF, A THIRD, OR A FOURTH? HOW MANY FOURTHS MAKE ONE-HALF?

FOLLOW-UP: Provide models of other fractions for children who have a good understanding of thirds, fourths, and thirds.

FRACTIONS: PARTS OF GROUPS

OBJECTIVE: To divide groups of objects into subgroups of halves, thirds, and fourths

MATERIALS: Draw, color, and cut out a mother rabbit, four child rabbits, and twelve carrots. (See Figure 29–3 for patterns.)

ACTIVITY: Place the mother rabbit, two child rabbits, and four carrots in front of the children. MOTHER RABBIT HAS FOUR CARROTS AND WANTS TO GIVE EACH OF HER CHILDREN HALF OF THE CARROTS. HELP HER BY DIVIDING THE CARROTS BETWEEN THE TWO CHILDREN SO EACH HAS HALF. Give the child time to complete the task. NOW SUPPOSE SHE HAS FOUR CARROTS AND FOUR CHILDREN (bring out two more children). SHOW ME HOW THEY CAN SHARE THE CARROTS SO EACH HAS ONE-FOURTH OF THE CARROTS. As long as the children are interested, continue the activity using different numbers of child rabbits and different amounts of carrots. Emphasize that the carrots must be shared so that it is fair to everyone.

FOLLOW-UP: Make other sets of materials (such as children and apples, dogs and bones). Move on to other fractions when the children can do these problems easily.

Figure 29–3 Patterns for the PARTS OF GROUPS activity.

FRACTIONS: LIQUID VOLUME

OBJECTIVE: To compare fractional parts of liquid volume

MATERIALS: Several sets of standard measuring cups, color coded, if available, and a pitcher of water. If color-coded cups are not available, mark each size with a different color (i.e., 1 cup with blue, $1/2$ cup with red, $1/3$ cup with yellow, $1/4$ cup with green).

ACTIVITY: Give each child a set of measuring cups. Let them examine them and tell you what they notice. After discussion and examination, have everyone pick up their 1-cup size and their $1/2$ cup size. HOW MANY OF THESE SMALL CUPS OF WATER WILL FILL THE LARGE CUP? Let the children predict. FILL YOUR SMALL CUP WITH WATER. POUR THE WATER IN YOUR LARGE CUP. IS THE LARGE CUP FULL? POUR IN ANOTHER SMALL CUP OF WATER. IS THE LARGE CUP FULL NOW? HOW MANY OF THE SMALLER CUPS WERE NEEDED TO FILL THE LARGER CUP? Follow the same procedure with the $1/3$ and $1/4$ size cups.

FOLLOW-UP: Reverse the procedure. Starting with the full 1-cup measure, have the children count how many $1/2$, $1/3$, and $1/4$ cupfuls it takes to empty the full cup. Do the same activity using rice, birdseed, or sand instead of water.

FRACTIONS: LENGTH

OBJECTIVE: To compare fractional parts of lengths

MATERIALS: Cuisinaire(TM) rods

ACTIVITY: Provide the children with a basket full of Cuisinaire(TM) rods. After they have had an opportunity to explore the rods, suggest that they select a long rod and find out which lengths of rods can be placed next to it to show halves, thirds, and fourths (Figure 29–4).

FOLLOW-UP: Follow the same procedure with whole straws and straws cut into halves, fourths, and thirds. Try the activity with other materials such as string, ribbons, and paper strips.

Figure 29–4 The children explore the properties of cuisinaire rods.

FRACTIONS: EGG CARTON MODELS
OBJECTIVE: To find fractional parts using egg carton models
MATERIALS: Egg cartons cut in halves, thirds, and fourths
ACTIVITY: Let the children explore the materials. Show them a whole egg carton. FIND SOME PARTS THAT WILL MAKE A WHOLE CARTON LIKE THIS ONE. Let them compare and work with the cartons until they find all the combinations that make a whole carton. Then ask, HOW MANY CUPS ARE IN A WHOLE CARTON? IN ONE-HALF OF A CARTON? IN ONE-FOURTH OF A CARTON? IN ONE-THIRD OF A CARTON?
FOLLOW-UP: Cut some cartons into sixths and twelfths. Encourage the children to explore the properties of these parts relative to the whole carton.

FRACTIONS: STORY PROBLEMS
OBJECTIVE: To have the children create and solve fraction story problems
MATERIALS: One or more model problems; paper, pencils, crayons, markers, and scissors; chart paper for model problems
ACTIVITY: Have the students brainstorm real-life situations in which things have to be divided into equal parts. Encourage them to write or dictate their own problems, draw a picture of the problem, and write out the solution. If they cannot come up with their own problems, provide one or two models that they can work through with you. Models (have each problem written on chart paper):

1. TWO CHILDREN FOUND SIX PENNIES. ONE CHILD TOOK THREE PENNIES. WHAT FRACTIONAL PART OF THE PENNIES WAS LEFT FOR THE OTHER CHILD? Draw two stick figure children. Draw six pennies on another piece of paper. Glue three pennies by one child and three by the other. WHAT PART OF THE PENNIES DOES EACH CHILD HAVE? YES, ONE-HALF. Write: "Each child has one-half of the pennies," (Figure 29–5)
2. BRENT INVITES THREE FRIENDS OVER FOR PIZZA. IF EACH CHILD GETS A FAIR SHARE, HOW WILL THE PIZZA LOOK WHEN IT IS CUT UP? Give each child a paper pizza. Ask them to fold the pizzas into the right size and number of parts. Cut up one of the pizzas and glue the parts on the chart. Write: "Each child gets one-fourth." (Figure 29–6)

FOLLOW-UP: Encourage the children to create and illustrate their own fraction problems.

"Two children found six pennies. One child took 3 pennies. What fractional part of the pennies was left for the other child?"

"Each child has one-half of the pennies."

Figure 29–5 The teacher can write and illustrate a model story problem which involves the partitioning of a set.

"Brent and his three friends each get a fair share of the pizza."

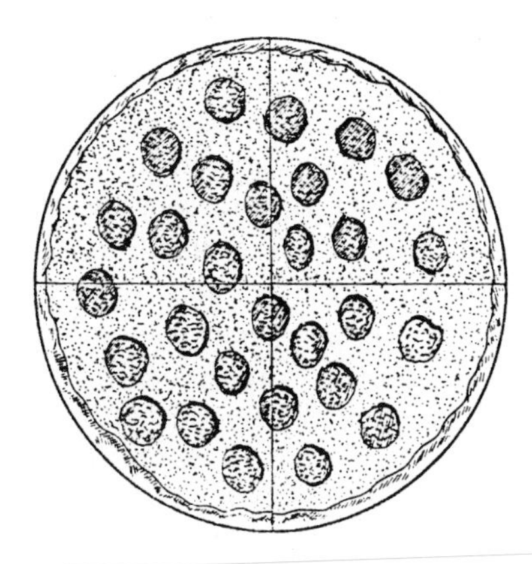

"Each friend gets one-fourth of the pizza."

Figure 29–6 The teacher can write and illustrate a model story problem which involves dividing a whole into equal parts.

FRACTIONS: GEOBOARD SHAPES

OBJECTIVE: To divide geoboard shapes into equal parts

MATERIALS: Geoboards, geoboard shape patterns, and rubber bands

ACTIVITY: Have the children make rectangles and squares on their geoboards with rubber bands, then divide them into equal parts using additional rubber bands.

FOLLOW-UP: Question the children regarding how they know that their shapes are divided into equal parts. Note if they use the number of geoboard nails as a clue to making their parts equal.

For the more advanced students, notation can be introduced.

Show them how one-half can be written ½ which means one part out of two. Then ask how they think they might write one-third as one part out of three, two-thirds as two parts out of three, and so on. Have them write the numerical fractions that match the parts of some of the materials suggested earlier in the unit.

Some children will enjoy using computer software that sharpens their knowledge of fractions. Third graders Vanessa and Jason enjoy racing cars and find the fraction problem-solving activities in *Grand Prix* (Random House) very challenging.

EVALUATION

Continue to note whether the children apply what they have learned about fractions during their everyday activities. Be sure to provide situations in which individual items or groups of items have to be shared equally with others. Note whether the children can use their concept of fractions in these situations. Also note if they apply their concept of fractions during measuring experiences that are a part of food preparation and science investigations. Administer the assessment tasks described in Appendix A.

SUMMARY

During the primary years, young children expand their informal concept of parts and wholes of objects and groups to the more formal concept of fractions, or equal parts. Primary children learn the vocabulary of fractions and work with fractions at the concrete level. Usually they do not go beyond halves, thirds, and fourths, and most of them are not ready to understand fraction notation until the latter part of the primary period. It is important not to rush young children into the abstract use of fractions until they are ready since fractions are much more difficult to work with than whole numbers.

FURTHER READING AND RESOURCES

Bezuk, N. S. 1988. Fractions in the early childhood mathematics curriculum. *Arithmetic Teacher. 35*: 56–60.

Burton, G. M. 1985. *Towards a good beginning.* Menlo Park, Calif.: Addison-Wesley.

Behr, M. J., and Post, T. R. 1988. Teaching rational number and decimal concepts. *Teaching mathematics in Grade K–8* 190–231. Boston: Allyn and Bacon.

Scott, L. B., and Garner, J. 1978. *Mathematical experiences for young children.* New York: McGraw-Hill.

Reys, R. E., Suydam, M. N., and Lindquist, M. M. 1984. *Helping children learn mathematics.* Englewood Cliffs, N.J.: Prentice-Hall.

RESOURCES FOR MATERIALS

Cuisinaire Company of America, Inc., 12 Church Street, Box D, New Rochelle, N.Y. 10802.

ETA, 199 Carpenter Avenue, Wheeling, Ill. 60090. (Ask for math catalog.)

Lakeshore Curriculum Materials Company, 2695 E. Dominguez Street, P.O. Box 6261, Carson, Calif. 90749.

Nienhuis-Montessori, 320 Pioneer Way, Mountain View, Calif. 94041.

SUGGESTED ACTIVITIES

1. Visit one or more primary classrooms during math instruction on fractions. Take note of the materials and methods used. Compare what you observe with what is suggested in the text.

2. Assess a primary child's level of development in the concept of fractions. Prepare an instructional plan based on the results. Prepare the materials and implement the lesson with the child. Evaluate the results. Did the child move ahead as a result of your instruction?
3. Be sure to add fraction activities to your Activities File.
4. Try out some of the activities suggested in this unit with primary children of different ages.
5. If the school budget allotted forty dollars for fraction materials, decide what would be your priority purchases.
6. Using the guidelines from Unit 9, Activity # 8, evaluate the following computer programs designed to reinforce the concept of fractions:

Fraction Factory. American Guidance Service. Circle Pines, Minn.: Apple Family.

Grand Prix. Part of *Basic Math Facts Drill.* Random House. New York, N.Y.

Tonk in the Land of Buddy-Bots. Mindscape, Inc. Norhtbrook, Ill.: Apple IIs, Atari, C64, IBM PC & PCjr.

REVIEW

A. Explain why primary children should not be rushed into working with fraction notation.

B. Identify the statements that are correct:

1. At the presymbolic level, work with fractions should be limited to halves, thirds, and fourths.
2. Fraction terminology should be introduced along with fraction notation.
3. At the presymbolic level, primary children can work with volume, regions, lengths, and sets.
4. Food experiences are excellent opportunities for working with volume.
5. Begin assessment with the tasks in this unit. It can be assumed that primary level children can already do the lower level tasks included in Unit 14.
6. Primary children still need exploratory experiences before being presented with structured fraction activities.

C. Label the following examples as A, partitioning (dividing) into sets; B, a whole object can be divided into equal parts; or C, a liquid can be divided into equal parts.

1. Liu Pei carefully folds rectangular pieces of paper in half and then in half again and cuts accurately on each line.
2. Ann has six crackers. She wants to be sure that she and both of her friends receive an equal amount. She decides that they should each have two crackers.
3. Jason has a one cup measure of milk. Ms. Hebert tells him that they will need one-half cup of milk for their recipe. Jason selects a smaller cup that says ½ on the handle and fills it with milk from the larger cup.
4. Trang Fung is experimenting with the Fit-a-Fraction set.

UNIT 30 Numbers Above Ten and Place Value

OBJECTIVES

After studying this unit, the student should be able to:

- Define place value, renaming, and regrouping
- Identify developmentally appropriate place value and two-digit whole number operations instruction
- Assess children's understanding of numbers above ten and of place value
- Provide developmentally appropriate place value and addition and subtraction instruction with two-digit numbers

During the latter part of the preoperational period (see Unit 25), children who are adept at manipulating quantities up to ten can move on to working with quantities above ten. Through manipulation of groups of ten and quantities between zero and ten, children move through the teens and up to twenty. Some will pick up the pattern of the twenties, thirties, and so on up through the nineties. As children enter concrete operations, they perfect their informal knowledge of numbers above ten and move on to whole number operations with numbers above ten. To fully understand what they are doing when they use whole number operations involving numbers above ten, they must be able to conceptualize *place value*. Place value pertains to an understanding that the same numeral represents different amounts depending on which position it is in. For example, consider the numbers 3, 30, and 300. In the first instance, *3* stands for three ones and is in the ones' place. In 30, *3* stands for three tens and is in the tens' place. In 300, *3* stands for three hundreds and is in the hundreds' place. In each of these cases, zero indicates there is no quantity in the place that it holds. In the number 32, *3* is in the tens' place and *2* is in the ones' place. An understanding of place value underlies the understanding of certain trading rules that govern place value and enable whole number operations to be accomplished. Examples of some trading rules are:

- Ten ones can be traded for one ten.
- One ten can be traded for ten ones.
- Ten tens can be traded for one hundred.
- One hundred ones can be traded for one hundred.

The place value concept enables us to represent any value using only ten digits (zero to nine).

Place value is one of the most difficult concepts for young children to grasp. Being able to rote and rational count above ten is only a beginning step on the way to an understanding of place value. Children need many counting experiences (as described in Unit 9) and many experiences with concrete models in order to develop the place value concept. All too often children are rushed into the place value operations involved in *regrouping* (what used to be referred to as borrowing and carrying) as a rote memory activity without the necessary underlying conceptualization. Through an understanding of place value, children will realize that when they take one from the tens' column they are actually taking one group of ten and that when they add numbers in the ones' column and arrive at a sum above nine that the amount they move to the tens, column represents one or more groups of tens. Understanding place value will also help them to see that the placement of numerals is critical in determining value. For example, sixty-eight can be written as 68, six tens and eight ones, and 60 + 8 but *not* 86. The sequence for writing numbers follows fixed rules just like a word sentence. That is, "Ball boy the throws" does not follow the conventions of correctly written English. In the same fashion, one hundred and twenty-one is not written as 10021. How to guide children to an understanding of this concept is the focus of this unit (Figure 30–1).

Figure 30–1 Place value problems are a difficult challenge for these primary grade children.

ASSESSMENT

Understanding place value is a difficult task for young children. They will normally flounder for a while, seeming to understand the concept in some situations and not in others. Teachers should be patient and accepting and give the children time and appropriate experiences. The following are examples of assessment tasks that can be used to discover where children are on the road to understanding two-digit numbers and the concept of place value.

SAMPLE ASSESSMENT TASK

Concrete Operational Ages 7–8

METHOD: Interview

SKILL: Child is able to count groups of eleven or more objects and tell how many tens are in the groups

MATERIALS: A container of 100 counters (e.g., chips, cubes, or sticks)

PROCEDURE: Place the container of counters in front of the child. Say, HERE ARE A BUNCH OF COUNTERS. COUNT OUT AS MANY OF THEM AS YOU CAN. If the child counts out eleven or more ask, HOW MANY TENS DO YOU THINK YOU HAVE? HOW MANY ONES?

EVALUATION: If the child answers correctly, then he probably has the concept of place value for tens. If he answers incorrectly, this indicates that although he may be able to rational count groups of objects greater than ten, he does not yet understand the meaning of each of the numerals in his response.

INSTRUCTIONAL RESOURCE(S):

Charlesworth, R., and Lind, K. 1990. *Math and Science for Young Children*. Albany, N.Y.: Delmar.

SAMPLE ASSESSMENT TASK

Concrete Operational Ages 7–8

METHOD: Interview

SKILL: The child is able to form two or more subgroups of ten objects each with some remaining from the original group and tell how many he has without counting each individual object.

MATERIALS: A container of 100 counters (e.g., chips, cubes, or sticks)

PROCEDURE: Place a pile of counters (start with about thirty-five) in front of the child. MAKE AS MANY GROUPS OF TEN AS YOU CAN. HOW MANY (counters) DO YOU HAVE ALTOGETHER?

EVALUATION: Note if the child can come up with the answer by counting the number of groups of ten and adding on the number of ones, or if he has to count each object to be sure of the total. If the child can determine the answer without counting by ones, this is an indication that he is developing the concept of place value.

INSTRUCTIONAL RESOURCE(S):

Charlesworth, R., and Lind, K. 1990. *Math and Science for Young Children*. Albany, N.Y.: Delmar.

On the average, first graders can learn to read, write, and understand two-digit numbers, second graders three-digit numbers, and third graders four-digit numbers. However, there will be a broad range of normal variation within any particular group. The best rule of thumb in assessment is to be sure children understand one-digit numbers before going on to two-digit, two before three, and so on.

ACTIVITIES

The following activities are adapted from the selection of resources listed at the end of the unit. Young children need many experiences in manipulating objects relative to numerals greater than ten and place value before proceeding to whole number operations with two-digit numbers. Start with counting activities such as those suggested in Unit 9. Then move on to the kinds of activities described in the following pages. The first two activities focus on constructing an understanding of the properties of amounts greater than ten (Figure 30–2).

Figure 30–2 A primary child mulls over a large number conservation problem.

NUMERALS GREATER THAN TEN: CONSERVATION OF LARGE NUMBERS

OBJECTIVE: To understand, when given a group of more than ten objects, that the number of objects remains the same no matter how they are arranged.

MATERIALS: Each child will need a container with fifty or more counters (i.e., cubes, chips, or sticks) and a place value board (as suggested by Richardson in *Developing number concepts using Unifix® cubes*, p.212). A place value board is a piece of paper divided into two sections so that groups can be placed on one side and loose counters on the other side. To make a place value board, take a 9″ × 12″ piece of paper or tagboard and glue or staple a 6″ × 9″ piece of colored paper on one half. Draw a picture in the upper right-hand corner of the white side so the child will know how to place the board in the correct position.

Unifix® cubes are excellent for this activity because they can be snapped together. However, cubes or chips that can be stacked or sticks that can be held together with a rubber band may also be used.

ACTIVITY: Tell the children to put some amount of counters greater than ten on the white side of their place value board. Suggest the children work in pairs so they can check each other's counting. When they agree that they have the same amount: MAKE A GROUP OF TEN (counters) AND PUT IT ON THE (color) SIDE OF YOUR BOARD. HOW MANY LOOSE (counters) DO YOU HAVE LEFT? HOW MANY (counters) DO YOU HAVE ALTOGETHER? Note how many children realize that they still have the same number of counters. Ask, DO YOU HAVE ENOUGH (counters) TO MAKE ANOTHER TEN?

Note if they respond correctly. If they do make another ten, ask again how many loose cubes they have and if the total is still the same. To connect their arrangements to number symbols, write the arrangements on the board:

3 tens and 4 ones
2 tens and 14 ones

FOLLOW-UP: The same activity could be done using containers to put the counters in. Each container could be labeled 10. If small counters such as tiny chips, bottle caps, or beans are used, this would be a more efficient way to keep the groups separated. See Richardson's book for additional variations on this activity.

NUMERALS GREATER THAN TEN: MEASURING

OBJECTIVE: To work with large numbers through nonstandard measurement activities (see Unit 18 for background information)

MATERIALS: Small objects that can be used as nonstandard units, paper and pencil to record measurements, and measurement cards (suggested by Richardson). Each measurement card depicts something in the classroom that can be measured.

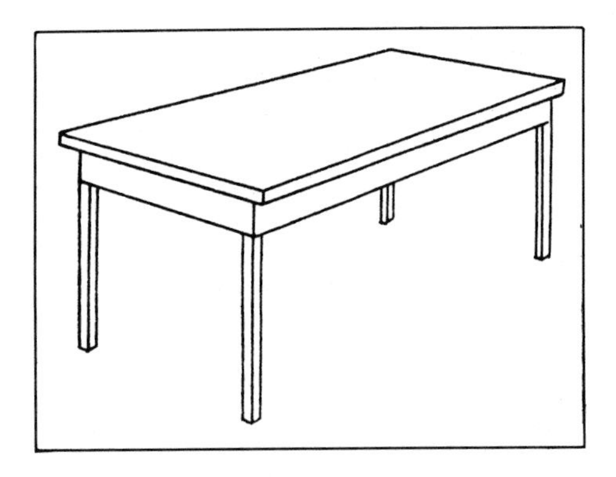

ACTIVITY: The children measure the items with the objects, group the objects into tens, then figure out how many they have used (Figure 30–3). They record the results on their paper.

The chair is 24 paperclips tall.

The table is 36 paperclips wide.

FOLLOW-UP: Have the children measure the same items using different nonstandard units and compare the number of units. See Richardson's book for further activities.

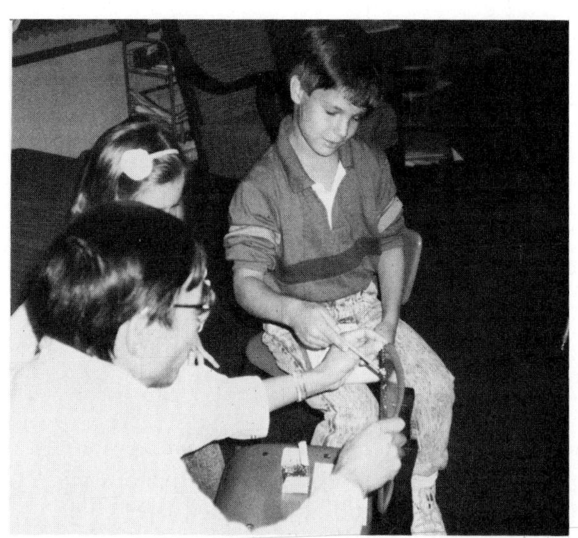

Figure 30–3 The teacher helps the children use paperclips as an informal unit of measure.

Once the children have a good understanding of counting and subdividing groups greater than ten, they are ready to move on to activities which gradually move them into the complexities of place value.

PLACE VALUE: CONSTRUCTING MODELS OF TWO-DIGIT NUMBERS

OBJECTIVE: To develop models of two-digit numbers using the place value board

MATERIALS: Place value board (described in previous activity) and a supply of cards with individual numerals zero to nine written on each card. Provide the child(ren) with at least two sets of numerals and a supply of counters that can be readily stacked, snapped, or bundled into groups of ten. One child working alone will need 100 counters. If children are working in small groups and sharing the counters, add fifty counters per child.

ACTIVITY: Have the children put the place value boards in front of them. Provide them with an ample supply of numerals in a small container and place the container of counters where it is convenient for everyone to reach. CLOSE YOUR EYES AND PICK TWO NUMERALS. PUT ONE ON YOUR BOARD ON THE (color) SIDE AND ONE ON YOUR BOARD ON THE WHITE SIDE. MAKE A MODEL OF THE NUMERAL YOU HAVE SELECTED.

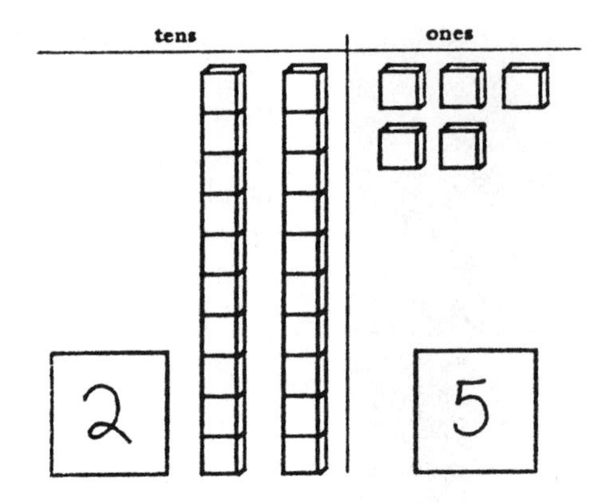

FOLLOW-UP: Have the students repeat this activity with different number combinations. Have them work alone, in pairs, and/or small groups. Have them do the same activity using Base Ten Blocks after they have been introduced. (Base Ten Blocks will be described later.) The same activity can be adapted to work with hundreds and thousands when the children are ready. The activity could also be done using spinners: one for each place value.

PLACE VALUE: ESTIMATION WITH COMMON ITEMS

OBJECTIVE: To estimate amounts and apply knowledge of place value to finding out how close the estimate is

MATERIALS: A container of common items such as a bag of peanuts, dry beans, or cotton balls; a box of paper clips, rubber bands, or cotton swabs; a jar of bottle tops, clothes pins, pennies, or other common items and some small cups, bowls, or plastic glasses. Flip chart numerals or paper and pencil to write results. (Flip chart numerals can be made by writing numerals zero to nine on cards, punching two holes at the top of each card, and putting the cards on rings.)

ACTIVITY: Estimating is the math term for guessing or predicting the answer to a math problem. Show the children the container of objects. Let them hold it, examine it, but not open it. LOOK CLOSELY AT THIS (item). HOW MANY (items) DO YOU THINK ARE IN THIS (container)? You record or have the children record each child's guess. Open the container. COUNT OUT GROUPS OF TEN (item) AND PUT THEM IN THE (cups, glasses, and so on). FIND OUT HOW MANY GROUPS OF TEN THERE ARE ALTOGETHER. EACH TIME YOU COUNT OUT TEN, TURN OVER A NUMBER ON YOUR FLIP CHART, or if the children can write, WRITE THE NUMBER THAT TELLS HOW MANY TENS YOU HAVE.

FOLLOW-UP: Once the children catch on, put out the estimate of the week. Children can record their guesses on a chart. Then on Friday they can open the container and find out the actual amount. A graph can be made showing the distribution of estimates:

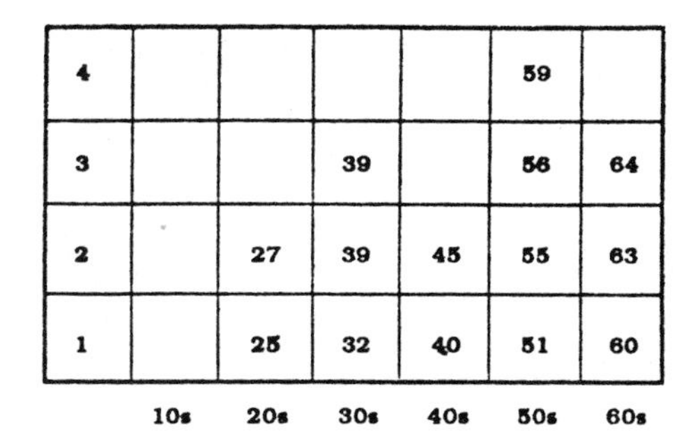

See *Math Their Way*, Chapters 11 and 12, for more ideas.

Base Ten Blocks provide a different kind of model for working with the place value concept. It is important that children work with a variety of types of materials in model construction so that they do not think there is just one way to view place value with concrete materials. So far we have described making models with discrete items. Base Ten Blocks depict each place with a solid model. Units (or ones) are individual cubes, rods (or tens) are the equivalent of ten unit cubes stuck together in a row, flats (or hundreds) are the equivalent of ten rods stuck together, and cubes (or thousands) are the equivalent of ten flats stacked and glued together. See the illustration to the right.

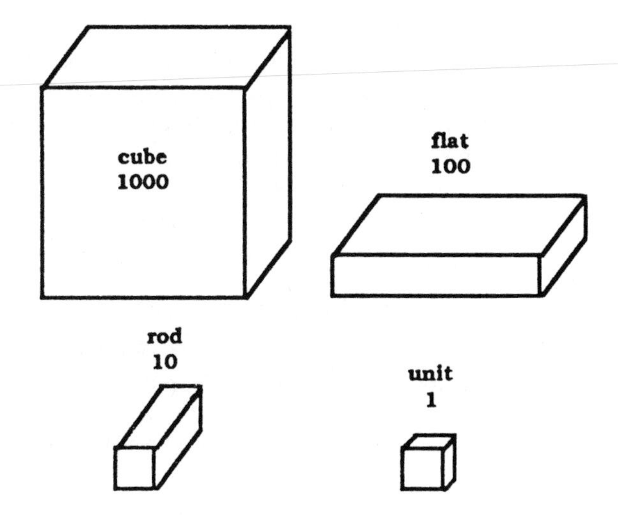

PLACE VALUE: BASE TEN BLOCKS

OBJECTIVE: To work with place value using Base Ten Blocks.

MATERIALS: A classroom set of Base Ten Blocks.

ACTIVITY: The Base Ten Blocks can be used for the activities above but adding on another step. Each time the child makes a group of ten, he can record this by selecting a rod. He can then count the rods at the end to find out how many groups of ten he has found.

FOLLOW-UP: Many auxiliary materials can be purchased that provide activities with Base Ten Blocks besides those that come with the Teacher's Guide. Base Ten Blocks can be purchased from ETA and Kaplan. Examples of auxiliary materials are:

- Base Ten Activity Book (ETA)
- Place Value Activity Cards (DLM)

Trading is another procedure for working with place value. Primary children need many experiences counting piles of objects, trading for groups of ten, and describing the results. Once they can do these activities with ease, they can move on to regrouping and renaming. Too often they are pushed into regrouping and renaming without an adequate conceptual base built on counting and constructing many groups of ten and relating them and any ones remaining to written numerals. *Regrouping* happens when one or more items are added or taken away so that an amount moves to the next ten, next hundred, next thousand, and so on. That is, a group might break down into two tens and six units, or twenty-six. Five more units are added so there are now eleven units. Ten units are then moved to the tens' place, leaving one in the ones' place. There are now three tens and one unit, or thirty-one. The group has now been *renamed* thirty-one.

Reverse trading would take place if units were removed. For example, if seven units were to be taken away from the three tens and one unit, a ten would be moved over to the units, making eleven units, and seven units could then be removed, leaving two tens and four ones, or twenty-four. Primary grade children need to do many trading activities with concrete materials before moving on to paper and pencil computations. These trades can be practiced with concrete items such as cubes and chips or the beads on an abacus or with solids such as Base Ten Blocks. A Chip Trading Set can be purchased (ETA), which includes a teacher's guide, or the guide can be purchased separately and counting disks or squares of paper or tagboard used for trading (Figure 30–4).

Figure 30–4 Constructing models builds the basic concept of place value. This student is working on a two-digit number.

PLACE VALUE: TRADING ACTIVITIES

OBJECTIVE: To construct the concepts of regrouping and renaming through trading activities

MATERIALS: A supply of paper squares: 100 reds (units) and thirty blues (tens) and a place value board with a ones' and tens' place

ACTIVITY: Have the children put their place value boards in front of them. Place a supply of red and blue paper squares where they can be easily reached. THE RED SQUARES ARE TENS AND THE BLUE SQUARES ARE ONES. Hold up a large 27. SHOW ME HOW YOU CAN MAKE A MODEL OF THIS NUMBER ON YOUR BOARD. Have the children try several examples. When you are sure that they understand that the reds are ones and the blues are tens, go on to regrouping. Go back to 27. MAKE TWENTY-SEVEN ON YOUR BOARDS AGAIN. NOW, SUPPOSE SOMEONE GIVES YOU FIVE MORE

ONES. TAKE FIVE MORE. WHAT HAPPENS TO YOUR ONES? Encourage them to describe what happens. Remind them that there cannot be more than nine in the one's place. Eventually someone will realize that 32 should be modeled with three ten's and two ones. Have the children discuss what they might do to get a model that has three tens and two ones by trading. SUPPOSE YOU TRADE TEN BLUES FOR ONE RED. WHERE SHOULD THE RED BE PLACED? Once everyone has the red in the ten's place, ask, SUPPOSE SOMEONE NEEDS FOUR ONES. HOW COULD YOU GIVE THEM FOUR? Encourage them to discuss this problem with each other and ask you questions until someone discovers that another trade will have to be made. Have everyone trade in a blue for ten reds, take four away, and note that they now have two blues and eight reds (28).

FOLLOW-UP: Create some story problems and have the children solve them using trading. Then move on to adding and taking away two-digit quantities.

When the children practice trading to regroup and rename on their place value boards they are actually adding and subtracting informally. Richardson's book explains how to carry this activity over to addition and subtraction of two-digit numbers. Richardson (*Developing number concepts using Unifix® cubes*) believes that it is confusing to begin two-digit addition and subtraction with numbers that do not have to be regrouped such as:

$$\begin{array}{r}22\\+35\\\hline 57\end{array}\qquad\begin{array}{r}53\\+14\\\hline 67\end{array}\qquad\begin{array}{r}46\\-34\\\hline 12\end{array}\qquad\begin{array}{r}18\\-13\\\hline 5\end{array}$$

This type of addition and subtraction may lead children to believe that adding or subtracting with two digits is exactly the same operation as with one digit and result in reponses such as those below:

$$\begin{array}{r}25\\+16\\\hline 311\end{array}\qquad\begin{array}{r}48\\+34\\\hline 712\end{array}\qquad\begin{array}{r}72\\-35\\\hline 43\end{array}\qquad\begin{array}{r}37\\-28\\\hline 11\end{array}$$

Note in these examples that the children have added or subtracted each column as though it were an individual one-digit problem. Introduce two-digit addition as follows:

PUT TWENTY-SIX CUBES ON YOUR BOARD.

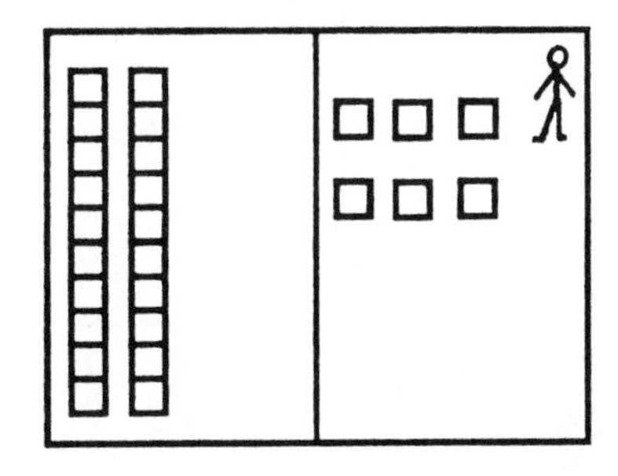

NOW GET EIGHTEEN CUBES AND PUT THEM NEXT TO YOUR BOARD.

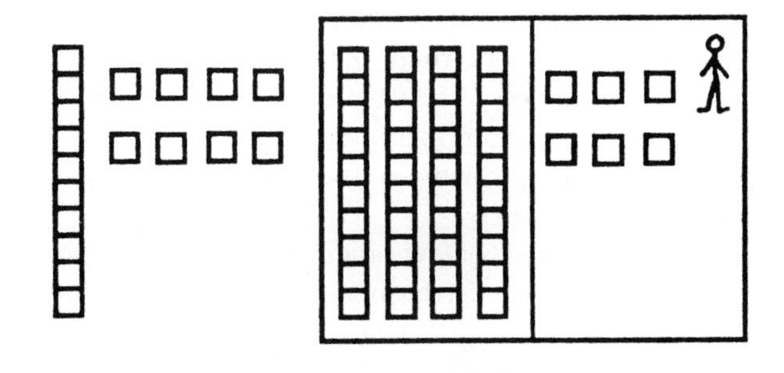

NOW PUT THEM TOGETHER.
HOW MANY TENS? HOW MANY ONES?
YES, WE HAVE THREE TENS AND FOURTEEN ONES.
DO WE HAVE ENOUGH TO MAKE ANOTHER TEN?

(Give them time to move ten cubes.) NOW WE HAVE FOUR TENS AND FOUR ONES.
HOW MANY IS THAT? YES, THAT IS FORTY-FOUR.

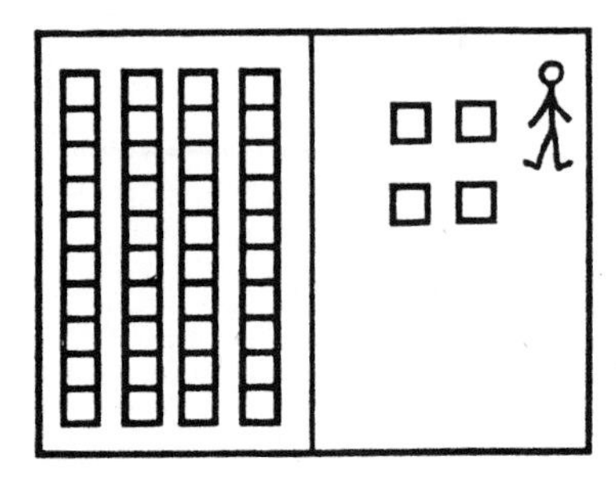

Repeat this process several times with different pairs of two-digit numbers: somewhere the sum contains more then nine units and somewhere it does not. When the children understand the process without symbols, connect the symbols by writing them on the board as you go through the process.

Subtraction of two-digit numbers can be introduced in parallel fashion. For example, have the children put out forty-two cubes, then tell them they need to take away twenty-five. They will discover that they have to break up a ten to accomplish this task. After doing many examples without written numbers, go through some problems in which you connect the quantities to numbers at each step, thus gradually introducing the notation. Then have them do mixed sets of problems. Finally, move on to story problems.

KAMII'S APPROACH

The activities suggested in this unit follow a fairly structured sequence while promoting construction of concepts through exploration. Constance Kamii (February 1988, *Arithmetic Teacher*) has been working with primary children using open-ended activities that provide for more child trial and error and self-sequencing. Interviewing primary students who had been through conventional workbook/textbook instruction, she discovered that they were able to do regrouping and renaming as a rote process without really knowing the meaning of the numbers they were using. For example, when asked to do a two-digit problem such as 28 + 45, they could come up with the correct answer:

$$\begin{array}{r} 1 \\ 28 \\ +45 \\ \hline 73 \end{array}$$

However, when asked what the 1 in thirteen means they said it meant one rather than ten. Kamii has had greater success in getting this concept over to primary children using games and letting them discover the relationship of the digits on their own. No workbooks or ditto sheets are used and neither are the kinds of concrete activities described in this unit. Problems are written on the board, children contribute answers, and every answer is listed. Then the children give their rationales for their answers. When working with double-digit addition, their natural inclination is to start on the left. They add the tens, write the answer, add the ones, and, if necessary, move any tens over, erasing the original answer in the tens' column. Through trial and error and discussion, they develop their own method and construct their own place value concept. Place value is not taught as a separate skill needed prior to doing double-column addition. This method sounds most intriguing and is more fully described in a book entitled *Young Children Continue to Reinvent Arithmetic, 2nd grade.*

CALCULATORS

When children explore calculators for counting, they notice that the number on the right changes every time, whereas the other numbers change less frequently. Calculators provide a graphic look at place value and the relation of each place to those adjacent. Suggest that the children try + 10 = = = = =, and note what happens, that is, which place changes and how much each time? This activity will assist children in seeing what the tens' place means. Concepts constructed using manipulatives can be reinforced with calculator activities. By adding one to numbers that end in nine and subtracting one from numbers that end in zero, students can see immediately what happens. Suggest that they guess which number follows nine, twenty-nine, and fifty-nine. Then have them use their calculators to check their predictions. Use the same procedure, except have them predict what will happen if they subtract one from twenty, forty, and seventy (Figure 30–5).

Figure 30–5 Hand calculator activities can support concepts constructed with manipulatives.

EVALUATION

An evaluation technique suggested by Kamii shows if children really understand place value in two-digit numbers:

> Show the child a 3″ × 5″ card with sixteen written on it. Ask, WHAT DOES THIS SAY? After the child says sixteen, count out sixteen chips. With the top of a pen circle the six of the sixteen, What does this *part* (the six) mean? Show me with the chips what this *part* (the six) means. Circle the one of the sixteen. What does this *part* (the one) mean? Show me with the chips what this *part* (the one) means.

Kamii says that after conventional instruction with workbooks and possibly some manipulatives, all first and second graders can answer correctly regarding the six, but of the primary children Kamii interviewed, none of the end-of-first-graders, thirty-three percent of end-of-third-graders, and only fifty percent of end-of-fourth-graders said that the one means ten. Children who learned about place value through constructing it themselves using Kamii's method did considerably better. At the end of second grade sixty-six percent said that the one means ten and seventy-four percent said that the five in fifty-four means fifty.

Whichever instructional method you use, be sure to observe very carefully the process each child uses. Be sure to question them frequently about what they are doing to be sure they really understand the concepts and are not just answering in a rote manner.

SUMMARY

Learning about place value and working with two-digit whole number operations that require regrouping and renaming is one of the most difficult challenges faced by the primary level child. Conventionally, it is taught from a workbook approach with few if any manipulatives to support the instruction. Also, it is conventionally introduced too early and becomes a rote memory activity for those who have the facility to remember the steps necessary to come up with correct answers. Less capable students flounder in a lack of understanding.

In order to understand place value and the processes of regrouping and renaming, most math educators believe that children develop these concepts through practicing counting and regrouping and renaming using models made with concrete materials. Kamii takes a different approach, using no concrete materials but guiding children through trial and error and discussion.

FURTHER READING AND RESOURCES

Balka, D. 1987. *Unifix mathematics activities, Book 1.* Peabody, Mass.: DIDAX Educational Resources.

Baratta-Lorton, M. 1976. *Mathematics their way.* Menlo Park, Calif.: Addison-Wesley.

Baroody, A. J. 1987. *Children's mathematical thinking.* New York: Teachers College Press.

Burton, G. M. 1985. *Towards a good beginning.* Menlo Park, Calif.: Addison-Wesley.

Early childhood mathematics. 1988. *Arithmetic Teacher.* 35 (6).

Labinowicz, E. 1985. *Learning from children: New beginnings for teaching numerical thinking.* Menlo Park, Calif.: Addison-Wesley.

Reys, R., and Suydam, M., and Lindquist, M. M. 1984. *Helping children learn mathematics.* Englewood Cliffs, N.J.: Prentice-Hall.

Richardson, K. 1984. *Developing number concepts using Unifix® Cubes.* Menlo Park, Calif.: Addison-Wesley.

Taverner, N. 1982. *Unifix teacher's manual.* Andover Hants, England: Philograph.

OTHER MATERIALS

DLM Teaching Resources, P.O. Box 4000, One DLM Park, Allen, Tex. 75002. (800-527-4747; In Texas, 800-442-4711).

ETA, 199 Carpenter Avenue, Wheeling, Ill. 60090. (800-445-5985).

Kaplan School Supply Corp, 1310 Lewisville-Clemmons Road, Lewisville, N.C. 27023. (800-334-2014; in North Carolina, 800-642-0610).

Lakeshore Curriculum Materials Co., 2695 E. Dominguez St., P.O. Box 6261, Carson, Calif. 90749. (800-421-5354; in California, 800-262-1777; in Los Angeles, 213-537-8600).

SUGGESTED ACTIVITIES

1. Visit one or more primary classrooms. Observe the math instruction. Describe the methods observed. Is place value and/or two-digit addition and subtraction being taught?

How? Is the method developmentally appropriate? Are the children ready for these concepts? Do they seem to understand?

2. Using Kamii's technique, interview one first grader, one second grader, and one third grader. Describe the results; compare the performance of the children you interviewed with those interviewed by Kamii. What did you learn from this experience?
3. Review the first-, second-, and third-grade levels of two or more elementary math textbook series. At what level are place value, two-digit addition and subtraction, regrouping, and renaming introduced? What methods of instruction are used? Compare the textbook methods with those described in this unit. What are the similarities and differences? Explain why you would or would not like to use these textbooks.
4. Prepare materials for one of the activities suggested in this unit. Try out the activity with a child or small group of children you believe should be at the right stage for it. Write a report explaining what you did, who you did it with, what happened, and your evaluation of the activity and the children's responses.
5. Keep your Activities File up to date.

REVIEW

A. Define the following terms:

1. Place value
2. Regrouping
3. Renaming

B. Identify which, if any, of the following are descriptions of place value being taught as described in this unit.

1. A kindergarten teacher is drilling her class on ones, tens', and hundreds' places using a worksheet approach.
2. A second grade teacher gives one of her students thirty-five Unifix® cubes and asks him to make as many groups of ten as he can.
3. A primary teacher is beginning instruction on two-digit addition with simple problems such as 12 + 41 so the students will not have to regroup and rename right away.
4. Some primary children are exploring with calculators. Their teacher has suggested that they take a list of the numbers 19, 29, 39, and 49, and predict what will happen if one is added to each. Then they are to try these operations with their calculators, write down what happens, and share the results with the other children and the teacher.

C. Look back at question B. Explain why you answered as you did.

D. List three kinds of materials that can be used for making place value models.

E. In what ways does Kamii's approach to teaching place value differ from what is conventionally done?

F. Check yourself on place value.

1. Identify the number of 100's, 10's, and 1's in each numeral.

	100s	10s	1s
37			
4			
276			

2. In the numeral 3,482, the 4 means _______, the 8 means _______, and the 2 means _______.

3.

$$\begin{array}{r} 1 \\ 67 \\ +17 \\ \hline 84 \end{array}$$

The 1 above the 6 means: ____________________

UNIT 31 Geometry, Graphs, Charts, and Tables

OBJECTIVES

After studying this unit, the student should be able to:

- List the basic concepts of geometry that young children learn informally at the primary level
- Assess children's readiness for primary level geometry and graphing experiences
- Plan and carry out developmentally appropriate primary level geometry and graphing instruction
- Be able to construct and interpret graphs and tables

Weather is the current science topic in Ms. Hebert's third grade class in South Louisiana. This morning the children are huddled over copies of the past week's weather forecasts that they clipped out of the local morning paper. They are reading the forecast section to find out what kind of information is included and to discuss how they might organize some of the data presented. They note that the day's forecast is included along with the normal highs and lows for that date. There is a regional forecast map for the southeast states and a larger forecast map and description of the weather for the entire United States. Selected national and global temperatures along with precipitation and outlook are included in a table. The coastal tide predictions and the nation's highs and lows are reported. The children compile a list of this information and discuss what they can learn and what information might be interesting to record.

Chan lived in Hong Kong for a few months before coming to the United States. He decides to record the high temperatures in Hong Kong for one week and compare them with Baton Rouge. He notices that Baton Rouge temperatures have been above normal during the past week, so he also records the normal highs. Vanessa decides to compare the highs for Baton Rouge with the highs for the nation. She will mark the highs on a map of the United States. She was disappointed to discover that the nation's highs were only included for four days. The children first made a chart and then a graph to depict the information from the chart. Then they wrote a description of the information obtained from the graphs. To complete this activity, they applied their mathematical knowledge (measurement, counting, graph making) to an activity that integrates science (the topic of weather), social studies (geography), and reading and language arts (reading and comprehending the article and writing about the information obtained). Figures 31–1 and 31–2 depict what Chan and Vanessa might have produced.

Groundwork for this type of activity was laid in several previous units: Unit 12 (Shape), Unit 13 (Space), Unit 20 (Graphs), and Unit 25 (Higher-Level Spatial Relations and Graph-Making Experiences). Primary children must have this groundwork before moving on to the activities described in this unit. Children need a basic understanding of shape and space, which

Cities	Dates						
	14	15	16	17	18	19	20
Hong Kong	75	81	77	81	82	86	88
Baton Rouge	92	90	92	93	93	93	90
Normal Baton Rouge	85	85	85	86	87	86	86

High Temperatures in Hong Kong and Baton Rouge, May 14 - 20

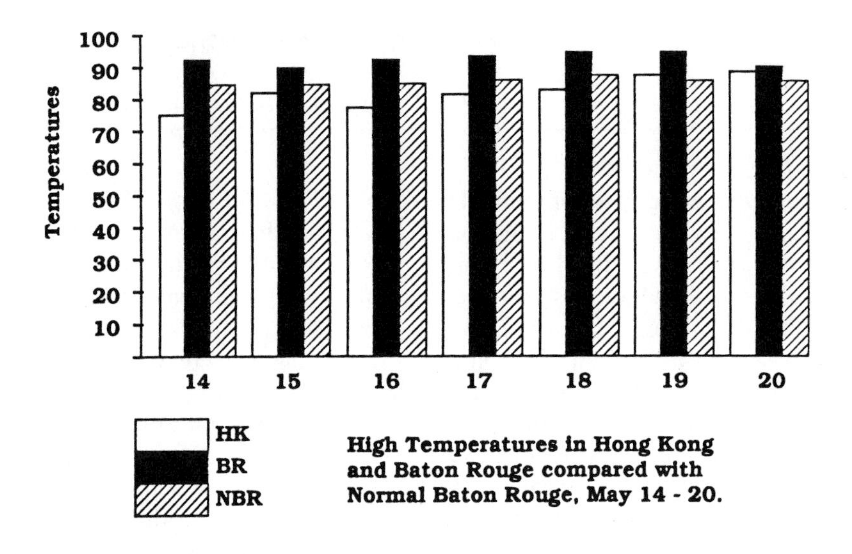

1. **Hong Kong started the week cooler.**
2. **Baton Rouge had a heat wave.**
3. **By Friday (May 20) they were almost the same.**

Figure 31–1 Chan's Data Table and Bar Graph

they apply to the early graphing and mapping experiences during the preoperational period, to help them to move on to higher-level graphing and geometry concepts. Children should know the basic characteristics of shape and be able to identify geometric shapes such as circles, triangles, squares, and rectangles when they enter the primary level. They should also have the spatial concepts of position, direction, and distance relationships and be able to use space for making patterns and constructions. During the primary years, they should continue with these basic experiences and be guided to more complex levels.

Geometry is studied in a very general, in-

Locations	Dates May 14	May 15	May 16	May 17	May 18	May 19	May 20
Nation	105	106	106	110	X	X	X
Baton Rouge	92	90	92	93	93	93	90

High Temperatures in Baton Rouge and the Nation, May 14 - 20.

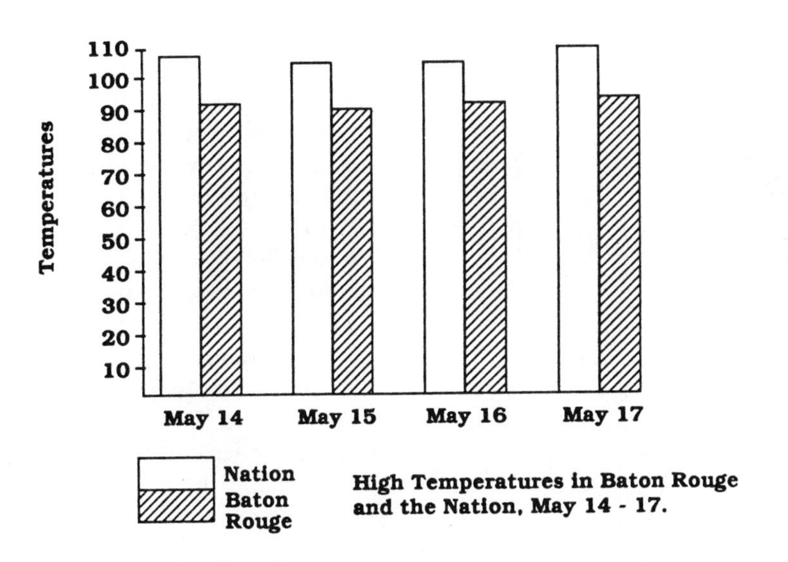

High Temperatures in Baton Rouge and the Nation, May 14 - 17.

1. **The hottest places were in Arizona and Texas.**
2. **Baton Rouge was hot but much cooler than the hottest places.**

Figure 31–2 Vanessa's Table and Bar Graph

formal way during the elementary grades. Spatial concepts are reinforced and the senses are sharpened. During the primary years, geometric concepts continue to be developed mainly at an intuitive level. However, geometric figures are used to teach other concepts, so children should be familiar with them. For example, multiplication is frequently illustrated in a rectangular grid:

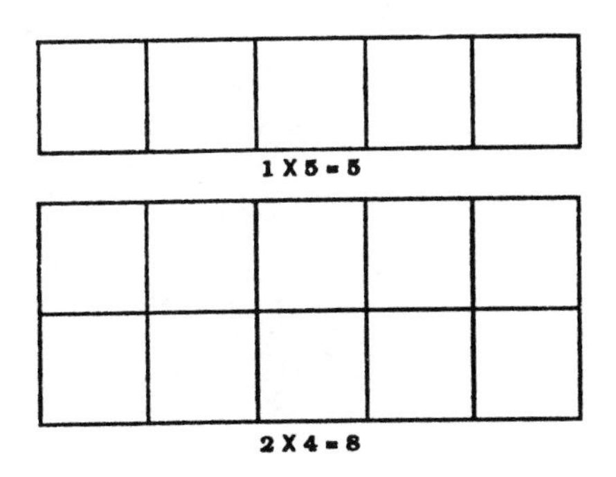

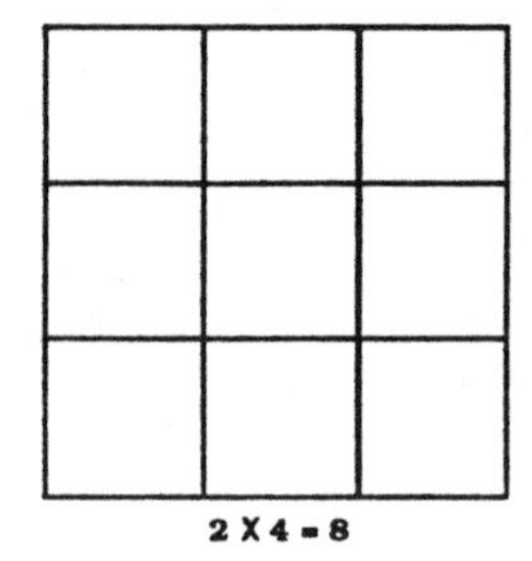
2 X 4 = 8

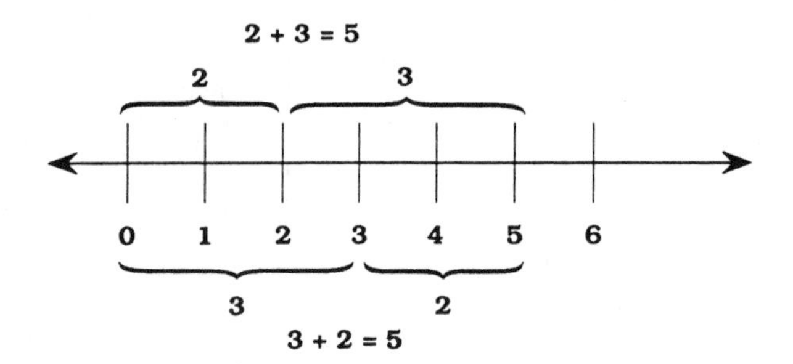

Fractions are commonly illustrated using geometric shapes (see Unit 29) such as:

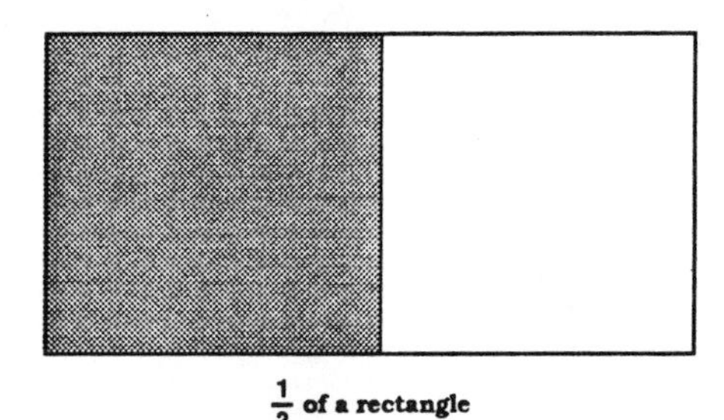
$\frac{1}{2}$ of a rectangle

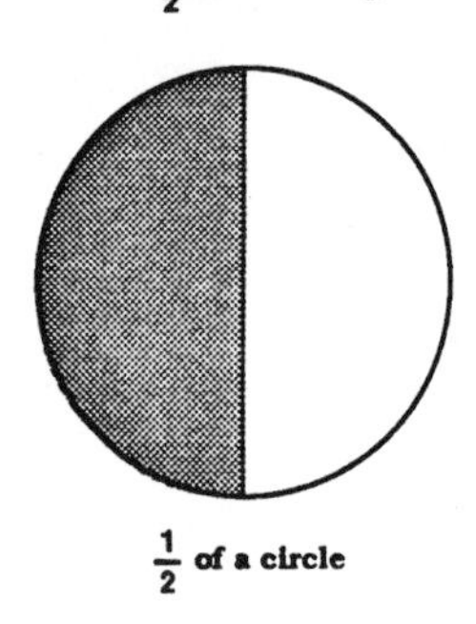
$\frac{1}{2}$ of a circle

Number lines (which will be described later) are conventionally used to help children visualize greater than, less than, betweenness, and the rules of addition and subtraction. For example the number line below is used to illustrate that $2 + 3 = 3 + 2$:

Graphing is closely related to geometry in that it makes use of geometric concepts such as line and shape. Note that the graphs depicted in this unit and in earlier units are based on squares and rectangles. More advanced graphs are based on circles, and others involve the use of line segments to connect points. Tables are the necessary first step in organizing complex data prior to illustrating it on a graph. Charts are closely related to graphs.

In summary, knowledge of geometry and construction of graphs, tables, and charts are closely related basic tools for organizing data. The remainder of the unit describes geometry, graphing, and the use of charts and tables at the primary level.

ASSESSMENT

Readiness for the following primary level activities should be assessed using the assessment tasks that accompany Units 12 and 13. Readiness is also measured by observing children's capabilities in accomplishing the graphing activities in Unit 20 and the higher-level graphing and spatial relations (mapping) activities in Unit 25. It is not safe to assume that children have had all the prerequisite experiences before they arrive in your primary classroom. You might have to start with these earlier levels before moving on to the activities suggested in this unit.

ACTIVITIES

This section begins with geometry, which includes mapping and LOGO computer applications to geometry. It then goes on to graphing, and finally to charts and tables.

Geometry

Primary children are not ready for the technicalities of geometry, but they can be introduced informally to some of the basic concepts. They can learn about *points* as small dots on paper or on the chalkboard. *Curves* are introduced as smooth but not straight paths that connect two points during a story or a mapping activity. *Lines* appear as number lines, in measurement activities, and as the sides of geometric figures. Children perceive *angles* (space made by the meeting of two straight lines) in geometric figures. *Congruency* or sameness of size and shape is what children deal with when they match and compare the size and shape of various figures such as when they sort attribute blocks or make collages from paper shapes. *Symmetry* (correspondence of parts of a figure on opposite sides of a point, line, or plane) is what children are working with when they do the paper folding suggested in the unit on fractions. The terms *point(s)*, *line(s)*, and *curve(s)* may be used with young children without going into the technicalities. The terms *congruency*, *symmetry*, and *angle* will be introduced beyond the primary level and are not essential to working with the concepts informally. The readings and resources at the end of the unit contain a multitude of ideas for activities that will lay the basis for the formal study of geometry. The following are some examples.

GEOMETRY: GEOBOARD ACTIVITIES

OBJECTIVE: To provide experience exploring the qualities of plane figures

MATERIALS: Geoboards and rubber bands. Geoboards may be purchased or made. A geoboard is a square board with nails placed at equal intervals so that it appears to be made up of many squares of equal size. Commercial geoboards have five rows of five nails each.

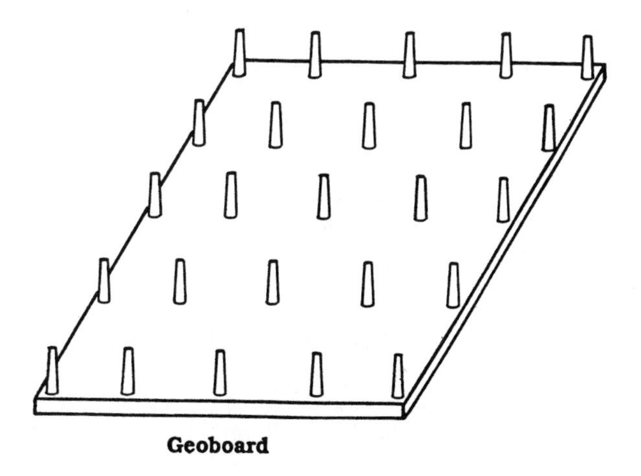

Geoboard

ACTIVITIES: 1. Put out the geoboards and an ample supply of rubber bands of different sizes and colors (special rubber bands may be purchased with the geoboards). Allow plenty of time for the children to explore the materials informally. Suggest that they see how many different kinds of shapes they can make.

2. Give each child a geoboard and one rubber band. MAKE AS MANY DIFFERENT SHAPES AS YOU CAN WITH ONE RUBBER BAND. Encourage them to count the sides of their shapes and to count the number of nails in each side. Suggest that they make a drawing of each shape.

3. Give each child a rubber band and an attribute block. MAKE A SHAPE JUST LIKE THE BLOCK'S SHAPE. Start with squares and rectangles, then triangles and hexagons.

4. On graph paper made to match the geoboard, draw patterns that the children can copy with their rubber bands. For example:

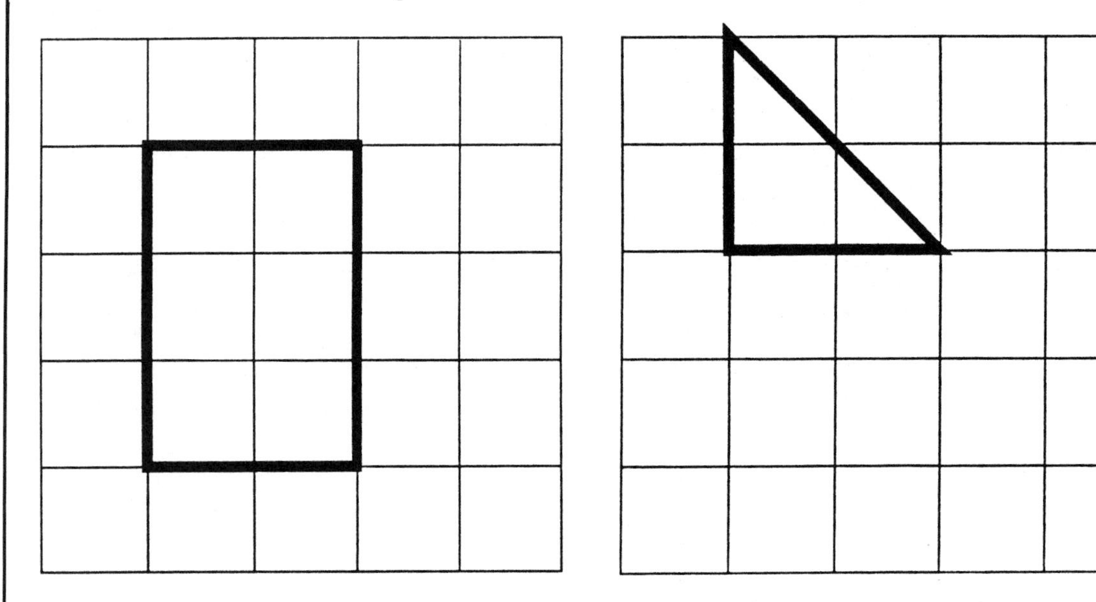

(Commercially made patterns can be purchased also.)

5. Have the children draw patterns on graph paper that matches the size of the geoboards. Demonstrate on the chalkboard first to be sure they realize that they need to make their lines from corner to corner. Have them try out their capabilities on some laminated blank graphs first. Then have them copy their own patterns on the geoboards. Encourage them to exchange patterns with other children.

FOLLOW-UP: Provide more complicated patterns for the children to copy. Encourage those who are capable of doing so to draw and copy more complicated patterns, possibly even some that overlap. Overlapping patterns can be drawn with different colored pencils or crayons and then constructed with rubber bands of matching colors.

GEOMETRY: ACTIVITIES WITH SOLIDS

OBJECTIVE: To explore the characteristics of solid geometric figures

MATERIALS: A set of geometric solids (available from ETA, Kaplan, Nienhuis-Montessori, etc.)

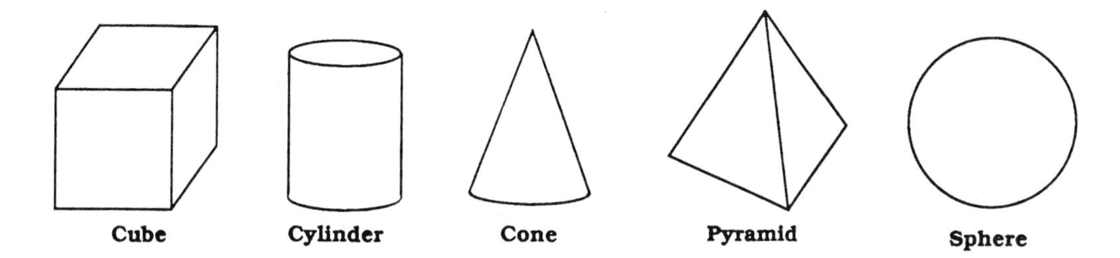

ACTIVITIES: Let the children explore the solids, noting the similarities and differences. Once they are familiar with them, try the following activities:
Put out three of the objects. Describe one and see if the children can guess which one it is.

- IT IS FLAT ALL OVER AND EACH SIDE IS THE SAME. (cube)
- IT IS FLAT ON THE BOTTOM. ITS SIDES LOOK LIKE TRIANGLES, (pyramid)
- IT IS FLAT ON THE BOTTOM, THE TOP IS A POINT, AND THE SIDES ARE SMOOTH. (cone)
- BOTH ENDS ARE FLAT AND ROUND AND THE SIDES ARE SMOOTH. (cylinder)
- IT IS SMOOTH AND ROUND ALL OVER. (sphere)

FOLLOW-UP: Have the children take turns being the person who describes the geometric solid.

GEOMETRY: SYMMETRY

OBJECTIVE: To provide experiences for exploration of symmetry

MATERIALS: Construction paper symmetrical shapes. (See Figure 31–3 for some patterns.) The following are some suggested shapes. Use your imagination to develop others.

ACTIVITY: Give the children one shape pattern at a time. Have them experiment with folding the shapes until the halves match.

FOLLOW-UP: 1. Have the children use the shapes they have folded to make a three-dimensional collage. PUT GLUE ON JUST ONE HALF OF YOUR FOLDED PAPER.

2. Give the children paper squares and rectangles. Show them how they can fold them in the middle and then cut the sides so they come out with a figure that is the same on both halves. These can also be used to make three-dimensional collages.

GEOMETRY: NUMBER LINES

OBJECTIVE: To apply the concept of a line as a visual picture of addition and subtraction and more than and less than

MATERIALS: A laminated number line for each child. A large laminated number line that can be used for demonstration and/or a permanent number line on the chalkboard. Markers (i.e., chips) to mark places on the number lines.

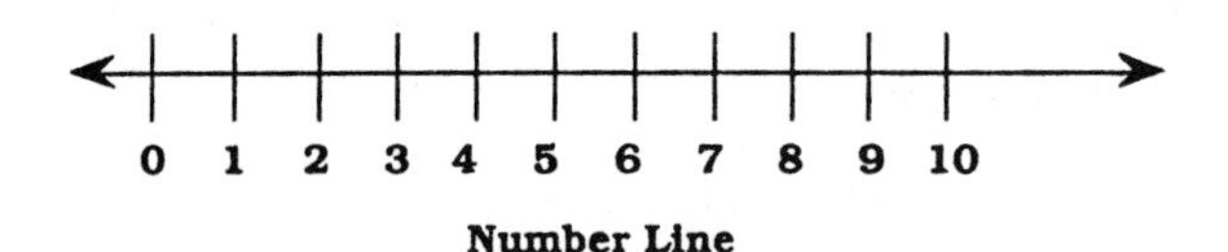

Number Line

ACTIVITIES: The number line can be used to illustrate the following:

1. Greater and less than. PUT A MARKER ON THE FOUR. FIND A NUMBER THAT IS GREATER THAN FOUR. PUT A MARKER ON IT. Discuss which numbers were selected. How did they know which numbers were greater than four. Go through the same procedure with the other numbers. Then go through the procedure looking for numbers less than a given number.

Figure 31–3 Patterns for Exploring Symmetry

2. Addition. HOW CAN WE SHOW [2 + 4] ON THE NUMBER LINE? Encourage the children to try to figure it out. If they can't, then demonstrate. FIRST I'LL PUT A MARKER ON TWO. NOW I'LL COUNT OVER FOUR SPACES. NOW I'M AT SIX. HOW MUCH IS TWO PLUS FOUR? Have the children try several problems with sums of ten or less.

3. Subtraction. HOW CAN WE SHOW 5 − 2 ON THE NUMBER LINE? Encourage the children to try to figure it out. If they cannot, then demonstrate. FIRST I'LL PUT A MARKER ON THE 5. NOW WHICH WAY SHOULD I GO TO FIND 5 − 2? Have them try several problems using numbers ten or less.

FOLLOW-UP: Have the children illustrate equivalent sums and differences on the number line (i.e., 3 + 4 = 1 + 6; 8 − 4 = 7 − 3. The children could also make up number line problems for themselves. Suggest that when they need help with a one-digit problem that they use the number line to find the answer.

GEOMETRY: COMPARING ROAD AND STRAIGHT LINE DISTANCES

OBJECTIVE: To see the relationship between a direct route and the actual route between points on a map. (This activity would be for the more advanced level primary students who have learned how to use standard measurement tools.)

MATERIALS: Maps of your state for everyone in the class, foot rulers, and marking pens

ACTIVITIES: Have the students explore the maps. See if they can find the legend and if they can tell you what the various symbols mean. Be particularly sure that they know which kinds of lines are roads, how you find out the mileage from one place on the map to another, and how many miles there are to the inch. Spend some time finding out how far it is from your town or city to some of the nearby towns and cities. Have the class agree on two places in the state they would like to visit. Using their rulers and marking pens. have them draw a line from your city to the nearest place selected, from that place to the other location selected, and from there back home. They should then have a triangle. Have everyone figure out the mileage by road and then by direct flight by measuring the lines. Add up the three sides of the triangle. Add up the three road routes. Find the difference between the road trip and the direct route. Discuss why the roads are not as direct as the lines (Figure 31–4).

FOLLOW-UP: Encourage interested students to compare road and direct distances to other points in the state.

LOGO computer language can provide experience with geometry at a number of levels. With just a few simple commands and minimal instruction, children can explore, play, and create an infinite number of geometric shapes and designs (Silvern, 1988). With a little more structured approach, they can learn how to plan out patterns ahead of time and use more complex

Figure 31–4 These children discover that a straight line is the shortest distance between two points on a map.

instructional commands (Campbell, 1987). The cursor, referred to as the *turtle* in LOGO, can be moved about in many directions, at different angles, to make straight or curved lines. Problem-solving skills are developed when children work on figuring out how they will make the turtle go just where they want it to in order to come up with a particular design or figure.

Making Graphs

Graphing includes constructing graphs, reading information on graphs, and interpreting what the information on a graph means. The data used for making graphs needs to be something of interest to the students. In Unit 20, there is a list of possible graphing subjects that young children might enjoy working. Other subjects will grow out of their current interests and activities.

The four most popular types of graphs are picture graphs, bar graphs or histograms, line graphs, and circle or pie graphs. The graphs described earlier fall into the first two categories and are the easiest for young children to construct and interpret. Circle or pie graphs are beyond the primary level. Some primary level children can begin to work with line graphs.

Line graphs demand concrete operational thinking because more than one aspect of the data must be focused on at the same time. Line graphs are made on a squared paper grid and apply the basic skills that children would learn by first doing the squared paper activities with

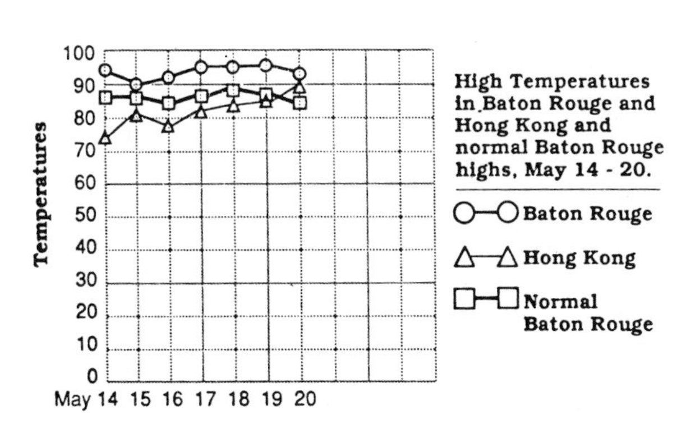

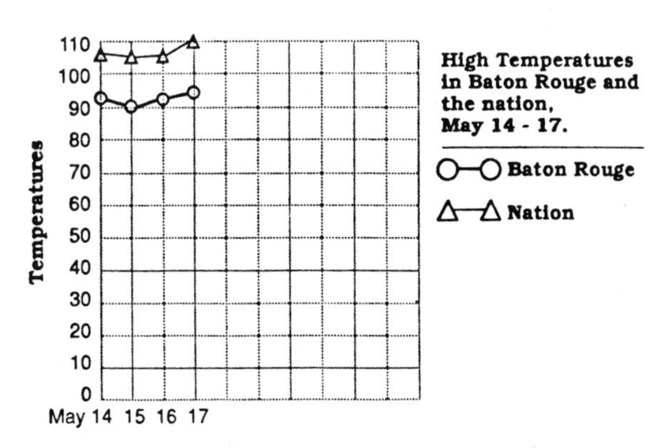

Figure 31–5 Chan's and Vanessa's data depicted in line graphs.

the geoboard. They are especially good for showing variations such as rainfall, temperature, and hours of daylight. In Figure 31–5, Chan's temperature data is translated from the bar graph to the line graph. Note that the left side and the bottom are called the axes and each must be labeled. In this case, the left side is the temperature axis and the bottom is the days-of-the-week axis. To find the correct point for each temperature on each day, the child has to find the point where the two meet, mark the point, and connect it to the previous point and the next point with a line. If two or more types of information are included on the same graph, then geometric symbols are frequently used to indicate which line goes with which set of data (see Figure 31–5). In her article "Coordinate graphing: Shaping up a sticky situation," Jeanne M. Vissa suggests some creative ways to introduce the use of coordinate (or line) graphing to young children.

GRAPHING: INTRODUCING COORDINATES

OBJECTIVE: To introduce finding coordinates on a graph

MATERIALS: A large supply of stickers of various kinds. On the bulletin board, construct a large 5 × 5 square coordinate graph. The grids can be made using black tape. Place stickers at the intersections of various coordinates (Figure 31–6).

ACTIVITIES: THIS IS THE CITY. DRIVING INTO THE CITY THE CORNER IS HERE AT 0,0. I WANT TO GO TO (name one of the stickers). TELL ME HOW MANY BLOCKS OVER AND HOW MANY BLOCKS UP I WILL HAVE TO GO. Suppose that the sticker is on 2, 3. YES, I HAVE TO GO OVER TWO BLOCKS AND UP THREE. THIS POINT IS CALLED 2, 3. Point out how the numbers on the bottom and the sides correspond to the point. Go back to 0, 0 and have the children direct you to other points on the graph. Let the children take turns telling you the coordinates of a sticker they would like to have. When they are able to give the correct coordinates, they get a matching sticker to keep.

FOLLOW-UP: During Center Time, encourage children to explore the coordinate map on their own or with a friend. Suggest they trace trips to different "corners" with their fingers. The children who understand the concept of coordinates can use coordinate paper to complete symmetrical shapes (Figure 31–7) and name the coordinates.

Charts and Tables

Charts and tables are contructed to organize data before it is graphed. A simple chart consists of tic marks such as depicted in the chart on floating and sinking objects in Unit 10 (see Figure 10–5). This information could be translated into a single variable graph showing frequency of floating or sinking for each object or into a double variable graph showing both (that is, a double-bar or a double-line graph). Simple tables are represented in Figure 31–1. These tables were used to organize the temperature data prior to constructing the graphs.

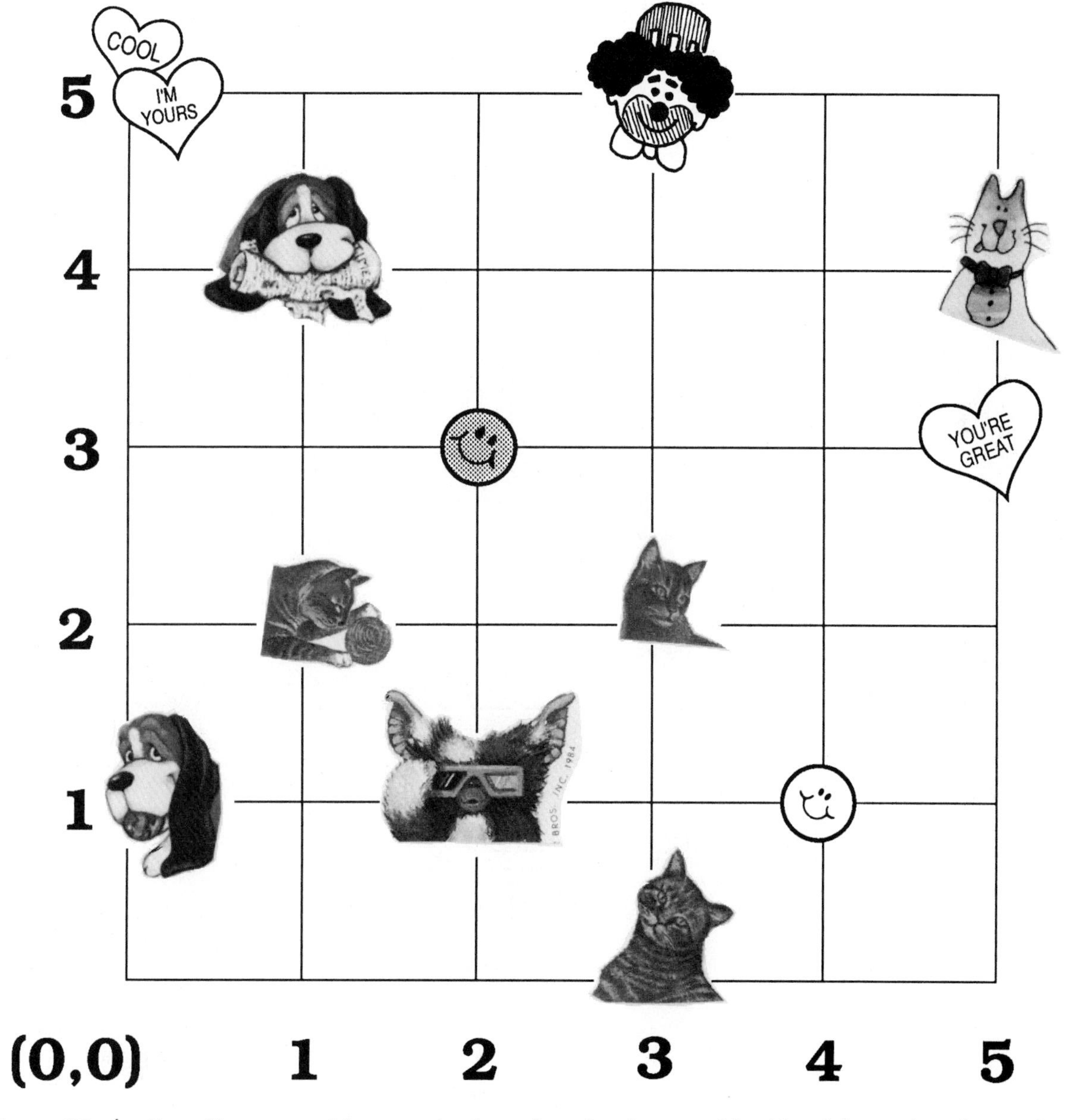

Figure 31–6 Coordinate graphing can be introduced using a grid with stickers placed at points to be identified.

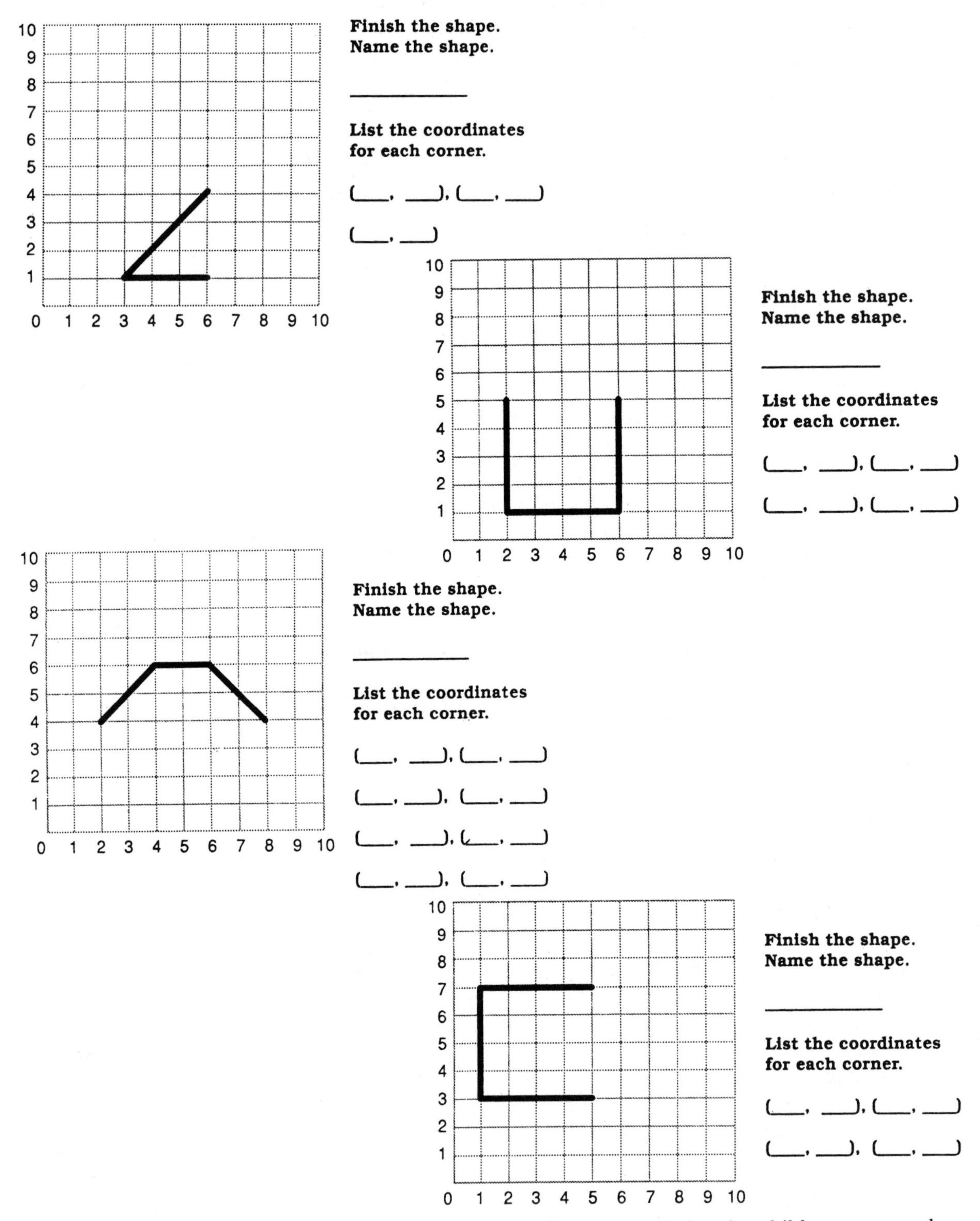

Figure 31–7 Half of a symmetrical shape can be drawn on a grid for the children to complete.

EVALUATION

Note whether children can follow directions and maintain involvement in the activities. Observation is critical for observing the process in these activities. When children are not able to do an activity, it is important to note where the process breaks down. Does the child have the basic idea but just need a little more practice and guidance? Does the activity seem to be beyond the child's capabilities at this time? These activities require advanced cognitive and perceptual motor development, so children should not be pushed beyond their developmental level. If children work in pairs or small groups of varied ability, the more advanced can assist the less advanced.

SUMMARY

Primary experiences with geometry, graphs, tables, and charts build on preprimary experiences with shape, space, and simpler graphs and charts. Primary level geometry is an informal, intuitively acquired concept. Children gain familiarity with concepts such as line, angle, point, curve, symmetry, and congruence. Geoboard activities are basic at this level.

Geometric and number concepts can be applied to graphing. Advanced children can develop more complex bar graphs and move on to line graphs. Charts and tables are used to organize data, which can then be visually depicted in a graph.

FURTHER READING AND RESOURCES

Balka, D. 1987. *Unifix® mathematics activities, Book 1.* Peabody, Mass.: DIDAX Educational Resources.

Bell, S. 1988. Let's go, LOGO. *Creative Classroom.* 2 (6): 62–63.

Burton, G. M. 1985. *Towards a good beginning.* Menlo Park, Calif.: Addison-Wesley.

Campbell, P. F. 1988. Microcomputers in the primary mathematics classroom. *Arithmetic Teacher.* 35 (6): 22–30.

Hoffer, A. R. 1988. Geometry and visual thinking. *Teaching mathematics in grades K-8.* Boston: Allyn & Bacon.

Reys, R., Suydam, M., and Lindquist, M. M. 1984. *Helping children learn mathematics.* Englewood Cliffs, N.J.: Prentice-Hall.

Scott, L. B., and Garner, J. 1978. *Mathematical experiences for young children.* St. Louis: McGraw-Hill.

Silvern, S. B. 1988. Creativity through play with LOGO. *Childhood Education.* 64: 220–224.

Vissa, J. 1987. Coordinate graphing: Shaping a sticky situation. *Arithmetic Teacher.* 35 (3): 6–10.

SUGGESTED ACTIVITIES

1. Assess some primary children's readiness for the types of geometry and graphing activities described in this unit. Describe the results and your evaluation of their degree of readiness.

2. Plan some activities that would be appropriate for the children you assessed. If possible, use these activities with the children. Evaluate the results. Did the children respond as you expected? Was your assessment accurate? What modifications, if any, would you make the next time? Why?

3. Provide the students in this course with some data. Have them develop some charts or ta-

bles using the data. Then have them make some line graphs. Do they have any problems in developing graphs?

4. Add geometry and graphing activities to your Activity File.

REVIEW

1. List the concepts and experiences that are prerequisite to the geometry and graphing concepts and activities described in this unit.
2. Make three diagrams that illustrate how fractions are visually depicted using geometric shapes.
3. Show on a number line how $4 + 5 = 3 + 6$.
4. Match the terms in column I with the definitions in column II.

Column I	Column II
1. Point	a. Correspondence of parts of a figure on opposite sides of a point, line, or plane
2. Curve	b. The space made by the meeting of two straight lines
3. Line	c. An idea that is represented on paper by a dot
4. Angle	d. Sameness of size and shape
5. Congruency	e. Represented on paper by using a straightedge ruler and pencil
6. Symmetry	f. A smooth but not straight line

5. Make a sketch of a geoboard and explain its purpose.
6. Find the following points on the graph: (0, 0), (3, 2), (3, 4), (0, 3), (4, 1), (1, 2).

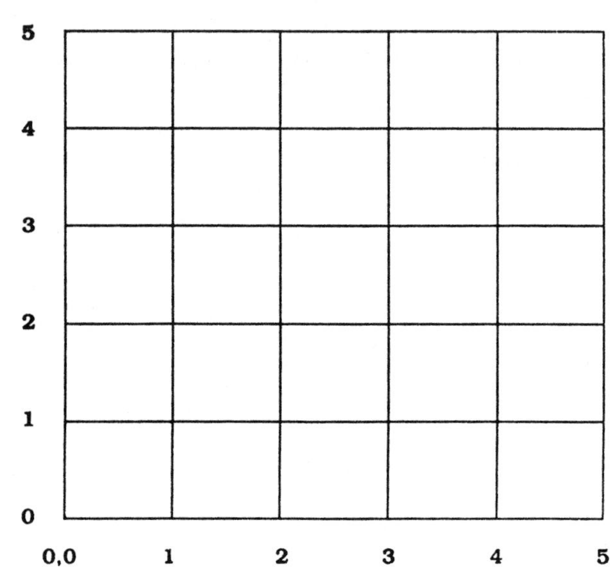

7. Make a line graph using the following data.

Cities High Temperatures, May 15–20

	Sat.	Sun.	Mon.	Tues.	Wed.	Thurs.
New York City	73	72	74	64	61	60
Cairo, Egypt	102	111	107	95	93	102

What conclusions can you make from an examination of the graph?

UNIT 32 Measurement with Standard Units

OBJECTIVES

After studying this unit, the student should be able to:

- List the reasons that measurement is an essential part of the primary mathematics program
- Know when to introduce standard units of measurement for length, time, volume, area, temperature, and money
- Name the two types of standard units of measurement that we use in the United States
- Plan and carry out developmentally appropriate primary level measurement instruction

Measurement is an extremely important aspect of mathematics. It is a practical activity that is used in everyday life during experiences such as cooking, shopping, building, and constructing. In the primary curriculum, it is essential to data gathering in science and can also be applied in other areas. Measuring is also a vehicle for reinforcing other math skills and concepts: the number line is based on length, a popular multiplication model is much like area, and measurement is an area that lends itself naturally to problem-solving activities. Counting, whole number operations, and fractions are used to arrive at measurements and report the results. This unit will build on the basic concepts of measurement described in Units 18, 19, and 22. The focus of this unit is instruction and activities for introducing the concept of standard units and applying the concept to length, volume, area, weight, temperature, time, and money measurement (Figure 32–1).

In Unit 18 (see Figure 18–1), five stages in the development of the concept of measurement were described. During the sensorimotor and preoperational periods, children's measurement activities center on play and imitation and making comparisons (e.g., long/short, heavy/light, full/empty, hot/cold, early/late, rich/poor). During the transition period from ages five to seven, they enjoy working with arbitrary units. During concrete operations (which an individual usually enters at age six or older), children can begin to see the need for standard units (stage 4) and begin to develop skills in using them (stage 5). Standard units are not introduced for each concept at the same time. In general, the following guidelines can be observed:

- Length (linear measure): The units inch/foot and centimeter/meter are introduced in the beginning of primary and for measurement during second grade.
- Area: Is introduced informally with non-standard units in grade one and ties in with multiplication in grade three.
- Time: Time measurement devices and vocabulary were introduced prior to primary, but it is generally the end of

Figure 32–1 The lengths of common objects are compared as a first step in linear measurement.

primary before conventional time is clearly understood and a nondigital clock can be read with accuracy.

- Volume (capacity): Done informally during pouring activities and accuracy stressed during preprimary cooking. Concept of units of volume is usually introduced in grade two.
- Weight: Standard measurement for weight is usually introduced in third grade.
- Temperature: Temperature units are identified and children may begin to read thermometers in the second grade, but it is usually beyond primary when children really measure temperature with accuracy and understanding.
- Money: Coins and bills are identified prior to primary, symbols are associated in early primary, but value does not begin to be understood until the end of primary.

The goal in the primary grades is to introduce the meaning of measurement, needed terminology, important units, and most common measurement tools.

Both English (customary in the United States) and metric units are introduced during the primary years. Although the metric system is much easier to use because it is based on tens and is used as the principal system in most countries, it has not been adopted as the official measure in the United States. In the 1970s there was a movement toward adoption of the system in this country, but it died out and the U.S. Metric Commission was abolished in 1984. However, children must learn the metric system because it is so widely used around the world as well as in industry and science. (See Unit 43 for metric charts.)

ASSESSMENT

Concrete operational thinking is essential for an understanding of the need for and the use of standard units. Conservation tasks for length (Figure 18–4), weight (Figure 18–4), and volume (Figure 1–10) were illustrated earlier in the text. In Unit 18, observational assessment guidelines are suggested for finding out what children at the early stages of understanding of measurement know about volume, weight, length, and temperature. Interview tasks for time can be found in Unit 19. Be sure that children can apply nonstandard measure before moving on to standard measure.

INSTRUCTION

The concept of measurement develops through measurement experiences. Lecture and demonstration is not adequate for supporting the development of this concept. Also, it is important to take a sequenced approach to the introduction of standard units. Adhere to the following steps:

1. Do comparisons that do not require numbers (see Unit 11).
2. Use nonstandard arbitrary units (see Unit 18).
   a. Find the number of units by counting.
   b. Report the number of units.
3. Compare the thing measured to the units used (e.g., a table's width is measured with paper clips and drinking straws).
4. Introduce standard units appropriate for the same type of measurement.
   a. Find the number of units using standardized measuring instruments (i.e., ruler, scale, cup, liter, thermometer).
   b. Report the number of units (Figure 32–2).

Introduction of new standard measurement techniques and instruments should always be preceded by comparisons and nonstandard measurement with arbitrary units. Naturalistic and informal measurement experiences should be encouraged at all levels.

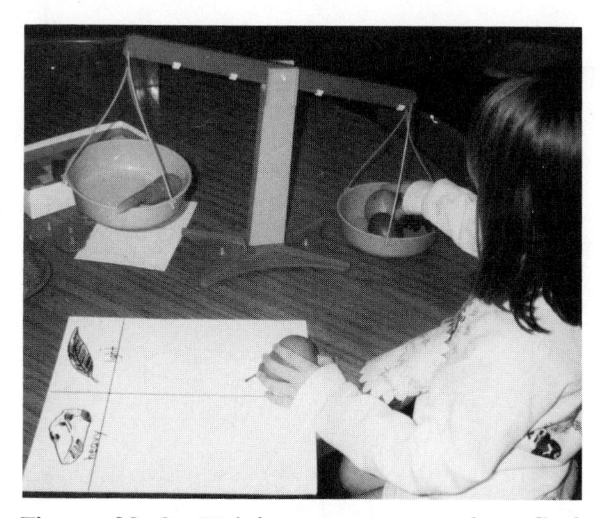

Figure 32–2 Weights are compared to find objects that are heavy and objects that are light before moving into arbitrary and standard units of weight measure.

The Concept of Unit

Children's ability to measure rests on their understanding of the concept of *unit*. Many children have difficulty perceiving that units can be other than one. That is, one-half foot could be a unit, three centimeters could be a unit, two standard measuring cups could be a unit, one mark on a thermometer equals two degrees, and so on.

By using nonstandard units of measure first, the concept of unit can be developed. Children learn that measurement can be made with an arbitrary unit, but the arbitrary units must be equal to each other when making a specific measurement. For example, when measuring with paper clips, each one must be the same length. Paper clip is not the unit. A paper clip of specific length is the unit. Through the use of arbitrary but equal units used to measure objects, children construct the concept of a unit. The concept is reinforced by using different arbitrary units (one kind at a time) and then comparing the results in terms of the number of units. For example, the children measure Lai's height using Unifix® Cubes, identical drinking straws, and the class math textbook. Soon they realize that measurement with smaller units requires more units than with larger units. When the students move on to standard units, they can compare the number of units needed to measure using teaspoons versus a standard cup measure, inches versus a yardstick, and so on.

Children should be aware that when they use units (arbitrary or standard) they must be accurate. For example, there cannot be gaps or spaces between units when measuring length. That is why it is a good idea to start with Unifix® Cubes, Lots-a-links, or some other units that can be stuck together and easily lined up. Once children are able to measure using as many units as needed to measure the whole length, capacity, and so on, then they can advance to using one or more units that must be moved to make a complete measurement. For example, they could make a ten Unifix® Cube length measure, place it on the item to be measured, mark where it ends, move the measure to that point, keep going until finished, and then add the tens and the remaining cubes to arrive at the length in cubes. For capacity, individual measuring cups could be filled and then the number of empties counted after a larger container is filled, or one cup could be used and a record kept of how many cupfuls filled the larger container. As children discover these shortcuts to measurement, they will be able to transfer this knowledge over to standard unit measure and understand the rationale behind foot rulers, meter and yardsticks, and quart and liter measures.

Measuring Instruments

With the introduction of standard units comes the introduction of measuring instruments. Rulers, scaled instruments (scales, graduated cylinders, thermometers), and clocks are the tools of standard measurement. Children have problems with these instruments unless they understand what they are measuring and what it means to measure. It is wise to begin with simple versions of the instruments, which are marked only with the unit being used. For example, if the unit is the centimeter, use a ruler that is marked only with centimeters (no millimeters). If the unit is an inch, use a ruler marked only with inches (no half, quarter, and/or eighth inches). Be sure the children understand how units are marked. For example, on a ruler, the numbers come after the unit, not before. Even most nine-year-olds will say that the ruler illustrated is five inches, rather than six inches long.

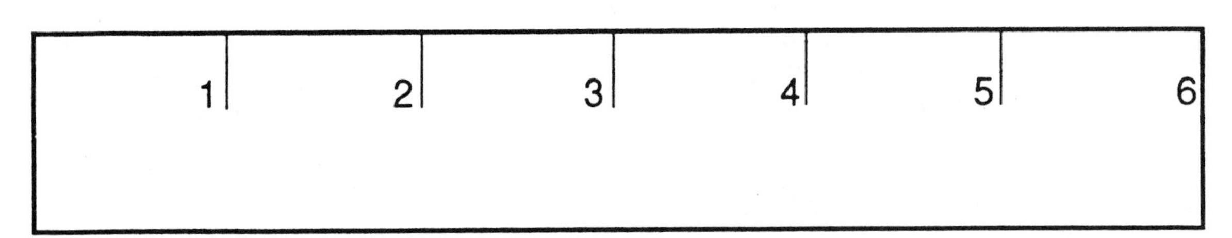

You must be sure children understand each number means that many units have been used. Children need many experiences measuring objects shorter than their ruler before they move on to longer objects with which they will have to measure, mark, and move the ruler. They also will need to be able to apply their addition skills. For example, if they are using a twelve-inch ruler, and they measure something that is twelve inches plus eight inches, can they add 12 + 8? To measure to the nearest 1/4, 1/8, or 1/16 of an inch, they must understand fractions.

Scaled instruments present a problem because every individual unit is not marked. For example, thermometers are marked every two degrees. A good way to help children understand this concept is to have them make their own instruments. They can make graphs using different scales or make their own graduated cylinders. The latter can be done by taking a large glass and putting a piece of masking tape down the side as shown in the following illustration.

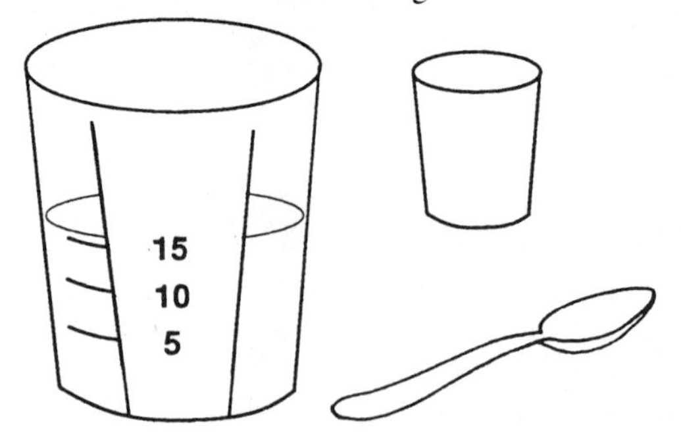

Take a smaller container and fill it with spoonfuls, count how many it holds, and empty it into the glass. Mark the level of the water and the number of spoonfuls. Fill the small container, empty into the glass, and mark again. In the example, the small container holds five teaspoons of water. This measure can be used to find out how much other containers will hold.

Clocks are one of the most difficult instruments for children to understand. Although there are only three measures (hours, minutes, and seconds), the circular movement of the hands makes reading the face difficult. Children vary greatly as to when they are finally able to read a clock face accurately. There is no set age for being able to tell time. Skills needed to tell time must be learned over many years and through practice with clock faces with movable hands. Digital clocks are easier to read but do not provide the child with the visual picture of the relationship between time units.

Money also offers difficulties because the sizes of the coins do not coincide with their value. That is, the dime is smaller than the penny and the nickel. Bills provide no size cues but do have numerical designations that relate one bill to the other. Relating the coins to the bills is a difficult task for young children.

Figure 32–3 These children are ready for linear measurement using standard units.

MEASUREMENT ACTIVITIES

This section includes activities that will help children construct the concepts of linear, volume (capacity), area, weight (mass), temperature, time, and money measurement using standard units. Always refer to earlier units for the comparison and arbitrary (nonstandard) unit experiences that need to be done prior to introducing the standard units. Also, remember to work on the concepts of the units used for each type of measurement as well as actual measurement (Figure 32–3).

STANDARD MEASUREMENT: LINEAR

OBJECTIVE: To be able to use standard units of measure to compare lengths of objects; to discover that the smaller the units, the more will be needed to measure a distance; and to learn to use a ruler to measure objects.

MATERIALS: Rulers (inch/foot, yardstick, centimeter, meter stick), tape measures, paper, pencils, crayons, markers, poster board.

ACTIVITIES: These activities are introduced following many exploratory experiences with comparing visually and with arbitrary units. The same objects can be measured using both methods to emphasize the need for standard units. The first activity is designed to develop an understanding of this need.

1. YOU HAVE MEASURED MANY THINGS AROUND THE ROOM, INCLUDING YOURSELVES, AND MADE COMPARISONS. WHAT YOU COMPARE AGAINST IS CALLED YOUR *UNIT OF MEASUREMENT*. WHAT ARE SOME UNITS OF MEASUREMENT YOU HAVE USED? (Children should name items used such as paper clips, Unifix® cubes, books, etc.) WHEN YOU TELL HOW LONG OR HOW TALL SOMETHING IS, YOU HAVE TO TELL WHICH UNIT OF MEASURE YOU USED. WHY? WHAT PROBLEM COULD OCCUR IF YOU ARE TELLING THIS INFORMATION TO SOMEONE WHO CANNOT SEE YOUR UNIT OF MEASURE? Encourage discussion. For example, they might suggest that if measuring with Unifix® cubes you send the person a cube or if measuring with a piece of string you send the person a piece of string of the same length. Suggest that a unit of measure might be more useful if EVERYONE IN THE WORLD KNEW EXACTLY WHAT IT IS BECAUSE IT IS ALWAYS THE SAME. THAT IS WHY WE HAVE *STANDARD* UNITS OF MEASURE. WE USE INCHES, FEET, AND YARDS. (Pass around foot rulers and yardsticks for children to examine and compare.) IN MOST PLACES IN THE WORLD, IN SCIENCE LABORATORIES, AND IN FACTORIES, CENTIMETERS, DECIMETERS, AND METERS ARE USED. Pass around meter sticks. Point out that the Base Ten Blocks they have been using are marked off in centimeters (ones units) and in decimeters (tens units), and that ten tens placed end to end is a meter. Have them test this with the meter sticks and the base ten blocks. THESE UNITS ARE THE SAME EVERYWHERE AND DO NOT CHANGE. GET INTO PAIRS AND MEASURE EACH OTHER'S BODY PARTS USING YOUR RULERS AND TAPE MEASURES. BEFORE YOU START, MAKE A LIST OF THE PARTS YOU PLAN TO MEASURE. Have the students call out the names of the parts they plan to measure. When they finish have them compare their results.
2. Have the children use their foot rulers and two other units (such as a shoe and a pencil) to measure the same objects. Have them record the results in a table.

Object	**Length**			
	Foot ruler		*Shoe*	*Pencil*
	Feet	*Inches*		
Table				
Windowsill				
Book shelf				

Discuss the results and what they mean.

3. Have the children make their own inch and centimeter rulers. Provide them with strips of cardboard six inches long and have them mark them off as indicated on the next page.

   Have them make comparative measures around the classroom. Then send the rulers home along with a note to the parents.

January 22

Dear Parents,
We are working with different ways of measuring length. Your child is bringing home an inch ruler and a centimeter ruler and is supposed to measure six things at home with each ruler and record the results. Please help your child if necessary. Have your child share the results with you and explain the differences in the number of inches versus the number of centimeters of length found for each object. Thanks for your help with this project.
Sincerely,
Jon Wang, Second Grade Teacher

Object	**Length**	
	Inch ruler	*Centimeter ruler*
1.		
2.		
3.		
4.		
5.		
6.		

FOLLOW-UP: Do many more linear measurement activities using standard units as suggested in the resources at the end of the unit.

STANDARD MEASUREMENT: VOLUME

OBJECTIVE: To be able to use standard units of measure to compare volumes of materials; to learn that the smaller the units, the more will be needed to measure the volume; to learn how to use standard measures of volume such as teaspoons, tablespoons, cups, pints, quarts, and liters (Figure 32–4).

MATERIALS: Containers of many different sizes (boxes, baskets, buckets, jars, cups, bowls, pans, bottles, plastic bags, and so on) and standard measures of volume (set of customary and set of metric measuring cups, liter and quart measures, customary and metric measuring spoons)

ACTIVITIES: These activities are introduced following many exploratory experiences with comparing visually and with arbitrary units. The same container's capacities can be measured with arbitrary and with standard units to emphasize the need for standard units. The first activity is designed to develop an understanding of this need.

1. Discuss volume following the same format as was used for linear measurement. YOU HAVE EXPLORED THE VOLUME OR SPACE INSIDE OF MANY CONTAINERS BY FILLING AND EMPTYING CONTAINERS OF MANY SIZES AND SHAPES. WHAT KINDS OF UNITS HAVE YOU USED? Encourage them to name some of the smaller units they have used to fill larger containers. See if they can generalize from the length discussion that there are problems in communication when standard units are not used. Pass around the standard measurement materials. Encourage the children to talk about the characteristics of these materials. Do they recognize that they have used these types of things many times to measure ingredients for cooking? Do they notice the numbers and scales marked on the materials?
2. Using the standard units, have the children find out the capacities of the containers previously used for exploration. Have them use materials such as water, rice, beans, and/or sand for these explorations. They should record their findings in a table such as the one below:

NUMBER OF UNITS TO FILL

Container	***Measure Used***	
	Customary Standard Cup Measure	*Small Plastic Juice Glass*
1.		
2.		
3.		
4.		
5.		
6.		

FOLLOW-UP: Continue to do exploratory measurement of capacity. Also, continue with cooking activities giving the children more responsibility for selecting the needed measurement materials.

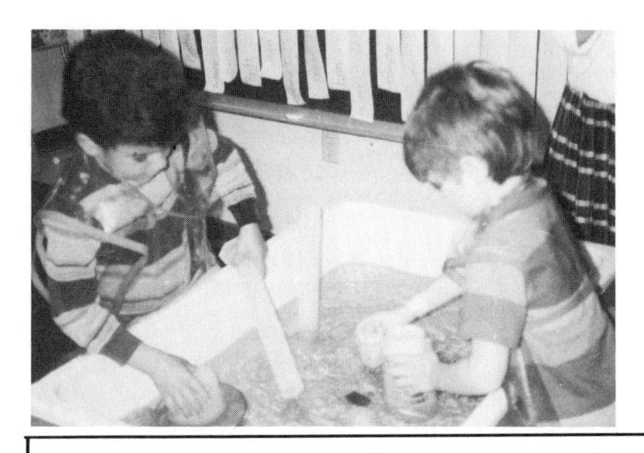

Figure 32–4 These boys are exploring volume using standard measuring tools. Behind them are height strips for each child in the classroom.

STANDARD MEASUREMENT: AREA

OBJECTIVE: To explore area in a concrete manner and its relationship to linear measurement

MATERIALS: Inch and centimeter cubes and two-dimensional patterns with squares and without squares; paper grids and paper squares. (Or purchase *Learning measurement inch by inch* from Lakeshore.)

ACTIVITIES: Area can be explored with squared paper and cubes long before formal instruction. Remember that area-type activities are frequently used as the visual representations of multiplication.

1. On poster board, make some shapes such as those shown that are in inch or centimeter units. Have the children find out how many inch or centimeter cubes will cover the whole shape.

   SHAPES WITH SQUARES MARKED

   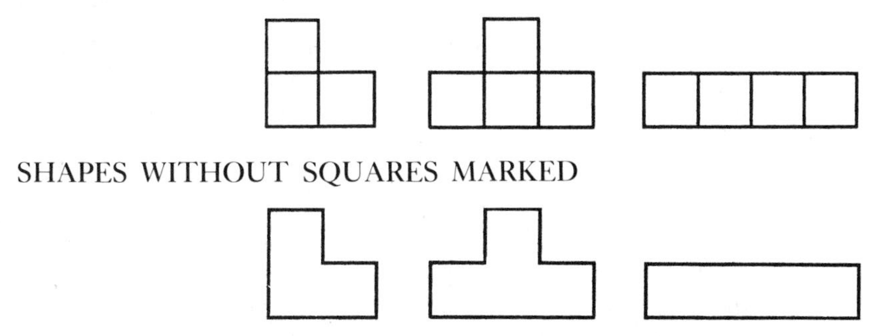

   SHAPES WITHOUT SQUARES MARKED

2. Make a supply of paper grids (inch or centimeter squares, about 5×4 squares). Make a supply of construction paper squares. Have the children make up their own areas by pasting individual squares on the grid. Have them record on the grid how many squares are in their area. (Or use *Learning measurement inch by inch*.)

FOLLOW-UP: If rulers have been introduced, have the children measure the lines on the grids and patterns with their rulers. Discuss how they might figure out how many square inches or centimeters are on a plane surface using their rulers.

STANDARD MEASUREMENT: WEIGHT

OBJECTIVE: To be able to use standard units of measure to compare the weights of various materials and to use balance and platform scales (Figure 32–5).

MATERIALS: Balance scales with English and metric weights, a set of platform scales, and a metric/and or customary kitchen balance scale, paper and pencil for recording observations, and many objects and materials that can be weighed.

ACTIVITIES: The children should have already explored weight using comparisons as depicted in Figures 32–2 and 32–5.

1. WHEN YOU LOOK AT TWO OBJECTS, YOU CAN GUESS IF THEY ARE HEAVY OR LIGHT, BUT YOU DON'T REALLY KNOW UNTIL YOU LIFT THEM BECAUSE SIZE CAN FOOL YOU. YOU COULD EASILY LIFT A LARGE BALLOON, BUT A ROCK OF THE SAME SIZE WOULD BE TOO HEAVY TO LIFT. (Have a balloon and a rock available for them to lift, if possible.) A MARSHMALLOW WOULD BE EASIER TO LIFT THAN A LUMP OF LEAD OF THE SAME SIZE. THE LEAD AND THE ROCK HAVE MORE STUFF IN THEM THAN THE BALLOON AND THE MARSHMALLOW. THE MORE STUFF THERE IS IN SOMETHING, THE HARDER A FORCE IN THE EARTH CALLED GRAVITY PULLS ON IT. WHEN SOMETHING IS WEIGHED, WE ARE MEASURING HOW HARD GRAVITY IS PULLING ON IT. YOU HAVE COMPARED MANY KINDS OF THINGS BY PUTTING ONE KIND OF THING ON ONE SIDE OF A BALANCE SCALE AND ANOTHER ON THE OTHER SIDE. NOW YOU WILL WORK WITH CUSTOMARY AND METRIC WEIGHTS IN YOUR PAN BALANCE ON ONE SIDE AND THINGS YOU WANT TO WEIGH ON THE OTHER SIDE. Discuss the sets of pan balance weights and have them available for the children to examine. Explain that they are all made of the same material so that size is relative to weight. Have the children weigh various objects and materials (water, rice, beans, and so on). Compare the weights of one cup of water versus one cup of rice or one cup of beans. Which weighs more? How much more? Remind the children they will have to add the amounts for each weight to get a total. Suggest that they work in pairs so they can check each other's results.
2. Show the children how to read the dial on the kitchen balance. Provide a variety of things to weigh.
3. If a platform scale for people is available, have everyone in the class weighed, record the results, and have the children make a graph that depicts the results.
4. Have the children go through newspaper grocery store advertisements and cut out pictures of items that are sold by weight and have to be weighed at the store to find out the cost.
5. Have the children weigh two objects at the same time. Remove one and weigh the remaining object. Subtract the weight of the single object from the weight of the two. Now weigh the other object. Is the weight the same as when you subtracted the first weight from the total for both objects?

FOLLOW-UP: Continue putting out interesting things to weigh. Make something with a European recipe that specifies amounts by weight rather than volume.

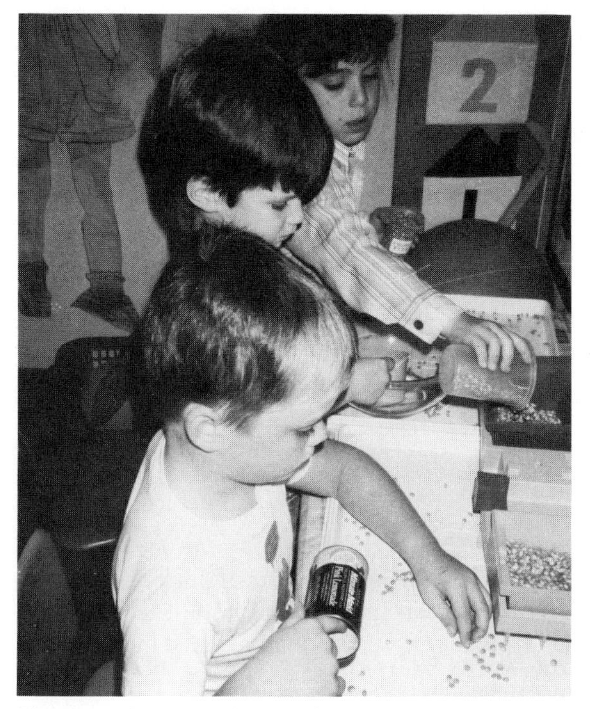

Figure 32–5 These boys are exploring weight using a pan balance.

Figure 32–6 A model thermometer is an important instructional tool for temperature.

STANDARD MEASUREMENT: TEMPERATURE

OBJECTIVE: To be able to use standard units of measure to compare temperatures and to learn how to use a thermometer to measure temperature

MATERIALS: Large demonstration thermometer, small thermometers for student use, outdoor thermometer

ACTIVITIES: These activities assume that children have talked about and have had experiences with hot, warm, and cold things and hot, comfortable, and cold weather.

1. Have the children examine the demonstration thermometer. Have them decide why there are two scales (Fahrenheit and Celsius) and how each is read. Discuss their experiences with thermometers (when they are ill or go to the doctor for a checkup, for measuring the outdoor and indoor air temperature, for controlling the thermostat on their furnaces and air conditioners) (Figure 32–6).
2. Provide some hot water and ice cubes. Have the children measure the temperature of the water and record the result. Have them add an ice cube, let it melt, and measure again and record the result. Keep adding ice cubes and recording the results. Have the children make a line graph to illustrate how the ice affects the temperature. Is any other factor affecting the

water temperature? (answer: the air temperature) Compare the results with the temperature of tap water.

3. If possible, post an outdoor thermometer outside the classroom window and have the children record the temperature each day in the morning, at noon, and at the end of the day. After a week, have them make some graphs depicting what they found out. Have interested students write daily weather reports to post on the bulletin board.

FOLLOW-UP: Have interested children record the daily weather forecasts from the radio, TV, or newspaper. Compare the forecast temperatures with those recorded at school.

STANDARD MEASUREMENT: TIME

OBJECTIVE: To be able to use standard units to measure time and to read time accurately from a conventional clock

MATERIALS: Large clock model with movable hands (such as the well-known Judy Clock); miniature model clocks that can be used individually during small group activities; a sixty-minute timer that can be used to help develop a sense of time duration; a class monthly calendar (teacher-made or purchased).

ACTIVITIES: 1. Children should have some sense of time sequence and duration by the time they reach the primary level. A timer is still useful to time events such as "Five minutes to finish up" or "Let's see if anyone can finish before the ten-minute timer rings." The major task for the primary child is understanding the clock and what it tells us, and eventually learning how to work with time in terms of the amount of time from one clock reading to another. Children can work together with the Judy Clock or with their smaller models, moving the hands and identifying the time (Figure 32–7). Much of clock knowledge comes from everyday activities through naturalistic and informal experiences. You can support these experiences by having a large wall clock in your classroom and having visual models of important times during the day that the children can match to the real clock. For example, the daily schedule might be put on the wall chart with both a conventional clock face and the digital time indicated for each major time block such as:

2. Clock skills may be broken down as follows:

a. Identification of the hour and the minute hands and the direction in which they move
b. Being able to say the time on the clock at the hour and being able to place the hands of the clock for the hour. Knowing that the short hand is on the hour and the long hand is on twelve.
c. Identify that it is after a particular hour.
d. Count by fives
e. Tell the time to the nearest multiple of five.
f. Count on from multiples of five (10, 11, 12, . . .)
g. Write time in digital notation (3:15).
h. Tell time to the nearest minute and write it in digital notation.
i. Match the time on a digital clock to a conventional clock's time.
j. Identify time is before a particular hour and count by fives to tell how many minutes it is before that hour.

3. Each child can make a clock to take home. Use a poster board circle or paper plate. Provide each child with a paper fastener and a long and short hand. Have them mark the short hand with an **H** and the long hand with an **M**. Send a note home to the parents suggesting some clock activities they can do with their child.
4. Provide the children with blank calendars each month that they can fill in with important dates (holidays, birthdays, etc.).

FOLLOW-UP: Continue to read children stories that include time concepts and time sequence. Have more advanced students who understand how to read clocks and keep track of time keep a diary for a week recording how much time they spend on activities at home (eating, sleeping, doing homework, reading, watching TV, playing outdoors, attending soccer practice, going to dancing lessons, and so on). Have them add up the times at the end of the week and rank the activities from the one with the most time spent to the least time spent. They might even go on to figure out how much time per month and year they spend on each activity if they are consistent from week to week.

Figure 32–7 The large clock model can be explored by children as they seek to understand how clocks tell us the time.

STANDARD MEASUREMENT: MONEY

OBJECTIVE: To be able to tell the value of money of different denominations; to associate the symbol "¢" with coins and "$" with dollars; find the value of a particular set of coins; be able to write the value of particular sets of coins and bills.

MATERIALS: Paper money and coin sets; pictures of coins; *Money Bingo Game* (Trend Enterprises, Inc.)

ACTIVITIES: Money is always a fascinating area for young children. Money activities can be used not only to learn about money but also to provide application for whole number and later for decimal skills (Figure 32–8).

1. Dramatic play continues to be an important vehicle of learning for primary children. First graders enjoy dramatic play centers such as those described in Unit 22. Second and third graders begin to be more organized and can design their own dramatic play activities using available props. They enjoy writing plays and acting them out. Play money should always be available in the prop box.
2. Make some price tags (first with amounts less than one dollar, later with more). Have the children pick tags out of a box one at a time and count out the correct amount of play money (Figure 32–9).
3. Have the children go through catalogs and select items they would like. Have them list the items and the prices and add up their purchases. Younger children can write just the dollar part of the prices.

FOLLOW-UP: *Play Money Bingo.* The bingo cards have groups of coins in each section. Cards with different amounts of cents are picked, and the children have to add up their coins to find out if they have a match.

EVALUATION

Evaluation of children's progress with measurement should be done with concrete tasks. The following are some examples. Give the children a list of three items in the classroom to measure. Arrange a set of measuring cups, material to measure, and a container to measure into and have each child turn in his answer. Set up a scale with three items of known weight, have each child weigh each item individually, and record the amount. Put out three model thermometers with different temperatures, have each child in turn tell you the readings, and if they indicate hot, comfortable, or cold. Show the child the time on a model clock, ask him to tell you the time, and explain how he knows. Have each child identify coins and then put them together to make various amounts, making the amounts appropriate to the child's level at the time.

SUMMARY

Measurement skills are essential for successful, everyday living. People need to know how to measure length, volume, area, weight, temperature, time, and money. These concepts develop gradually through many concrete experiences from gross comparisons (i.e., long/short, a lot/a little bit, heavy/light, hot/cold, early/late, rich/poor) to measurement with arbitrary units, to measurement with standard English (customary)

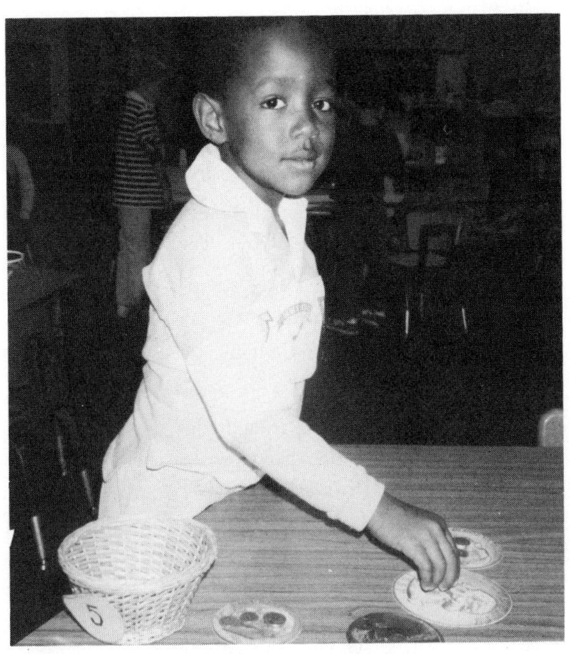

Figure 32–8 An understanding of money begins with identifying and matching real coins to pictures of coins.

Figure 32–9 Working with the value of money applies the concepts of number and counting.

and/or metric units. Measurement activities are valuable opportunities for applying whole number skills, fraction knowledge (and later decimals), and obtaining data that can be graphed for visual interpretation. Measurement concepts are acquired through practice with real measuring tools and real things to measure. Lecture and demonstration alone are not adequate methods of instruction.

FURTHER READING AND RESOURCES

Aho, C., Barnett, C., Judd, W. P., and Young, S. 1976. *Measure Matters*. Palo Alto, Calif.: Creative Publications. (Out of print, but an excellent resource if you can locate a copy.)

Bendick, J. 1971. *Measuring*. New York: Franklin Watts.

Higgins, J. L. 1974. *A metric handbook for teachers*. Reston, Va.: National Council of Teachers of Mathematics.

Johnson, G. L. 1987. Using a metric unit to help preservice teachers appreciate the value of manipulative materials. *Arithmetic Teacher*. 35 (2): 14–20.

Mathematic curriculum guide: Grades K-8, Revised. 1986. Baton Rouge, La.: State of Louisiana Department of Education.

Nelson, D., and Reys, R. E. 1976. *Measuremant in school mathematics*. Reston, Va.: National Council of Teachers of Mathematics.

Reys, R., Suydam, M. N., and Linquist, M. M. 1984. *Helping children learn mathematics*. Englewood Cliffs, N.J.: Prentice-Hall.

Wilson, P. S., and Osborn, A. 1988. Foundation ideas in teaching about measure. *Teaching mathematics in grades K-8*. Boston: Allyn & Bacon.

SUGGESTED ACTIVITIES

1. On your Activity File cards for this unit, make note of the prerequisite activities in Units 18, 19, and 22.
2. Locate the article by Gretchen L. Johnson on helping preservice teachers understand metrics. Prepare the materials for one or more of the suggested activities and try them out with the other members of your class. Have the students in your class evaluate whether or not the experience was helpful in providing a better understanding of the metric system and for suggesting some appropriate instructional ideas.
3. Assess one first-grade, one second-grade, and one third-grade student on their readiness for using standard units of measure. Write a report describing what you did, recording the childrens' responses, and comparing the three childrens' readiness.
4. Prepare materials for the suggested evaluation tasks. Evaluate one first-grade, one second-grade, and one third-grade student on their understanding of standard units of measure. Write a report describing what you did, recording the childrens' responses, and comparing the three childrens' levels of understanding.
5. Plan instructional activities based on the results of your interviews with the primary children. If possible, use the activities with the children.
6. Develop a plan for a measurement center for a first-, second-, or third-grade classroom. Include one station for each type of measurement discussed or five stations for one type of measurement. If possible, procure the materials and set up the center in a classroom for a week. Keep a record of what the children do in the center. At the end of the week, evaluate the center. Was it appropriate? Were the materials set up in such a way that the children could work independently? Did children return again? Did they try the activities at each station? Would you do it the same way next time? Why? What changes would you make?

REVIEW

A. List two reasons that measurement is an essential part of the primary mathematics curriculum.

B. There are five stages children pass through on their way to understanding and using standard units of measure. Put the steps in the correct developmental order:

1. Understands the need for standard units
2. Works with arbitrary units
3. Applies standard units of measure
4. Makes gross comparisons (i.e., heavy/light, hot/cold)
5. Play and imitation

C. Children are not ready to understand all the different units of standard measure at the same time. Match the measurement concepts on the left with the usual time of readiness on the right.

Measurement Concept	Time When Ready for Standard Units
1. Length	a. In grade two
2. Time	b. Introduced in grade two but well beyond primary when accuracy is achieved
3. Volume	
4. Area	

5. Weight
6. Temperature
7. Money

c. Introduced gradually, starting in preprimary, but not done accurately until the end of primary and even later. Instrument is difficult to understand.
d. Value is not really understood until the end of primary.
e. Units introduced in grade one; measurement in second grade
f. Introduced in grade three
g. Mainly informal during primary but important for visual representation of multiplication

D. Name the two types of units used in the United States. Which is the conventional standard?

E. Why do we need to learn metric if it is not the standard in our country?

F. Which cognitive developmental stage do children need to be in to be able to understand and apply standard units of measure?

G. List the most important factors in the comprehension of the concept of *unit*.

H. Provide two or more examples of the difficulties associated with using measuring instruments. Explain how these difficulties can be eased.

I. Select one measurement area and describe how you would introduce measurement in that area following the sequence of instruction described in this unit.

UNIT 33 Problem Solving

OBJECTIVES

After studying this unit, the student should be able to:

- List the National Council of Teachers of Mathematics' goals for mathematics instruction
- Recognize routine and nonroutine problems
- Explain the term *heuristic* and its significance for mathematics problem solving
- Understand the value of estimation techniques
- Implement developmentally appropriate primary level problem-solving assessment and instruction

As mentioned in Unit 1, the National Council of Teachers of Mathematics (NCTM) has published goals and standards for mathematics instruction in grades K–12 (NCTM, 1989). These new standards were developed in response to the back to basics movement of the 1970s and 1980s, which moved instruction back to the 1930s. This backward movement has had a negative effect in that national assessments show mathematics scores to have decreased, rather than increased. These new NCTM standards are designed to move instruction forward into the twenty-first century. As previously mentioned, the standards emphasize five general goals:

1. Children should become mathematics problem solvers.
2. Children should be able to communicate mathematically.
3. Children should learn to reason mathematically.
4. Children should value mathematics.
5. Children should feel confident in their ability to do mathematics.

This text has geared the curriculum and instruction in the direction designated in the NCTM standards. It has emphasized problem solving, communications, reasoning, number sense and numeration, concepts of whole number operations, applications through measurement, geometry, and data collection (introduction to later study of statistics and probability), fractions, and patterns and relations. It has also emphasized computation as a means to an end, rather than an end in itself. Problem solving is now considered the most important focus of the curriculum, with paper and pencil skills taking a much lower priority than in the past. This unit focuses specifically on problem solving.

OVERVIEW OF PROBLEM SOLVING

It has been shown that young children can solve simple word problems even before they learn formal computation (Unit 25). Before they are provided with adult-generated problems, they generate their own problems during their daily activities. The importance of exploration as a vehicle for constructing concepts and skills has been stressed throughout this text. A major aspect of exploration is meeting problems and developing solutions. Once children reach the stage at which they begin to compute, they can handle more complex problems.

Consider the following children as they work on three kinds of problems:

- Brent and the other children in his class have been given the following problem by their teacher, Mr. Wang: "Derick has ten pennies. John has sixteen pennies. How many pennies do they have together?" Brent notes the *key* words "How many altogether?" and decides that this is an addition problem. He adds 10 + 16 and finds the answer, twenty-six pennies.
- Mr. Wang has also given them the following problem to solve. Juanita and Lai want to buy a candy bar that costs thirty-five cents. Juanita has fifteen pennies and Lai has sixteen pennies. Altogether do they have enough pennies to buy the candy bar? Brent's attention is caught by the word "altogether," and again he adds 15 + 16. He puts down the answer, thirty-one pennies.
- Brent has five sheets of 8 ½ by 11-inch construction paper. He needs to provide paper for him and six other students. If he gives everyone a whole sheet, two people will be left with no paper. He then draws a picture of the five sheets of paper. Then he draws a line down the middle of each. If he cuts each sheet in half, there will be ten pieces of paper. There are seven children. That leaves three extra pieces. What will he do with the extras? He decides that it would be a good idea to have the three sheets in reserve in case someone makes a mistake (Figure 33–1).

The first problem is a *routine problem*. It follows a predictable pattern and can be solved correctly without actually reading the whole question carefully. The second is called a *nonroutine problem*. There is more than one step, and the problem must be read carefully. Brent has centered on the word "altogether" and stopped with the addition of the two girls' pennies. He misses the question asked, which is, Once you know there are thirty-one pennies, is that enough money to buy a thirty-five-cent candy bar? (Figure 33–2). The current focus in mathematics problem solving is on providing more opportunities for solving nonroutine problems, including those that occur in

Figure 33–1 This student has figured out how to give each person the same number of pieces of paper.

Figure 33–2 These children use chips to represent pennies. They are trying to figure out if they have enough money to buy a 35-cent candy bar if one child has 16 cents, another 10 cents, and a third 5 cents.

naturalistic situations such as the problem in the third example. Note that the third problem is multistepped: subtract five from seven, draw a picture, draw a line down the middle of each sheet, count the halves, decide if there are enough, subtract seven from ten, and decide what to do with the three extras. This last problem really has no one right answer. For example, Brent could have left two of the children out, or he could have given three children whole sheets and four children halves. Real problem-solving skills go beyond simple one-step problems.

Note that when dealing with each of the problems, the children went through a process of self-generated questions. This process is referred to as *heuristics*. There are three common types of self-generated questions:

- Consider a similar but simpler problem as a model.
- Use symbols or representations (draw a picture or a diagram; make a chart or a graph).
- Use means-ends analysis such as: identifying the knowns and the unknowns, working backward, setting up intermediate goals, analyzing the situation.

We often provide children with a learned idea or heuristic such as a series of problem-solving steps. Unfortunately, if the rules are too specific, they will not transfer (note Brent's focus on the key word), but if they are very general, how will you know if the idea is mastered? It is important to note that to apply a heuristic such as developing the relevant charts, graphs, diagrams, pictures, or doing needed operations requires a strong grounding in the basics such as counting, whole number operations, geometry, and so on.

Researchers have been concerned with whether heuristics can be taught and with what successful problem solvers do that leads to their success. It has been found that general heuristics cannot be taught. When content is taught with heuristics, content knowledge improves, not problem-solving ability. From studying successful problem solvers it has been discovered that they know the content and organize it in special ways. Therefore, content and problem solving should be taught together, not first one and then the other. Good problem solvers think in ways that are qualitatively different from poor problem solvers. Children must learn how to think about their thinking and manage it in an organized fashion. Heuristics is this type of learning, not just learning some lists of strategies that might not always work. Children have to learn to consciously

1. Assess the situation and decide exactly what is being asked.
2. Organize a plan that will get them to their goals.
3. Execute the plan using appropriate strategies.
4. Verify the results: behaviors that evaluate the outcomes of the plans.

It is important that children deal with real problems that might not have clearly designated unknowns, that might contain too much or too little information, that can be solved using more than one method, have no one right answer (or even no answer), that combine processes, have multiple steps that necessitate some trial and error, and that take considerable time to solve. Unfortunately, most textbook problems are of the routine variety where the unknown is obvious, only the necessary information is provided, the solution procedure is obvious, there is only one correct answer, and the solution can be arrived at quickly. Children must learn that their first approach to solving a problem might be erroneous but that if they keep trying, a solution will materialize. The best instructional strategy is to begin with small group problem solving. This method forces children to look at other points of view (expert problem solvers develop multiple points of view on their own). The steps the group should go through would be:

1. Look for a similar problem that might provide a clue.
2. Develop representations.
3. Identify the knowns and unknowns. Describe them to a peer.
4. Develop solutions.
5. Compare solutions.

Explaining problems to someone else will aid in learning to explain problems to yourself. Disagreements will support cognitive development.

Assessment

Assessment of the children's problem-solving expertise is not an easy task. It demands that teachers be creative and flexible. Development of problem-solving skills is a long-term activity; it is not learned in one lesson. It must be focused on the process of problem-solving, not just the answers. Therefore, you must provide children with problem-solving situations and observe how they meet them, interview students, have small groups of children describe how they solved problems, and have students help each other solve problems. Reyes, Suydam, and Linquist suggest several questions the teacher can consider while walking around the room observing the children as they work:

- Is there evidence of careful reading of the problem?
- Do individual children seem to have some means of beginning to attack a problem?
- Do they apply a strategy, or do they try to use the last procedure you have taught?
- Do they have another strategy to try if the first one fails?
- How consistent and persistent are they in applying a strategy?
- Are careless errors being made, and if so, when and why?
- How long are they willing to keep trying to solve a problem?
- How well are they concentrating on the task?
- How quickly do they ask for help?
- What strategies does each child use most frequently?
- Do they use manipulative materials?
- What do their behaviors and such factors as the expressions on their faces indicate about their interest and involvement?

Write anecdotal records describing these factors.

Interviews have been emphasized as a format for assessment throughout the text. The interview is also an excellent way to look at problem-solving behavior. Present the child with a problem and let the child find a solution, describing what he's thinking as he works. Make a tape recording or take notes.

Checklists can be developed to check off mastery of particular strategies or particular parts of the process. Paper and pencil tests should have good problems that the child will find interesting and challenging.

Instruction

Researchers agree that children should be taught a variety of problem-solving strategies so that they do not approach every problem in the same stereotyped way. They need to be given problems that are at their developmental level. Natural informal methods of instruction should begin the exploration of problem solving just as in other areas. If you look back through the earlier units in this text, you will find these kinds of problems presented to children in the sections on naturalistic and informal instruction. For example, asking how many children are in the classroom today, how many glasses of juice we will need at Kate's table, and so on.

To be effective problem solvers, children need time to mull over problems, to make mistakes, to try various strategies, and to discuss problems with their friends. When teaching in the kindergarten and primary grades, check your grade level textbook. If you find that most of the problems are routine, you will have to devise some nonroutine problems. Use the following criteria:

- Devise problems that contain extra information or that lack necessary information.
  1. George bought two bags of cookies with six cookies in each bag for ten cents a bag. How many cookies did George buy? (Price is extra information.)
  2. John's big brother is six feet tall. How much will John have to grow to be as big as his brother? (We don't know John's height.)
- Devise problems that involve estimation or that do not have clearly right or wrong answers.
  1. Vanessa has one dollar. She would like to buy a pen that costs forty-nine cents and a notebook that costs thirty-nine cents. Does she have enough money? (Yes/no rather than numerical answer)
  2. How many times can you ride your bike around the block in ten minutes? In one hour? In a week? In a month? (estimation)
- Devise problems that apply mathematics in practical situations.
- Base problems on things children are interested in, or make up problems that are about students in the class (giving them a personal flavor).
- Devise problems that require more than one step.
- Ask questions that will require the children to make up a problem.

Collect resources that children can use for problem solving. Gather statistics that children can work with (such as the weather information from the daily newspaper that was used in a previous unit). Use children's spontaneous questions (How far is it to the zoo?). Have children write problems that other children can try to solve. Calculators can be very helpful tools for problem solving. Children can try out more strategies because of the time they save that might have been spent in tedious hand calculations.

Microcomputers can also be used for problem solving. LOGO programming is a problem-solving activity in itself. Remember, have children work in pairs or small groups (Figure 33–3).

The conventional strategies taught have been:

- Understand the problem.
- Devise a plan for solving it.
- Carry out your plan.
- Look back and check your procedure and your answer.

However, children need more specific alternative strategies to try when they meet a problem. Reys et al suggest several such strategies.

1. Act out the problem. That is, use real objects or representations to set up the

Figure 33–3 These boys study the weather page in order to find data that they need to record high and low temperature.

problem and go through the steps for solution. This is the type of activity used to introduce whole number operations.

2. Make a drawing or a diagram. Stress that these pictures need to be very simple, only including the important elements. For example:

George wants to build a building with his blocks that is triangular shaped with seven blocks in the bottom row. How many blocks will he need? Drawing:

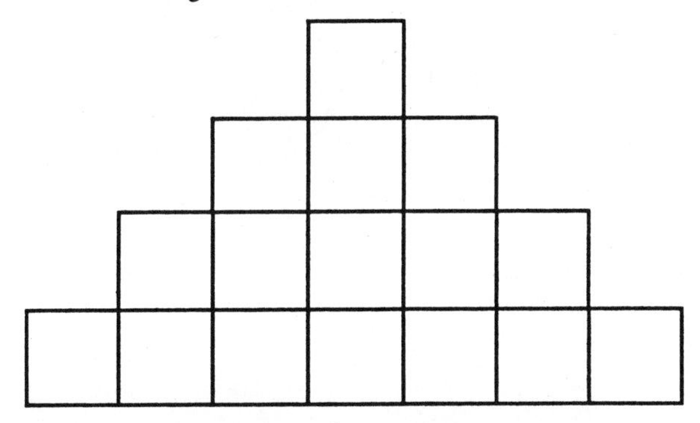

George will need 16 blocks

Theresa's mother's van has three rows of seats. One row holds three passengers, the next two, and the back row holds four. Can ten passengers and the driver ride comfortably?
Drawing:

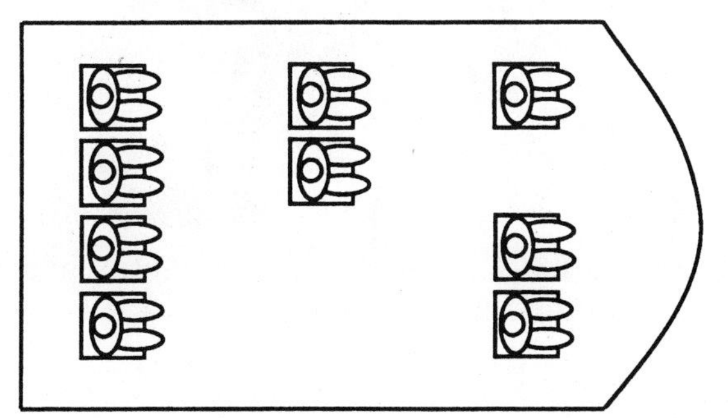

The van holds eight passengers and the driver. Ten passengers and the driver would be crowded.

---

3. Look for a pattern. (See Unit 28.)
4. Construct a table. (See Unit 31.)
5. Account systematically for all possibilities. That is, as different strategies are tried or different calculations made, keep track of what has been used.

Below is a map showing all the roads from Jonesville to Clinton. Find as many different ways as you can to get from Jonesville to Clinton without ever backtracking.

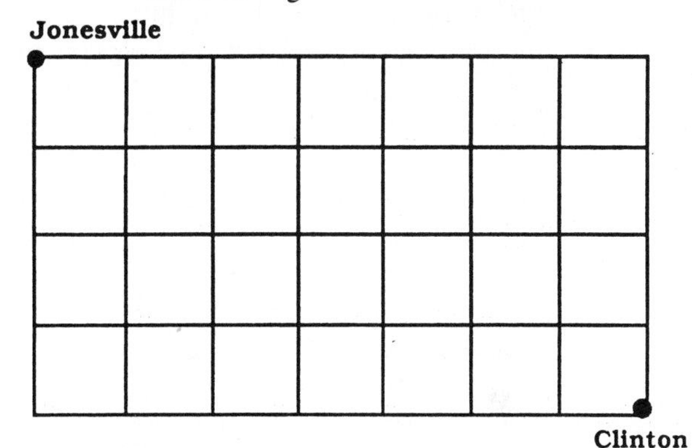

---

6. Guess and check. Make an educated guess based on attention to details of the problem end past experience. Some problems demand trial and error and best guess to solve.

Using only the numbers 1 through 9 fill the squares so that the sum in every row and column is 15.

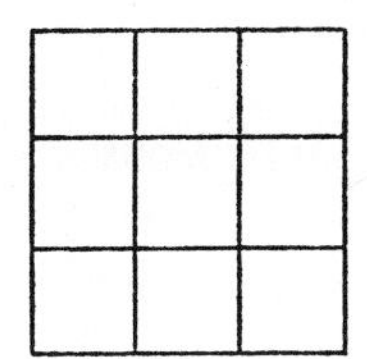

7. Work backward. In some problems, the end point is given and the problem solver must work backward to find out how the end point was reached. A maze is a concrete example of this type of problem.

Chan's mother bought some apples. She put half of them in her refrigerator and gave two to each of three neighbors. How many apples did she buy?

---

8. Identify wanted, given, and needed information. Rather than plunging right into calculations or formulating conclusions, the problem solver sorts out the important and necessary information from the extraneous factors or has to collect additional data. Taking a poll is a common way of collecting data to make a decision.

Trang Fung says that most of the girls would like to have pepperoni pizza at the slumber party. Sara claims that most of the girls prefer hamburger. To know how much of each to order, they set up a chart, question their friends, and tally their choices.

---

9. Write an open sentence. This process is not easy and is too frequently the only strategy included in some textbooks.
10. Solve a simpler or similar problem. Sometimes large numbers or other complications get in the way of seeing how to solve a problem. Sometimes making a similar problem can help the child discover the solution. For example, in the problem below, the amounts could be changed to Derick has $4 and Brent $6.

If Derrick has saved $4.59 and Brent has saved $6.37, how much more money has Brent saved?

Sometimes problems have to be broken down into smaller parts. If a problem is put in the child's own words, sometimes a strategy will be clarified.

11. Change your point of view. Is the strategy being used based on incorrect assumptions? Stop and ask, "What is really being said in this problem?"

All these strategies will not be learned right away. They will be introduced gradually and acquired throughout the elementary grades (Figure 33–4).

Figure 33–4 Using toy cars, these children act out different ways to get from Jonesville to Clinton.

ESTIMATION

Estimation is arriving at an approximation of the answer to a problem. Estimation should be taught as a unique strategy. Estimation should be mental and not be checked. Later on, children can apply estimation after computation to help decide if a computed answer is a reasonable one. First, however, the concept must be developed. At the primary level, the most common problems for applying estimation involve length or numerosity and are solved through visual perception. Children might guess how wide the rug is or how many objects are in a container. Computational estimation is usually introduced near the end of the primary period, that is, in the third grade.

There are a number of strategies that can be used for estimation. The *front-end* strategy is one that young children can use. This strategy focuses on the first number on the left when developing an estimate. For example:

37 43 +24	To estimate the sum, focus on the left column first. Note that there are nine tens, which would be ninety. Then look at the right column. Obviously, the answer is more than ninety and noting that the right column adds up to more than ten, an estimate of 90 + 10 = 100 is arrived at.

Another strategy is called *clustering*. Clustering can be used when numbers are close in value. For example, estimate the total attendance in class for the week.

Class attendance

Monday	27	1. There were about thirty students each day.
Tuesday	29	
Wednesday	31	
Thursday	32	
Friday	20	2. 5 × 30 = 150, the estimated total for the week

Rounding is a strategy that is helpful for mental computation. Suppose you wondered how many primary children had eaten lunch at school this week. You found out that there were forty-three first graders, thirty-eight second graders, and fifty-two third graders:

Number of Primary Students Eating Lunch		Round	Add
First graders	43	40	40
Second graders	38	40	40
Third graders	52	50	50
			130 (est.)

There are two additional strategies that are much more complex and would be used by more advanced students beyond the elementary grades. *Compatible numbers* strategy involves more complex rounding. *Special numbers* strategy is one that overlaps several strategies. For the most part, primary children will be using only the noncompu-

Figure 33–5 Estimating the number of beads in the jar is a challenge.

tational and the front-end strategies. The important point is that children begin early to realize that mathematics is not just finding the one right answer but can also involve making good guesses or estimates (Figure 33–5).

EVALUATION

Evaluation focuses on advancements in the processes of problem solving. Look back at the section on assessment. Evaluation should focus on the same aspects of the problem-solving process. Observation is critical for looking at the developing problem-solving skills during the school year. An interview is recommended at the end of the year.

SUMMARY

Problem solving is a process that should underlie all instruction in mathematics. It is at the top of the list of goals developed by the National Council of Teachers of Mathematics. Paper and pencil skills have lower priority. Problem solving emphasizes the process rather than the final product (or correct answer). The important factor is that during the elementary years, children gradually learn a variety of problem-solving strategies and when and where to apply them. For young children, problems develop out of their everyday naturalistic activities. It is critical that children have opportunities to solve many nonroutine problems. That is, problems that are not just simple and straightforward with obvious answers, but those that will stretch their minds. Both assessment and evaluation should focus on the process rather than the answers. Both observation and interview techniques may be used.

Problem-solving instruction for young children begins with their natural explorations in the environment. It requires time and careful guidance. During the primary grades, teacher-initiated problems can be gradually introduced. A variety of strategies for problem solving can be introduced one by one throughout the elementary years.

FURTHER READING AND RESOURCES

Baroody, A. J. 1987. *Children's mathematical thinking*. New York: Teachers College Press.

Burns, M. 1986. Those pesky word problems. *The Instructor*. November/December.

Burton, G. M. 1985. *Towards a good beginning*. Menlo Park, Calif.: Addison-Wesley.

Dougherty, B. J., and Crites, P. 1989. Applying number sense to problem solving. *Arithmetic Teacher*. 36 (6): 22–25.

Kamii, C. K. 1985. *Young children reinvent arithmetic*. New York: Teachers College Press.

Labinowicz, E. 1985. *Learning from children: New beginnings for teaching numerical thinking*. Menlo Park, Calif.: Addison-Wesley.

Lesh, R., and Zawojewski, J. S. 1988. Problem solving. *Teaching mathematics in grades K–8*. Boston: Allyn & Bacon.

National Council of Teachers of Mathematics. 1989. *Curriculum and Evaluation Standards for School Mathematics*. Reston, Va.: Author.

Nelson, D., and Worth, J. 1983. *How to choose and create good problems for primary children*. Reston, Va.: National Council of Teachers of Mathematics.

Platts, M. E. 1964. *Plus: A handbook of elementary mathematics*. Educational Service, Inc., Stevensville, Mich.

Reys, B. 1988. Estimation. *Teaching mathematics in grades K–8*. Boston: Allyn & Bacon.

Reys, R. E., Suydam, M. N., and Lindquist, M. M. 1984. *Helping children learn mathematics*. Englewood Cliffs, N.J.: Prentice-Hall.

Rosenbaum, L., Behounek, K. J., Brown, L., and Burcalow, J. V. 1989. Step into problem solving with cooperative learning. *Arithmetic Teacher*. 36 (7): 7–11.

Stochl, J. 1988. Curriculum reform in mathematics. *National Association of Early Childhood Teacher Educators (NAECTE) Newsletter*. 9 (2): 8–10.

Van de Walle, J. A. 1988. Problem solving tips for teachers: Hands-on thinking activities for young children. *Arithmetic Teacher*. 35 (6): 62–63. (Note: *Arithmetic Teacher*, includes a monthly column on problem-solving tips.)

Whitin, D. J. 1987. Problem solving in action: The bulletin-board dilemma. *Arithmetic Teacher*. 35 (3): 48–50.

COMMERCIAL MATERIALS

The following can be purchased from ETA, 199 Carpenter Avenue, Wheeling, Ill. 60090:

Building thinking skills primary activity kit. (K–4) Also can be purchased in separate parts (K–2 and 2–4).

Tops beginning problem solving (K–2). Uses story themes to present problems for solution.

SUGGESTED ACTIVITIES

1. Start a section on problem solving in your Activity File. Look through previous units for activities that involve problem solving. If you find some routine problems, rewrite them as nonroutine problems. Look through other resources for additional problem ideas. Write some problems of your own. Look through your favorite science and social studies units and develop some problems that fit the unit themes.

2. Select or create three problems: one you think is appropriate for a first grader, one for a second grader, and one for a third grader. Interview a child from each of the three grade levels using all three problems. Write a report describing the problems, the children's responses, and your evaluation of the children and the appropriateness of the problems. Note the strategies they use.

3. If possible, work on problem solving with the same group of primary children for five to ten sessions. Record their responses. Use the list in the assessment section of this unit as a guide to behaviors to be noted. Summarize the children's responses and note any changes you observe during the time you work with them.

4. Look through the problem-solving activities in a first, second, or third grade mathematics textbook. Categorize the problems as *routine* or *nonroutine*. Report your findings to the class.

5. Select or create three nonroutine math problems. Have the students in the class try to solve them. Have them write a description of the steps they take and the strategies they use. Have them compare their stategies with those described in the unit and with each other.

REVIEW

A. List the National Council of Teachers of Mathematics' five general goals for mathematics education.

B. Put an **R** for routine problems and an **N** for nonroutine problems.

_____1. Larry has four pennies and his dad gave him five more pennies. How many pennies does Larry have now?

_____2. Nancy's mother has five cookies. She wants to give Nancy and her friend Jody the same number of cookies. How many cookies will each girl receive?

_____3. Larry has four small racing cars. He gives one to his friend Fred. How many does he have left?

_____4. Nancy has three dolls. Jody has six. How many more dolls does Jody have?

_____5. Tom and Tasha each get part of a candy bar. Does one get more candy? If so, which one?

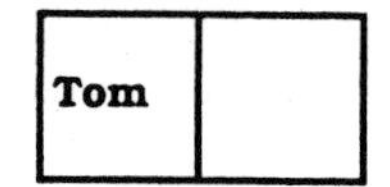

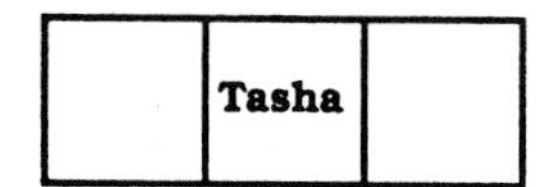

C. Explain what a heuristic is.

D. Put an X by the statements that are correct.

_____1. If children have the heuristics mastered, mastery of the content is not important.

_____2. Good and poor problem solvers have about the same heuristic skills.

_____3. Expert problem solvers are skilled at thinking about thinking. They can assess the situation, decide what is being asked, organize a plan, carry out the plan, and verify the results in an organized fashion.

_____4. Most textbooks today include plenty of nonroutine problems to solve.

_____5. Having children work in small groups to solve problems is the most productive instructional approach.

_____6. If a child has to explain a problem to someone else, it will clarify the problem for him.

E. List three techniques that can be used to assess children's problem-solving skills.

F. Critique the following instructional situation:

"We are going to do some more math problems today. I will pass out the problems to you. You will have fifteen minutes to complete them. Do not talk to your neighbors. Follow the steps on the board." The following steps are listed on the board:

1. Understand
2. Plan
3. Follow the plan
4. Check

There are no manipulatives in evidence and there is no room on the paper for drawing pictures or diagrams.

G. In the math center there are two glass jars filled with marbles. One jar is tall and thin. The other jar is short and fat. There are some 3″ × 5″ cards and a shoebox next to the jars. There is a sign that says:

How many marbles in the BIG jar?

How many marbles in the SMALL jar?

Does one jar have more or do they both have the same amount?

On a card write:

1. Your name
2. Number of marbles in the BIG jar
3. Number of marbles in the SMALL jar
4. Put your card in the shoebox.

What is the term used for this type of activity?

SECTION VI

Using Skills, Concepts, and Attitudes for Scientific Investigations in the Primary Grades

UNIT 34 Overview of Primary Science

OBJECTIVES

After studying this unit, the student should be able to:

- Develop appropriate science learning experiences for the primary age child
- Design lessons that guide students in primary science investigations
- Incorporate process skills into science investigation lessons
- Guide students in collecting, observing, sorting, and classifying objects

This unit relates the skills needed for primary science investigations and the fundamental process skills to science lessons. Children in the primary grades continue to be avid explorers. Even though they are beginning to refine their inquiry skills, to identify changes in observed events, and to understand relationships among objects and events, they still require time to interact and manipulate concrete objects.

As children leave kindergarten and enter the primary level, they are also leaving the preoperational level and entering concrete operations. They begin to be able to use abstract symbols such as numbers and written words with understanding if they are tied to concrete experiences such as science investigations. They are also entering a period of industriousness in which they enjoy long-term projects, building things, making collections, and playing games that require turn-taking, learning systems of rules, and making predictions. Peers are becoming increasingly important, thus working in small groups becomes a basic instructional strategy (Figure 34–1).

This unit begins with examples of how the love of collecting and the ability to play games can be applied in the science curriculum. Next, planning for investigations is described, followed by suggestions on how to manage the classroom. Finally, examples of how precomputer and computer activities can be used to enhance logical thinking are described.

COLLECTING

Primary children love to collect. They are increasingly aware of details, and their ability to compare and categorize objects is developing. They are apt to begin collecting pocketfuls of

Figure 34–1 Classifying and sorting shells.

small and portable objects that they see around them. Use this natural inclination to encourage children to observe, compare, sort, and classify. Collections can consist of many things such as plants, animals, feathers, fur, rocks, sand, sea shells, soils, and anything else that interests children.

However, whatever the composition of the collection, it should be viewed as a means of encouraging inquiry, not as an end in itself. In this way, simple identification of objects does not become the focal point of collecting. Instead, children will learn the basic steps in scientific inquiry (Figure 34–2).

Getting Started by Using Magnifiers

Magnifiers are useful for both collecting and classifying. Hand-held plastic magnifiers are perfect for all ages, are inexpensive, and have the advantage of being mobile for outdoor explorations. Magnifying boxes are hollow plastic boxes with a removable magnifier at the top. They are ideal for observing small treasures and animals such as live insects. When using magnifiers, carefully catch the insects, observe them, then return them to their environment without injury. In this way, you will encourage humaneness as well as observation.

Students can also use the jumbo-sized magnifier mounted on a three-legged stand. There is no need to hold objects with this type of magnifier, plus objects of different sizes can be examined at the same time. Although some dexterity is required for adjusting most magnifiers, the effort is important because the magnifiers are a bridge to using microscopes. In the following scenario, second graders are introduced to magnifiers before collecting (Figure 34–3).

Mrs. Red Fox introduces magnifiers by letting the children explore for a period of time on their own. She does not tell them what to look at, rather she gives them time to "mess around" with the magnifiers. "Wow!" observes Trang Fung. "Look how big the hair on my arm looks." As she continues to look around at objects close to her, Dean motions her over to his table.

Figure 34–2 A rock collection brought form home.

Figure 34–3 Introduce children to magnifiers.

"Look, Trang. Look at the sleeve of my shirt." "It looks different, something is in my shirt." "Mrs. Red Fox," asks Sara. "Do things always look bigger with magnifiers? Can we always see more?" "Let's look through them and see," suggests the teacher.

Mrs. Red Fox plans several opportunities to view objects in different ways. She asks the children to describe the object before viewing with a magnifier, to describe what they see while they are viewing, to compare how objects look under different powers, and to compare and contrast appearances of objects after viewing is completed. She asks, "How does the object appear under the magnifier? Why do you think it looks different?"

The teacher groups the children and has them examine different areas of the room, their clothes, lunch, and a spiderweb. She explains that the magnifying glass itself is called a *lens* and confirms that things look different under a magnifier. She explains that this is because of the way the magnifier is constructed, but does not suggest technical explanations. Primary children will enjoy noticing details not seen with the naked eye (Figure 34–4).

Figure 34–4 Sara observes a chemical reaction with a box magnifier.

Focusing the Collecting

Practice collecting on the school grounds or in the neighborhood. Help children focus on their collections by giving them suggestions. After collecting, suggest classification systems, let children come up with their own, or try sorting objects in different ways. The primary purpose of collecting is not identification at this age. Rather, collecting should be viewed as an opportunity to encourage inquiry and become aware of the variety of similarities and differences in nature. As children collect, they observe, compare, classify, and begin to think as a scientist might think. The following collecting ideas will get the class started.

1. *Leaves.* Collect and sort leaves by color, shape, vein patterns, and edges, and so on. Ask, HOW MANY RED LEAVES CAN YOU FIND? CAN YOU FIND LEAVES THAT ARE SMOOTH? DO SOME OF THE LEAVES FEEL DIFFERENT? TRY PUTTING ALL OF THE LEAVES THAT SMELL THE SAME IN A PILE. Children will want to associate the leaf with its name on a label. Suggestions for displaying collections are found in Unit 42 (Figure 34–5).
2. *Shells.* Collect different types of empty shells; for example, nut shells, eggshells, snail shells, seashells. Ask, WHAT KIND OF OBJECTS HAVE A SHELL? HOW DO YOU THINK THEY USE THEIR SHELL?
3. *Litter.* Collect litter around the school ground. Ask, WHAT TYPE OF LITTER DID YOU FIND MOST OF-

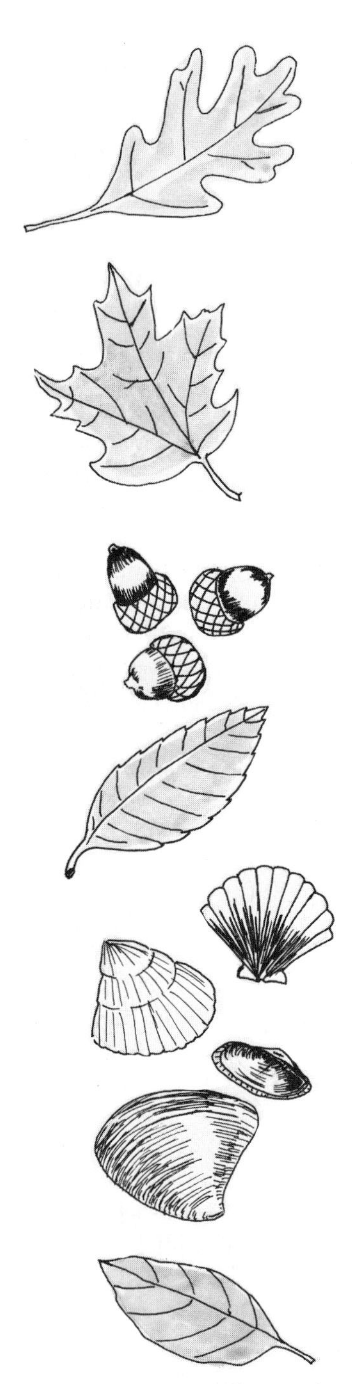

Figure 34–5 Leaves, shells, and seeds are fun to collect, sort, and label.

TEN? WHERE DID YOU FIND IT?

4. *Seeds*. Seeds can be found on the ground or flying in the air. Walk through a field and examine the seeds clinging to your trouser legs and socks. Sort the seeds by size, color, and the way they were dispersed. Ask, WHAT TYPE OF SEED DID YOU FIND MOST OFTEN? Suggestions for setting up a center on seeds are found in Unit 40.
5. *Spiderwebs*. Spiderwebs are all around us, but children will need practice and patience to collect them. Spray powder on the web, then put a piece of dark paper on one side of it. Hold the web in place with hairspray. Ask, HOW ARE THE SPIDERWEBS THE SAME? HOW ARE THEY DIFFERENT? Children might enjoy pulling twine through white craft glue to duplicate the way a spider forms its web. Let dry and hang.
6. *Feathers*. Feathers can be found at home, at school, and on the way to school. Children will enjoy examining the feathers with their magnifiers. Point out the zipper-like barbs that open and close the feathers. Ask, HOW ARE THE FEATHERS ALIKE? HOW DO THEY DIFFER? If children are studying birds, identify the function of the different types of feathers (Figure 34–6).
7. *Rubbings*. Another way to collect is to collect impressions of objects. Rubbings of bark and fossils are made by holding one side of a piece of paper against an object and rubbing the other side with a dark crayon. Mount the resulting patterns on colored construction paper and display. Have the children compare and classify the patterns.
8. *Modeling clay*. Modeling clay can be used to create an impression of fossils, bark, leaves, and seeds. Simply press the clay

Figure 34–6 Zipper-like barbs can be seen with the help of a mangifier.

Figure 34–7 Making a plaster mold of raccoon tracks. "What animal made these tracks?"

against the object, remove, and compare impressions.

9. *Plaster molds.* Footprints can be preserved by forming a plaster mold. Mix plaster, build a small cardboard rim around the footprint, and pour plaster into the impression that you have found. Carefully remove the plaster when dry and return the cast to the room for comparison (Figure 34–7).

Collecting Small Animals without Backbones

Mr. Wang asks his students, "What is your favorite animal?" "That's easy," Derick says. "I like lions, seals, and cats." "Me, too," Theresa says. "But I really like horses the best." Brent chimes in that his favorite pet is a gerbil, and Liu Pei insists that dogs are the best animals because they guard your house.

A few students mention reptiles as a favorite animal, but most of the children name mammals. To introduce invertebrates as animals and clear up a common misconception, the teacher has the children to run their hands down their back. He asks, "What do you feel?" "Bones," says Theresa. "I feel my backbone." Mr. Wang asks the children if their favorite animals have backbones. "Yes, of course they do," Brent answers.

Mr. Wang explains that not all animals have backbones; these animals are called invertebrates. He mentions a few such as worms, sponges, mollusks, starfish, crayfish, spiders, and insects and is surprised at the children's interest.

The teacher decided to plan a collecting trip to a nearby pond and field. Each student was to wear old clothes, long-legged pants, and bring a washed peanut butter sized jar. Mr. Wang had prepared insect sweepers and a few catch jars (Figure 34–8). (See Unit 35 for directions for making these items.)

Before leaving on the trip, Mr. Wang gave instructions. "We will need to be quiet and to look high and low." Derick cut in, "And under

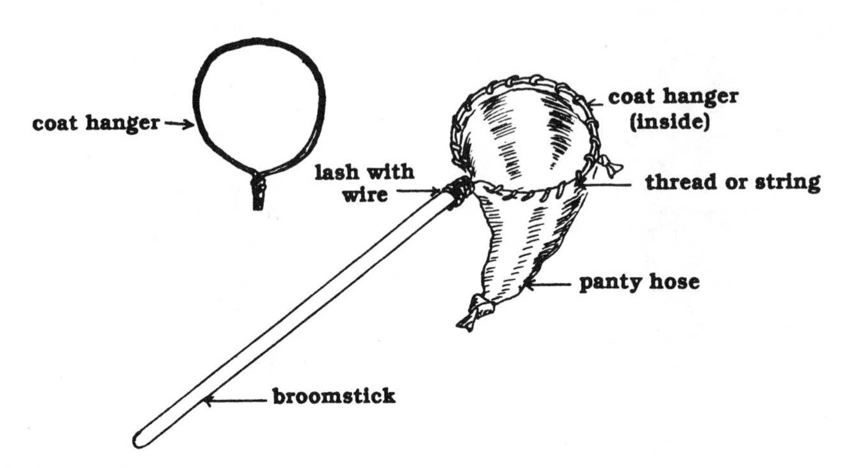

Figure 34–8 Making an insect sweeper from discarded nylons and a hanger.

rocks, bushes, and fallen branches." "Yes," the teacher said. "Look carefully where you walk."

The class worked in teams during the hunt, and each team tried to capture only one kind of animal. "Remember where you captured your animal," Mr. Wang reminded the children. "We will turn the insects loose in the same habitat that we found them."

After returning to the room, the teams used nature books to find out how to keep their captives alive. Each animal was displayed with an index card giving its name, habitat where it was found, and the collector's name. The animal was on display for two days and then released to its original habitat. Many questions were raised. "Does my worm like to eat raisins?" Derick wanted to know. Brent was interested in knowing if his worm was able to see, and Liu Pei wondered if her sow bugs would dig in the dirt of their container.

Mr. Wang encouraged close observation and investigation with such questions as, "Does the animal like rough or smooth surfaces? Does it spend most of its time in the light or in the dark?"

The next day, Liu Pei brought in animals she had found on her way to school. The display was growing. Children classified the animals by color, number of legs, and where they were found. "Let's group the animals into those found on land and those found on water," Theresa said. "Don't forget the air," Brent suggested. "Mine were found in the air."

Figure 34–9 After observing ants, the class enjoys dramatizing the way ants move to the song, "The ants go marching . . ."

The invertebrate display was an excellent way to observe animals and their habitats. The students had an opportunity to classify, investigate characteristics, study life cycles, and expand their definition of animals. Some students developed a skit involving metamorphosis, others worked on a wall chart, and all gained an appreciation of the little creatures and their worlds (Figure 34–9).

GAMES

Primary age children enjoy games. "Put the Lobster into the Chest" is a game that offers an opportunity for spatial awareness exploration and verbal communication, while reinforcing science content. Refer to Unit 16 for preliminary strategies for using this science board game.

Put the Lobster into the Chest

To play this game, which is similar to Battleship, you must first create two identical game boards and sets of playing pieces. Correlate the activity to a science topic of your choice. For instance, if you are doing a unit on farm animals, a pasture or a farmyard would be the obvious choice. For example, imagine that you are off the coast of Maine. For the game pieces, select animals commonly found in this setting. In this case, you might include lobsters, crabs, eels, clams, whales, dolphins, and seals. You will need two of each animal. Or, use three-dimensional game pieces (Figure 34–10).

Figure 34–10 Make sets of all figures and place them on identical game boards.

It is best to play this game as a class before having the students break into pairs or small groups. When you give the instruction, "Put the lobster into the chest," students can consult with each other and ask you questions to make sure they know which is the lobster. Separate the players with a box or similar barrier so they cannot see each other's board (Figure 34–11). A typical beginning interchange might go like this:

Figure 34–11 Put the lobster under the plant.

INITIATOR: Take the long thing with claws and put it beneath the floating green stuff.

FOLLOWER: Just on the sand?
INITIATOR: No. Put it on the sand under the plant.
FOLLOWER: Under the floating stuff near the brown fish?
INITIATOR: That's right.

The follow-up discussion is crucial to this game. After playing, always have the players compare their boards and talk about any differences between them.

DEFINING INVESTIGATIONS

You know that hands-on manipulative activities promote literacy skills and the learning of science concepts. Yet, many teachers shy away from including children in the active participation needed to develop these skills and concepts. Messy science investigations are avoided for several common reasons. One reason is that some teachers do not feel comfortable teaching science. Other teachers find managing children and materials an overwhelming task, and still others believe that investigations should be reserved for older children.

These fears should be put to rest. You do not need an extensive science background to guide children in science investigations. What you do need is instruction in how to do it. Explore strategies for teaching investigations to primary age children and try the suggested strategies for managing children and materials. Early science experiences provide the necessary background for future, more sophisticated skill and concept development.

Teaching Science Investigations

Science investigations usually begin with a question. For example, ask, HOW LONG CAN YOU KEEP AN ICE CUBE FROM MELTING? Children can initiate investigations by asking their own questions, such as HOW LONG WILL THIS ICE CUBE LAST IF I LEAVE IT ON MY PLATE?

After the initial question or questions, you usually predict what you think is going to happen. In the ice cube example, a child might predict, "I think the ice cube will last until lunch."

Determining variables simply means taking all the factors into account that might affect the outcome of the investigation. Although most third-grade students can handle the term *variable*, it is not advisable to introduce this term to kindergarten through second-grade children. This component of investigation is usually too difficult for primary age children to understand. Instead, ask questions that help children consider variables in the investigation. For example, ask, WHAT ARE WE TRYING TO FIND OUT? WHAT SHALL WE CHANGE? The idea of keeping something constant can be understood by talking about the idea of fairness. It just would not be fair for everything not to be the same.

Keep records of observations, results, procedures, information obtained, and any measurements collected during the investigation. Conclusions, of course, are statements that tell if the original prediction, hypothesis, was rejected. Ask, WHAT HAPPENED? IS THIS WHAT YOU THOUGHT WOULD HAPPEN? DID ANYTHING SURPRISE YOU?

This procedure resembles the scientific method for a reason. This is because children are unconsciously using the scientific method as they observe, predict, and reach conclusions. Encouraging them to investigate capitalizes on their natural interest.

In fact, we could call investigations student research. Basically, children are trying to find the answers to questions for which they do not know the answers. As they investigate, they find answers to their questions in the same way as a scientist does. They are thinking like a scientist.

MANAGING THE CLASS

Distributing Materials and Working Together

A lack of classroom management can wreck your best plans. In addition to planning your lesson, give a few moments to consider how you will organize your children and the materials they will work with. The following suggestions will start you thinking.

Organize Children for Learning

Organize the class into teams of no more than four children. It is essential that each child have responsibilities on the team. Designate a team leader, who is responsible for seeing that all materials are correctly obtained, used, and put away. The recorder is responsible for obtaining and putting away, recording and writing books, and reporting results to the class. Appoint a judge, who has the final word in any science activity related disputes, and an investigator, who is responsible for seeing that the team follows the directions when conducting the science activity. Make sure that the children understand their responsibilities.

Organize Materials for Learning

Materials can be organized several ways. One possibility is to distribute materials from four distribution centers in your room. Each center should be labeled 1, 2, 3, 4 and contain all of the equipment needed for an exploration. Then, number the children in the team 1, 2, 3, 4. The 1's are responsible for acquiring the materials at distribution center 1, the 2's at center number 2, and so on. Locate the distribution sites in separate areas of the room to reduce confusion, give a time limit for collection of materials, and provide a materials list so that the team leaders can check to see that everything needed is on their table.

If you are teaching a structured lesson, you may want to make a list that indicates what each individual will do during the exploration. For example, if you want the children to explore concepts of surface tension with soap and pepper, have the 1's pour the water into the cup, the 2's tap the pepper in the cup, the 3's coat the toothpick with liquid detergent, and the 4's plunge the toothpick into the pepper.

Science explorations are fun. To maintain discipline while conducting complex activities with the whole class, an organization system for children and materials is needed. Establish simple and clear rules for classroom operation. Once the children are comfortable with the rules and organization system, you will only need to periodically review what is expected to keep them on task.

Pocket Management Strategy

Primary teacher Maureen Awbrey prepares for managing children in learning areas by making a personalized library card pocket for each child in her class. Each pocket has the child's picture on it and a distinguishing symbol such as an orange triangle or a green square. Children learn their classroom jobs and become familiar with symbols as they review their job list each morning.

After the children are comfortable with their personal symbols and pockets, Mrs. Awbrey introduces them to the learning areas of the room. Each area is designated by a pocket and symbols that tell the children how many individuals may work in an area. For example, four circles mean that four children are permitted in the area. If Mrs. Awbrey does not want the children to use certain equipment, she places a "closed" sign in the area.

Colored strips of laminated paper containing the child's name and that of an available learning area are kept in the child's pocket. These strips are called *tickets*. When Scott wants to construct a zoo in the block center, he takes his block ticket to the block area and puts it in the pocket. When he is finished working in the block area, he removes his ticket and puts it in a basket. In this way, children rotate through the room and are exposed to a variety of learning experiences (Figure 34–12).

SAMPLE INVESTIGATIONS

The following investigations allow children to develop science processes as they conduct scientific investigations.

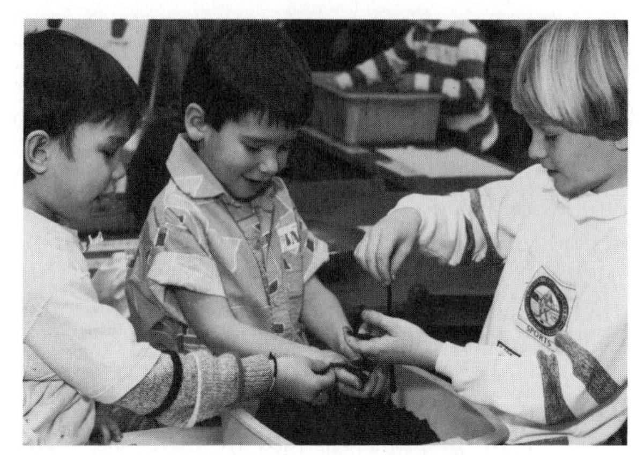

Figure 34–12 The boys cooperate as they explore worms.

HOW LONG DOES IT TAKE FOR A PAPER TOWEL TO DRY?

CONCEPT: Water disappears into the air during evaporation.

OBJECTIVE: Children should be able to simulate clothes drying on a line by investigating the length of time it takes for paper towels to dry.

MATERIALS: Bowl of water, paper towels, cardboard, pie plates

PROCEDURE: Ask children how their clothes get dry after they are washed. Are they put in a dryer? Are they hung on a line? Say, LET'S PRETEND THESE PAPER TOWELS ARE CLOTHES AND SEE HOW LONG THEY WILL TAKE TO DRY.

1. Soak a paper towel under water.
2. Squeeze out all the water that you can.
3. Open the towel and lay it on a pie plate.
4. Leave the plate on a table. Have children check and record the time.
5. Feel the towel at thirty-minute intervals to see if it is dry. When it is dry, record the time.
6. How long did the towel take to dry? Have the groups share their findings on a class chart.

EXTENSIONS: Do you think it will make a difference if you put the towels in the sun or in the shade? Which wet towel do you think will dry first? Do you think wind will make a difference in how fast a paper towel dries? Set up a fan and create wind. Measure the difference in drying time.

CAN GOLDFISH BE TRAINED?

CONCEPT: Animals can be trained to respond to light and other signals.

OBJECTIVE: Children should be able to train a goldfish to respond to light.

MATERIALS: Goldfish (at least two), tank, flashlight, fish food

PROCEDURE: Ask, "HAVE YOU EVER TRAINED A PET TO DO SOMETHING?" Children usually have a dog that sits or stays. "DO YOU THINK THAT GOLDFISH KNOW WHEN IT IS TIME TO EAT?" "HOW COULD WE TRAIN THE GOLDFISH TO GO TO A CORNER OF THEIR TANK TO EAT?" Discuss possibilities.

Shine the flashlight into a corner of the tank. Ask, "DO THE GOLDFISH SWIM TOWARD THE LIGHT?" (no) Each day sprinkle a little food in the water as you shine the flashlight in the corner (Figure 34–13). Ask, "WHAT ARE THE FISH DOING?" (swimming toward the light)

Do this for several days in a row. Then, shine the light without adding food. Ask, "WHAT HAPPENS WHEN YOU ONLY SHINE THE LIGHT?" (Fish come to the light, but only for a couple of days if food is not offered.)

EXTENSIONS: Ask, "WILL THE FISH RESPOND TO DIFFERENT SIGNALS?" "DO BOTH FISH RESPOND IN THE SAME WAY?" "WILL DIFFERENT TYPES OF FISH RESPOND IN THE SAME WAY?" Have children record their attempts and successes. They may want to write stories about their fish, tell others, and invite other classes to see their trained fish.

Figure 34–13 Can a goldfish be trained to respond to light?

Examples of Topics to Investigate

1. Can we design a container to keep an ice cube from melting?
2. Will mold grow on bread?
3. Can we get a mealworm to change direction?
4. What objects in our classroom will a magnet attract?
5. Can a seed grow without dirt?
6. Which part of a wet spot dries faster: the top, middle, or bottom? Or, does it all dry at the same time (Figure 34–14)?

COMPUTER ACTIVITIES THAT ENHANCE LOGICAL THINKING

Children like to have fun when they learn. Why not make precomputer activities fun while developing early problem-solving skills? Have

Figure 34–14 Use a chalkboard for investigating the drying time of wet spots.

students look for patterns, guess the rules that might govern those patterns, and apply the rules to a situation.

The Floor Turtle

In the following scenario, Mrs. Red Fox is aware that young children learn best by becoming physically involved in activities. The teacher selects *Valiant Turtle* by Harvard Associates, Inc., to spark interaction among her students. Commands typed into Terrapin LOGO move the cordless computerized floor turtle forward or backward and turn as far to the right or left as the child desires.

Sara types TFD 10 (turtle forward ten steps). "Oh, look, the turtle went forward," she says. Sara then types TBK (turtle backward), TRT, and TLT (turtle right or left turn, respectively), and a number of steps. The class is delighted with the floor turtle's response and everyone wants a turn.

Trang Fung gets ready to make a prediction. She asks, "What commands are you going to give the turtle, Sara? Okay, I think the floor turtle will stop, right here." Trang Fung places a piece of masking tape with her name on it on the predicted spot. She waits as Sara gives commands. "Wow, I was really close. Let's try it again, Sara" (Figure 34–15).

As the children play with the floor turtle, they observe unit amounts and estimate distances. Steps and direction are predicted. The following are suggested floor turtle activities:

1. Put a sticker on the floor where the floor turtle will start and another where it stops. Measure the distance.
2. Have a child stand with feet apart in the path of the turtle. Ask, HOW MANY STEPS WILL THE TURTLE NEED TO TAKE TO PASS UNDER DEAN'S LEGS?
3. Have the turtle pass over two different types of floor surfaces. Have student's investigate, WILL THE TURTLE

Figure 34–15 Sara at the computer *(From Jane Ilene Davidson, CHILDREN & COMPUTERS TOGETHER IN THE EARLY CHILDHOOD CLASSROOM, Copyright 1989 by Delmar Publishers, Inc.)*

TRAVEL THE SAME DISTANCE OVER CARPET AS IT WILL OVER TILE? This activity will also help children begin to understand friction.

4. Send the floor turtle over a certain path. Have students list the way it will have to move. Ask, WHICH WAY WILL INVOLVE THE FEWEST NUMBER OF STEPS? Children can interact with each other, form and test hypotheses, list a sequence of steps that the floor turtle will follow, and become aware that they control the turtle (Figure 34–16).

Recognizing Rules for Patterns and Shapes

Elements of a pattern can be recognized and rules developed which apply to those patterns. *Gertrudes Secret* by the Learning Company gives children a chance to keep patterns going by offering attribute games. Before learning the pattern games, introduce children to the spacebar and the up, down, left, and right keys on the computer. Mrs. Red Fox designs an off-computer activity that prepares the students for the software.

Figure 34–16 Children working with a floor turtle. *(From Jane Ilene Davidson, CHILDREN & COMPUTERS TOGETHER IN THE EARLY CHILDHOOD CLASSROOM, Copyright 1989 by Delmar Publishers Inc.)*

Mrs. Red Fox makes squares, diamonds, and hexagons with three different colors of construction paper. She shows the children three pieces together that would fit a certain rule; for example, all shapes held up are red. She asks, "Can you tell me a rule for this group?" "Yes," says Dean. "Those shapes are all red." "Good," answers Mrs. Red Fox.

She continues to play games with a secret rule. For example, she holds up cutouts that are all diamonds or all yellow, and the children guess what the rule is. Then she asks, "If all of these pieces are yellows, what other pieces will also fit in this group?" When the children master this game they are ready to play the game "Loop" on *Gertrude's Secret*.

The cutout shapes can also be used to make an attribute train. A piece is selected at random to make the train's "engine." "Cars" are made by attaching a shape that is different in only one way, either shape or color, to the previous one. The game continues until all the shapes have joined the train. Play a game called "Trains" on *Gertrude's Secret* when children fully understand the off-computer train game.

SUMMARY

When children investigate, they develop science processes and science concepts. Science concepts are the "big ideas" in science; they explain the way the world is viewed.

Primary age children continue to be avid investigators. At this age, they refine their inquiry skills and begin to understand abstract symbols that are tied to concrete experiences. The unit suggests science learning experiences that enhance the primary child's special fondness for collecting and playing games. Ideas for classroom management and use of simple classroom investigations are also suggested.

FURTHER READING AND RESOURCES

Campbell, S. 1987. A playful introduction to computers. *Science and Children*. 24 (8): 38–40.

Green, B., and Schlichting, S. 1985. *Explorations and investigations*. Idea Factory, Inc.

McClurg, P. 1984. Don't squash it! Collect it! *Science and Children*. 21 (8): 8–10.

McIntyre, M., 1984. *Early childhood and science*. Washington, D.C.: National Science Teachers Association.

Smith, E., Blackmer, M., and Schlichting, S. 1987. *Super science sourcebook*. Idea Factory, Inc.

VanDeman, B. 1984. The fall collection. *Science and Children*. 27 (1): 20–21.

Zeitler, W. R., and Barufaldi, J. P. 1988. *Elementary School Science Instruction—A Perspective For Teachers*. White Plains, N.Y.: Longman.

SUGGESTED COMPUTER PROGRAMS

Gertrude's Secrets. The Learning Company. (Grades K–4).

Mystery Objects. MECC (#A-211).

Moptown Parade. The Learning Company (Grades 1–3).

LOGO Revised. MECC (#A-775). (Grades K–3).

USING WORD PROCESSORS

Children can use word processors or speech synthesizers to write about science topics. They can then listen to output, or print out and illustrate what they have created. Some good word processing programs for children are:

Magic Slate (20 Col.). Sunburst.

Milliken. Scholastic.

Bank Street Writer. Scholastic.

SUGGESTED ACTIVITIES

1. Observe differences in the ability to classify. Design lessons that take advantage of childrens' abilities to observe detail.
2. Begin a collection of twenty objects from nature. How could you use this collection in your teaching? Use the collection in this way and share children's response with others.
3. Select a collecting topic such as leaves. Describe naturalistic, directed, and structured collecting activities that might take place with leaves. Observe children as they collect and identify each of these three types of learning.
4. Identify the process skills and science concepts emphasized in the investigation activities in this unit. Design an investigation and introduce it to primary age children. Record children's responses. What type of questions did children ask? How did they go about conducting the investigation?

REVIEW

1. Give two differences in the way that primary age children learn and the way that kindergarten age children learn.
2. How are primary grade science experiences similar to kindergarten lessons?
3. Why encourage children to collect? When collecting, what preparations and procedures will you follow before, during, and after collecting?
4. What are invertebrates and why should they be mentioned in the early grades?

UNIT 35 Life Science

OBJECTIVES

After studying this unit, the student should be able to:

- Develop an understanding of life science topics that are appropriate for primary grade children.
- Develop structured and unstructured life science learning activities with primary age children.
- Describe appropriate conditions for the care of animals in the classroom.
- Apply unit webbing strategies to life science content.

Whether they live in large cities or small towns, children display an eagerness to learn about the living things around them. The countless opportunities that exist to acquire firsthand knowledge of plants and animals make learning fun for both children and teacher. This unit begins with life science concepts that are basic to primary grade learning; it then presents an example of a seed unit planned with the webbing strategy discussed in Unit 7. Next, guidelines for animal care and a variety of investigations and learning experiences with animals are described. Finally, sources for teaching about plants, animals, and the environment are suggested.

LIFE SCIENCE CONCEPTS

A knowledge of science concepts is basic to planning and teaching life science to primary age children. Remember, the concepts are the "big ideas" from science that we want students to understand. Thus, learning experiences are planned around them. For example, Figure 35–1 shows a unit planning web, as described in Unit 7, that focuses learning activities around three basic concepts from the topic of seeds. The following concepts are basic to understanding plants, animals, and all living things (Figure 35–1):

Living Things

1. Living things can be distinguished from nonliving things.
2. Plants and animals are living things.
3. Animals and plants affect one another.
4. Living things have unique characteristics.

Seeds and Plants

1. Seeds differ in size, shape, color, and texture.
2. Seeds germinate and grow into a specific type of plant.
3. Some seeds grow inside fruits.
4. Some seeds grow into flowers, shrubs, and trees.
5. Some seeds grow into food that we eat.
6. Seeds are dispersed in several ways.
7. Seeds need water, light, and warmth to grow.
8. Seeds and plants grow and change.
9. Leaves tend to grow toward light and roots tend to grow into the soil.
10. Plants grow from seeds, roots, and stems.
11. Some plant forms do not have seeds, roots, or stems.

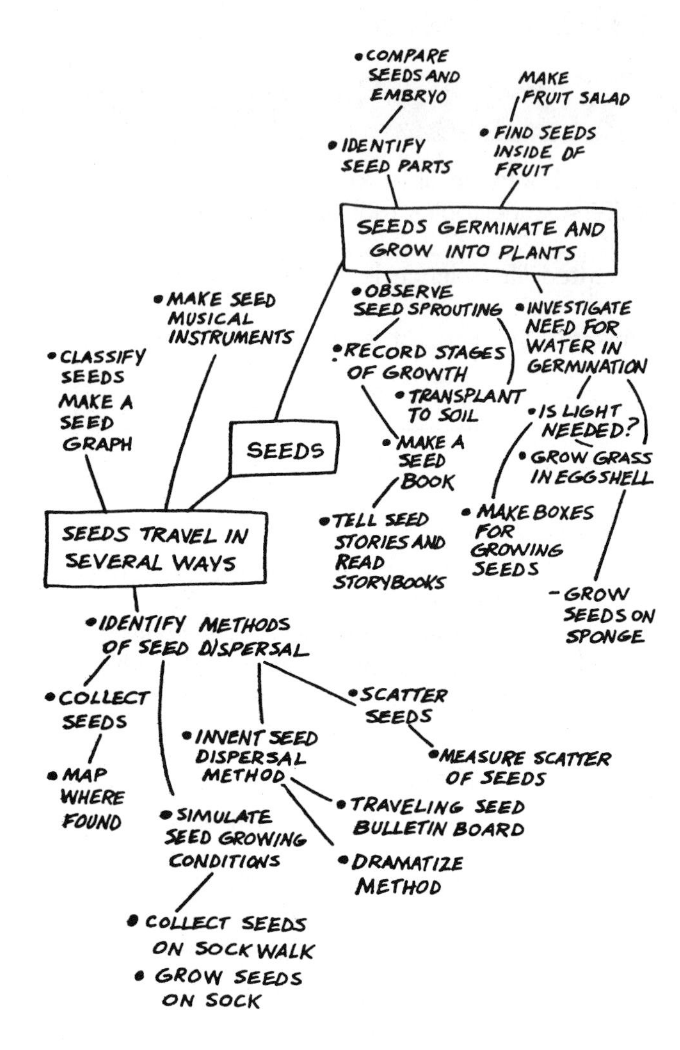

Figure 35–1 A planning web for a seed unit

12. Some plants grow in the light and some plants grow in the dark.
13. Some plants change in different seasons.

Animals

1. Animals need food, water, shelter, and a unique temperature.
2. Animals have individual characteristics.
3. Animals have unique adaptations. They move, eat, live, and behave in ways that help them survive.
4. Animals go through a life cycle.
5. Pets are animals that depend on us for special care. We love and take care of our pets.
6. There are many kinds of pets.
7. Different kinds of pets need different types of care to grow and be healthy.
8. Aquariums are places for fish and other living things to grow.

PLANNING AND TEACHING A SEED UNIT

The learning experiences suggested in this unit follow the format described in Unit 7 and are designed to meet the needs, interests, and developmental levels of primary age children described in Unit 34. Each lesson states a concept, a teaching objective based on what the child should be able to do, materials needed, and suggestions for teaching the concept. Extensions, integrations, and possible evaluation procedures are indicated in the body of the lesson when appropriate and at the end of the unit.

SEEDS

CONCEPT: Seeds germinate and grow into plants; seeds contain a baby plant.

OBJECTIVE: Discover that a seed has three parts. Identify the embryo inside of the seed.

MATERIALS: Lima beans that have been soaked overnight (if you soak them longer they may rot), paper towels, a variety of seeds

PROCEDURE: Show children a lima bean seed. Ask, WHAT DO YOU THINK MIGHT BE INSIDE OF THIS SEED? CAN YOU DRAW A PICTURE OF HOW YOU THINK THE INSIDE OF THE SEED LOOKS?

Have children open the lima bean seed with their thumbnails. Ask, WHAT DO YOU NOTICE ABOUT YOUR SEED? HOW DOES THE SEED FEEL? IS THERE A SMELL? DOES THE INSIDE OF THE SEED LOOK LIKE THE PICTURE YOU HAVE DRAWN? Discuss similarities and differences. Have students draw a picture of the bean after it is opened.

Point out that there are three basic parts in all seeds. The seed cover (for protection), food for the baby plant (cotyledon), and the baby plant itself (embryo). Introduce the term *germinate*. (This is when the seed grows—it sprouts.) Have students paint a picture of the plant as they think it will look when it has grown.

EXTENSION: Soak different types of seeds and have children compare the inside and outside of the seeds to the lima bean seeds. This could be done in an unstructured way as part of a Seed Center. In beans and peas, there are two seed halves. In seeds such as corn and rice, there is only one seed half or cotyledon (Figure 35–2).

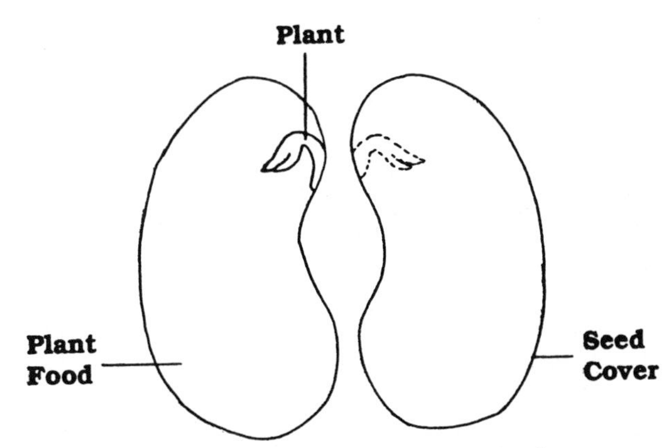

Figure 35–2 All seeds have three basic parts

BABY PLANTS

CONCEPT: Seeds germinate and grow into plants.

OBJECTIVE: Observe and describe how seeds grow into plants. Describe how the embryo grows into a plant.

MATERIALS: Lima beans, paper towels, water, cotton, clear containers or glass tumblers

PROCEDURE: Say, YESTERDAY WE SAW A BABY PLANT INSIDE OF A SEED. DO YOU THINK IT WILL GROW INTO A BIG PLANT? LET'S FIND OUT.

Soak lima beans overnight. Have children line the inside of a drinking glass with a wet, folded paper towel. Then, stuff cotton into the glass. This holds the paper towel in place.

Put soaked seeds between the paper towel and glass. Then, pour water to the edge of the bottom seed in the glass (the paper towel and cotton will absorb most of the liquid). The sprouting seeds will be easy for the children to see through the clear container. Have the children draw the sprouting lima bean and write a story about the investigation.

Ask, WHAT DO YOU SEE HAPPENING? WHAT COLORS DO YOU SEE? WHICH WAY ARE THE ROOTS GROWING? WHERE IS THE EMBRYO GETTING ITS FOOD? (from the cotyledon) As the plant gets larger, the seed becomes smaller. In fact, the plant will begin to die when the cotyledon is used up.

When the food provided by the seed is used up, ask, WHAT OTHER TYPE OF FOOD CAN WE GIVE THE PLANT? Discuss the possibility of transplanting the young plants to soil as a continuation of the project.

EXTENSION: This concept can be extended throughout the year. It is a good science activity with the advantage of going from simple observation to experimentation. The concept lends itself to a short study of seeds or an extended study of how light, moisture, heat, color, soil, air, and sound affect the growth of plants.

GLASS GARDEN

CONCEPT: Moisture is needed for seeds to sprout.

OBJECTIVE: Observe the growth of seeds

MATERIALS: Mung beans, cheesecloth, jar

PROCEDURE: Soak a handful of mung beans in a jar and cover the jar tightly with cheesecloth. Protect the jar from light by wrapping it in a towel. Then, have children simulate rain by rinsing and draining the seeds three times a day. Have children predict what will happen to the seeds. Ask, DID THE SEEDS SPROUT EVEN THOUGH THEY WERE IN THE DARK? (yes) DO YOU THINK THE SEEDS WILL SPROUT IF WE KEEP THEM IN THE DARK BUT DO NOT WATER THEM? (no) Repeat the activity and investigate what will happen if the seeds are kept in the dark and not watered.

EXTENSION: You may want to make a salad and top it off with the mung sprouts. Several variables can be tried, one at a time of course, so that children can investigate variables in sprouting seeds and growing plants (Figure 35–3).

TEACHING NOTE: Lima beans usually work well, but mung beans show the speedy growth that impatient young children may demand. Mung beans will not grow as tall as others, but they will show growth. Corn, on the other hand, takes longer than lima beans to germinate.

Figure 35–3 Make a salad and top it off with mung sprouts.

WHAT'S INSIDE?

CONCEPT: Seeds come from the fruit of plants.

OBJECTIVE: Discover that seeds develop inside of fruit.

MATERIALS: A variety of fruits, pinecones, flowers, plastic knives, paper towels

PROCEDURE: Select several fruits and place them on a table with plastic knives and paper towels. Have children examine the fruits and ask, WHAT DO YOU THINK IS INSIDE OF THE FRUIT?

Invite children to cut the fruits open. Ask, WHAT DID YOU FIND? HOW MANY SEEDS DID YOU FIND? ARE ALL OF THE SEEDS THE SAME COLOR? WHAT DO YOU THINK THE SEEDS ARE DOING INSIDE OF THE FRUIT?

Explain that a fruit is the part of a flowering plant that holds the seeds. Ask, CAN YOU NAME ANY FLOWERS IN YOUR BACKYARD THAT HAVE SEEDS? (sunflowers, dandelions) ARE THEY FRUITS? (yes) Point out that not all fruits are edible.

EXTENSION: Show a picture of a pine tree. Ask, DOES THIS TREE HAVE FLOWERS? (no) DOES IT HAVE SEEDS? (yes) WHERE ARE THE SEEDS? (cones) Have children examine pinecones to find out how the seeds are attached. Compare the seeds enclosed in fruits to those attached to cones. Point out that seeds can grow inside of a flower, surrounded by fruit, or in a cone.

TRAVELING SEEDS

CONCEPT: Seeds need to travel to grow.

OBJECTIVE: Investigate why seeds travel by simulating growing conditions.

MATERIALS: Two containers, soil, seeds

PROCEDURE: Have children bring in seeds from the school grounds or their neighborhood. Examine the seeds and discuss where they were found. Hold up a seed and ask, DO WE HAVE THIS KIND OF PLANT ON OUR SCHOOL GROUNDS? (no) HOW DID THE SEED GET HERE? (wind) I WONDER WHY IT BLEW AWAY?

After discussing ask, DO YOU THINK THAT SEEDS WILL GET A CHANCE TO GROW IF THEY ALL FALL AT THE BOTTOM OF THE PARENT PLANT? LET'S SEE WHAT HAPPENS WHEN SEEDS FALL IN ONE PLACE.

Fill two containers with seeds. Have children help plant seeds close together in one container and far apart in the other. Ask, WHICH CONTAINER DO YOU THINK WILL GROW THE MOST PLANTS? Discuss and list suggested reasons. Have children take turns giving all of the seeds water, light, warmth. Watch them for many days. Have children measure the growth of the plants and record what they see. Then ask, IN WHICH CONTAINER DO SEEDS GROW BETTER? (spaced apart) In this way, children will see a reason for a seed to travel (Figure 35–4).

SCATTERING SEEDS

CONCEPT: Seeds are adapted to disperse in several ways.

OBJECTIVE: Identify several ways of dispersing seeds.

MATERIALS: Seeds, magnifying glasses, mittens

PROCEDURE: Arrange seeds from weeds, grasses, and trees in a science center. Have a magnifying glass and mitten available for students to examine seeds. Give basic directions and let children "mess around" and explore the seeds. Ask, WHICH SEEDS SEEM TO CATCH IN THE MITTEN? (hairy ones, or ones with burrs) Have children predict and draw what they think they will see when they observe a burr through a magnifying glass. Say, LET'S LOOK THROUGH THE MAGNIFYING GLASS AND DRAW WHAT WE SEE. Discuss the before and after pictures and describe the tiny hooks and how they are used. Introduce the term *disperse* and use it in sentences and stories about the burr with tiny hooks. Refer to Unit 34 for suggestions for the use of magnifiers.

EXTENSION: Take children outside to explore how different seeds are dispersed by shaking seeds from pods, beating grass seed spikes to release grains, brushing hairy seeds against clothes, and releasing winged seeds. Compare what happens to each seed.

Figure 35–4 In which container do the seeds grow better?

INVENT A SEED

CONCEPT: Seeds are modified to travel in a specific way.

OBJECTIVE: Modify a seed for travel.

MATERIALS: Dried bean or pea seeds, a junk box (colored paper, glue, rubber bands, tape, cotton, popsicle sticks, balloons, pipe cleaners, paper clips, string) scissors

PROCEDURE: Show children a coconut (the largest seed) and ask, HOW DO YOU THINK THIS COCONUT TRAVELS? (water) Display pods such as milkweed to demonstrate a pod that bursts and casts its seeds to the wind. HOW WILL THE BURR TRAVEL? (catch on things) Birds and other animals eat fruit such as berries. Then, they digest the fruit and leave the seeds somewhere else. WHY DID THE BIRD WANT TO EAT BERRIES? (They were good to eat.) Children will conclude that seeds have specific modifications to travel in specific ways. To reinforce the term *modify*, have students modify their clothes for different weather or activities.

Ask, IF YOU WERE A SEED, HOW WOULD YOU LIKE TO TRAVEL? Discuss preferences and then assign one of the following means of travel to each small group of children: attract animal, catch on fur, pops or is shot out, floats on water, or carried by wind. Give each group seeds and junk box materials and ask them to invent a way for their seed to travel. When the children have completed their creations, have them demonstrate how their seed travels (Figure 35–5).

EXTENSION: Make a bulletin board that displays the modified seeds traveling in the way the students intended. Have children examine seeds and think of other objects that work in the same way. (For example, the burr is the inspiration for the development of velcro.)

HOW SEEDS TRAVEL

SLING SHOT

witch-hazel

HELICOPTER

maple

HITCH HIKER

burdock (burr)

ANIMAL EXPRESS

cherries

BOATS

coconut

Figure 35–5 Seed dispersal methods

1. *Sock walk.* Drag a sock through a field. Then cover with soil and water and wait for a variety of plants to grow from the seeds caught on the sock. Or, put the sock in a plastic bag, shake the seeds out of the sock, and examine them (Figure 35–6). Ask, WHAT TYPES OF SEEDS DID THE SOCK ATTRACT?
2. *Egghead hair.* Draw a face on half of an eggshell. Place the egg in an egg carton and fill the eggshell with moist soil. Sprinkle grass seed on top, then water. Grass will grow in five days. Attach construction paper feet with clay to display (plastic eggs also work well). To show the benefit of light on leaves, first grow the grass hair in the dark. Discuss the resulting pale, thin grass and ask, HOW CAN WE MAKE THE GRASS GREEN AND HEALTHY? Place the egg man in the light and be prepared to trim the healthy green hair (Figure 35–7).
3. *Collections.* Seeds make an easy and fun collection. Collect from your kitchen, playground, or neighborhood.

Investigation Questions for Growing Seeds

1. How deep can you plant a seed and still have it grow?
2. If you crush a seed, will it still grow?
3. What direction will the sprouts in the glass garden grow if the glass is turned upside down?

Figure 35–6 A sock garden

Figure 35–7 Can you grow healthy grass hair for your eggshell people?

SUBJECT INTEGRATIONS

Science and Math

1. *Seed walk.* Take a seed walk around the school. Ask, WHERE SHALL WE LOOK FOR SEEDS? WHAT DO YOU THINK WE MIGHT SEE? Gather seeds and use as materials to sort, match, count, and weigh in the science center.
2. *How many ways.* Provide a box of seeds for students to classify. Compare size, shape, color, and texture of the seeds. Construct a seed graph.
3. *Greenhouse.* Simulate a greenhouse by placing a plastic bag over a container of germinating seeds. Place seeds in a jar on a moist paper towel. Put the jar inside of a plastic bag and record the amount of time needed for sprouting. Compare sprouting time with seeds that are not placed in a plastic bag (Figure 35–8).
4. *Cooking.* Create a fruit salad with the fruits gathered for seed activities. Or, measure ingredients and follow a recipe to make apple

Figure 35–8 Make your own greenhouse

pie or other tasty dishes. Nut butter can be made by passing the seeds or nuts through a nut grinder. Store in refrigerator and use as spreads.

Science and Social Studies

1. *Seeds swell.* Years ago, seeds were used to stretch tight leather shoes. To simulate this, fill a small bottle with beans, cover beans with water, close the bottle with plastic wrap, and secure with rubber bands. Beans will swell and lift the plastic.
2. *Neighborhood map.* Have children bring in seeds that they have found. Make a map of where the seeds might be found.
3. *Seeds can be edible.* Try to bridge the gap between the processed food that the children eat and its raw form. Show pictures of wheat, fruit trees, and vegetables growing. Discuss edible seeds such as pumpkin, sunflower, peanuts, and those that are not eaten such as watermelon and apple seeds.
4. *Careers.* Discuss different jobs associated with growing plants. Make hats and act out these jobs.
5. *Woodworking.* Create boxes and planters for growing seeds.

Science and Language Arts

1. Have students remove the "s" from seed and make new words (weed, feed, need, bleed).
2. Pretend you are a seed. Draw a picture and write a story about how you will look and where you might travel.
3. Dramatize the growth of an embryo into a plant.
4. Make a seed book (in the shape of a lima bean) and fill it with bean stories and drawings.
5. Read, tell, and dramatize stories about seeds such as *Jack and the Beanstalk*, *The Story of Johnny Appleseed*, or *Popcorn* by Tony DePaola.

Science and Music

1. Sing songs about seeds and plants from your music book.
2. Use seeds to create musical instruments such as maracas and other types of "shaking" instruments in tubes with plastic lids.
3. Some children can make whistles from acorn caps. Hold thumbs in a V shape, with the top of the thumbs forming the V, under the hollow side of the cap. Blow gently at the base of the V to make a sound.

Science and Art

1. Create landscapes that would be favorable for seed growth.
2. Make faces on moist sponges with bird-

seed. Trim the green growth into different patterns.

3. Design seed mosaics. Glue seeds to paper or cardboard to make pictures.
4. Jewelry can be made by stringing seeds with thread and needle. You may need to boil seeds and wait for them to soften. Some children might want to paint the seeds and use them to decorate costumes.
5. Stick burrs together to make a basket. What else can you make from burrs? Pinecones and walnut shells can become birds and other creatures by adding glue and construction paper.

ADDITIONAL PLANT ACTIVITIES BASED ON SCIENCE CONCEPTS

When your unit on seeds is completed, you might want to develop additional concepts about plants. Here are a few suggestions.

Concept: Plants Grow from Roots and Stems

Growing plants from cuttings can be an exciting experience for primary age children. They will enjoy observing familiar products from a grocery bag doing unfamiliar things. Investigations with cuttings are long term and provide opportunities for record keeping, process skill development, and subject integration. You will need water, light, and common vegetables to reinforce that seeds are not the only way that plants reproduce.

1. *Dish garden.* To make a dish garden, cut off the top inch of carrots, beets, or white turnips. Keep the tops in dishes of water while the children observe the roots growing. Shoots will usually appear in a week to ten days. Ask, CAN YOU TELL WHAT THIS PLANT IS BY THE LEAVES?
2. *Hanging around.* Suspend a yam or a fresh sweet potato, tapered end down, in a jar filled with water. Insert toothpicks so that only one-third of the sweet potato is in the water (Figure 35–9). Put the jar in a warm, dark place until buds and roots grow. Then put it in a sunny place and prepare a string trellis for the upcoming foliage to climb. Children can chart the number of days for root and foliage growth and observe and record changes. Ask, WHAT HAPPENS TO THE SWEET POTATO AS THE VINE GROWS?
3. *Pineapple tops.* Cut a two and one-half-inch section of pineapple fruit below the leaves. Put the pineapple top in a dish with water. When the roots develop, put the plant in potting soil and make a greenhouse by covering it with a plastic bag. Keep the pine-

Figure 35–9 Will a sweet potato grow in a glass of water?

apple greenhouse warm but not in direct sunlight. In about three weeks, take the new plant out of the greenhouse, add water, and place it in the sun. Eventually, tiny pineapples may form (six to twelve months).

Concept: Molds Grow in Dark, Moist Conditions

Children have probably seen mold in their own refrigerators or bread baskets, but do not realize that mold is actually a tiny plant. Observing a mold garden gives children an opportunity to focus on recording observations, keeping records, and writing predictions as they investigate the world of tiny plants.

1. To begin the lesson, show the children a molding orange and ask, WHAT IS ON THE ORANGE? IS THE ORANGE GOOD TO EAT? WHERE HAVE YOU SEEN MOLD? Make a list of the conditions of the places where mold has been found. There should be replies of "moist," "dark," "types of materials such as fruit and bread."
2. To make the mold garden, fill a glass jar about one-third full of sand. Sprinkle water on the sand and place items that the children bring in on top of the sand. Screw the lid on the jar and begin observations, discussion, and predictions. WHICH OBJECTS DO YOU THINK MOLD WILL GROW ON FIRST? WILL ANY OBJECTS NOT GROW MOLD? Place the mold garden in a dark place and prepare the children to observe. Changing colors, shapes, and sizes should prove interesting (Figure 35–10).
3. You might want to mention the important role that fungus plays in breaking down materials and in returning the components to the soil.

Figure 35–10 A mold garden

Teaching Notes

1. Most mold of this type is harmless, but do not take any chances. KEEP THE LID ON THE JAR. In addition, have children wash their hands if any moldy items are touched, do not sniff molds (check for mold allergies among students), and when the activity is completed, throw away all mold gardens without opening them and in a tightly sealed bag.
2. Growing mold in soil is speedy because it is rich in organic materials and for that reason is recommended by some science educators. However, a teacher does not know what else might be in the soil sample, ready to grow. Thus, play it safe and slower—stick to sand.
3. Mold is a type of fungus which lacks chlorophyll. Although warmth, darkness, and moisture are ideal conditions for mold growth, neither warmth nor darkness is necessary for growth. Further investigations of conditions for mold growth may be appropriate in your classroom.

ANIMALS IN THE CLASSROOM

Caring for and studying living things in the classroom and outdoors can be an excellent way to develop a respect and knowledge of the daily requirements of all forms of life. As children care for a living thing, they seem to develop a sensitivity and sense of responsibility for the life around them. As they maintain the living organism, they become aware of the conditions under which that animal (or plant) survives as well as the conditions under which it will perish. You hope that these understandings and attitudes will carry over into the child's life and the human condition in general.

Before you think about caring for any living organisms, take the precautions listed in this unit. It is vital that living things do not suffer from too much care, such as overfeeding and handling, and too little care, such as improper diet, water, temperature, and shelter. In other words, before allowing an animal in your classroom, be sure you know how to take care of it.

Care of the animal should begin before the animal arrives. It is recommended that the entire class has an opportunity to help prepare the cage or environment. As preparations are made, discuss why you are doing so to develop the children's understanding of the specific needs of a species.

When the animal arrives in class, give the new visitor time to become acclimated to its new surroundings. The expected enthusiasm and interest that an animal visitor is likely to generate may overwhelm the newcomer. Instruct children to quietly observe the animal in pairs for brief periods of time. Then, small groups can watch, always quietly, as everyone (animal and children) adjusts to the environment (Figure 35–11).

When planning for an animal guest, ask yourself and your students the following questions:

1. What type of cage or environment does this animal require?

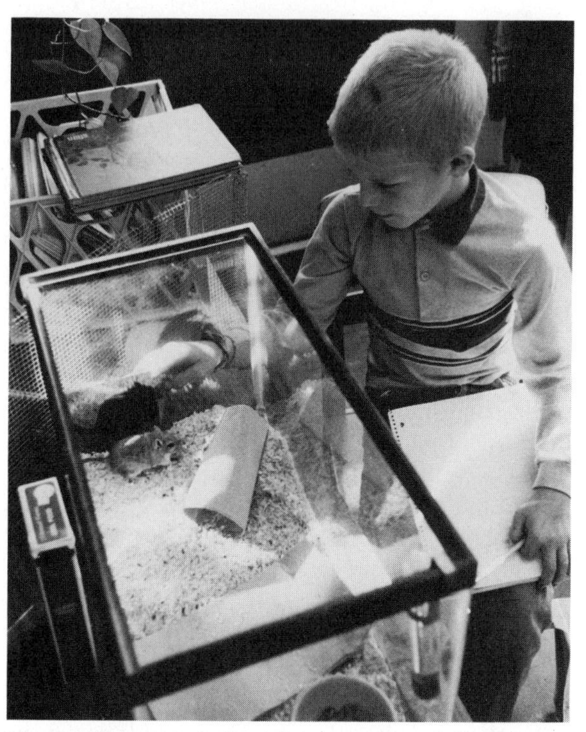

Figure 35–11 John observes a gerbil as it acclimates to its new environment.

2. What temperature must be maintained?
3. What type of food is needed and how should that food be presented?
4. Will the animal be able to live in the room over weekends? What will happen to the animal during vacations?
5. Is this an endangered animal? Has this animal been illegally captured or imported? For example, for each parrot of many species that a child encounters, about ten parrots have perished in capture or transport. Or, as in the case of a species such as the ball python, some animals will never eat in captivity. Limit animal use to those bred in captivity.
6. Does the animal need special lights? Many reptiles must have ultraviolet light, a warm and cool area of their cage, and live food.

7. Will this animal make a good classroom visitor? Should it be handled? Avoid impulse purchases that will lead to future problems.

Tips for Keeping Animals in the Classroom

When animals are in the classroom, care should be taken to insure that neither the children nor the animals are harmed. Mammals protect themselves and their young by biting, scratching, and kicking. Pets such as cats, dogs, rabbits, and guinea pigs should be handled properly and should not be disturbed when eating. For example, rats, rabbits, hamsters, and mice are best picked up by the scruff of the neck, with a hand placed under the body for support (Figure 35–12). In addition:

1. Check school district procedures to determine if there are any local regulations to be observed. Personnel at the local humane society or zoo are often very cooperative in assisting teachers to create a wholesome animal environment in the classroom. However, zoos receive numerous requests for adoption and usually do not want classroom animals for their collection or "feeder stock."

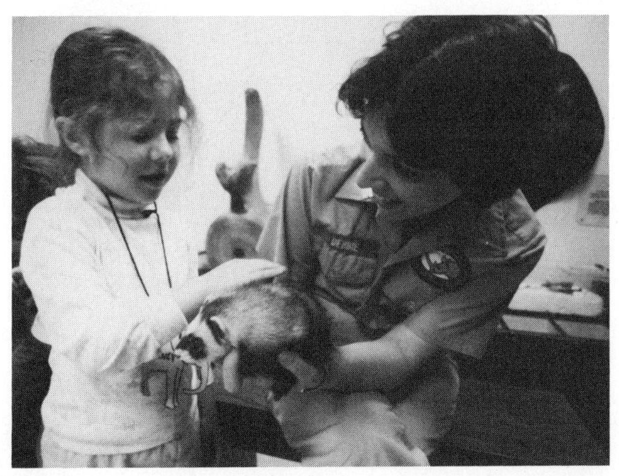

Figure 35–12 The teacher holds the rabbit for the children to touch.

2. Caution students never to tease animals or to insert their fingers or objects through wire mesh cages. Report animal bites and scratches immediately to the school's medical authority. Provide basic first aid. Usually children follow the example set by the teacher. If you treat the animals with respect, the children will, too.
3. Purchase fish from tanks in which all fish appear healthy.
4. Discourage students from bringing personal pets to school. If they are brought into the room, they should be handled only by their owners, provided a place for fresh water, a place to rest, and then sent home.
5. Refer to Appendix D, the "Code of Practice on Use of Animals in Schools," before beginning animal observations and activities. Guidelines for collecting invertebrates and caring for them in the classroom are provided in Unit 34.

Teaching with Animals

Entire teaching units can be designed around observing animals in the classroom. Science concepts and subject integrations occur naturally as children observe, categorize, and communicate their experiences with animals. The benefits of observing animals far outweigh the cautions and are well worth your time and energy. Here are guiding questions and integrations that lead to investigations.

1. Describe how the animal moves. Do you move that way? See if you can figure out why the animal moves in the way that it does. Does anything make it move faster or slower? Add your observations to the Observation List (Figure 35–13).
2. How does the animal eat its food? Does it use its feet or any other part of its body to help it eat? Does it prefer a specific type of food? Make a chart or graph showing preferences.

Michael Parker	Gerbils
They play.	
The gerbils drink water.	
They are hiding in their house.	
He's lookin, playin.	
He jumped back down in the house.	

Figure 35–13 Michael observes and records some of the daily activities of gerbils.

3. What do the animals do all day? Keep a record of the animal's activities or pinpoint several specific behaviors such as recording the type of food eaten. Are there times when the animal is more or less active? Take black-and-white polaroid pictures of the animal at the same time of the day. Chart the results.
4. How do animals react to different objects in their environment? Offer gerbils paper towel tubes and watch them play. Write a story that sequences the actions of the gerbils.
5. What type of sounds do the animals make? Identify different taped animal sounds and decide if you can tell the size of an animal by its sound. Have children tape record different sounds made by the animal such as sleeping, eating, playing, drinking. Play the tape and have children make up stories about what they think is happening.
6. How can we keep earthworms in the classroom? Have children research what is required for keeping earthworms healthy. Earthworms make good pets because they can be handled with less stress to the animal than mammals, birds, reptiles, and amphibians.
7. Describe the animal's body covering. Why do you think it has this type of covering? Compare the body coverings of different animals. Design a center that allows children to feel objects that simulate animal coverings (refer to Unit 40). Observe these coverings on a trip to the zoo.
8. How does the animal drink? Can you think of other animals that drink in the same way? Categorize ways that animals drink (a cat laps water, a snake sucks water).
9. Seven- and eight-year-olds are not ready for formal outlining yet, but building a web is a good way to organize what they are learning about animals. This type of organization refreshes memory for what children have seen or studied and relates any information they are collecting. Figure 35–14 presents their ideas in a loosely structured way. The teacher selects the main divisions and writes them on the chalkboard. Then, the children give the

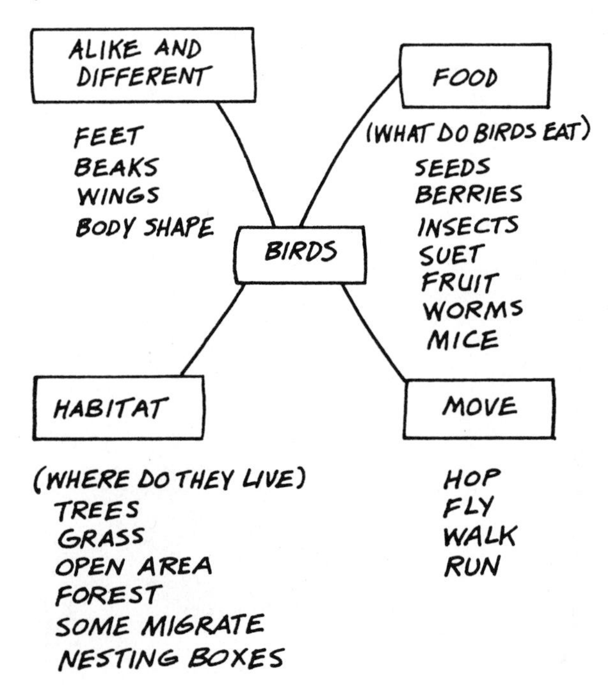

Figure 35–14 Webbing to review what we have learned.

main points. If the children are working in committees to research different aspects, this strategy illustrates the relatedness of the entire project.

A TRIP TO THE ZOO

A trip to the zoo is a terrific way to offer children opportunities to explore the world of animals. Observing animals in a zoo setting is a high interest activity that helps children learn about animals and their needs (Figure 35–15). The following suggestions are designed to help maximize the unique setting of a zoo.

Before, During, and After

To make the most of your zoo visit, plan learning experiences that will be taught before going to the zoo, during the zoo visit, and after returning to the classroom. For example, you might want to focus your plans on the similarities and differences between reptiles and amphibians. In the following scenario, Mrs. Red Fox decides to relate class discussions and observations to turtles.

Before going to the zoo, Mrs. Red Fox prepares learning stations for her class. The stations will prepare the children for their zoo visit.

Figure 35–15 The Louisville Zoo has a children's zoo.

Station 1. Mrs. Red Fox prepares items that simulate body covering for the children to feel. To represent the dry, scaly skin of most reptiles, she rolls shelled sunflower seeds in clay. Oiled cellophane represents the moist, glandular skin of most amphibians. As children visit this station, they are invited to touch the simulated coverings and match picture cards of animals that might feel dry or slimy.

Station 2. To emphasize the role of camouflage in both reptiles and amphibians, the teacher prepares two pieces of black construction paper. A yellow pipe cleaner is taped to one piece of paper and a black pipe cleaner is taped to another. Ask, WHICH IS HARDER TO SEE? HOW DO YOU THINK THIS HELPS THE ANIMAL SURVIVE? DESIGN CAMOUFLAGE FOR A GREEN PIPE CLEANER THAT LIVES IN A DESERT.

Station 3. Some reptiles and amphibians rely on vibrations to "hear" because they do not have outer ears. Mrs. Red Fox has the children tap a tuning fork on a surface and hold it near their ear. She says, DESCRIBE WHAT HAPPENS. TAP THE FORK AGAIN AND PRESS IT AGAINST YOUR CHEEK. WHAT HAPPENS THIS TIME? The class discusses how vibrations tell the animal what is happening. Some of the children begin comparing how birds and mammals hear and pick up vibrations.

Station 4. Mrs. Red Fox has the children draw what they think a turtle will look like and what kind of home it will have.

At the Zoo

Mrs. Red Fox divides her class into small groups and assigns a chaperone to each small group. Each group has a turtle task card. When the groups come to the turtle exhibit, the chaperone helps the children complete the questions (Figure 35–16). The children are asked to observe and answer:

Figure 35–16 Can you see the animal?

- How many turtles do you see?
- What are they doing?
- Are turtle toes like your toes?
- What are the turtles eating?
- Try walking around the areas as slowly as the turtle is moving. How does it feel?
- How are you and the turtle alike?
- How are you different?
- What other animals are living with the turtle?
- Is it easy to find the turtles in the exhibit?

As the children move to an amphibian exhibit such as frogs, toads, and salamanders, they make comparisons between the turtles and the amphibians. Mrs. Red Fox makes sure that the exhibit signs are read and questions are asked to gather additional information.

After the Zoo Visit

Mrs. Red Fox reinforces what the children have learned with a variety of learning experiences. The class makes turtle candy from caramels, nuts, and chocolate bits, and the children create a play about turtles and how they live.

Each child creates a shoebox diorama of the turtle habitat. The teacher asks, DID THE TURTLE LIVE IN THE SAME WAY THAT YOU THOUGHT IT WOULD? WHAT WAS DIFFERENT? WHAT WAS THE SAME? She has the children compare what they knew about turtles before the visit to the diorama that they have created. Mrs. Red Fox has storybooks and reference books available for children to find out more about reptiles and amphibians.

Additional Animal Activities

The following questions and activities can be used before, during, or after a trip to the zoo:

1. HOW MANY TOES DOES A GIRAFFE HAVE? WHAT COLOR IS A GIRAFFE'S TONGUE? CAN YOU TOUCH YOUR NOSE WITH YOUR TONGUE? Explain that giraffes use their tongues the way we use forks and spoons.
2. A snake is covered with scales. Ask, ARE THE SCALES ON A SNAKE'S STOMACH THE SAME SHAPE AND SIZE AS THOSE ON ITS BACK? DOES A SNAKE HAVE FEATHERS? DOES A SNAKE HAVE FUR?
3. People have several different types of teeth in their mouth. After looking in a few mouths, have the children look at a crocodile's teeth. ARE THE TEETH ALL THE SAME TYPE? WHY DO YOU THINK SO?
4. WHAT DO WE USE TO PROTECT OUR FEET? HOW ARE THE FEET OF THE SHEEP AND DEER ADAPTED TO WHERE THEY LIVE? When in front of the hoofed animal display, have students stretch out their arms and then slowly move their arms together while looking directly ahead. WHEN CAN YOU SEE YOUR HANDS? Direct the children's attention to the hoofed animals. WHERE ARE THE DEER'S EYES LOCATED? Discuss the location of the eyes

and name some advantages of having eyes that can see behind you.

5. In front of the aquarium ask, HOW DO YOU THINK FISH BREATHE? DO FISH HAVE NOSES? FIND A FISH THAT FEEDS FROM THE TOP OF THE TANK. FIND ONE THAT FEEDS FROM THE BOTTOM OF THE TANK. DO ANY OF THE FISH FEED IN THE MIDDLE OF THE TANK?
6. DO ANY OF THE ANIMALS HAVE BABIES? Refer to these questions when observing animal babies. DO THE BABIES LOOK LIKE THE ADULT? HOW ARE THEY DIFFERENT? DOES THE BABY MOVE AROUND ON ITS OWN? IS ONE OF THE ADULTS TAKING CARE OF THE BABY? WHAT IS THE ADULT DOING FOR THE BABY?

Learning at the zoo can be an exciting experience. For a successful trip, plan activities for before, during, and after a zoo visit. If this is your first trip to the zoo, concentrate on familiar body parts and activities. For example, how does the animal move, eat, hear, see, or protect itself? Use these questions to encourage discussion of the special abilities and adaptations of animals (Figure 35–17).

Figure 35–17 Touching a snake is fun in the zoo class.

SUMMARY

Teaching life science concepts on a first-hand basis is essential to planning and teaching science to primary age children. Children display an eagerness in learning about the living things around them. This unit provides activities about seeds, plants, and molds and contains suggestions for caring and studying living things, both in the classroom and outdoors, and keeping animals in the classroom.

Animal observations capitalize on the interest of children. The benefits of these observations far outweigh any cautions. A trip to the zoo can be a meaningful experience if specific plans are made for the visit. Prepare activities to do before, during, and after the trip.

RESOURCES FOR ANIMAL CARE

Science and Children, a journal of the National Science Teachers Association, has an informative monthly column on the care and maintenance of specific living organisms. Each month a different plant or animal receives in-depth treatment.

Dolensek, E. *A Practical Guide to Impractical Pets.* New York: Bronx Zoo

Hampton, Hampton, Kammer, 1988. *Classroom creature culture.* Washington, D.C.: National Science Teachers Association.

Orlans, F. B. 1977. *Animal care from protozoa to small animals.* Menlo Park, Calif.: Addison-Wesley.

Nichol, J. 1989. *The Animal Smugglers.* New York: Facts on File Inc.

Smith, H. M. 1988. Snakes as pets. *Science and Children.* 25 (6): 36.

———. 1980. *Snakes as pets.* 4th ed,. New Jersey: T.F.H. Publications, Inc. Ltd.

FURTHER READING AND RESOURCES

Althouse, R., and Main, C. 1975. *Science experiences for young children*. New York: Teachers College Press.

Cobb, V. 1972. *Science experiments you can eat*. New York: J. P. Lippincott Company.

Council of Elementary Science International, 1984. *Outdoor Learning Experiences*. Columbus, Ohio: SMEAC Information Reference Center.

Dean, R. A., Dean, M. M., and Motz, L. *Safety in the Elementary Classroom*. 1987. Washington, D.C.: National Science Teachers Association.

DeVito, A., and Krockover, G. H. 1980. *Creative sciencing: Ideas and activities for teachers and children*. Boston: Little, Brown and Company.

Gega, P. 1986. *Science in elementary education*. New York: John Wiley and Sons.

Harlan, J. 1988. *Science experiences for the early childhood years*. 4th ed. Columbus, Ohio: Charles E. Merrill Publishing Co.

McIntyre, M. 1984. *Early Childhood and Science*. Washington, D.C.: National Science Teachers Association.

Stein, S. 1979. *The science book*. New York: Workman Publishing.

Tolman, M. N., and Morton, J. 0. 1987. *Life science activities for grades 2–8*. West Nyack, N.Y.: Parker Publishing.

Wassermann, S., and Ivany, J. W. 1988. *Teaching elementary science*. New York: Harper & Row.

SUGGESTED COMPUTER PROGRAMS

Fish Scales. DLM. (Grades 1–2). Subtitles:

Fish Jump.

Today's Catch.

Look and Hook.

Which Fish?

Fishing Dock.

Fishing Derby.

Elementary Volume 4 Math/Science. MECC (#705). (Grades 2–up).

Odell Lake.

Odell Woods.

The Pond. Sunburst. (Grades 2–up).

Plant Doctor. Scholastic, Inc. (Grades 2–up).

Bumble Games. The Learning Company. (Grades K–up). Programs emphasize graphing, number lines, and using coordinate positions. Subtitles include:

Find Your Number.

Find the Bumble.

Butterfly Hunt.

Visit From Space.

Tic Tac Toc.

Bumble Dots.

SUGGESTED ACTIVITIES

1. Use a web to plan a unit centered around two concepts important in teaching about animals to primary age children. Groups could brainstorm different concepts and put a web of the results on a large sheet of paper. Teach this unit to primary age children.
2. Would you web the example seed unit in the same way? There are countless ways to web a seed unit. Select different seed or plant concepts around which to focus your planning and share them with the class. State your rationale.

3. Develop questions and evaluation strategies for the seed unit.
4. Design an activity for teaching with a classroom animal.
5. How many subject integrations will you use in your animal activity?
6. Identify the teaching strategies that were discussed in Units 5, 6, and 7. Compare lists.

REVIEW

1. List six life science concepts that are appropriate for primary age children.
2. What planning strategy can you use to develop a life science teaching unit?
3. Answer *True* or *False*:

   ______ a. Animals can suffer from overfeeding and too much care.

   ______ b. Wait until the animal arrives before preparing a home.

   ______ c. Children should be encouraged to bring personal pets to the classroom.

   ______ d. Most school systems do not have rules about animals in the classroom.

4. What are questions you should ask yourself as you prepare for an animal in your classroom?

UNIT 36 Physical Science

OBJECTIVES

After studying this unit, the student should be able to:

- Design physical science investigations appropriate for primary age children
- Develop an understanding of physical topics that are appropriate for primary grade children
- Develop structured and unstructured life science learning activities with primary grade children
- Apply unit webbing strategies to physical science content
- Integrate science with other subject areas

Physical science experiences are fun and exciting for primary age children. Even though these experiences can be dramatic and astounding for children, teaching the concepts is more foolproof than you might think. This is true for several reasons.

First, the physical sciences abound with opportunities for using discrepant events (discussed in Unit 6) that put students in disequilibrium and readies them for learning. Thus, they are eager and ready to explore. Second, most experiences are instantly repeatable. In this way, the child can continue exploring without elaborate preparation. And finally, the hands-on nature of physical science explorations make them ideal for use with primary age children.

This unit begins with the physical science concepts that are basic to primary grade learning and presents an example of ways that the topic of air can be integrated with science concepts of movement and changes in matter, gravity, light, and color. The subject integrations and lessons are part of the webbing scheme depicted in Figure 7–7 of Unit 7. Suggestions for the teaching of sound are also included (Figure 36–1).

Figure 36–1 "I can play a tune."

PHYSICAL SCIENCE CONCEPTS

The following concepts are basic to understanding the physical sciences. Use this list to identify concepts around which you can develop a planning web such as those described in Units 7 and 35.

- Air takes up space.
- Air has weight.
- Air is all around us.
- Sound moves in the air.
- Moving air pushes things.
- Air slows moving objects.
- Gravity causes objects to fall.

- Everything is made from material called *matter*.
- All matter takes up space.
- Matter can change into a solid, liquid, or gas.
- Matter can be classified according to observable characteristics.
- Physical and chemical changes are two basic ways of changing things.
- In a physical change, appearances change, and the substance remains the same. (tearing paper, chopping wood)
- In a chemical change, the characteristics change so that a new substance in formed. (burning, rusting)
- A mixture consists of two or more substances that retain their separate identities when mixed together.

- Temperature tells how hot or cold an object is.
- There are many types of energy—light, heat, sound, electricity, motion, magnetic.
- Magnets attract materials made of iron, steel, and nickel.
- Static electricity is produced when two different materials are rubbed together.
- There are three parts to a simple electric circuit: A source of electricity, a part for electricity to travel, and something that uses the electricity (a light bulb or bell).
- Some materials allow electricity to pass through them.
- To see color, light is needed.

- Pushing and pulling are forces.
- Machines make it easier to move things.
- Things move in many ways.

PLANNING AND TEACHING A UNIT ABOUT AIR

The learning experiences suggested in this unit follow the format described in Unit 7 and used in the seed unit (refer to Unit 35). Each lesson states a concept, an objective, materials, and suggestions for teaching the concept. Extensions, integrations, and possible evaluation procedures are indicated in the body of the lesson, when appropriate, and at the end of the unit. The topic of air is explored with bubble and sound lessons (Figure 36–2).

Exploring Bubbles

Children love to play with bubbles. They love to create bubbles, chase bubbles, and pop bubbles. Why not take advantage of this interest to teach science concepts with a subject children already know something about?

Figure 36–2 "What is inside of the bubble?"

THE BUBBLE MACHINE

CONCEPT: Bubbles have air inside of them.

OBJECTIVE: The child will construct a bubble machine by manipulating materials to produce bubbles.

MATERIALS: Mix a bubble solution of eight teaspoons (one-half cup) liquid detergent and one quart water. Punch holes about halfway down the side of enough paper cups for each child. A box of drinking straws will also be needed. Write each child's name over the hole.

PROCEDURE: Assemble a bubble machine by inserting a straw through a hole in a paper cup and filling with detergent to just below the hole. Have the children observe the machine as you blow bubbles. Ask, WHAT DO YOU THINK IS HAPPENING? Have the children assemble bubble machines. Each child should insert a straw into the hole in the side of the paper cup and then pour some detergent solution into the cup just below the hole. Give children time to try bubble blowing and ask, WHAT DO YOUR BUBBLES LOOK LIKE? DESCRIBE YOUR BUBBLES. HOW MANY BUBBLES CAN YOU BLOW? WHAT IS INSIDE OF THE BUBBLE? (Figure 36–3).

EXTENSIONS: Make bubble books with drawings that depict the bubble machine experiences. Encourage children to write about their pictures. Read the books to the class then put them in the library center for perusal. (Refer to Figure 7–7 for additional extensions.)

Stephanie Karinn Anderson **October 12, 1988**

We had bubble machines and we made lots of bubbles. They were big and little, soapy and soft, all kinds of shapes and colorful but most of all they popped easily! They were lots of fun though.

Figure 36–3 Stephanie describes the bubbles she made with her bubble machine.

INVENTING BUBBLE MACHINES

CONCEPT: Many different shapes will create bubbles.

OBJECTIVE: Make a shape that will create a bubble. Observe air entering bubbles and taking up space.

MATERIALS: Mixed bubble solutions soft, bendable wire, pans for bubble solution

PROCEDURE: YOUR BUBBLE MACHINE WORKS WELL. CAN YOU BEND WIRE INTO SHAPES THAT WILL MAKE BUBBLES? Have children "mess around" with wire to create a variety of shapes for making bubbles. Have them test the shapes by blowing air into the bubble film or moving the wire shape through the air. Ask, DID YOU MAKE A BUBBLE? WAS YOUR BUBBLE THE SAME SHAPE AS THE WIRE? (Figure 36–4).

EXTENSION: Have other materials available for bubble machine making such as a spatula, turkey baster, berry basket bottom, sieve, and other kitchen items. Ask, WHAT SHAPE DO YOU THINK THESE BUBBLES WILL BE? WHAT DO YOU NOTICE ABOUT THE SIZE OF THE BUBBLES? (different sizes) Have children record predictions and results for the different bubble blowers. (Figure 36–5).

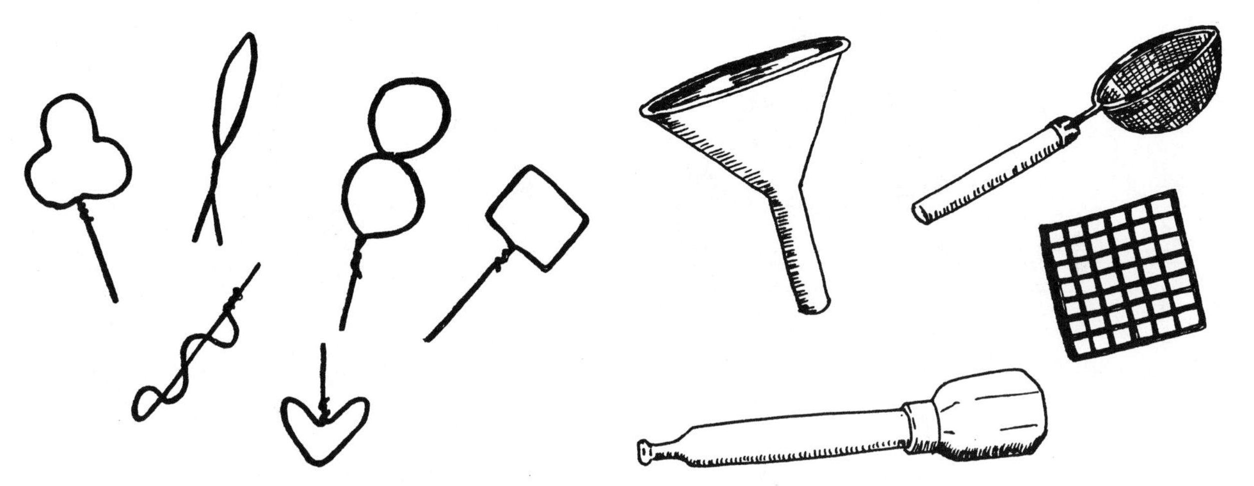

Figure 36–4 Wire shapes for blowing bubbles

Figure 36–5 Kitchen shapes make interesting bubbles.

TABLETOP BUBBLES

CONCEPT: Air takes up space.

OBJECTIVE: Use air to create bubble sculpture. Predict how the bubbles will act on top of tables.

MATERIALS: Bubble solution, straws, cups, flat trays

PROCEDURE: Have children dip out a small amount of bubble solution on a tray and spread it around. Ask, WHAT DO YOU THINK WILL HAPPEN IF YOU BLOW AIR INTO THE BUBBLE SOLUTION? (bubbles will form) Instruct children to dip a straw into the soap on the tray and blow gently. Give them time to "mess around" with tray top bubble blowing. Present challenges. CAN YOU BLOW A BUBBLE THE SIZE OF A BOWL? Have several children work together to make a community bubble. One child begins blowing the bubble and others join in by adding air to the inside of the bubble. (The straws must be wet to avoid breaking the bubble.) (Figure 36–6). Bubble cities of the future can be made by adding more solution to the tray and blowing several bubble domes next to one another.

EXTENSION: Bubbles inside of bubbles. CAN YOU FIT A BUBBLE INSIDE OF A BUBBLE? Have children blow a bubble. Then insert a wet straw inside of the bubble and start a new bubble. WHAT HAPPENS IF THE SIDES TOUCH? (Figure 36–7). Make bubble chains. Have children blow a bubble with a straw. Then, wiggle the straw and blow another one. Soon, a chain of bubbles will form. Ask, HOW MANY BUBBLES CAN YOU MAKE?

Figure 36–6 Children working together to create bubbles.

Figure 36–7 "Can you fit a bubble inside of a bubble?"

Investigation Questions for Exploring Air and Bubbles

1. Will other liquids create bubbles?
2. Will adding a liquid called *glycerin* to the bubble solution make the bubbles stronger? (yes) Can you think of other things to add to the bubble solution to make the bubbles last longer? Test your formulas.
3. Are bubbles always round? (Yes, unless some of them stick together. Forces acting inside and outside of the soap film are the same all over it.)

SUBJECT INTEGRATIONS

Bubbles and Science

1. *Observe bubbles.* Describe color, movement, size, shape. How many different colors do you see in the soap bubble? How do the colors move and change? Look at the bubbles through different colors of cellophane or polarized sunglasses. Do the bubbles change?
2. *Classify bubbles.* Classify by color, size, shape (bubble machine bubbles will stick together and take different shapes), location, how the bubble was made.
3. *Changes in matter.* Ask children to describe some changes that they notice in bubbles. Ask, DID ANYTHING CHANGE? (yes) DID YOU SEE A PHYSICAL CHANGE IN THE BUBBLE SOLUTION? (Yes, matter only changes form in a physical change.)

Bubbles and Art

1. *Bubble painting.* Mix bubble solution with liquid poster paint. Put the mixture in a bowl. As the child blows into the paint with a straw, a paint bubble dome forms. Lay construction paper over the bubble. As

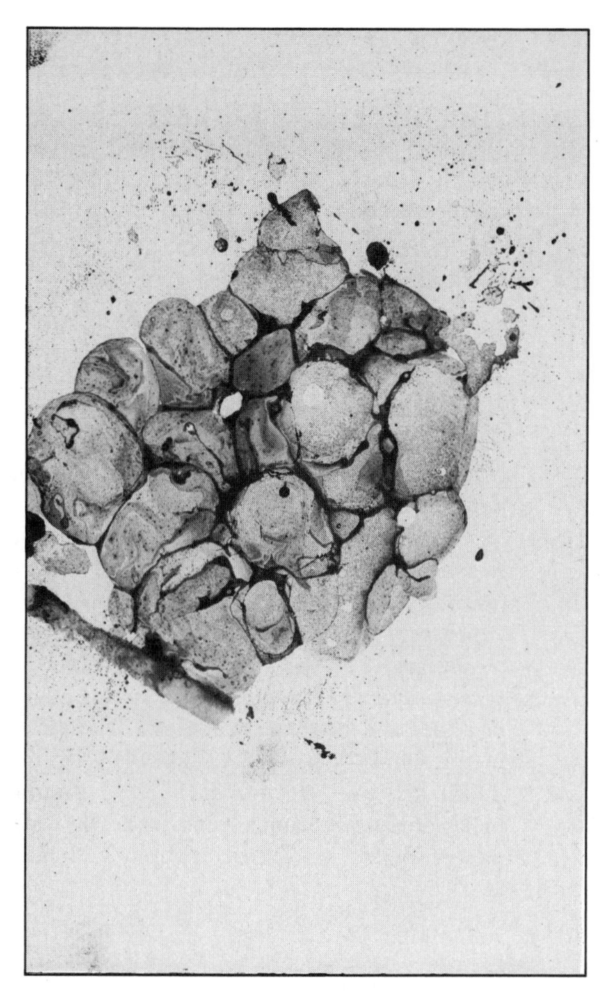

Figure 36–8 Christy's bubbles splat and make a pattern on the paper.

Figure 36–9 George measures the height of a large bubble.

the paint bubble bursts, it splatters paint on the paper. (Figure 36–8).

2. *Bulletin board.* Make a background of large construction paper bubbles. Display children's writing and art activities in this way.

Bubbles and Math

1. *How high?* Measure the height of the bubbles on a tray top. Measure how long a bubble lasts, predict and measure how many bubbles can be blown, and how far they can float. (Figure 36–9).
2. *Lung capacity.* Make a graph of who has the largest lung capacity. Have children discover a way of determining lung capacity by blowing bubbles.

Bubbles and Language Arts

1. Science experiences can provide inspiration for important first steps in writing.
2. *Bubble books.* Provide paper for children to record their bubble explorations. These can be bubble shaped and illustrated.
3. *Books about bubbles.* Read storybooks about bubbles.
4. *Chart story.* Have children write or dictate stories about what they observed when they examined air inside of bubbles (Figure 36–10).

Bubbles and Food Experiences

Children can observe the tiny air bubbles that go into various whipped mixtures. For example, whipped milk topping can be made for

```
Matthew                                              October 12, 1988

To day we were sintest again  it was a lot of fun.  We were seeing
if air took up space.  Some people thoght that it did and some
thoght it didn't.  I blew a very very very very big bubble but it
popped in my face.  There were different colors and shapes.  But
they didn't taste good.
```

Figure 36–10 Matthew has made any observations about bubbles.

cookies. You will need one-half cup of instant dry skim milk, and one-half cup cold water.

Beat with an electric mixer at high speed for four minutes. Have children watch the mixture change. Then, add two tablespoons of sugar and one-half teaspoon vanilla and beat at high speed until the mixture stands in peaks. You may want to add the ingredient "air bubbles" to your recipe. Remind children that air went into what they are eating.

Concept: Air Can Move Things and Slow Things Down

The following suggestions extend science concepts:

1. *Moving bubbles.* Have children move bubbles in the direction they want them to go by fanning them with paper (Figure 36–11).
2. *Paper fans.* Make paper fans to move air. These can be accordian-pleated sheets of paper or small paper plates stapled to popsicle sticks.
3. *Glider designs.* Make gliders and find out how far they can fly.
4. *Straw painting.* Blow paint with plastic straws. Pick up the paint with air pressure (as with the bubble solution), drop it on paper, and blow the puddles of paint into a design.
5. *Make pinwheels.* Have children fold squares of paper and hold them in front of a fan or in the wind. Ask, WHAT HAPPENS TO THE PAPER? (It moves.) Help children make pinwheels from typing paper or wallpaper (Figure 36–12). A pushpin will hold

Figure 36–11 "Can you make bubbles move?"

Figure 36–12 Making pintwheels

the pinwheel to a pencil eraser and allow it to turn freely. Ask, HOW DID WE PRODUCE MOTION?

6. *Air walk.* Take children on a walk to find things that are moved by air. Discuss and write about what you have seen. Local windmills are ideal, but weather socks, flags, trees, seeds, clothes on a clothesline, and other moving things work well.
7. *Exploring parachutes.* Have children make parachutes out of squares of cloth. Tie each corner with string, then thread the four strings through a wooden bead or washer. Drop the parachute from various heights and predict what will happen. Cut a hole in the top of the parachute and observe any differences. Make parachutes out of different materials, such as plastic or cupcake liners, and compare their flight (Figure 36–13).
8. *Baby seeds.* Open milkweed pods in the wind and watch the wind disperse the seeds. Refer to the seed lessons in Unit 35 for further investigations.

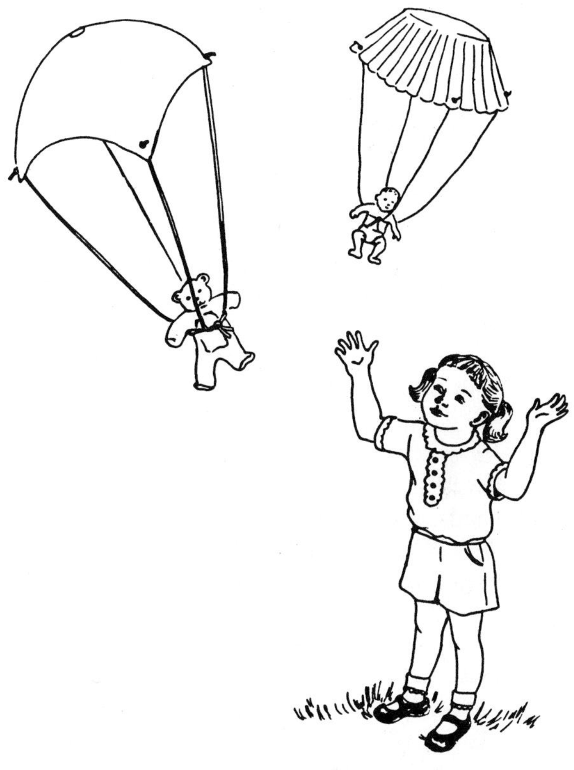

Figure 36–13 Kate compares the flight of two parachutes.

Exploring Sound

Children love to make music, but they probably do not know anything about the nature of sound. Help them understand that sound is caused when something vibrates by using the following lessons to teach children to see, hear, and feel the sound vibrations around them. Then, make musical instruments that reinforce the concepts.

GOOD VIBRATIONS

CONCEPT: Sound is caused when something vibrates.

OBJECTIVE: Observe objects vibrating and making sounds. Construct musical instruments based on concepts of sound.

MATERIALS: Paper plates, rubber bands, bottle tops, hole punch, paper cups, waxed paper, coffee cans, flexible tubing

PROCEDURE:

1. Have children begin vibration observations by gently resting their fingers over their vocal chords and saying "Ahhhh" and "Eeee." Ask, WHAT DO YOU

FEEL? (something moving) Explain that they are feeling vibrations occurring in their vocal chords, where all sounds they make with their voices come from.

2. Then, have them place their fingertips against their lips while they simultaneously blow and hum. Reinforce the connection between vibration and sound by touring your classroom. Ask children to identify things that vibrate—the aquarium pump, kitchen timer, air vent, overhead lights. If possible, let them feel the vibrations in these objects and listen to the sounds each makes (Figure 36–14).
3. *Drum.* Make a simple drum out of a coffee can that has a plastic top. Place grains of rice on the can lid and tap lightly, while your students watch and listen. Ask, WHAT HAPPENS TO THE RICE WHEN YOU DO THIS? (It bounces up and down to the sound of the drum.)
4. *Drumsticks.* Make drumsticks from tennis balls and dowels. Punch a hole in the ball, apply glue to the end and edges of the dowel or stick, and push it into the center of the ball. Or, use a butter brush, wadded up rubber bands, or a plastic lemon (Figure 36–15).
5. *Humming cup.* Make a humming cup by cutting the bottom out of a small paper cup. Then have the children cover the newly opened ends with waxed paper and secure with rubber bands. Say, PLACE YOUR LIPS LIGHTLY AGAINST THE WAXED PAPER END OF THE CUP AND HUM. They will get another feel for sound. Paper towel cups can be used in a similar way to make horns (Figure 36–16).
6. *More horns.* A yard of flexible tubing will provide a chance to whisper back and forth and feel sound traveling to their ears through a talking tube. Then, tape a funnel to one end and make a bugle-like horn (Figure 36–17).
7. *Tambourines.* You can construct this favorite instrument by decorating and shellacking paper plates in which you have already punched holes. Distribute bottle caps that also have holes punched in them and have children tie two or three caps to each hole in their plates. This is a noisy but fun way to show the relationship between vibrations and sound.

EXTENSIONS: *Build a Band.* Making musical instruments will give children an opportunity to observe vibrations in other ways. Make a variety of instruments and decorate for a special occasion.

Figure 36–14 Derrick feels the air vibrate in a tube as Linwood whispers to Brent.

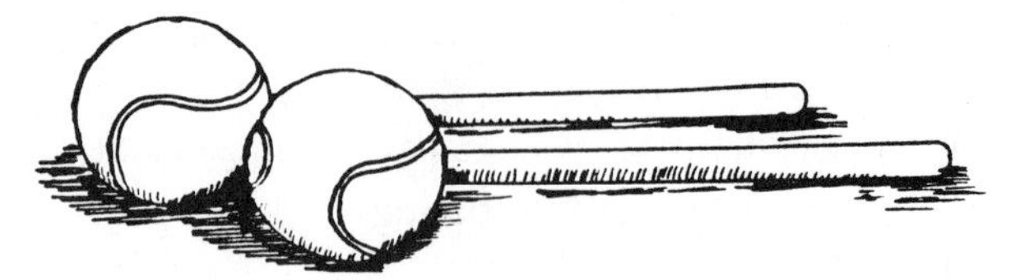

Figure 36–15 George makes drumsticks from tennis balls.

Figure 36–16 "Can you feel the vibrations of a humming cup?"

Figure 36–17 Sam made a bugle horn from a funnel and a piece of a plastic tubing.

Wind Instruments

Wind instruments such as flutes, whistles, and panpipes depend on vibrating columns of air for their sound. The longer the column of air, the lower the pitch. (The highness or lowness of a sound is its pitch. The more vibrations per second, the higher the tone that is produced; the fewer vibrations, the lower the tone.) The following lesson demonstrates the relationship between the length of an air column and musical pitch.

BOTTLES, WATER, AND AIR

CONCEPT: Pitch is the highness or lowness of a sound.

OBJECTIVE: Recognize differences in sound. Construct wind instruments.

MATERIALS: Ten bottles of the same height, water

PROCEDURE: Prepare by filling glass soda bottles to varying levels with water. (Plastic bottles are too easily knocked over.) Hold up an empty bottle and ask, WHAT DO YOU THINK IS IN THE BOTTLE? WHAT DO YOU THINK WILL HAPPEN IF THE AIR IN THE BOTTLE VIBRATES? Demonstrate by blowing across the top of the empty bottle. Teach children to direct air across the mouth of a bottle. This is worth taking the time to do. Instruct children to press the mouth of the bottle lightly against their lower lip and direct air straight across the mouth of the bottle. (Do not blow into the bottle. The secret is blowing straight across. Give them time to "mess around" with their new skill (Figure 36–18). Then ask, WHAT DO YOU THINK WILL HAPPEN IF YOU BLOW ACROSS A BOTTLE WITH WATER IN IT? Pour water into bottles and have children find out what will happen. Children will discover that the bottles make sounds that vary in pitch. They might even play a tune (Figure 36–19).

EXTENSION: Emphasize the meaning of pitch by reading a familiar tale in which high- and low-pitched sounds figure. For example, ask children to imitate the three billy goats gruff. Dramatize differences in pitch with voice and body activities.

Figure 36–18 Marie has learned to make music with a bottle.

BOTTLES AND WATER

Collect eight bottles to make a scale—your job will be easier if they are all the same kind and size. To tune the bottles, you'll need to fill them to varying heights with water.

Start by dividing the volume of water that one of the bottles could accommodate by the number of notes you wish to produce. For example, if you want a one octave scale, divide the volume by eight. Then, leaving bottle one empty, put $1/8$ of the volume into bottle two, $1/4$ into bottle three, $3/8$ into bottle four, and so on until bottle eight contains $7/8$ of its volume of water. Adjust the amount of water up and down in the bottles until the scale intervals sound (more or less) true. Use masking tape to mark the correct level on each bottle.

Figure 36–19 Formula for bottle music.

COMPUTER SUGGESTIONS

Wood Car Rally. MECC (#A-214), (Grades 3–up).

Miner's Cave. MECC (#A-213).

Mystery Matter. MECC (#A-212).

SUMMARY

When children explore physical science concepts, they are learning basic skills, which develop future understandings. To be effective, these lessons must emphasize process skills, concrete experiences, and integrations. Thus, the children see a relationship between the concepts and their world.

Children are not expected to understand distinctions in all physical science concepts. Many of the concepts are explored to prepare them for future learnings. Physical science concepts can be presented in many ways and with a variety of instructional strategies.

FURTHER READING AND RESOURCES

Abruscoto, J. 1988. *Teaching science to children*. Englewood Cliffs, N.J.: Prentice Hall.

Allison, L., and Katz, D. 1983. *Gee, wiz!* Boston, Mass.: Little, Brown and Company.

Althouse, R., and Main, C. 1975. *Science Experiences for young children*. New York: Teachers College Press.

Harlan, J. 1988. *Science experiences for early childhood years*. Columbus, Ohio: Charles E. Merrill.

Hawkinson, J., and Faulhaber, M. 1969. *Music and instruments for children to make*. Chicago: Whitman.

Lind, K. K. 1986. The beat goes on. *Science and Children* 23(7). 39–41.

______ 1985. The beat of the band. *Science and Children*. 23 (3): 32–33.

Council for Elementary Science International Sourcebook V. 1988 *Physical science activities for elementary and middle school*. Columbus Ohio: SMEAC Information Reference Center.

Perez, J. 1988. *Explore and experiment*. Bridgeport, Conn.: First Teacher Press.

Victor, E. 1985. *Science for the elementary school*, 6th Ed. New York MacMillan Publishing Company.

Zubrowski, B. 1979. *Bubbles*. Boston, Mass.: Little, Brown and Company.

SUGGESTED ACTIVITIES

1. Which process skills are used in the bubble unit? Identify each skill and how it is incorporated into the study of air inside of bubbles. Teach the unit to children. Which integrations and activities did you use?
2. Interview teachers of primary age children to find out what types of physical science activities they introduce to children. Do any of these activities include the exploration of materials? Record your observations for discussion.
3. Brainstorm strategies to prepare children to understand physical science topics such as magnets, light, and electricity. Pick a topic and prepare a learning experience that will aid in future concept development.
4. Select a physical science concept and prepare a learning cycle to teach that concept. Teach it to children. Record your observations for discussion.
5. Present a lesson about sound to primary age children. What types of learning experiences will you design for your students? Explain how you will apply Piaget's theory of development to this topic.
6. Go to a bookstore or library and identify books to enhance the teaching of physical science. Present an evaluation of these materials and how you might use them to reinforce science concepts.

REVIEW

1. Explain why primary age children should explore physical science activities (give at least two reasons).
2. List six physical science concepts that are appropriate for primary grade science lessons.
3. What type of activities should be used to teach physical science concepts?

UNIT 37 Earth Science

OBJECTIVES

After studying this unit, the students should be able to:

- Design earth science investigations for primary age children
- Develop an understanding of earth science topics that are appropriate for primary age children
- Develop structured and unstructured earth science learning experiences with primary age children
- Apply unit planning webbing strategies to earth science content
- Integrate science with other subject areas

Teaching children about rocks, fossils, weather, and space exposes them to new vocabulary and helps them gain experience with earth science concepts. It is not meant that children memorize words and ideas, but rather "mess around" in preparation for future learnings. In this way, when they are introduced to the same concepts at more advanced stages of development, they will have familiarity with the ideas and be open and receptive to learning.

This unit begins with appropriate earth science concepts that are basic to primary grade learning and suggests a way that the topic of rocks can be developed into a unit and integrated into the classroom. The rock lessons are included to help you get started, you may find other concepts and lessons useful. This is simply a place to begin. (Figure 37–1).

EARTH SCIENCE CONCEPTS

- Rocks are formed in different ways.
- Rocks are nonliving things.
- Sand is made up of tiny pieces of rock. Weathering is the action of wind and water on the surface of the earth.
- Mountains and land are made of rocks and soil.
- Water can evaporate. Evaporation means a liquid changes to a gas.
- Condensation means a gas changes to a liquid. Water vapor is an example of condensation.
- The atmosphere is the air around us. It has conditions such as wind, moisture, and temperature. Weather is what we call the conditions of the atmosphere. Temperature is how hot or cold something is. Thunder is caused by lightning.

Figure 37–1 Looking for a special rock

- There are four seasons, and each has unique weather.
- The earth has a layer of air and water around it. Water is a liquid and is called ice when it is solid. Moving air is called wind.
- The quality of air, water, and soil is affected by human activity.
- The sun gives off light and heat and warms the earth. Clouds are objects that have different shapes and move in the sky.
- The sun sets each evening and changes position in the sky. The moon changes in size and shape and position in the sky. The sun and moon repeat cycles.
- Objects can stop light and make a shadow. When light bounces off objects, it is called reflected light.
- The dinosaurs are the largest land animals that ever lived on earth.
- They are extinct, but fossil evidence tells us about them. Our only knowledge of some plants and animals is through the fossils that we have found.

PLANNING AND TEACHING A UNIT ON ROCKS

The learning experiences suggested in this unit follow the format described in Unit 7 and used in Units 35 and 36. Each lesson states a concept, an objective, materials, and suggestions for teaching the concepts. Extensions, integrations, and possible evaluation procedures are indicated in the body of the lesson, when appropriate, and at the end of the unit.

The following basic geology lessons introduce fundamental concepts and stimulate children's curiosity about rocks, minerals, and Earth processes. The strategies will acquaint primary age children with the nature of rocks in a way that is immediate, exciting, and fun.

A CARTON OF ROCKS

CONCEPT: There are many kinds of rocks.

OBJECTIVES: Observe different types of rocks. Classify rocks in different ways.

MATERIALS: An egg carton for each child or group, small labels that can be glued on the box, pencils or pens, a handful of rocks

PROCEDURE: Show the children the rocks in your hand. Ask the children, WHAT ARE THESE AND WHERE ARE THEY FOUND? Give each small group of children a few rocks and encourage them to develop classification schemes for them, such as flat, speckled, animal-shaped. Younger children will notice size and color first, then shape and the way the rock feels. Have the children develop labels for the categories.

Take the children to any place where there are rocks. Set a generous time limit and have them search for different types of rocks to fill their egg cartons. Then, have the children share and classify what they have found. Use labels to identify rocks in the egg carton collections (speckled, round, and so on) (Figure 37–2).

EXTENSION:

1. Older children are able to compare rock characteristics with pictures in rock identification handbooks.

2. Use various rocks to create a rock collage.
3. Read *Sylvester and the Magic Pebble* by William Steig. A round red pebble makes a nice prop.

Figure 37–2 Kate and Sara sort the rocks they have collected.

THINKING LIKE A GEOLOGIST

CONCEPT: Some rocks are harder than other rocks.

OBJECTIVE: Examine rocks and minerals to determine how hard they are. Classify rocks by characteristics.

MATERIALS: An assortment of rocks (a rock and mineral set, if possible) roofing nails, pennies

PROCEDURE: Give groups of children an assortment of rocks to observe. After they have had time to observe the rocks, ask, CAN YOU SCRATCH ANY OF THE ROCKS WITH YOUR FINGERNAIL? As they notice a difference in rocks, ask, CAN SOME ROCKS BE SCRATCHED IN THIS WAY? (yes) WHAT ELSE COULD YOU USE TO SCRATCH ROCKS?

After children try to scratch rocks with different objects such as the penny and fingernail, explain that finding out if a rock crumbles easily is one of the things a geologist does. (Even though children may not remember the term, the labeling is important.)

Tell the children that they are going to think like a geologist and determine how hard a rock is. Make a list of ways that the children will test the rocks. For example:

Can be rubbed off on fingers
Can be scratched with fingernail
Can be scratched with a penny
Hard to scratch

Have the children test the rocks and place them with the statement that best describes the rock. Ask, WHICH ROCKS DO YOU THINK ARE THE HARDEST? (those that are hard to scratch) (Figure 37–3).

Then, take the children outside to see which rocks can be used to draw streaks or lines on the sidewalk. Children might even see distinctive colors. Ask, WHAT COLORS ARE THE STREAKS? Have children compare the colors in the rocks with the colors of the streaks. Ask, CAN WE CLASSIFY ROCKS IN THIS WAY? (yes) You may want to review some of the ways that geologists classify rocks (appearance, texture, hardness, color) (Figure 37–4).

EXTENSION: As children wash their hands, compare wet and dry rocks. What differences do they notice? Geologists use hardness and streak tests to identify minerals. Introduce the term "minerals" when children notice bits of color in the rocks. Explain that rocks are made up of many different ingredients (minerals).

Figure 37–3 Labeling rocks by hardness

Figure 37–4 The sidewalk makes a good place to make streaks with your rocks.

EXPLORING COOKIES

CONCEPT: Rocks are formed in different ways.

OBJECTIVE: Identify similarities between rocks and cookies. Stimulate curiosity about Earth processes.

MATERIALS: Oatmeal chocolate chip cookie recipe on chart and on index cards, ingredients for making cookies (two for each child)

PROCEDURE: Pass out recipe cards to the small groups. Show the children what the various ingredients look like as you discuss those used in making oatmeal chocolate chip cookies. Give each small group of children samples of the cookie ingredients. Then, give each child two cookies (one to examine and one to eat).

Have the children explore the cookie and compare the ingredients and the information on the recipe card with the final product. Ask, CAN YOU SEE THE INGREDIENTS? (some of them) WHICH INGREDIENTS DO YOU SEE? CAN YOU TASTE ANY OF THE INGREDIENTS? (Figure 37–5). DO THE INGREDIENTS LOOK THE SAME? (No, some have changed.) Once the children recognize that some ingredients are still recognizable after baking and some are no longer in the same state, ask, IN WHAT WAY HAVE THEY CHANGED? (melted, dissolved, and so on) WHAT CHANGED THE INGREDIENTS? (mixing, pressing, heat) Discuss heat and chemical changes, if appropriate (Figure 37–6).

Give children an assortment of rocks and tell them that rocks have ingredients, too. The Earth acts like a baker and turns ingredients into rocks. (Rock ingredients, minerals, are turned into rock by pressure, heat, and moving around.) Let children handle mineral samples and try to find evidence of these samples in the rocks.

EXTENSION: Take a rock field trip. Have each child find an egg-sized rock. Wrap the rock in cloth and have an adult hit it with a hammer. Use hand lenses to examine the pieces of broken rock. Ask, HOW MANY PIECES OF ROCK ARE THERE NOW? Have the children explore the shape, size, and color of the fragments and describe the inside and outside appearance of the rock. The inside of a dull-looking rock is usually filled with pattern, texture, and color. Use a small light to help compare the shine and luster of various rocks. Have the children draw or color what they see through the hand lens.

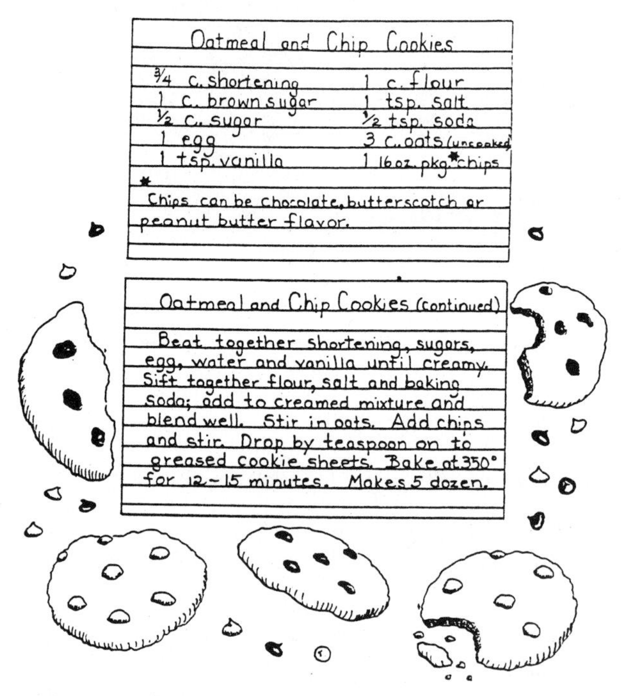

Figure 37–5 Compare the ingredients for making cookies with the cookies.

Figure 37–6 Exploring oatmeal chocolate chip cookies

HOW ROCKS ARE FORMED

You can make the study of rocks more meaningful for your students if you know the basic way rocks are formed and a few common examples of each.

Igneous Rocks

Igneous rocks are formed through the cooling or magma or lava. Examples of igneous rocks formed by the rapid cooling of surface lava are pumice, obsidian and basalt. Granite is the most common example of slow, below-surface cooling of molten rock. Refer to the Igneous Rock Fudge cooking experience integration.

Sedimentary Rocks

These rocks are formed from eroded rocks, sand, clay, silt, pebbles, and other stones. Compressed skeletons, shells, and dissolved chemicals also form sedimentary rocks. These rocks are usually deposited in layers and are compacted and cemented by the pressure of the overlying sediments. Examples of sedimentary rocks are conglomerate, sandstone, shale, *and* limestone. Fossils are frequently found in sedimentary rock. Refer to the Pudding Stone cooking experience integration.

Metamorphic Rocks

The term *metamorphic* means "changed in form." Metamorphic rock is formed when sedimentary and igneous rocks are completely changed in form through pressure and heat. For example, limestone becomes marble, shale becomes slate, and sandstone may become quartzite. These rocks are harder than most rocks.

Show the effect of pressure by pressing a bag of marshmallows or several spongy rubber balls under books on a flat surface. Have the children describe the change in the physical appearance of the objects and relate the change to the metamorphic process.

To simulate the effect of heat on the rocks, melt several different color marshmallows together into one interlocking marshmallow. Have children observe and describe the changes.

A CLASSROOM CAVE MODEL

Making a model is a way to simulate something that is difficult to observe firsthand. Keep in mind that the effectiveness of a model depends on the observations, inferences, and predictions that children make. The following lesson is an example of model building that will help children gain understandings about caves and how some rocks are formed. Many additional questions, integrations, and strategies can be developed to design a cave unit.

ICICLE STALACTITES

CONCEPT: Stalactites are formed as mineral-rich water drips from the ceilings of caves.

OBJECTIVE: Observe the formation of stalactites in a simulated cave. Make a model of a cave.

MATERIALS: Cardboard box, pictures of caves, cotton string, small cardboard box, water, food coloring, eye dropper.

PROCEDURE: Prepare a model of a cave by cutting a large hole in one side of a cardboard box. Make a hole in the top of the box, tie heavy cotton string to a stick, and hang the string through the hole. The string should be about an inch above the bottom of the box. You may want to add a few rocks to the bottom of the box.

The box should be attached to the outside of a classroom window with the hole in the side of the box facing the inside of the classroom. The little cave should be placed in the shade so that the icicle stalactite will not melt. In this way, children can observe the formation of an icicle "stalactite." On a cold day, begin forming the stalactite by dripping water down the string a few drops at a time. Each day add more drops of water. This model will take a long time to develop, much like the development of a cave stalactite. In this way, children learn about the formation of stalactites and that they take a long time to form (Figure 37–7).

Show children pictures of caves and discuss features of caves. Explain that *stalactites* develop from the ceiling of a cave and *stalagmites* develop from the floor of the cave.

As the icicle stalactite forms, begin adding food coloring to the drops of water to simulate the minerals found in the dripping cave water. Ask, WHAT CHANGES DO YOU SEE IN THE STALACTITE?

Have children predict if a stalagmite might form in their cave. Then, as the stalagmite begins to grow, ask, WHAT CAUSES THE STALAGMITE TO FORM? (some water will drip on the bottom of the cave and begin building up to meet the stalactite) Have children record and draw the changes inside of the classroom box cave.

A large calendar chart will emphasize the length of time required for a stalactite to develop. Have children take turns drawing the developing cave. Ask, WHAT CHANGES DO YOU SEE? WHAT DO YOU THINK WILL HAPPEN NEXT?

Children love caves and will enjoy making up stories about them. To encourage thinking, ask, WHAT KIND OF ANIMALS MIGHT LIVE IN OUR CAVE? COULD YOU LIVE IN A CAVE?

EXTENSION: Cover a table with blankets and have children pretend they are cave explorers. (Or, connect a series of large boxes to make a cave.) Papier-mâché hats and flashlights make good props for exploration. Help children decorate the cave with pillow rocks and paper stalactites and stalagmites. The children will enjoy discussing what they might see in a cave before they enter.

Investigation Questions for Exploring Caves

1. Can seeds break up rocks?
2. How does freezing water break up rocks?
3. How does weathering change a rock?

SUBJECT INTEGRATIONS

Rocks and Science

1. *Smooth rocks.* Additional ideas about rocks and their formation can be introduced by walking along a beach or stream and observing the smooth stones at the bottom of the stream or on the shore. Children can collect rocks and begin to draw comparisons about what makes the rocks smooth. Have them watch the action of the waves or stream. Ask, WHAT IS HAPPENING? Compare using sandpaper with the action of sand against the rocks.
2. *Sand from rocks.* Have children rub two rocks together to simulate wear on them. Discuss how this wear might take place in nature (water current, waterfalls, wind, and so on).

Figure 37–7 "What causes the stalactite to form?"

Rocks and Art

1. *Rock gardens.* Construct small rock gardens. Use pictures of Japanese gardens as inspiration.
2. *Rock necklaces.* Glue yarn to the back of flat, round rocks to make necklaces.
3. *Vegetable fossils.* Make mold fossils of vegetables. (See Additional Lesson Ideas for directions.)

Rocks and Language Arts and Reading

1. *Storybooks.* Read *Stone Soup* and make vegetable soup. See Unit 26 for suggestions.
2. *Rock shape book.* Make-rock shaped covers for this bookmaking activity from brown construction paper. Inside pages should also be in the shape of the rock cover. Punch two holes in the top of the book and fasten with metal rings or yarn. You may want to start the book with phrases like "A rock is a . . . " (Figure 37–8).

Figure 37–8 "Record observations in your rock shape book."

3. *My rock.* Have each child select a rock and describe it. The children can record or dictate descriptions. Encourage them by asking, WHAT DOES YOUR ROCK FEEL LIKE? DOES YOUR ROCK HAVE A SMELL? Stimulate thinking with, "My rock is as sparkly as a . . . ." Then, have children write a biography of their rock.

Rocks and Math

1. *Ordering rocks.* Put rocks in order of size, shape, and color, then from smoothest to roughest. Try doing this activity blindfolded.
2. *Weighing rocks.* Use a small balance to weigh rocks. Predict which rocks are the heaviest.
3. *Flannel board rocks.* Make flannel board rock shapes for math problem solving. Different sizes, colors, and shapes can be put in order, organized into patterns, sorted, and counted.
4. *Rock jar.* Fill a jar with rocks and have the children predict how many rocks are in the jar. Verify the number of rocks in the jar by counting by groups of ten.
5. *Dinosaur rocks.* Have children predict the

number of rocks needed to build a dinosaur cave. Children can determine the needs of a dinosaur, such as enough rocks to shelter it and keep larger dinosaurs out of the cave.

Rocks and Cooking Experiences

1. *Igneous rock fudge.* After showing children pictures of volcanoes, find out what children know about them. Have pumice available for the children to examine. Some children might notice that the stone is light and has air bubble spaces. Explain that igneous means fire and that the pumice stone is an igneous rock that was once hot, liquid iron in a volcano that hardened as it cooled.

   To demonstrate how lava cools, make fudge in a medium-sized pan. Mix 1/3 cup water, 1 cup sugar, a pinch of salt, 3 tablespoons of cocoa, and 1 teaspoon of vanilla. Boil the mixture for 3 minutes and cool in one of two ways. To cool the fudge fast and produce a shiny, smooth surface, pour some of the mixture into a pie plate that is resting in a bowl of ice. For rough and lumpy granite-like fudge, cool at room temperature in another pie pan. Ask, WHICH WAY DO YOU THINK THE PUMICE WAS COOLED? (Fast, because it was blown out of the volcano. The black glassy obsidian is another example.) (Figure 37–9).
2. *Pudding stones.* To make the conglomerate rock called pudding stone, have the children mix plaster and add various types of gravel and weathered stones to the mixture. Then, make pudding and add raisins and other food bits for a tasty pudding stone snack (Figure 37–10).

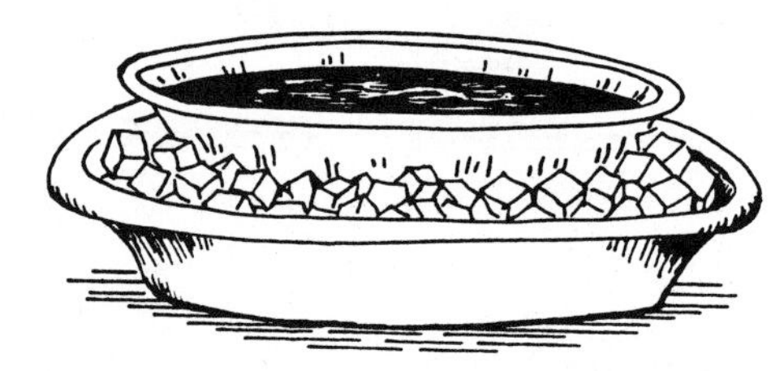

Figure 37–9 Making igneous rock fudge

ADDITIONAL LESSON IDEAS

Fossils

Fossils are the remains or signs of animals or plants that are most likely to be found in metamorphic rock. To make a fossil, give the children pieces of plants and have them coat the plant with petroleum jelly. Next, mix plaster of paris with water—according to the directions on the package—in the bottom of an aluminum foil

Figure 37–10 Making a conglomerate rock

pie plate. Instruct the children to gently press the plant into the surface of the mixture. After the plaster dries, remove the plant and observe the imprint. Explain that this is one way a fossil is formed.

Children can press clay against the imprint, and the surface of the clay will take the shape of the original plant. Try making mold fossils of vegetables. Children can form a fossils collage with the results (Figure 37–11).

Weather

Weather lessons are especially appropriate for primary age children. In addition to discussing and studying the daily weather, take the children on a field trip to observe weather equipment in action. Local trip opportunities might include airports, television and radio stations, or high school or college weather stations. Visiting a weather station familiarizes children with the instruments and information they will study as they follow and record changes in weather and differences in seasons. Include constructing weather instruments, graphing, storybooks, subject integrations, and hands-on experiences as you teach children about weather.

Space

Space travel, the moon, sun, and stars intrigue young children. They will not be able to comprehend the vastness of space or the enormity of the sun, moon, and Earth, yet, there are aspects of space science that they can readily observe.

Figure 37–11 Making a mold fossil

The sun is the most observable object in the sky. (CAUTION: children should never look directly into the sun.) Ask, ON WHAT SIDE OF THE BUILDING DOES THE SUN SHINE IN THE MORNING? Then, ON WHAT SIDE OF THE BUILDING DOES THE SUN SHINE WHEN WE LEAVE FOR THE DAY? Take the children for a walk around the building and note where the sun's rays are shining at different times of the day. Draw these changes on a chart. Ask, WHY DO YOU THINK THAT THE SUN'S RAYS SHINE IN DIFFERENT PLACES?

Take the children outside of the building in the morning and have them draw the school building and the location of the sun. Begin a bulletin board mural. Make the school building and surrounding features and have the children place a construction paper sun where it belongs in the morning sky. Then make paper sun rays shining on the school building.

Shadow play is a natural for young children as they learn about the sun. Draw chalk outlines of the shadows cast and then try casting shadows on a cloudy day. Ask, DO YOU HAVE A SHADOW TODAY? WHY NOT? Discuss the color of the sun and the moon. Ask, IS THE MOON THE SAME COLOR EVERY NIGHT? CAN YOU SEE THE MOON DURING THE DAY? (yes) Many children think that the moon goes to bed at night. Help them speculate on why the moon does not look as bright during the day.

If children think that the stars are held up with tape, do not discourage them. They are in good intellectual company. Aristotle, for example, thought that stars were embedded in concentric crystalline spheres. A primary age child cannot understand such concepts. Help the child notice that things in the sky look different at various times of the day and night (Figure 37–12).

Figure 37–12 Children construct and dramatize a space mission.

COMPUTER SUGGESTIONS

How to Start a Rock Collection. Danville, Ky.: Teacher Video Production Corporation.

Dinosaur Construction Kit—Tyrannosaurus Rex. D. C. Heath and Co. Subtitles:

Animal Watch—Tracks
Animal Watch—Whales
Animal Watch—Wolves
A Closer Look—The Desert

Space Subtraction. MECC (#A-145). (Grades 1–3). Subtitles:

Cosmic Creature
Blast Off
Zemoon Walk
Space Match
Shuttle Trip

SUMMARY

Rocks and minerals are not simply objects laying on the ground. They are part of the continually changing process of the Earth. As children study earth science topics such as rocks, minerals, weather, and space, they study the conditions and forces that affect their planet. An early introduction to these difficult concepts exposes children to their world and helps ready them for future understandings.

FURTHER READING AND RESOURCES

Gega, P. 1986. *Science in elementary education*. New York: John Wiley & Sons.

McBiles, J. L. 1985. *Mining, minerals, and me*. Nashua, N.H.: Delta Education, Inc.

McCormack, A. 1979. *Outdoor areas as learning laboratories*. Council for Elementary Science International Sourcebook. Columbus, Ohio: SMEAC Information Reference Center.

McIntyre, M. 1984. *Early childhood and science*. Washington, D.C.: National Science Teachers Association.

Perez, J. 1988. *Explore and experiment*. Bridgeport, Conn.: First Teacher Press.

Renner, J. W., and Marek, E. A. 1988. *The learning cycle and elementary school science teaching*. Portsmouth, N.H.: Heinemann.

Williams, R. A., Rockwell, R. E., and Sherwood, E. A. 1987. *Mudpies to magnets*. Mt. Rainer, Md.: Gryphon House, Inc.

SUGGESTED ACTIVITIES

1. Recall your rock collecting experiences in elementary school. Where did you find your rocks? Did you display your rocks? Did you know what your rocks were or how they were formed? What do you think the formation of rocks meant to your teachers?
2. Make your own teaching web. Select lessons

from the rock unit and design a unit that is appropriate for a group of children you are teaching. You should add ideas to the ones presented. Teach the lesson to children and record your observations for class discussion.

3. Which process skills are used in the rock unit? Give an example of each skill and when it was used. Can you think of additional ways to integrate process skills into the unit on rocks?

4. Present a lesson on weather to primary age children. What types of learning experiences will you design for your students? How do you plan to apply what you know about Piaget's theory of development to the teaching of weather?

5. Examine a teacher's edition of an elementary school science textbook. Identify a lesson on rocks and one on weather. Do you agree with the way the lesson is presented? Explain why or why not. How would you teach the lesson?

6. Select an earth science topic and concept and prepare a learning cycle lesson to teach that idea. Teach it to children.

7. Imagine that your class has just returned from collecting rocks. What types of questions will you ask to help students generate a list of characteristics that can be used to classify rocks into different groups? Your questions could take a variety of forms. Use your question list with children and note the interactions for class discussion.

REVIEW

1. What is the purpose of introducing young children to earth science concepts?

2. List six earth science concepts appropriate for primary age children.

3. Match the following words with their proper definition.

_____ a. weathering	1. conditions of the atmosphere
_____ b. evaporation	2. a gas changes to a liquid
_____ c. condensation	3. ancient evidence of plants and animals
_____ d. reflected light	4. a liquid changes to a gas
_____ e. weather	5. action of wind and water on the surface of the earth
_____ f. fossils	6. when light bounces off an object

4. Name the three classifications of rocks. How will you teach these types of rocks to children?

UNIT 38 Health and Nutrition

OBJECTIVES

After studying this unit, the student should be able to:

- Design health and nutrition investigations for primary age children
- Develop an understanding of health and nutrition topics that are appropriate for primary age children
- Develop structured and unstructured learning experiences in health, nutrition, and the human body
- Integrate science in the primary grades with other subjects

Optimum growth and development depends on good nutrition, adequate experience, plenty of rest, and proper cleanliness. Concepts in health and nutrition need to be introduced to young children so they learn how to take care of their bodies and develop beneficial lifelong habits at an early age. This unit suggests experiences that help children learn about themselves through exploring the human diet, major food groups, and the human body. Lesson ideas for promoting good health habits are also presented.

The following health, nutrition, and human body concepts are basic to primary grade learning (Figure 38–1).

- A balanced diet consists of carbohydrates, fats, proteins, vitamins, minerals, and water.
- The Four food groups are: meat and meat alternates, vegetables and fruits, breads and cereals, and milk and milk products.
- A diet that is a balance of the Four food groups, plus water, makes up a healthful regimen for normal people.
- A good diet is necessary for the growth and development of the human body.
- You can help yourself stay healthy.
- The human body can be affected by a variety of diseases.
- The human body carries on life processes.
- Food is the fuel and the building material of the body.
- The human body must take in and digest food.
- The human body consists of a number of groups of organs that work together to perform a particular function.
- Movement of the human body is made possible by the skeleton and muscles.
- The human body takes in oxygen and gives off carbon dioxide.

Figure 38–1 How is a rock like our bones?

- The human body reacts to stimuli.
- The circulatory system is the transportation system of the body.
- Humans reproduce and give off waste products.

- Humans observe and learn by seeing, hearing, feeling, tasting, and smelling.
- The senses of smell, sight, and taste help us identify foods and the world.

EXPLORATIONS IN HEALTH AND NUTRITION

The learning experiences suggested in this unit introduce fundamental concepts that acquaint children with their bodies and how to stay healthy. Topics include good health habits, treatment of cuts and scrapes, how the body gains energy, the food groups and food activities, and bones, teeth, vitamins, and minerals.

Health Habits

Good health habits begin early in a child's life. Encourage children to begin an understanding of what healthy living really means by providing a variety of instructional techniques and subject integrations that relate to the child's life. Here are a few to get you started.

ARE YOUR HANDS DIRTY?

CONCEPT: Hands that look clean can still be dirty.

OBJECTIVE: Observe the growth of mold on potatoes. Compare the effect of mold growth resulting from handling potatoes with washed and unwashed hands.

MATERIALS: Two small potatoes, potato peeler, two clean jars with lids that seal, labels, marker

PROCEDURE: Have children examine their hands. Ask, ARE YOUR HANDS CLEAN? DO YOU SEE ANY DIRT ON THEM? Discuss how hands look when they are dirty. WHAT IS THE DIRTIEST YOUR HANDS HAVE EVER BEEN? THE CLEANEST?

Peel two potatoes, ask a child to handle one of them, and put it in a jar. (Select a child who has not washed his or her hands for several hours.) Point out that the child's hands appear to be clean.

Label the jar with the name of the child and the words "unwashed hands." Instruct a child to wash his or her hands, handle the second potato, and put it in the remaining jar. Label the jar with the child's name and the words "washed hands." Discuss the relative cleanliness of the "washed hands" potato.

Place the two jars in a warm place where the children can observe and record what happens. Compare what is happening to the potatoes after a day or two (mold is likely to form on the "unwashed" potato). Compare the potatoes and discuss how the mold got on the potato (Figure 38–2). (Refer to Unit 35 for additional mold lessons.)

EXTENSION: Have children describe and draw what the unwashed and washed hands looked like. Ask, DID THE UNWASHED HANDS LOOK DIRTY? (no) This would be a good time to discuss germs and the importance of washing hands before handling food. Have children design posters to encourage hand washing before meals.

Good health habits chart. Keep a record of your health habits. Answer questions about yourself for a week. (For example: I brushed my teeth after every meal, I took a bath or shower, I washed my hands before meals today. I exercised today.) At the end of the week, look at all of your answers. What can you do to improve your health habits?

Figure 38–2 What is happening in the unwashed hands jar?

Figure 38–3 Apples with puncture wounds.

APPLES AND ME

CONCEPT: Microorganisms (germs) can enter our body through cuts and scrapes in the skin.

OBJECTIVE: Compare the healing of injuries on an apple with injuries on our skin. Describe how antiseptic keeps germs out of our body.

MATERIALS: Three apples, soil or other type of dirt, antiseptic, two sewing needles

PROCEDURE: Lead the children in a discussion of current injuries that they have received. Such incidents are a common happening in the primary grades. Encourage children to tell how the injury occurred and how it was treated.

Then hold an apple in your hand. Ask, HOW IS THE COVERING OF AN APPLE SIMILAR TO YOUR SKIN? HOW IS IT DIFFERENT? LET'S SEE WHAT HAPPENS WHEN THE SURFACE OF AN APPLE IS INJURED.

Label one apple the "control apple." You will not injure this apple. Label the second apple "no antiseptic," and the third apple "antiseptic." Then, stir one of the needles in dirt and puncture the second apple three times. Stir the other needle in dirt, sterilize the needle with antiseptic, and puncture the third apple. Place the three apples in a sunny place and have the children keep a record of changes in the apples. Ask, WHAT DO YOU THINK THEY WILL LOOK LIKE IN ONE DAY? As children record the progress, discuss possible reasons for differences in the appearance of the apples. (The apple punctured by the unsterilized needle will rot fairly quickly.)

After several days, cut all of the apples in half, including the control apple. Ask, WHAT

CHANGES DO YOU SEE IN THE APPLES? WHICH APPLE LOOKS THE WORST? DID STERILIZING THE NEEDLE MAKE ANY DIFFERENCE? WHAT DO YOU THINK WOULD HAPPEN IF THE SURFACE OF YOUR BODY WERE PUNCTURED? Discuss the best possible action to take in case of an injury (Figure 38–3). (Refer to the mold activities in Unit 35.)

EXTENSION: Sterilize the needles in different ways, such as using a candle flame to simulate heat sterilization or various commercial antiseptics to compare products.

Children will be interested in trying out this experiment with different types of foods and different types of injuries (scraping, cutting, slicing, and so on). Ask, IS THERE ANY SPECIAL WOUND THAT WILL LET THE MOST MICROORGANISMS FIND THEIR WAY THROUGH YOUR SKIN AND INTO YOUR BODY? (puncture wounds)

Test types of dirt. Fingernail dirt, shoe dirt, desk dirt. Find out if some dirt is "dirtier" than others.

Germ story. Write a story about the invading germs and what can be done to stop them. Children will enjoy acting out the invasions of germs into the body and their subsequent destruction by antiseptic.

FOODS WE EAT

PLANT POWER

CONCEPT: We get energy from plants.

OBJECTIVE: Identify the parts of plants that store food and the parts we eat.

MATERIALS: A variety of foods, toothpicks

PROCEDURE: When teaching about foods, it is important to stress that they are necessary for growth and development. One way to do this is to compare children's bodies to the bean seeds in Unit 35. Ask, WHERE DID THE BEANS GET THE ENERGY TO SPROUT? (from the food stored in it) Remind children that young plants used the food stored in the seed to grow. Ask, CAN YOU GET ENERGY FROM SEEDS? (Yes, when you eat them.)

Show the children different plants and ask, WHERE DO YOU THINK GREEN PLANTS STORE THEIR FOOD? (leaves) Hold up carrots and turnip plants and ask, WHERE DO THESE PLANTS STORE THEIR FOOD? (roots) Show a picture of sugarcane plants (or a piece, if available) and ask, WHERE DO YOU THINK THE SUGARCANE STORES ITS FOOD? (stem, which is between the root and the leaves) Remember that the white potato is really a stem plant and the sweet potato is a root plant.

After you assess student progress, you may want to continue the lesson by introducing them to fruits, which store food for plants and flowers such as broccoli and cauliflower. Ask, WHERE DO YOU THINK THESE PLANTS STORE FOOD? (flowers)

EXTENSION: Reinforce the idea that energy comes from plants and their stored food with a plant-tasting party. You will need enough food samples for the entire class to taste. (You could ask each child to bring a sample.) Before the tasting party, have the children group plants by the parts that store food, wash the plants, cut them into bite-size parts, and put a toothpick in each piece. Ask, DO WE USE ALL OF THESE PLANT PARTS FOR FOOD? (yes, taste, seeds, leaves, roots, stems, fruits, and flowers and discuss favorite plant parts.)

FOOD GROUPS

CONCEPT: Foods from the four basic groups provide the nutrients necessary for good health.

OBJECTIVE: Identify the four groups as dairy products, meat and egg, breads and cereals, and fruits and vegetables. A possible fifth food group could contain foods that are high in fats and carbohydrates and contain little nutrients.

MATERIALS: Magazine pictures of food

PROCEDURE: Have children cut out magazine pictures from each food group. Mount the pictures on construction paper, add identifying food group labels, and pin to a bulletin board. Let the children decide if the pictures are labeled correctly. Move misplaced pictures to the correct categories.

EXTENSION: *Bag it.* Label lunch bags for each food group. Then, cut out and laminate pictures of food. Have children sort the cutouts into the correct bags. You can use this idea in a center or as an interactive bulletin board.

Food group biography. Choose a group and write your biography. If you were a food group, what one would you be? What would it be like to be this food group? Does your food group have a future?

Sell your food group. Make a television or radio commercial that will persuade the class that your food group is worthy of being eaten every day.

ADDITIONAL FOOD ACTIVITIES

1. *Fatty foods*. Ask, HOW CAN YOU FIND OUT IF FOOD HAS FAT IN IT? Place a drop of liquid fat on a piece of brown paper. Hold the paper up to the light and ask, DOES THE LIGHT COME THROUGH WHERE THE FAT IS? After a few hours ask, IS THE SPOT DRY OR DOES IT STAY THE SAME? (stays the same) Test the fat in foods such as peanut butter, nuts, olives, and milk by leaving the food on brown paper for a few hours. Then ask, IS THE SPOT DRY OR DOES IT STAY THE SAME? After testing several foods ask, HOW DO YOU KNOW THERE IS FAT IN THESE FOODS? (fatty oils stay on the paper)
2. *Sugar time*. Show the children a variety of food containers (jars, boxes, and so on) Have them predict which foods contain sugar. After they write their predictions on a prediction sheet, have them read the ingredients on the food containers to check their predictions. Explain that the ingredients are listed in sequence from greatest amount to smallest. Also note information on vitamin, mineral, and calorie content. Compare results. Then, write the word "sugar" on the chalkboard. Ask, DOES THE LIST ON THE LABEL SAY

SUGAR? (not always) Introduce the children to some of the ways sugars are listed: corn syrup, maltose, sucrose, fructose, corn sweetener, syrup, dextrose, glucose, lactose, molasses. Discuss the amount of sugar in foods and decide if eating a lot of sugar is a good idea.

3. *Classify foods.* Give children practice in identifying which foods belong in which food group by making charts and food group books. This can be done by using pictures, classifying individual children's diets, analyzing school lunch menus, the contents of bag lunches, and the evening meal.
4. *Finding out.* Divide the class into small groups and give each questions to explore about different food groups. Make a web of their findings as they share with the class (Figure 38–4).
5. *Foods I like chart.* Have children write the names of the foods they would like to eat for breakfast, lunch, and dinner on a chart. Then, make a chart and list the foods that they actually eat at these meals. Compare the charts and identify foods that are high in fats, carbohydrates, and sugar (potato chips, etc.). Discuss the nutritional value of the foods. Children might want to keep a log of foods that they eat for breakfast over a week's time (or food eaten for lunch or dinner). Compare class eating habits.
6. *Visitors.* Invite a cafeteria worker to talk about cleanliness and a nutritionist to explain how the lunch menus are created.
7. *Tasting party.* One effective method of evaluating food learning is to have a tasting party in which the children select their meal from a variety of foods. Simply observe which foods the children select to make up the meal.

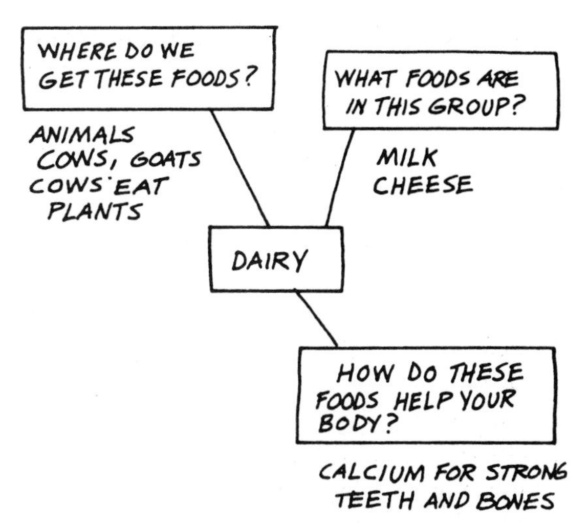

Figure 38–4 Web

8. *Recipes.* Integrate social studies by tasting foods from different countries or from different areas of town. Families also cook in different ways. Discuss similarities and differences and make recipe books of favorite recipes. Have parents or other guests from a variety of ethnic backgrounds make one of their special dishes or share some special food with the class (refer to the many food activities in previous units).

We Are What We Eat

Study vitamins and minerals as far as the developmental level of your students permits. Here are a few ideas to get you started.

KEEPING BONES AND TEETH STRONG

CONCEPT: Food supplies the body with important minerals. Calcium is a mineral that is used to make bones and teeth strong.

OBJECTIVE: Test rocks for the presence of calcium. Observe evidence of calcium.

MATERIALS: Pieces of limestone, chalk, marble, vinegar, hand lenses, medicine droppers

CAUTION: Hydrochloric acid is usually used for this test. However, when working with young children, use vinegar.

PROCEDURE: Show children the rocks and ask, WHAT ARE ROCKS MADE OF? (Many answers; one will probably be minerals.) Tell the children that they can test for the presence of a mineral called calcium by dropping acid or vinegar on a rock.

Give children samples of limestone. Ask, WHAT DO YOU THINK WILL HAPPEN IF WE DROP VINEGAR ON A ROCK? Instruct the children to fill the medicine dropper and carefully drop the vinegar on the rocks. Have the children observe and record what happens to the rocks (bubbles form). Generate a list of words that describe children's observations (fizz, bubble, smell, and so on). Explain that the bubbles indicate calcium is present (Figure 38–5).

EXTENSION: Children enjoy comparing the hard parts of their bodies to rocks. Ask, HOW ARE BONES AND TEETH LIKE ROCKS? (They both contain minerals.) Explain that two minerals, calcium and phosphorus, combine to make bones and teeth hard. Good sources include milk and other dairy foods, fruits, and vegetables. Other foods, such as eggs, leafy green vegetables, meats, grains, and nuts, have some calcium. Bring an end to the lesson by asking, WHERE DO WE GET THE CALCIUM TO MAKE OUR BONES AND TEETH STRONG? (from our foods, calcium)

Chicken bones. Testing the brittleness of chicken bones can help children understand that bones are more likely to break if they do not have the necessary minerals. To conduct this test, scrub two large chicken bones thoroughly. Soak one in water and one in vinegar for several days. Have the children compare the bones with hand-held lenses and note any differences. Then, predict which of the two bones will break more easily. (vinegar dissolves calcium and phosphorus) Compare the chicken bones to human bones.

Tear or crush? Have the children examine their teeth with mirrors and identify different sizes and shapes of teeth. How are the front teeth shaped? Ask, WHEN DO YOU USE YOUR FRONT TEETH? HOW DO YOUR FRONT TEETH HELP YOU EAT? WHAT TYPES OF FOOD DO YOU NEED TO CRUSH TO EAT? WHICH TEETH HELP YOU CRUSH FOOD?

This would be a good time to discuss carnivorous and herbivorous animals. Show pictures of the animals eating. Ask, WHAT TYPE OF TEETH DO THESE ANIMALS HAVE? After categorizing animals (and dinosaurs) by what they eat and discussing the shape and use of teeth, ask WHICH CATEGORY DO WE BELONG IN? (both) WHY? (We have several types of teeth. We eat both plants and animals.)

How do minerals get into food? To show how the roots of vegetables take in minerals from the soil, mix red food coloring and water. Have children observe the small rootlets that grow from the core of a carrot. Then, submerge the carrot in dyed water for twenty-four hours. Ask, WHAT HAPPENED TO THE RED COLOR? (passes to the core of the carrot) HOW DID IT GET THERE? (through the rootlets)

You may want to set up this observation with several different vegetables. Children will discover that dissolved minerals come into plants through plant roots. Let them decide that as we eat the plants, we take in the same minerals (Figure 38–6).

Then ask, IF PLANTS GET MINERALS FROM SOIL WATER, CAN WE GET MINERALS FROM WATER TOO? (yes) Minerals in water are sulfur, chlorine, sodium, calcium, magnesium, potassium, fluorine. Discuss how we get minerals from water (drink, eat, plants, animals).

Figure 38–5 What will happen when vinegar is dropped on limestone?

Figure 38–6 Why is the celery turning red?

ADDITIONAL MINERAL ACTIVITIES

Iron. Iron is a mineral that you need in your blood. The iron combines with protein to form a compound that makes blood red. This compound also carries oxygen to all parts of your body. Meat, beans, peas, grains, and leafy vegetables are good sources of iron. Children will enjoy acting out the oxygen processing function of iron. Ask, HOW DO WE GET IRON INTO OUR BODIES?

Children can "become" blood platelets, oxygen, and food by wearing appropriate labels and drawings. Make a vein path with red yarn for the children to follow as they trace the journey of a blood platelet. Have children create a script about how iron enters the body and what it does to help them stay healthy.

Anemia. You can get sick if you do not get enough iron. Ask, DO YOU EAT FOODS THAT GIVE YOUR BODY IRON? Discuss foods that contain iron. Explain that when you have had a good night's sleep but still are tired all the time, you may be anemic. People with anemia get tired easily because their blood does not carry enough oxygen. Ask, WHAT CAN YOU DO TO AVOID GETTING ANEMIA? (Eat foods that contain iron.)

Plan a meal. Have children plan a meal for someone with anemia. If you are studying vitamins, children could plan a meal for someone with beriberi. (People with beriberi lack essential vitamins and minerals and are too weak to use their muscles to do work.)

Mineral people. Study the labels of common foods. Categorize cans and boxes of food by the minerals they contain. Then make mineral people such as an Iron Person or Calcium Person. These figures can be made from the actual foods or boxes or pictures of these foods (Figure 38–7). Then, decide what foods are needed to create Vitamin People.

STRATEGIES FOR TEACHING THE HUMAN BODY

The internal anatomy of the human body is a difficult concept for young children to understand. Concrete experiences on which to base understandings can be difficult to provide. Refer to Unit 26 to review the garbage bag strategy for introducing children to the inside of the human body. In addition to the suggestions in Unit 26, the following suggestions will help you provide the children with concrete learning strategies.

Inside of Me

Although it is difficult to observe the inside of the human body, children can infer that muscles, bones, and organs are inside of them. For example, bones give the body shape, help the body move, and protect the organs. Preliminary activities should include body awareness activities in which children explore, feel, and identify some of the major bones. Observation of X-ray films can further enhance the children's concept of bones. Then, trace a Halloween skeleton, cut out the bones, and assemble the skeleton using paper fasteners as joints. Try some of the following strategies for teaching about bones with children:

Our skeleton has joints. Have the children move their arms and legs in a way that explores the way joints work: marching, saluting, swinging arms and legs, and acting like a windmill. Ask, DO YOUR LEGS BEND WHEN YOU MARCH? CAN YOU MARCH WITHOUT BENDING YOUR LEGS? IS IT EASY TO DO? Point out and explore the joints that help your body move. Discover the difference between hinge joints, which bend one way, and ball and socket joints, which can bend, twist, and rotate (Figure 38–8).

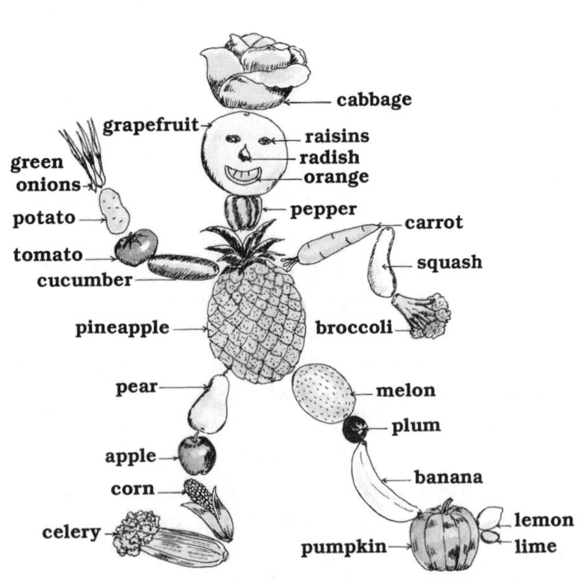

Figure 38–7 Create a mineral person

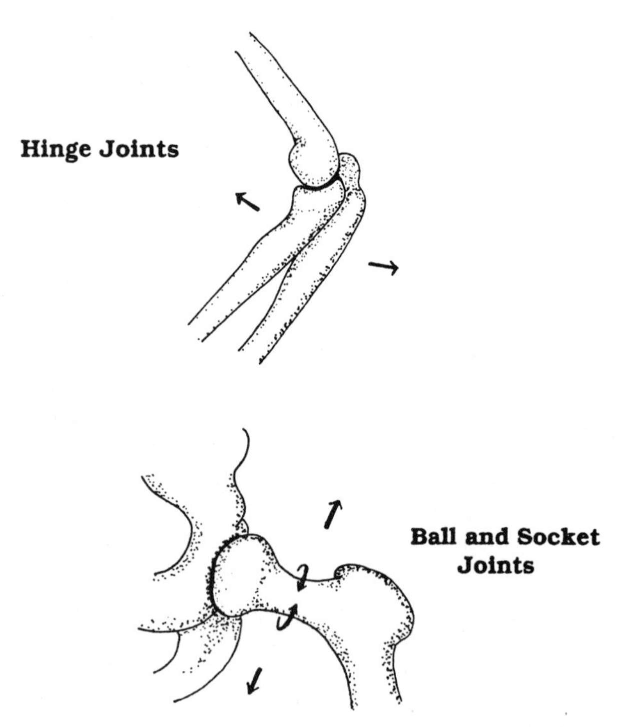

Figure 38–8 A ball and socket joint and a hinge joint

Find the joints. Have children work in pairs to make a paper body and locate joints. Have one child lie on a piece of butcher paper and the other child draw the outline of the body. Then, have the children locate joints on the paper body. Discuss why the type of joint used in the body is the most appropriate. Ask, WHAT WOULD HAPPEN IF WE HAD A BALL AND SOCKET JOINT IN OUR KNEES?

Writing about bones. Explore the advantages and disadvantages of different types of joints by having children write stories about children with joints in different places. For example, if a child had a hinge joint where a shoulder joint is needed, how would the child move? What would a day in the life of that child be like?

Do chickens have joints? Find out if the children think that a chicken has joints. Then, make chicken bones available to demonstrate how joints work. After children have had time to explore the chicken bones, ask, HOW IS YOUR SKELETON LIKE A CHICKEN SKELETON? (both have joints and other similarities)

Finger bones. Have children place their fingers over the light of a flashlight. Ask, WHAT DO YOU SEE? HOW CAN YOU TELL THAT YOU HAVE BONES INSIDE OF YOUR FINGERS? Have a doctor visit your class and discuss the use of X rays. The children will enjoy seeing X-ray pictures of bones, especially broken bones. This would be a good time to discuss first aid for broken bones.

Bone music and art. Sing "Dry Bones." (The toe bone's connected to the foot bone, the foot bone's connected to the ankle bone, and so on.) Children can compare skeleton poster bones as they sing, or a model of the backbone can be made with yarn and pieces of macaroni.

Make a muscle. Young children enjoy flexing their arm muscles and feeling the muscles they can make, but they might think that arm muscles are the only ones they have. Encourage them to discover other muscles by raising their heels off the ground in a tiptoe position. Ask, CAN YOU FEEL YOUR MUSCLES MOVE AS THEY DO THIS? Have the children lie on mats, stretch out like a cat, then curl up into a ball. Ask, WHICH MUSCLES CAN YOU FEEL NOW?

Explore facial muscles by wiggling noses and raising eyebrows. Ask, DOES YOUR TONGUE HAVE MUSCLES? (yes) Give children time to find unexpected muscles in their bodies. Then, apply some of the bone and joint learning experiences to teaching about muscles.

All About Me

Children like to know that they are the same as others and that they are different, too. Discuss the similarities and differences in people that make them who they are. Give children opportunities to explore their bodies and what makes them an individual.

Exploring the body: Using hands and feet to help us move. Ask children to find out how many things they can do with their hands such as wave, rub, scratch, or make a fist. Then, find out how many things can be done with their feet: skip, hop, walk, tiptoe. Have children write or dictate a list of their findings. Ask, CAN YOU THINK OF SOMETHING THAT YOU CANNOT DO WITH YOUR HANDS OR CANNOT DO WITH YOUR FEET? HOW IMPORTANT IS YOUR THUMB? Have the children do an activity without using their thumbs.

Paper bodies. Full-sized paper outlines of a child's body can be used as a starting point for many learning experiences. Partners take turns tracing around each other on butcher paper with a crayon (Figure 38–9). Then, each child cuts out his or her body by following along the lines. Examples include:

- *My body.* After the paper body is made, have the children study their faces in a mirror and draw their features and clothes on one side of the paper body. The other side of the paper can be used to express yet

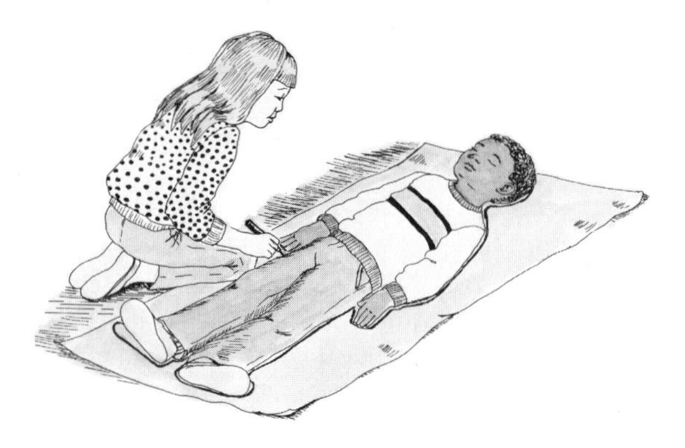

Figure 38–9 Make a paper body

a different look. Children will enjoy creating a fantasy body or storybook character on their own body shape.

- *Math body.* The cutout paper body can be an easy way for children to measure parts of their body. They can measure length of body, arms, and so on, and add weight and physical characteristics on a chart.
- *Action body.* After children are familiar with body tracing, let them trace each other in action poses and guess who is doing what and with what. Have magazines available for references of how a dancer uses muscles, how joints bend as an athlete runs, and how joints are used to sit in a chair.
- *I am curious.* Have children fill the paper body with magazine pictures depicting things that they like to do, are curious about, or want to learn about. Be careful not to limit paper body collages or subsequent "All About Me" books to simple biographies or listing of facts. This is not just a consumer activity or one about children's own acquisitiveness, but one that extends the exploration to include activities that enhance self-esteem. Children's interest in the world around them is worth learning more about. Ask, WHAT WOULD YOU LEARN MORE ABOUT? FIND THINGS THAT YOU ARE CURIOUS ABOUT. Make I Am Curious bodies, and the like.
- *Body mobile.* Separate this body at the joints and attach the body parts with yarn. These body mobiles can hang from the ceiling.
- *Body parts.* Cut up paper bodies can be added to a bulletin board or center. The body part bulletin board features an envelope of body parts that are assembled with thumbtacks. Labels may be added to name the parts, bones, joints, or organs that are being emphasized (Figure 38–10).

Figure 38–10 A body part bulletin board

Senses

Studying the senses and how they can be used to gather information is a natural exploration for young children. Refer to Units 16, 24, and 26 for exploration ideas and to the senses learning center in Unit 40 for suggestions. The following sense walk experience is a useful beginning or extension for an ongoing study of the senses.

Sense walk. Ask the children to be absolutely silent as they take a walk around school. Plan to pass the lunchroom, office, custodian's area, gymnasium, other classrooms and special area rooms, such as art and music. Instruct children to use their senses to notice as many things as they can. When you return to class, make a class chart or web of all the things noticed on the walk, such as different types of noises, smells, textures, sounds, and activity. Ask questions to encourage thinking, such as, DID THE FLOOR OF THE OFFICE FEEL LIKE THE FLOOR OF THE CLASSROOM? On a different day, take the class on an outside walk and brainstorm an experience web with the children to reinforce observations (Figure 38–11).

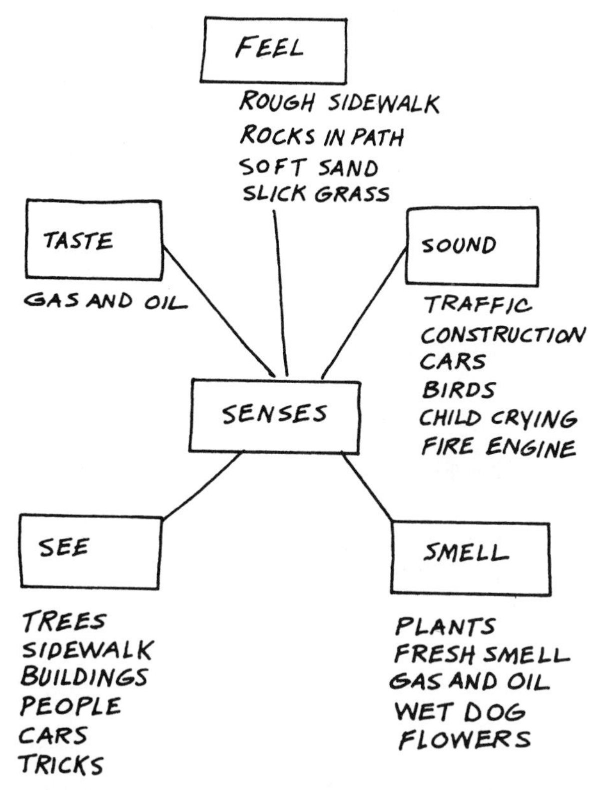

Figure 38–11 Brainstorm with an experience web

SUMMARY

Good health and nutrition habits begin early in a child's life. In order for children to begin to understand what it means to be healthy, they need to be exposed to a variety of learning experiences. These experiences need to be integrated into the child's life to have meaning and be as concrete as possible for better understanding.

Although children are curious about their bodies and how they work, their knowledge about internal anatomy is limited and especially difficult for children to understand. Concrete experiences on which to base learnings about bones, muscles, and organs must be provided to ensure understanding.

MEDIA RESOURCES

Body Movement, Health and Your Body. Encyclopedia Britannica Educational Corporation, Chicago, Ill. Each set contains six films with posters and teacher's guide.

Slim Goodbody: Your Body, Health, and Feelings Module. A multimedia kit from the Society for Visual Education, Chicago, Illinois.

COMPUTER SUGGESTIONS

Bones and Muscles: A Team to Depend On. Scholastic, Inc. (Grades 2–up).

Body-Awareness. Learning Well. (Prekindergarten–grade 2).

Incredible Lab. Sunburst. (Grade 3).

FURTHER READING AND RESOURCES

Abruscato, J. 1988. *Teaching children science.* Englewood Cliffs, N.J.: Prentice Hall, Inc.

Allison, L. 1976. *Blood and guts.* Boston: Little, Brown, and Company.

Bershad, C., and Bernick, D. 1979. *Bodyworks: The kids' guide to food and physical fitness.* New York: Random House.

Carter, C. D., and Phillips, F. K. 1979. *Activities that teach health.* New York: Instructor Publications, Inc.

Harlan, J. 1988. *Science experiences for the early childhood years.* Columbus, Ohio: Charles E. Merrill Publishing Co.

Hendry, L. 1987. *Foodworks.* Menlo Park, Calif.: Addison-Wesley.

Lind, K. K. 1985. The inside story. *Science and Children.* 22. (4): 122–123.

——— 1983. Apples, oranges, baseballs and me. *Understanding the healthy body.* (pp. 123–125). Columbus, Ohio: SMEAC Information Reference Center.

Katz, L. G. 1988. *Early childhood education: What research tells us.* Bloomington, Ind.: Phi Delta Kappa Educational Foundation.

Science and Children. 1982. Skeleton Poster. Washington, D.C.: National Science Teachers Association.

Stronk, D. 1983. *Understanding the human body.* Council for Elementary Science International Sourcebook. Columbus, Ohio: SMEAC Information Reference Center.

Wassermann, S., and Ivany, J. W. G. 1988. *Teaching elementary science.* New York: Harper & Row.

SUGGESTED ACTIVITIES

1. Assemble materials for one of the suggested learning activities. Try out the lesson with a small group of primary age children. Report to the class on what you did and how the children responded. How did you evaluate the children's responses? Would you change the activity if you were to teach it again?
2. Design a nutrition learning strategy for your Activity File. Which process skills will you emphasize? How many areas of the curriculum can you integrate into this lesson?
3. Develop your own teaching web. Select lessons from this unit and design a teaching unit that is appropriate for a group of children you are teaching.
4. Prepare a learning cycle lesson to teach a health, nutrition, or human body concept. Teach it to a group of children and report to the class. What was the advantage of using the learning cycle? Would another teaching strategy have worked better with this concept or group of children?

REVIEW

1. List two reasons health and nutrition education should be an essential part of the science curriculum in the primary grades.
2. Define health education as it applies to primary grade science.

SECTION VII

The Math and Science Environment

UNIT 39 Materials and Resources for Math

OBJECTIVES

After studying this unit, the student should be able to:

- Categorize math materials in six levels from concrete to abstract
- Set up a math learning center
- Select appropriate math materials applying educational criteria
- List examples of basic math materials that can be used to support specific math concept development

A variety of materials is important to help children learn math. The teacher can select materials that are easy to find, are safe, and can be used in many ways. Commercial materials and math kits can be purchased. However, math materials do not have to be expensive to be useful. Many math materials are items found in the home or available free from a local resource.

In Unit 3, six categories of materials were described: real objects, real objects used with pictorial representations, two-dimensional cutouts, pictures, wipe-off folders, and paper and pencil. These categories follow a developmental sequence from the concrete manipulative to the abstract representational. Preoperational children work with only the first four types of materials. During the transition to concrete operations, the last two categories may be available for those children who can deal with them. All through the concrete operations period, new concepts and skills should be introduced with concrete manipulative and pictorial materials before moving on to the abstract representational (Figure 39–1).

Many kinds of concrete manipulative materials have been introduced throughout the preceding units. Some are very versatile and others serve specific functions. Pictorial manipulatives and other picture materials have also been suggested. Children's picture books are an especially rich source of pictorial and language information as was suggested in Unit 15. Stories, poems, and pictures enrich the math curriculum. These materials help teach math vocabulary, illustrate the use of math in a variety of settings, and expand children's ideas of how math can be used. Books should be carefully selected. Be sure the illustrations accurately portray the concepts the book purports to help teach. As noted in Unit 24, special care must be taken when selecting counting books because the illustrations are frequently inaccurate in their depiction of the set/symbol relationships. The teacher should ask, "Which concept or concepts are illustrated in this book? How will reading this story or poem help Richard, Liu Pei, or Mary to better understand this concept?" Books should have good artwork and be colorful and well written. For example, an excellent children's book that includes math concepts is *Over in The Meadow* by Ezra Jack Keats. The rhythm of

"Over in the meadow, in the sand, in the sun,
Lived an old mother turtle and her little turtle one.
"Dig!" said the mother.
"I dig," said the one.
So he dug all day,
In the sand, in the sun"

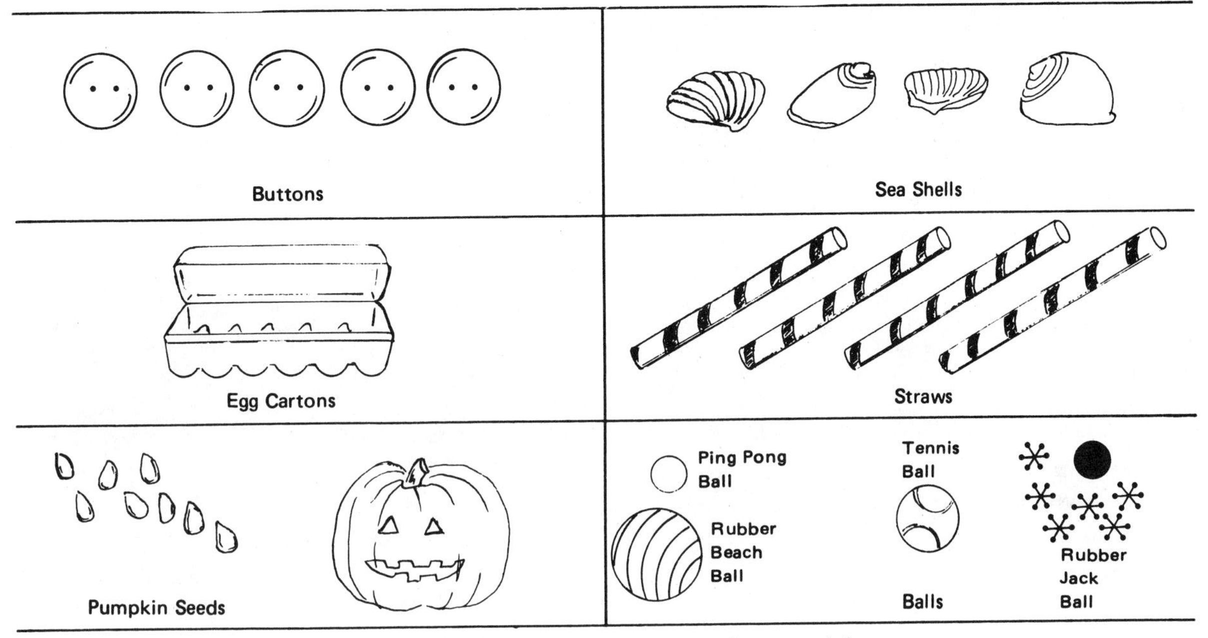

Figure 39–1 Inexpensive math materials

is appreciated by the teachers as well as the children. Appendix B contains a list of concept books suitable for young children.

The flannel board and various two-dimensional felt pieces are excellent vehicles for working with abstract symbols without having to be concerned with paper and writing implements. For example, Mrs. Jones is not sure if Mary understands sets matched to numerals. She cuts the numerals zero to nine from felt and forty-five felt pumpkins. On the flannel board, Mrs. Jones places the numerals 1, 2, 3, and 4. She gives Mary some pumpkins and tells her to put the appropriate number of pumpkins by each symbol. If Mary can match the correct amounts one to four, then Mrs. Jones can go on to five to nine.

MATH LEARNING CENTER

Math materials should be displayed so that children can easily see them, use them, and put-them away. Too many math materials stacked in an unorganized pile limits their use. It is better to place the materials side by side in baskets or other open containers on low shelves in the math learning center (Figure 39–2). This tells the child that these materials are available and can be used. This helps children learn proper replacement locations.

Teachers should change some of the math materials according to the needs of individual children. This may be done on a daily, weekly, or monthly basis. Rotating special materials is called creating a responsive environment. A responsive environment is one that continually challenges and interests children. Teachers should observe which math materials are being used. They should also take note of which materials are selected by specific children. New materials can be introduced as needed by individuals or small groups. Children can be encouraged to try out new materials if the teacher sits down and uses the material with them. Adult-child interaction can be the stimulus for child-child in-

Figure 39–2 Open containers and low shelves invite children to use the math materials.

Figure 39–3 This math center focuses on measurement activities and materials.

teraction. After the adult teaches one or more children to play a game or use a new type of material, these children can then teach others how to use the material or play the game.

Math centers can also focus on specific concepts when these concepts are being introduced. For example, Figure 39–3 shows materials in a measurement center. Focused centers can be set up for any of the skills and concepts in the text (Figure 39–3). Notice the signs that call attention to the print associated with measurement language. Math centers can be set up relative to available space and furnishings. A center might be a set of shelves with an adjacent table and chairs or carpeted area, a small table with a few different materials each day, or a space on the floor where math materials are placed. In one primary classroom in Guilderland, New York, a math bed was installed. A blackboard was at one end and shelves for books and materials at the other end: a comfortable and inviting math center.

The math center should be available to every child, every day. Too frequently, the math center is open only to those children who finish their workbook and other reproducible paper assignments. Unfortunately, it is probably the children who do not finish their paper and pencil work who are most in need of experiences with concrete manipulative materials. If workbooks are required, after they have been introduced, put them on a special shelf in the math center and let the children who are ready select them from among the other materials available.

Many teachers, especially in the primary grades, feel pressured to follow the textbook and teach whole group math lessons without the use of manipulatives to introduce concepts as described throughout this text. It is possible to find a compromise that is beneficial for both students and teacher. For example, one first-grade teacher, feeling discouraged with the progress her students were making in mathematics, decided to try a new approach: She took the objectives for the text assigned to her class and found developmentally appropriate activities that correlated with each one. She selected the activities from *Developing number concepts using Unifix® cubes*, *Mathematics their way*, *Workjobs*, and *Workjobs II*. She then assembled the necessary materials. For each concept, she introduced it to the whole class and then divided the class into small groups to work with the concrete materials. The children felt more satisfied, performed their tasks with enthusiasm, and even commented on how much fun it was not to have to write! The teacher had time to circulate and help individual children. A positive side effect emerged from the opportunity to 'talk' math; The children's language was extended and unclear points were clarified. Peer tutoring developed naturally, to the advantage of both the tutors and those who needed help. The games and manipulatives were available for further use during the children's free choice time. The teacher's enthusiasm and the enthusiasm of the students was picked up by the other first-grade teachers and the principal. The teacher was asked to inservice the other first-grade teachers so that they could adopt the same system.

SELECTING MATH MATERIALS

Each child has a variety of needs and a unique potential in math development. Every school and classroom includes children who are at various levels of concept development. It is important to provide a wide array of math materials and time to explore them freely.

Early math materials are those that provide firsthand sensory experiences for touching, tasting, hearing, and seeing. The importance of allowing young children to explore materials in a richly provided environment cannot be overemphasized. Exercise of small muscles and fine motor control occurs with peg boards, bead stringing, tinker toys, sequence cards, and stack clowns.

In Unit 3, three considerations regarding selection of math and science materials were discussed:

- The materials should be sturdy and versatile.
- The materials should fit the selected objectives.
- The materials should fit the developmental levels of the children.

In addition, the following should be considered:

- The materials should be safe.
- The materials should be easily supervised.

Materials for young children should be stimulating. This is facilitated through appropriate size, height, and weight of the equipment. They should be "child-size." The cost of a material is related to its use. Is the material a special piece of equipment used only once a year? Or is the material one that is used daily? Blocks, for example, are used daily by children. All equipment must be safe for children to use. Hazardous features, such as sharp or pointed edges that could injure a child, should be avoided. Painted surfaces should have a nontoxic and nonflammable paint. Paper materials should be laminated to strengthen them. Wooden items are desirable as long as they have smooth edges. Metal materials can rust. Plastic items can crack. Quality math materials should serve more than one purpose. For example, blocks can be used in many ways. Placement of materials is an important consideration for supervision. Climbing bars in the outdoor play space should be located where there is room for the children to explore.

Before deciding what new math materials to obtain, the teacher lists what is already available. She should make notes as to the condition of current math materials. Are Lego pieces missing? Do games need mending? After this inventory, a list is made of the most needed materials to collect or purchase first. The teacher now considers the children's needs, both individually and as a group (Figure 39–4). She also considers the indoor and outdoor learning environment and the amount of room available. She considers the math objectives of the program. She considers how the materials will be used. Are they for naturalistic, informal, or structured activities?

After these decisions are made, the teacher

Figure 39–4 Children help each other learn

is ready to purchase the math materials. In preprimary, when children have shorter attention spans, it is recommended that a variety of small sets of materials (enough for two or three children to share at one time) be purchased. In the primary grades, start with a classroom set of Unifix materials. These materials are among the most versatile because they can be used for teaching almost every primary mathematics concept. Many accessories are available and there are numerous resources, already mentioned in this text, from which to select activities. Gradually add Base Ten Blocks, fraction materials, and other manipulatives.

Another source of materials are those that can be made by teachers and/or volunteer parents. For example, beanbags are easy to make. Scrap materials can be used, as shown in Figure 39–5. Waste materials may also be donated. Parents can be asked to save egg cartons, buttons, boxes and other containers, bottle caps, yarn, ribbon, and other materials that can be used in the math program (see the list in Unit 43). Parents might also donate inexpensive items such as toothpicks, golf tees, playing cards, funnels, measuring cups, and so on. Lumber companies

1. Use scrap pieces of felt fabric or other sturdy material in bright primary colors of red, yellow, and blue.
2. Cut 2 each of the 4 basic shapes. Make the square 4" x 4". The circle, rectangle, and triangle should be similar in size.
3. Cut small black felt numerals 1, 2, 3, and 4. Cut small black felt shapes in numbers to match each of the number symbols (1 circle, 2 rectangles, 3 triangles, and 4 squares, for example).

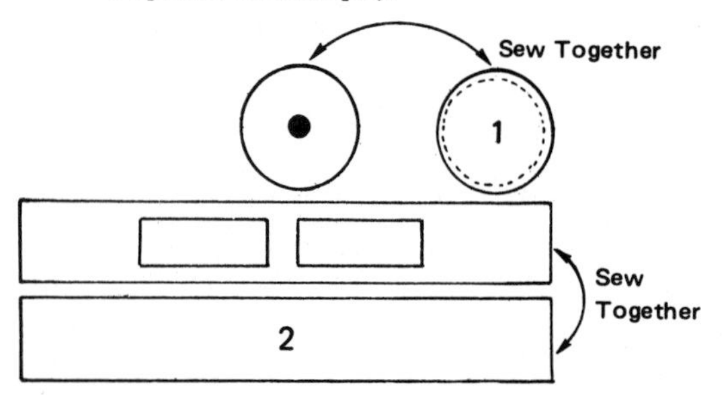

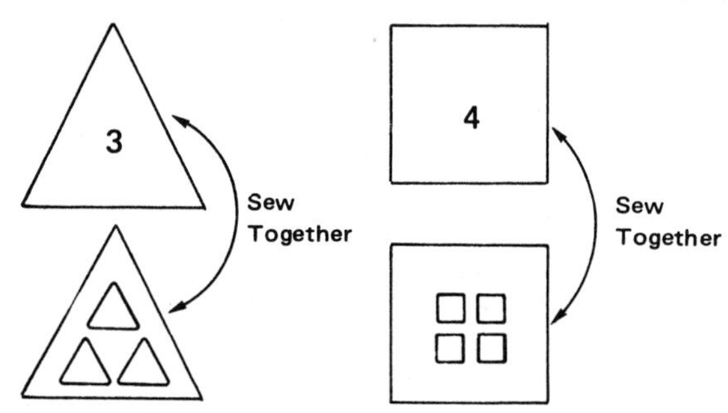

4. Machine sew one number symbol on one of each of the 4 basic shapes that were cut from felt earlier. Sew all the similar small felt shapes to each of the remaining 4 basic shapes, as shown.
5. Machine sew the matching shape pieces together. Stitch 1/4 inch from edge. Leave 1-inch opening. Stuff the bags with sawdust, cutup nylon hosiery pieces, or commercial pillow stuffings. Machine or hand sew the opening.

Figure 39–5 Number and shape beanbags

might donate scrap lumber. Restaurant supply companies will often sell teachers trays, various size containers, and the like at low prices. Try to convince your principal to let you spend your share of the supply money on concrete manipulatives and supplies for making two-dimensional manipulatives rather than investing in workbooks and ditto paper. Some resources for ideas for teacher-made materials are listed in Figure 39–6.

When the needs are clear and decisions have been made as to which math items can only be obtained by buying them new, the teacher looks through catalogs or visits local school supply stores. There are many quality resources. It is important to look at several catalogs. Although supply companies offer similar items, each company also has special math items. Some well-known catalogs are listed in Figure 39–7.

Baratta-Lorton, M. (1972) *Workjobs*. Menlo Park, CA: Addison-Wesley.
Baratta-Lorton, M. (1975) *Workjobs for parents*. Menlo Park, CA: Addison-Wesley.
Baratta-Lorton, M. (1976) *Mathematics their way*. Menlo Park, CA: Addison-Wesley.
Baratta-Lorton, M. (1978) *Workjobs II*. Menlo Park, CA: Addison-Wesley.
Carson, P. and Dellosa, J. (1977) *All aboard for readiness skills*. Akron, OH: Carson-Dellosa Publishing.
Carson, P. and Dellosa, J. (1979) *Holiday learning activity ideas*. Akron, OH: Carson-Dellosa Publishing.
Dellosa, J. & Carson, P. (1981) *Buzzing into readiness*. Akron, OH: Carson-Dellosa Publishing.
Englehardt, J.M., Ashlock, R.B., & Wiebe, J.H. (1984) *Helping children understand and use numerals*. Boston: Allyn & Bacon.
Kamii, C. (1985) *Young children reinvent arithmetic*. New York: Teachers College Press.
Platts, M.E. (1964) *Plus*. Educational Services, Inc., P.O. Box 219, Stevensville, MI 49127.
Richardson, K. (1984) *Developing number concepts using Unifix Cubes*. Menlo Park, CA: Addison-Wesley.
Scott, L.B., & Garner, J. (1978) *Mathematical experiences for young children*. St. Louis: McGraw-Hill.

Figure 39–6 Resources for ideas for teacher-made materials

- American Guidance Service, Publishers Building, P.O. Box 99, Circle Pines, MN 55014
- Childcraft; 20 Kilmer Road, Edison, New Jersey 08817
- Community Playthings; Rifton, New York 12471
- Creative Publications; P.O. Box 10328, 3977 East Bayshore Road, Palo Alto, California 94303
- Constructive Playthings, 1227 East 119th. Street, Grandview, MO 64030 (1-800-255-6124; in MO 816-761-5900)
- Dale Seymour Publications, P.O. Box 10888, Palo Alto, CA 94303 (outside CA, 1-800-USA-1100; in CA 1-800-ABC-0766)
- DLM Teaching Resources, P.O. Box 4000, One DLM Park, Allen, TX 75002 (1-800-527-4747; in TX 800-442-4711)
- Educational Teaching Aids (ETA), 199 Carpenter Avenue, Wheeling, IL 60090 (800-445-5985)
- Ideal School Supply Company; 11000 South Lavergne Avenue, Oak Lawn, Illinois 60453 (Early Learning Special Education) (1-800-323-5131; in IL 312-425-0800)
- Judy Co.; 310 N. second St., Minneapolis, Minnesota 55402
- Kaplan School Supply Corp., 1310 Lewisville-Clemmons Road, Lewisville, NC 27023 (East: 1-800-334-2014 in US; 1-800-642-0610 in NC) (West: 619-450-2172; 1-800-433-1591 in CA)
- Lakeshore Curriculum Materials Co., 2695 E. Dominguez St., P.O. Box 6261, Carson, CA 90749 (800-421-5354 in US; 800-262-1777 in CA; 213-537-8600 in Los Angeles)
- Nasco, 901 Janesville Ave., Fort Atchinson, WI 53538; Nasco West, 1524 Princeton Ave., Modesto, CA 95352 (1-800-558-9595)
- Nienhuis Montessori U. S. A., Inc., Mountain View, CA 94041 (415-964-2735)
- Rigby (For Math Big Books), P.O. Box 797, Crystal Lake, IL 60014 (1-800-822-8661)

Figure 39–7 Math materials catalogs

BASIC MATH MATERIALS

Basic math materials can be grouped according to their major use. However, as you have already seen, many are used to teach multiple concepts. Mathematics learning begins with play and exploration, proceeds to informal instruction, and then to structured activities. The following briefly summarize the materials suggested in each unit.

One-to-one correspondence (Unit 8) is the understanding that one group has the same number of things as another. Matching materials include wood inlay puzzles, play tiles, pegboards and pegs, picture lotto games, dominoes, fitting sets, snap beads, snap brads, wooden beads, and strings (Figures 39–8 and 39–9). Also, very useful are colored inch cube blocks and Unifix® cubes.

Figure 39–8 "Oh dear, will I get the right match?"

Figure 39–9 "Yeah, I did it!"

Number and counting (Unit 9) include the idea of what quantities are and rote and rational counting. Rote counting involves reciting the names of the numerals in order from memory. Rational counting involves attaching the name of each numeral in order to a series of objects in a group. Almost anything can be counted. Crayons, marbles, poker chips, sticks of all kinds, buttons, pegs, unifix cubes, shells, money, toothpicks, beads, dried beans or peas, popcorn kernels, small toys, counting frames, and inch cube blocks are examples. Flannel boards and flannel shapes or objects may be used for rational counting activities. Most of the above counting objects can be counted out in egg cartons or other containers that are marked with symbols.

Classifying (Unit 10) consists of sorting and grouping objects. Accessories that children like to classify are small farm animals, zoo animals, people, fruit, flowers, and play dishes. These can be made of wood, metal, rubber, or plastic. Buttons, hardboard geometric shapes, felt shapes, wooden beads, unit blocks, and color cubes also are good for classifying (Figure 39–10).

Comparing (Unit 11) is finding a relationship between two things or sets of things. All of the objects used for classifying can be used to make comparisons. Hoops or loops made from shoelaces, ribbon, rope, or string can be used to sort sets. Paper plates, pieces of construction paper, muffin tins, or boxes are also used to separate items.

Figure 39–10 "Teacher, these are wooden people."

Each thing the child meets in the environment has *shape* (Unit 12). Many commercial shape materials can be purchased. Shape sorting boxes and shape coordination boards are made by several companies. Clear shape stencils and size and shape puzzles are also available. Montessori materials include well-made shape cylinders and insets. Attribute blocks are excellent for shape sorting, matching, and labeling (Figure 39–11).

Space as a part of math and geometry was discussed in Unit 13. Construction materials help a child learn about space. Position, direction, and distance are manipulated by children when they build. Besides the basic unit blocks, which are discussed in Unit 41, there are many commercial

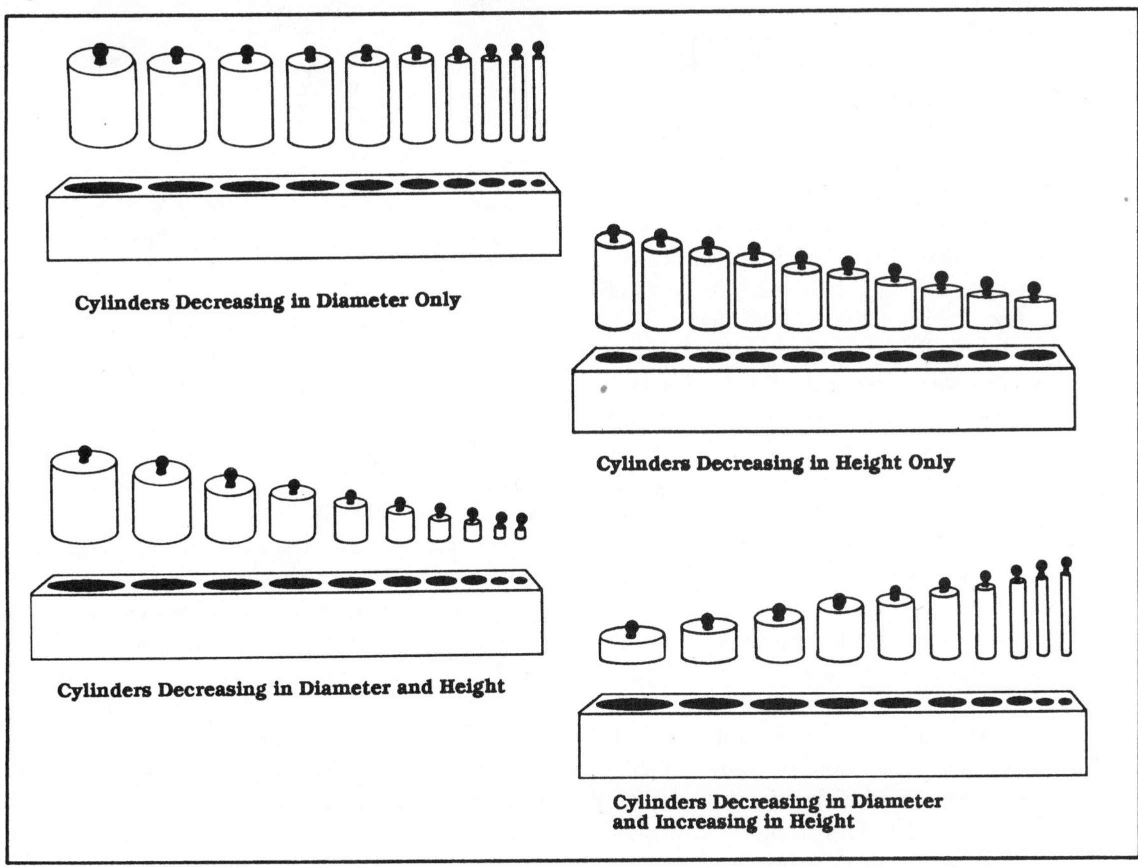

Figure 39–11 Montessori cylinder blocks with insets for ordering

construction sets for children. Space is also explored through organization and pattern materials. Some of these include geoboards, parquetry blocks, hexangle pattern set Peg-A-Pattern®, grid mosaic, play tiles, sewing boards, stringing rings, wooden beads and strings, or see-thru threading shapes.

The concept of *parts and wholes* is basic for learning about fractions (Unit 14). Many commercial companies produce lotto cards and puzzles that children can manipulate to learn about parts and wholes. Naturalistic experiences such as food preparation also help children understand this concept.

Ordering involves comparing more than two things or more than two sets. Ordering was discussed in Unit 17. There are many commercial ordering items. Nesting boxes, dolls, and boats are available. Color stacking discs, ring-a-rounds, a learning tower, pan pile-cups, handy boxes, and pagoda tower builder can be purchased. Home items are measuring cups and spoons as well as different sizes of containers.

Patterning (also discussed in Unit 17) is closely related to ordering. It involves making or discovering regularities. At the preprimary level, patterns can be made with color and/or number and various categories using cube blocks, Unifix® cubes, teddy bear, or other counters or almost any other type of objects (Figure 39–12). At the primary level, children move on to looking for more abstract patterns (Unit 28) using quantities and numerals. The same basic materials can be used in more complex ways (Figure 39–13).

Measurement was discussed in Units 18 and 19 at the exploratory levels, in Unit 25 at the level of nonstandard units, and in Unit 32 for the introduction of standard units. Many materials are discussed for nonstandard measure such as lengths of cardboard, ribbon, or string; objects such as paper clips, toothpicks, Unifix® cubes, books; containers of any kind; objects and substances that can be placed in a pan balance. For

Figure 39–12 Building with plastic bricks is a challenging activity.

standard measurement scales, tape measures, measuring cups, and other calibrated measures can be introduced. *Time* and *sequence* materials include a day-by-day calendar, clocks, timers, hourglass, and sundial. Sequence cards and seasons charts may be made or purchased. Play money and a cash register can be valuable for learning about the value of money.

At the primary level, the same basic materials are still used but for more complex, higher level activities. For constructing operations with *whole numbers* (Unit 27), many kinds of counters can be used. Teacher-made and commercial materials such as fraction pies can be used when working with *fractions* (Unit 29). Base ten blocks are excellent when students are introduced to place value (Unit 30), but other kinds of counters

Blockbusters	Lincoln Logs	Play Squares	Beam and Boards
Lego	Baufix	Octons	Habitat
Free Form Posts	Crystal Climbers	Girders	Busy Blocks
Sprocketeers	Rig-A-Jig	Play-Panels	Space Wheels
Tinkertoys	Color Cone	Wonderforms	Magnastiks
Toy Makers	Tectonic	Geo-D-Stix	Keeptacks
Cloth Cubes	Structo-Brics	Connector	Balancing H Blocks
Snap Wall	Giant Structo-Cubes	Wood'n Molds	Block Head
Lock & Stack Blocks	Floresco	Poki Blocks	Snap-N-Play Blocks
Giant Interlockers	Ring-A-Majigs	Multi-fit	Channel Blocks
Unifix Cubes	Crystal Octons	Disco Shapes	Wee Waffle Blocks
Unit Blocks	Ji-gan-tiks	Snap Blocks	Struts
Flexibricks	Mobilo	Bristle Blocks	Duplo
Play Shapes	Ring-A-Majig	Bristle Bears	Gear Circus
Stackobats	Locktagons	Brio Builder	Create It
Magnetic Blocks	Connect-A-Cube	Flexo	LASY Construction Kits
Form-A-Tions	Tuff Tuff Blocks	Giant Interlockers	Giant Double Towers
Poly-M	Polydron	Klondikers	

Figure 39–13 Construction materials for math.

can be used also. For *geometry* (Unit 31), geoboards provide many opportunities to explore geometry.

Hand calculators and computers have been recommended as support materials for many concepts. Calculators can be used to work on number patterns in early primary and later for making more complex calculations in a faster way. Computer software is becoming more available and more creative. It offers many opportunities for problem solving and reinforcement of basic concepts.

SUMMARY

Math materials can be classified into six categories that vary in degree of concreteness to abstractness. Math learning centers can be designed to fit the available space. The important features of the math center are that the materials should be developmentally appropriate and available for all the children in the class to use. Textbooks can be used as guides to scope and sequence without sacrificing developmentally appropriate instructional practices. Materials should be selected with care following the guidelines suggested in this unit. Materials can be purchased or made. The basic materials suggested and described in earlier units are briefly reviewed in this unit.

FURTHER READING AND RESOURCES

For Setting Up Centers

Carroll, J. 1983. *Learning centers for little kids.* Good Apple, Inc. Carthage, Ill.

Ziegler, N., Larson, B., and Byers, J. 1983. *Let the kids do it! Book 1: A manual for self-direction through indirect guidance.* Belmont, Calif.: Fearon Teacher Aids.

Ziegler, N., Larson, B., and Byers, J. 1983. *Let the kids do it! Book 2: Symbols and rebus charts.* Belmont, Calif.: Fearon Teacher Aids.

SUGGESTED ACTIVITIES

1. Visit a local educational materials store. Make a list of available math materials and their prices. Bring the list to class.
2. Participate in a small group in class and compare math materials lists. Make cooperative decisions as to what math materials should be purchased if a new prekindergarten, kindergarten, and/or primary classroom is to be furnished. Each group should consider cost as well as purpose.
3. Go to the library. Find and read at least ten children's picture books that contain math concepts (see the Appendix for suggestions). Write a description of each one. Tell how each book could be used with children.
4. Make two different "homemade" math resources that could be used with young children. Share the materials with the class. Be prepared to show the class how the resources can be made.
5. Visit two preschool, kindergarten, and/or primary classes. Ask to look at the math materials used. Diagram the math learning center or math shelves. Share the information with your class. Tell the strong and weak points of the math setup. List any changes which would make it better.
6. Add one file card that lists math materials to each of the activity units in your file.
7. Send for free commercial catalogs. Make a list of math materials that are new in each. Write down the descriptions presented in the catalogs. Share this information with your class.

REVIEW

A. Describe each of the following:

1. Concrete/manipulative math materials
2. Abstract/ representational

B. Respond to each of the following situations:

1. Teacher A has 6 lotto games, a geoboard with no rubber bands, and some odds and ends of blocks on a table in the corner of the nursery school. Teacher B has 2 lotto games, some buttons in a container, a stack clown, a wooden shape puzzle, and blocks. Which teacher has a better selection of math materials? Why?
2. Teacher C says she does not believe in teaching math to young children. Her preschool has a sand table, water table, and a woodworking set. Children are observed playing in the sand with measuring cups and bowls. Some children are also weighing plastic fruit on a balance scale in the housekeeping center. Is Teacher C teaching math? If you think so, explain how she really is teaching math to children.
3. Teacher D teaches first grade. She is going to set up a math center. She has been given \$250 to purchase basic materials. She asks you for help. What would you suggest?
4. Teacher E teaches second grade. She has

just read this book. She is trying hard to provide a developmentally appropriate math environment. However, she cannot decide how to begin. What suggestions would you give her?

C. Select appropriate math materials for the following.

1. A preschool center that already has picture books, pegboard and pegs, wooden beads and strings, unifix cubes, assorted blocks, a learning tower, and a woodworking area.
2. A kindergarten that already has picture books and ditto materials.

D. Match the correct items from Column II with each item in Column I.

I	II
1. Thermometer	a. Matching
2. Color stacking discs	b. Number and counting
3. Unifix® cubes	c. Sets and classifying
4. Dominoes	d. Comparing
5. Marbles	e. Shape
6. Plastic flowers	f. Space
7. Parquetry blocks	g. Parts and whole
8. Scale	h. Ordering
9. Geoboards	i. Measurement
10. Calendar	j. Whole number operations
11. Attribute shapes	k. Patterns
12. Boxes of different sizes	l. Fractions
13. Pictures of people in various jobs	m. Geometry
14. Big/small blocks	n. Measurement with standard units
15. A sandwich cut in half	o. Place value
16. Lego	
17. A meter stick	
18. Rectangular construction paper cut in three equal parts	
19. Base Ten Blocks	
20. Calculator	
21. Computer and software	

UNIT 40 Materials and Resources for Science

OBJECTIVES

After studying this unit, the student should be able to:

- Set up a science learning center
- Select appropriate science materials for teaching science
- List examples of basic science materials that can be used to support specific science concept development

Whether learning experiences are presented in an informal, unstructured, or structured approach (a learning cycle, discrepant event, or demonstration) or are used in learning centers or manipulated by an entire class at one time, manipulative science requires materials for the child to explore. Hands-on science requires that materials be handled, stored, distributed, and replaced whenever they are used. Do not be discouraged. In the long run, once the materials are accumulated and organized, less time is needed for teacher preparation because much of the classroom instruction will be carried out by the interaction of the child and the materials.

There are several general categories of science materials. The most complete listing of available materials appears each January in *Science and Children*, a journal of the National Science Teachers Association, available through NSTA Publications. This useful supplement answers the question, "Where do I go for help?" The publication is organized into four main sections: equipment/supplies, media producers, computers/software, and publishers. This unit offers suggestions for the selection of basic science materials and resources and preparing science learning centers. The guidelines for selecting math materials for young children can also be applied to selecting science materials. (Refer to Unit 39 for additional information.)

BASIC SCIENCE MATERIALS

There are two major types of science materials: those you purchase and those you "scrounge." Purchased materials include textbook publishers' kits, general kits, and items purchased at supply houses or local retailers. Materials that are scrounged or contributed by parents and other benevolent individuals are known by many teachers as "good junk." Regardless of how the materials are acquired, they must be organized and managed in a way that promotes learning. (Refer to Unit 34 for additional suggestions for classroom management.)

The Good Junk Box: Things to Scrounge

Many teachers rely on boxes of miscellaneous materials that have been gathered from many sources. Such a junk box comes in handy. Invite children, friends, business people, and others to add to your junk box. Once people know you collect odds and ends, they will remember you when they are ready to throw something away. For example:

- Glass containers and 2-liter bottles make good aquariums, terrariums, and places to display animals (Figure 40–1).
- Aluminum foil, pie plates, freezer food containers are useful for numerous activities.
- Film cans make smell and sound containers.
- Hardware supplies are always welcome for the tool center; plastic tubing, garden hoses, funnels are ideal for water play and making musical instruments.
- Candles, thumbtacks, paper clips, and sink and float items come in handy.
- Magnets from refrigerators for magnet experiences; old, leaky aquariums make good housing for small mammals or reptiles.
- Oatmeal containers make drums; shoe boxes are great for dioramas and general organization and storage.

Figure 40–1 A pop bottle terrarium *(Reprinted from* Science and Children *(1979). 16 (7), 47, with permission from the National Science Teachers Association, 1742 Connecticut Avenue, Washington, D.C. 20009. Gilmore, V. "Helpful hints-Coca-Cola bottle terrarium.")*

- Toys, clocks, and kitchen tools can be added to the machine center.
- Flashlights, batteries, and wire from telephone lines for electricity.
- Pipe cleaners are always useful for art; buttons and other small objects are needed for classifying and comparing.
- Straws, balloons, paper cups, pieces of fabric, and wallpaper are objects for the touch box.
- Some stores invite teachers to collect their old carpet and wallpaper sample books.
- Always keep an eye out for feathers, unusual rocks, shells, seed growing containers, plastic eggs—the list is endless.

Some teachers send home a list of "junk" items at the beginning of the year. Parents are asked to bring or send available items to school. Such a list will be easy to complete when you become familiar with "good junk" and have an idea of some of the items that you will use during the year. In addition, parents are usually responsive to special requests such as vegetables for the vegetable activities like those in Unit 26 or ingredients for the cooking activities in Unit 37.

Commercial Materials

Your school district might decide to purchase a kit from a publisher when selecting a textbook for teaching science. Publisher kits are available from most major companies and contain materials specifically designed to implement the activities suggested in the textbook. These kits can be helpful in providing the hands-on component of a textbook-based science program.

General kits are available from many sources and range in size from small boxes to large pieces of furniture with built-in equipment such as sinks and cabinets. General kits contain basic materials but may not be directed to your specific needs. An advantage of the boxes or rolling tables is that they are easy to circulate among teachers.

Specific topic kits such as "Mining, Minerals, and Me" from Delta Education (suggested in Unit 37) are boxed by topic and grade level. The idea is to provide the teacher with the materials necessary to teach a specific science topic. Teachers' manuals and materials allow teachers to use the kit as a supplement to a textbook or as the major means of teaching a science concept.

Advantages and Disadvantages of Kits

Most kits contain a limited amount of consumable supplies, usually just enough for a class to do all of the activities covered in the teacher manual. Reordering must be continuous for the kit to be used again. However, the kits also contain permanent supplies such as magnifiers, clay, small plastic aquariums, petri dishes, balances, and the like.

The major drawbacks to commercial kits of any type are maintaining the consumables in the kits and finding the money to purchase the kits. Kits are self-contained, complete, and ready for use; consequently, they are also expensive.

Purchased Equipment

There are some items that you might have to purchase to be an effective science teacher. This list includes magnifiers, eyedroppers, plastic tubing, mirrors, rock and mineral set, magnets, and batteries and bulbs. If you select just one of these items, do not hesitate to choose magnifiers as the most useful piece of science equipment for the early childhood classroom. (Refer to Unit 35 for suggestions on using magnifiers.)

Although they are not interactive in the same way as rollers, ramps, and constructions, magnifiers provide children with their first look at fascinating magnified objects. Piaget was a biologist and probably would have wholeheartedly approved. It is hard to believe that he did not eagerly explore his environment up close at an early age.

Organizing and Storing Materials

As you develop your own materials for teaching science, storage might become a problem. Most commercial kits have neat, ready-made labeled boxes, but the "junk box" system will need some organizing.

One way to manage a variety of materials is to place them in shoe boxes. The shoe boxes contain the equipment needed to teach a specific concept. If the boxes are clearly marked, they can be very convenient. The trick is to keep everything you need in the box such as homemade equipment materials, task cards, materials to duplicate, and bulletin board ideas. To be effective, the box, envelope, or grocery bag should display a materials list on the outside. In this way, you have a self-contained kit for teaching science (Figure 40–2).

Materials relating to a learning center can be stored in boxes under or near the science cen-

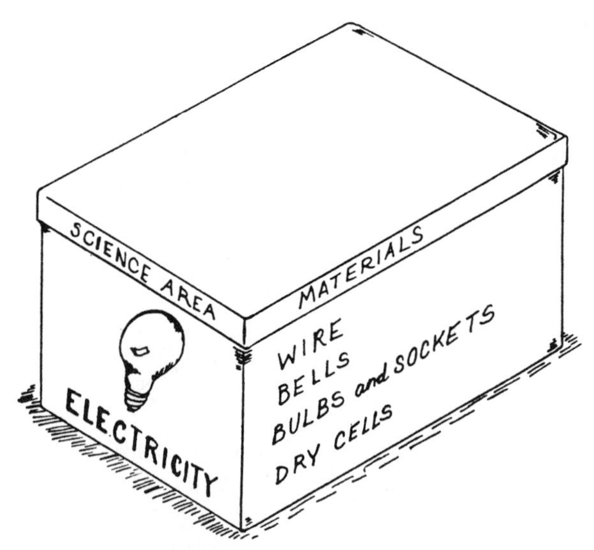

Figure 40–2 Store materials needed to teach concepts in a labeled shoebox.

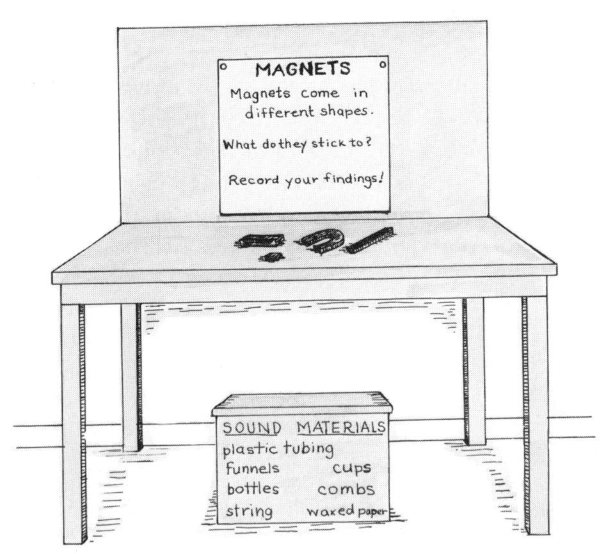

Figure 40–3 Labeled boxes can be stored under a learning center.

ter. Materials will be handy and older children will be able to get their own materials. Primary age children can be very effective organizers. They will enjoy and benefit from the job of inventorying science materials (Figure 40–3).

THE SCIENCE LEARNING CENTER

You know that the interests of children stem from the kinds of learning materials and experiences available to them. One way to provide stimulating explorations is by setting up learning centers appropriate for an individual child or for a small group of children. As children work in centers, they learn to learn on their own in a planned environment. In this way, instruction is individualized and children have time to explore science materials.

There are several different types of science learning centers, characterized by different purposes and modes of operation. The basic centers can be labeled Discovery Center, Open Learning Center, Inquiry Learning Center, and Science Interest Center.

Discovery Center

The word *discovery* implies that some action will be taken on materials and that questions and comments about what is happening takes place. All is not quiet in a discovery center. Thus, it should be located away from listening and literacy centers. Discovery centers can be located on large tables, desks against a wall, or any spot that has room. Mobile discovery centers made of trays, shoe boxes, or baskets can simply be picked up and taken to a designated spot on the floor. An area rug makes a good place to work without taking up too much space. (Figure 40–4).

Figure 40–4 Working with an activity tray

In Figure 40–5, activity trays are used to explore concepts about air. As with any learning center, the size of your center area and ages of the children must be taken into account. For the five activity trays described in Figure 40–5, it is recommended that the teacher use two trays with five-year-olds and, initially, one tray with younger children (Figure 40–5).

Open Learning Center

The science center in an open classroom is very creative and contains an abundance of manipulative materials, the majority of which are homemade. Using minimum directions, the child might be asked to "Invent something with these materials." Guidance is provided in the way of helpful suggestions or questions as the children

Activity Trays—Exploring Concepts About Air

TRAY	CONTENTS	CONCEPTS TO BE EXPLORED	INTRODUCTION	FURTHER EXPLORATION
1	2-4 each of several sizes and colors of paper bags Box of wire closures	Air can be felt even when it cannot be seen.	Open up bags and look inside. Do you see anything? Fold the bags up again. What do you feel as you do this? Try with several sizes and colors of bags.	How can you make the bag like a balloon? What is the closure used for?
2	12 plastic glasses, water pitcher, plastic drinking straws Red and blue food coloring, liquid soap	Bubbles are really air in water.	See how many different sizes of bubbles you can blow in a glass of water. What else can you do with the bubbles? What happens when you touch them?	Color the water with food coloring. What kinds of bubbles can you make now? What are the bubbles like when liquid soap is added?
3	Colored balloons, string, scissors, and felt-tip pens	Air can be felt even when it cannot be seen. Air makes a noise in some activities.	Blow up balloons. (A small air pump can be used for younger children.) Let the balloons go when they are large enough. What can you hear, feel, and see?	Draw faces and figures on your balloons. How do these change as the balloon is blown up? Tie balloons securely with a string and leave overnight. What changes are apparent the next day? What caused the changes?
4	Several sheets of heavy, colored paper, stapler, and crayons to decorate paper	Air can be moved with a fan.	Fold the colored paper accordion style to make a hand fan. Use a stapler to pull the bottom together. Use the fan to create air movement.	Try moving items from tray 5.
5	Feathers, cloth scraps, small plastic toy cars, plastic animals, packing forms, and cotton balls	Air moves many things.	Try to move these items using your fan. How many small items can you move across a table or floor space?	Try to guess which items will be more difficult to move. Why? Separate items that move easily from those that do not.

Figure 40–5 Activity trays—Exploring concepts about air
(Reprinted from Science and Children *(1984), with permission from the National Science Teachers Association, 1742 Connecticut Avenue, Washington D.C. 20009. Margaret McIntyre, "Discovery and Exploration Centers.")*

pursue and tinker with their invention.

An example of an open center is a sink and float center. In this center, reinforcing the concept of sink and float is the central objective. The center contains a plastic tub half full of water and a box of familiar materials. On the backboard are the questions, "What will float?" and "What will sink?" With no further direction, the children explore the questions with the materials.

Teacher evaluation is not formal; rather, the teacher visits the center and satisfies any concerns about children's progress. The teacher might find children making piles of things that do and do not float or trying to sink something that they thought would float. Some teachers suggest that the children communicate what happened at the center by drawing or writing a few sentences (Figure 40–6).

Figure 40–6 Construct a beach

Inquiry Learning Center

A more directed discovery approach focuses on a science concept or topic and contains materials to be manipulated by the children, directions for the investigation, and open-ended questions to be asked at the end of the inquiry. The objective is not to reinforce a concept but to engage the child in problem solving beyond what is known to gain new insights.

At an inquiry center, the child might find directions that say, "Using the materials in this box, find a way to light the bulb," or, "What kinds of materials are attracted to magnets?" Science concepts are turned into questions that are placed on activity cards. There might be a single task or a series of tasks that a primary age child can complete in fifteen or thirty minutes.

It would be foolish to send children to an inquiry center without adequate preparation. Discuss directions, procedures, and task cards with the children before beginning center work. The children are expected to achieve a desired learning outcome. So, if things do not go smoothly, ask yourself the following questions: Were the directions clear? Did the children know where to begin and end? Did they know what to do when they were finished? Is the center appropriate for the age group? Did the first group using the center know how to restore the materials for the next group? (Figure 40–7).

Science Interest Center

Science interest corners, tables, and centers are popular in many schools. This interest center reinforces, enriches, and supplements ongoing programs with materials that stimulate children's interest. The primary goal of the center is to motivate children to want to learn more about the subject at the center.

As general rules, children are not required to visit this center, and no formal evaluation of their activities is recorded. Many times the center repeats a lesson exploration. For example, if

Figure 40–7 An inquiry center

the class has done the "Thinking like a Geologist" lesson described in Unit 37, some children might want to try scratching minerals to test hardness or classifying the available rocks. Filmstrips, storybooks, and resource books on rocks and minerals should be available for the children to explore. Children could practice their measuring skills by using string to measure an assortment of objects or become aquainted with a balance scale by comparing the weight of different rocks.

Plan Your Center

State learning center objectives in such a way that you know what you want the child to learn. Know your children's developmental level. You can avoid inadvertently creating busywork for students by taking steps to carefully plan the concepts to be developed at the center. Start by writing a sentence that communicates what the child is expected to learn at the center. Children must be able participate in the activities and methods independently. Evaluate the center by asking yourself: Is the center effective in achieving my objectives? How can the center be improved? Figure 40–8 suggests a guide to planning a learning center.

SELECTING SCIENCE MATERIALS

Providing materials that encourage children to "mess around" and explore is the responsibility of the teacher. At the center, children use the process skills to observe, investigate, classify, and maybe hypothesize. Because this learning is not accidental, planning must go into setting up centers and selecting materials. After completing the planning guide suggested in Figure 40–8, consider the following criteria in selecting and arranging materials:

1. *Are the materials open ended?* That is, can they be used in more than one way? For example, water play provides the opportunity to explore measuring or floating.
2. *Are the materials designed for action?* In science, children do something to materials to make something else happen. If substances are to be dissolved, which will offer the best comparison, salt, sugar, or pudding mix?
3. *Are the materials arranged to encourage communication among children?* If appropriate, place materials to create cooperation and conversation. Arrange materials in categories such as pitchers of water in one section of the center, substances to be tested in another, and spoons and dishes for mixing in another. Children quickly learn to cooperate and communicate in order to complete the activity.
4. *Is there a variety of materials?* If the center is to be used over an extended period, some children will visit it many times. A variety of materials will prevent overexposure to the exploration.

A Simple Outline for Planning a Learning Center

Purpose of Center	Characteristics of Students	Concepts and Skills	Activities and Materials	Expected Learning Outcomes	Evaluations	Suggestions for Change
					Students:	
					Center:	

Figure 40–8 A simple outline for planning a learning center. *(Reprinted from* Science and Children *(1976) 14 (3) 12, with permission from the National Science Teachers Association, 1742 Connecticut Avenue, Washington, D.C. 20009. Geraldine R. Sherfey and Phyllis Huff "Designing the science learning center.")*

5. *Do the materials encourage "What if . . ." statements?* The sink and float activity described in the open classroom learning center invites children to predict what will happen if they try to float a marble, toothpick, or sponge.
6. *Are the materials appropriate for the maturity of the children?* Consider the maturity level of the class. Select materials that the children can handle safely and effectively.
7. *Do the materials allow for individual differences such as ability, interest, working space, and style?* After considering floating and sinking, some children will begin to consider size and other characteristics of the available objects. Have objects with a variety of textures and features available.
8. *How much direction do the materials require?* In giving directions, consider the age of the children. Four- and five-year-olds might receive directions from a cassette recording, but three-year-olds and young fours respond best to personal directions. Rebus-type directions are also appropriate.
9. *Do the materials stress process skills?* The process skills are the fundamental skills that are emphasized in science explorations with young children. These skills will come naturally from manipulating materials. However, a variety of appropriate materials is required for this to happen.

Sensory Learning Center

Children have a "sense-able" approach to the world around them. The approach that they use is as basic to science as it is natural to young children. Use the type of center you prefer to give children an opportunity to taste, smell, touch, observe, and hear their environment.

Thinking like a Criminologist

Skin prints are a way to take a closer look at the skin children live in. They enjoy examining their fingertips, taking their own fingerprints, and thinking like a criminologist at this center. Since criminologist investigates crimes by analyzing clues in a systematic way, studying fingerprints is an important part of this job.

You will need one No. 2 pencil, white scratch paper, plain paper, a hand-held lens, and a damp paper towel at this station. Show children how to rub soft pencil lead onto a sheet of paper and pick up a good smudge with one finger. Then, they carefully pick up the smudge with a piece of clear tape, pull it away, and press the fingerprint onto a clean paper.

Begin by having students make a set of their own fingerprints on a sheet of clean paper. If someone makes a mistake, simply peel off the tape and start over. Then create a classroom mystery for your students to solve. Ask, WHO WAS THE LAST PERSON TO USE THE PENCIL SHARPENER? Each group that visits the center can be a group of "suspects." Prepare prints in advance, then show prints that were "found" on the pencil sharpener. Have the children use the hand-held lens to compare their prints to the set belonging to the "criminal." (Figure 40–9).

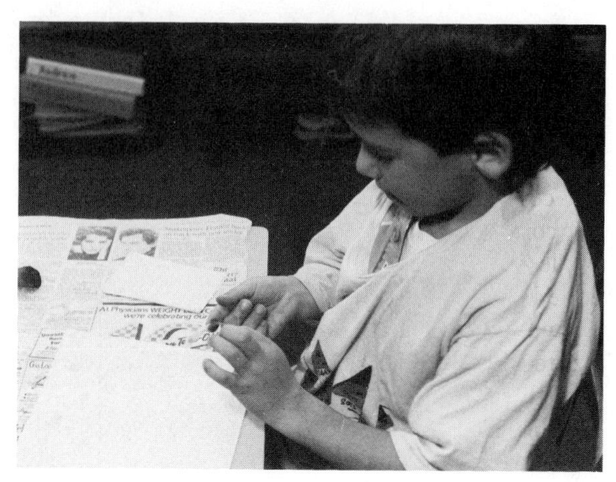

Figure 40–9 Carefully pull the tapes away from your fingertip.

Integrate fingerprinting into art by letting the children use ink pads to make prints from their hands or fingers. Children can bring the prints to life by adding legs, arms, and other features. Fingerprint stories can be developed, and you may even want to include toe prints as a homework assignment. You could go further and classify fingerprints into three basic patterns: whorl, arch, and loop. Refer to the further readings list for resources.

Do You Hear What I Hear?

Walk in the hallway with a tape recorder and record five interesting sounds. Bring the tape back to the classroom, provide earphones, and have the children draw a picture of the sounds that they hear. Children enjoy using a tape recorder. Let them take turns tape recording sounds during recess, assemblies, or the lunch period. Challenge them to record unusual sounds and infer what is making the sounds. Add further interest to this center by sending the tape recorder home with a child to tape "home sounds." As children identify the home sounds at the center, have them compare and contrast the sounds in their own home and write a story about the sounds and how they are created.

The tape recorder also makes a good listening center for a variety of tapes and records. Many of the recordings should be stories. Children will enjoy illustrating the stories. Some will enjoy taping a story of their own for use in the center or as background for plays, puppet shows, and radio programs.

Red, Yellow and Blue

The object of this center is to mix the primary colors and observe and record the results. Have the children record the color that results from dropping the correct amounts of color into the correct container (Figure 40–10).

Smells, Smells, Smells

Gather small amounts of familiar substances with distinctive odors, such as coffee, popcorn, orange extract, hot chocolate, onions, apples, and peanut butter. Put the items in jars, shoe boxes, plastic sandwich bags, or loosely tied brown paper bags—whichever seems appropriate—and leave small holes in the containers so that the odors can escape. As children take turns smelling the containers, ask them to describe the various odors and guess what might be producing these distinctive smells. Make a duplicate set of containers and ask the children to match the ones that smell alike. Be sure to show children the safe way to smell an unknown substance: They should gently fan the air between their noses and the container with one hand and breathe normally. The scent will come to them.

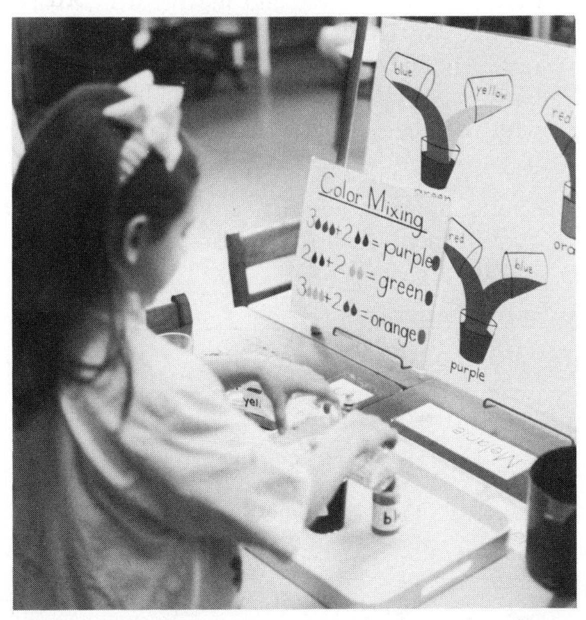

Figure 40–10 A rebus card gives the directions for mixing colors.

Apple or Potato?

Spread vegetable pieces on a plate for children to taste. They will have to work in pairs. One puts on a blindfold, holds his nose, and tries to guess what he is tasting. The other child keeps a record by writing down what the vegetable piece really is and what his partner thinks it is. Then, the children trade places, redo the test, and compare their scores.

The children will be surprised that they made a lot of wrong guesses. This is because it is hard to tell one food from another of similar texture. The secret lies inside the nose. Tongues only tell if something is sweet, sour, salty, or bitter; the rest of the information comes from the odor of the food. Lead children to the conclusion that they would have a hard time tasting without their sense of smell.

Tasting liquids that children drink frequently is fun. Pour some milk, orange juice, water, soda, or other popular drinks into paper cups or clean milk cartons and provide clean straws for tasting. Have the children describe the taste and guess what it might be. Some children will enjoy drawing a picture of their favorite flavor. A sip of water or bite of bread in between items helps clear their palates.

SUMMARY

Stimulating science lessons do not happen by accident. The materials selected to teach science and the format that they are presented in are essential for successful explorations. Whether materials are purchased or scrounged, they must be flexible and appropriate to the developmental age of the child and the type of science learning that is required. Learning centers are designed and used to meet specific teaching objectives and must be evaluated for their effectiveness.

Lists of resource books for teachers, books for children, and free and inexpensive materials are listed in the appendix and further reading list.

FURTHER READING AND RESOURCES

Carin, A. A., and Sund, R. B. 1989. *Teaching science through discovery.* Columbus, Ohio: Merrill Publishing Company.

Esler, W., and Esler, M. 1989. *Teaching elementary science.* Belmont, Calif.: Wadsworth Publishing Company.

Gilmore, V. 1979. Helpful hints—Coca-Cola bottle terrarium. *Science and Children.* 16 (7).

Houle, G. B. 1987. *Learning centers for young children.* (3rd edition) West Greenwich, R.I.: Tot-lot Child Care Products.

McIntyre, M. 1984. Discovery and exploration centers. *Early Childhood and Science.* Washington, D.C.: National Science Teachers Association.

Poppe, C. A., and Van Matre, N. 1985. *Science learning centers for the primary grades.* West Nyack, N.Y.: Center for Applied Research in Education.

Sherfey, G. R., and Huff, P. 1976. Designing the science learning center. *Science and Children.* 14 (3), p. 12.

Supplement of Science Education Suppliers. Published every January. Washington, D.C.: National Science Teachers Association.

SUGGESTED ACTIVITIES

1. Obtain a copy of the current NSTA Supplement of Science Education Suppliers. Select a science concept and list materials that could be used in teaching that concept.
2. Reflect on the nature of science materials. Then, make a list of materials needed for teaching a science unit that you have developed. Where will you get these materials? How will you organize them for hands-on teaching?
3. Construct a learning center. Use the center with a group of children and share the effectiveness of the materials with the class.
4. Compile a list of free and inexpensive science teaching materials. Send for some of these materials and evaluate if they are appropriate for young children. Share this information with the class.
5. Visit a preschool, kindergarten, and primary class. Compare the science teaching materials used in these classes. What are the similarities and differences?

REVIEW

1. Why are learning centers essential for science learning?
2. What are the two main types of science equipment?
3. List some useful items for a science junk box.
4. What are some of the advantages of science kits? What are some of the disadvantages?
5. Match the description with the center.

   ______ Discovery Center

   ______ Open Learning Center

   ______ Inquiry Learning Center

   ______ Science Interest Center

   a. Children explore a concept with as little direction as possible. A leading question.

b. Little direction is given. Children "mess around" with materials at small tables or trays.

c. A directed discovery approach focuses on a concept. This center includes directions and is used to encourage children to go beyond what they know about a concept.

d. This center motivates children to want to learn about its materials.

UNIT 41 Math in Action

OBJECTIVES

After studying this unit, the student should be able to:

- Plan and use woodworking for math experiences
- Plan and use blocks for math experiences
- Plan and use math games for math experiences
- Share math fingerplays and songs with children

Math goes on all the time in the classroom for young children. The young carpenter measures wood. The block builder engineers a building. This is math in action. Throughout the day the teacher of young children may use fingerplays and songs that contain math ideas. Many games, such as musical chairs, that young children like to play help them understand the use and the meaning of math. As children move into concrete operations, math in action includes more complex group games and the introduction of team sports and preplanned building projects.

WOODWORKING

Both boys and girls like to work with wood. It provides concrete experiences with measurement, balance, and spatial relationships. As children work with wood, they compare the size of one piece of wood to another. They learn to judge length, width, and thickness. Their conversations contain phrases such as "long enough," "about the same," and "I need more." As children move into the primary level, they can apply standard measurement: "I will need four pieces of 12 × 8 plywood for my bird house." The more advanced primary children can follow simple instructions and use patterns to make projects.

Effective woodworking requires a sturdy workbench, real tools of good quality, and an assortment of soft wood. The workbench should be large enough for two or three children to build at one time. It must be put in a place where it can be well supervised at all times. Work surfaces can be a purchased workbench designed for young children, an old sturdy table at children's height, or a large tree stump.

High quality tools should be purchased for woodworking. Figure 41–1 illustrates the basic carpentry tools for children ages four and five. As children become more proficient, additional woodworking and measuring tools can be added. The tools should be located where they are accessible and easy to keep in order. Tools should not be allowed to get rusty. Other accessories include short nails with large heads, coarse sandpaper, white glue, and wire.

A soft wood such as pine makes a good surface to pound. For the child who is learning to use the saw, lengths of two inches by one-half inch should be provided. The pine wood should be placed in a vise, so that it will be held steady while the child saws it.

Lumberyards often give away scraps of wood. Styrofoam, chunks of layered corrugated cardboard, or acoustical ceiling tile also can be used for carpentry experiences.

To help children explore space creatively, the adult should also provide containers of odds and ends. Bottle caps can be wheels. Straws can be cut and used for oars. Plastic lids can be windows. Sponges, spools, cloth scraps, ribbons,

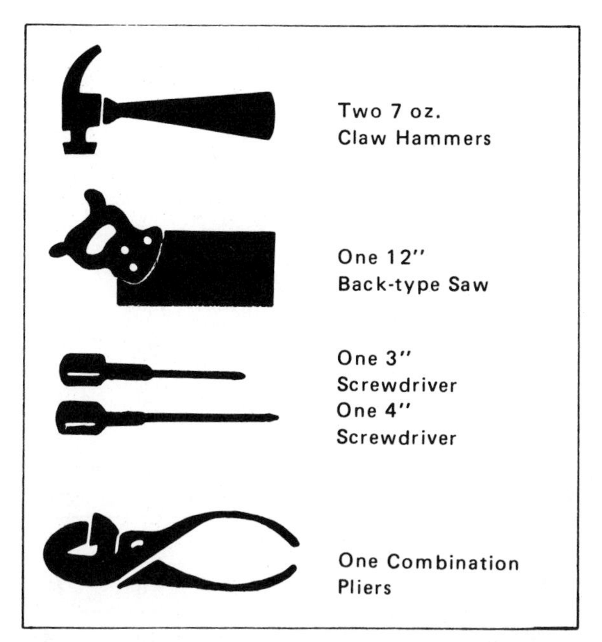

Figure 41–1 Basic woodworking tools for Four- and Five-Year-Olds: Start with these and add more as children become proficient.

and bows help the young woodworkers create all sorts of objects. Even feathers, flowers, and large weeds can be nailed or glued onto a piece of wood. Children get much satisfaction from their woodworking projects. In creating them, the children make plans and sequence their activity. When the project is finished, the children want to talk about what they have done. This is the time the alert adult can help a child use his math vocabulary.

BLOCKS

Blocks are probably the play material most used by young children. Unfortunately, blocks are seldom seen in classrooms beyond the kindergarten level, but they can also function as valuable concept-building materials for primary children. Basic math concepts are developed as children explore the relationship of unit block sizes and shapes (Figure 41–2). Math terms such as "one more" or "two wide" are used by children as they experiment with blocks. Likenesses and differences in form can be seen as well as how forms fit together. When children use blocks, they are working in space. They learn to manipulate length, width, and height. They measure. They discover that a certain number of blocks are required to equal the length or width of another block. As children take two half-circle blocks and put them together, they make a circle. They learn about parts and wholes. They learn about fractions.

The block area should be located where there is adequate floor space for building and away from other activities. Low, open shelved block cabinets should be provided so that children can easily reach the blocks and put them away. Shelves should be marked to help children store blocks. The teacher's role is to gradually introduce more blocks or a variety of shapes as children's needs grow. The teacher can ask questions to help children discover how things are different or how they are alike (Figure 41–3).

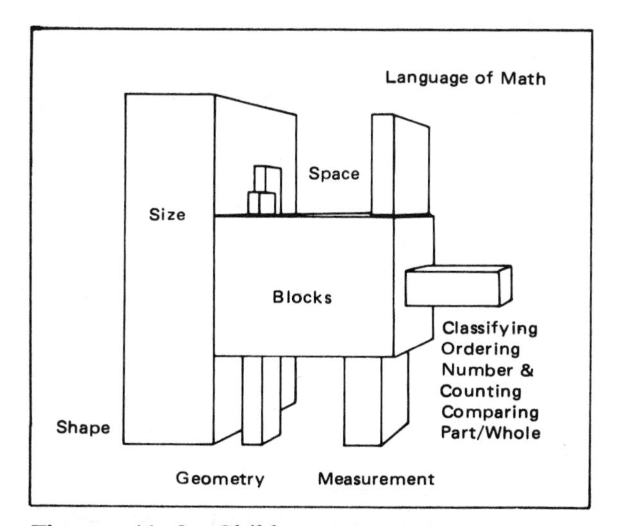

Figure 41–2 Children can construct many concepts as they work with blocks.

Figure 41–3 Marked shape pieces help children store blocks properly.

Mrs. Red Fox notes that Trang Fung has used all square blocks in her structure, whereas Sara has developed her structure with units and double units. "It looks like each of you have your favorite size blocks."

Blocks are available by sets. Each set is an assortment of shapes and sizes. Unit blocks are the basic forms. The basic unit is a brick-shaped rectangle (1 3/8″ × 2 3/4″ × 5 1/2″). Figure 41–4 lists the great variety of shapes combining straight and round surfaces.

Unit blocks should be of durable hardwood with all edges beveled to prevent wear and splintering. Smoothly sanded surfaces are important for safe handling. Precise dimensions are important for effective building.

Blocks are expensive. However, with proper care, blocks should last at least ten years. They should be kept dry and free of dust. If oil or wax is put on occasionally, the blocks will last longer. There are toys which may be used with blocks. Cars, trucks, farm and zoo animals, block people, and colored inch cubes add stimulation and variety to block play.

Teachers of young children have observed that all children pass through stages in block construction (Figure 41–5). When first introduced to

Name	Nursery	Kgn. & Primary
Square	40	80
Unit	96	192
Double Unit	48	96
Quadruple Unit	16	32
Pillar	24	48
Half Pillar	24	48
Small Triangle	24	48
Large Triangle	24	48
Small Column	16	32
Large Column	8	16
Ramp	16	32
Ellipse		8
Curve	8	16
¼ Circle		8
Large Switch & Gothic Door		4
Small Switch		4
Large Buttress		4
½ Arch & Small Buttress		4
Arch & ½ Circle		4
Roofboard		24
Number of Shapes	12	23
Number of Pieces	344	760

Figure 41–4 Child Craft Block Sets

A. Child handles blocks.

B. Child builds rows.

C. Child connects blocks.

D. Child builds enclosures.

E. Child makes decorative patterns.

F. Child names his structure. "Look, I made a fire station."

Figure 41–5 Stages of development in block construction: A. Child handles blocks; B. Child builds rows; C. Child connects blocks; D. Child builds enclosures; E. Child makes decorative patterns; F. Child names structure: "Look, I made a fire station."

blocks, a child carries the blocks from place to place. He handles them, but does not build with them. Second, the child makes rows. Third, he makes bridges. He connects two blocks with a third. In the fourth stage, the child places blocks in such a way that they enclose a space. This is called making enclosures. Fifth, the child uses the blocks to make decorative patterns. Much symmetry (balance) is observed. In the sixth stage, the child names his structures and dramatic play begins. Finally, the child makes buildings that represent actual structures he knows. This may be his school or his house or apartment building. After they reach this stage, children use their structures for dramatic play (Figure 41–6).

There are other kinds of building materials. Large hollow wood blocks have been used for a long time in preschools. There are also colorful plastic substitutes and corrugated board boxes available at a lower cost. Cardboard boxes are also used imaginatively by children. As they walk around, climb over, crawl into, and peek in and out of boxes, children learn math vocabulary. All of these building materials show differences in weight, size, shape, and texture. As children pick them up, move them, and build with them, they develop an awareness of comparisons between size, proportion, and volume.

Figure 41–7 "Who is in the big box?"

Figure 41–6 In which stage of block construction are these children?

They have the opportunity to manipulate their environment. Math is in action. (Figure 41–7).

MATH GAMES

Games become a part of a child's experience at an early age. Mother plays "peek-aboo" with the infant. As he gets older, his brothers and sisters play "going-in-the-car" and "let's find" games. As his father tosses him and catches him, the child experiences free space. Grandmother plays "how big" is Tommie today. By the time the child goes to school, he has played many informal games and has used his senses in them.

Games in the preschool should not be competitive. Young children do not understand competition. Games should be simple and contain only two or three directions. The teacher presents games with certain aims in mind to aid in understanding math ideas.

Board games provide an excellent way to teach math. *Candyland®*, *Numberland Counting Game®*, *Chutes and Ladders®*, *Fraction Brothers Circus®*, *Memory®*, *Picture Dominoes®*, *Picture Nines®* (domino game), *Candyland Bingo®*, *Tri-ominos®*, *Connect Four®*, *Count-a-Color®*, *Farm Lotto®*, and other bingo and lotto boxed games can be purchased. Some inexpensive board games are *Mailman®* (number recognition) and *Down On the Farm®* (set recognition) in *Readiness Gameboards®* from Frank Schaffer Publications, and *Dog Bones®* and *Match the Shapes®* from the Dellosa and Carson Publication *Fluttery Readiness*. For more advanced children, basic concepts can be practiced using board games such as *Multiplication/Division Quizmo®*, *Addition/Subtraction Mathfacts Game®*, *Multiplication/Division Mathfacts Game®*, *UNO®*, *UNO Dominoes®*, *Yahtzee®*, and *IMMA Whiz Math Games®*. *Pay the Cashier* and *Count your Change* are games that help children learn money concepts. A teacher may make a shape game as follows:

Make a large cardboard triangle, a large square, and a large circle. Place these in the middle of the floor. Make small shape forms that correspond to the large triangle, square, and circle. Make at least twenty of each. To play, children draw six shapes from a bag. The teacher calls the name of a shape. If a child has that shape, he places it on the large form. When one child uses up all his shapes, he gets to call the names. As soon as the next child uses all his shapes, he gets to call the names. The other child may draw six more shapes or stop playing.

Other basic board games were described in Unit 24. Card games enjoyed by primary children include those suggest by Kamii (see Unit 27) and those that are perennial favorites such as *Go Fish*, *Concentration*, *Crazy Eights*, *Old Maid*, *Flinch*, *Solitaire*, and *Fantan*. Look through catalogs and examine games at exhibitors' displays when you attend professional meetings. There is a vast selection available.

Aiming games such as bowling and drop the clothespins help children learn to count. Cardboard containers covered with Contac® paper with numerals or shape pieces attached to them make good tossing equipment to use with beanbags. Beanbags can be used in many ways. In Unit 40, directions for making beanbags were given. Figure 41–8 lists ideas for using beanbags.

Action games involve physical activity and mental alertness. For example, one child closes his eyes. The other children act out the following:

> I'm very, very tall,
> I'm very, very small.
> Sometimes tall,
> Sometimes small,
> Guess what I am now?

The child has to guess whether the other children are tall or small, as the last line indicates. This game can include other math ideas, also:

> I'm very very big; I'm very very thin
> I'm very very slow; I'm very very fast.

Another action game that twenty children can play requires running and catching.

> Ten little mice looking for cheese,
> They run on the floor as nice as you please.
> Ten little mice climb up high,
> They eat and eat a great big pie.
> Ten big cats out looking for fun,
> They chase the mice and make them run.

IDEAS FOR USING BEANBAGS

1. Balance a beanbag on your head. (Variation: Have a race.)
2. Walk a "balance beam" (2′ x 4′ board on some blocks) while balancing the beanbags. Walk backwards.
3. Set plastic kitchen bowls at varying distances. Try hitting the bowls.
4. Hang a balloon on a wall. Allow the child to try to hit the balloon with the beanbags. (Variation: Put several balloons on wall; assign numbers to each balloon. Hit the balloons in sequence. More advanced – add the numbers of the balloons hit.)
5. Play the "old-fashioned" game of catch, calling out the color and/or numbers of the bag being thrown. (Variation: In a group, children sit in a circle on the floor. The leader throws a bag and says one of the following: "Color, Shape, Number." The child responds, then throws to another in the group.)
6. Make a clown board to throw beanbags through, matching shapes, colors, or numbers.

Clown Throw: Purchase 4′ x 4′ x 1/2″ Masonite at the local lumberyard (about $3.00). Using a jigsaw, cut out eyes, ears, etc., in various shapes. Paint and decorate with leftover materials. (Instead of a clown, use any animal.)

7. Put masking tape on the floor, making various shapes. Then put letters and numbers within the shapes. Throw corresponding bags into the shapes. For more advanced children, throw the beanbag on a letter and have the child give a word using that letter as the initial sound (ending sound, or middle sound). Throw two (or more) beanbags on different numbers and have the children give the sum or the difference.
8. Using a large piece of oilcloth, mark on it a large square with a smaller square inside. (A third square inside it also possible.) Throw a bag in the "big" square or the "small" square. Teach the concept of big, bigger, biggest or small, smaller, smallest. (Variation: Have two sets of squares, stand the child in one little square, throw the beanbag into the other largest square.)
9. Make a basketball hoop with an old coat hanger. Hang it inside or outside. Play basketball.

Figure 41–8 Ways to use beanbags

Musical Chairs is a well known action game. Children and chairs are arranged in a circle. A record player and records are available. The record is played for ten or fifteen seconds. As the music plays, children walk around the chairs. The adult removes one chair. When the music stops, the children sit. One child is left standing. This child may remove the next chair. Children can count and match: Are there more chairs, more people, or one chair for each person? Musical chairs is a math game that reinforces the concept of comparison.

Another type of game that involves a musical tool, the triangle, is called *The Striking Clock*. In this action game, children stand in a circle with their feet spread apart and rock from side to side. The teacher stands in the center with a triangle to strike the hour when the poem has been said. Children count with the striking of the triangle.

We are swinging pendulums
Hanging from a clock;
As we count the hours struck,
We rock and tick and tock.

Primary children enjoy jumping rope. A popular jingle that requires counting is:

> **Mabel**
> Mabel, Mabel,
> Set the table,
> Don't forget the red hot pepper
> (turn the rope fast and count).

The children try to see who can jump the most red hot peppers. During the primary years, children are in the stage of industry versus inferiority. The struggle between these forces leads them into a natural interest in competitive activities, such as the games listed, and into team sports and races. Adults have to take care to find ways for each of the children to achieve so that they do not experience debilitating inferiority feelings. Primary children enjoy races that give them practice in time and distance relationships. Hurdle jumping can begin with high and low jumps then move into standard measures of height. Balls, beanbags, or frisbees can be thrown and the distances compared and measured. Team sports require score keeping and an understanding of more and less and ordinal relations (that is, who is up first, second, third, and so on).

Primary children also enjoy math puzzlers and brainteasers that give them practice in problem solving. Some of these types of problems were introduced in Unit 33. The following are additional examples:

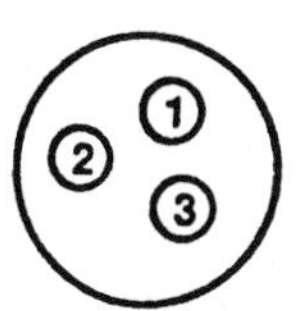

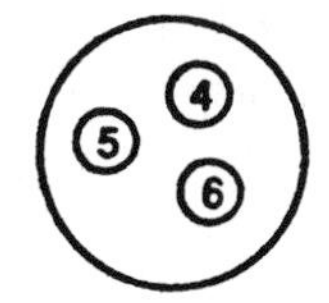

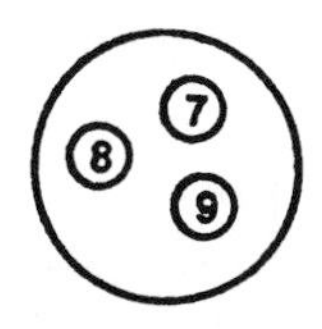

Move one so each set has a sum of 15.

MAGIC TRIANGLES

Write the numbers one through six in the circles of the triangle below in such a way so as to have a total of nine (or ten or eleven or twelve) on each side.

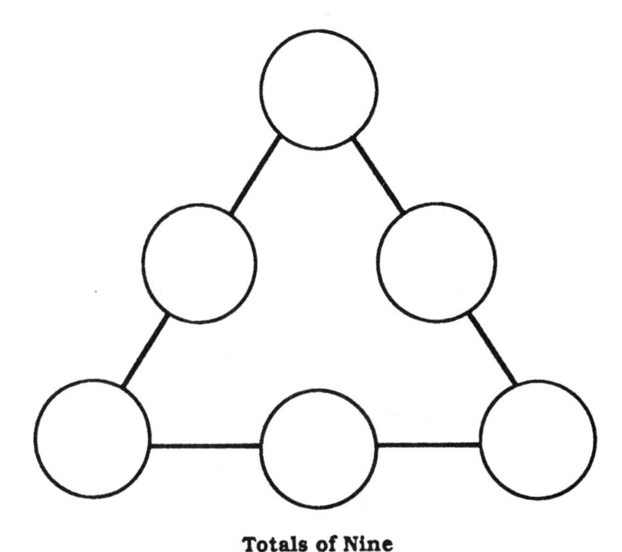

Totals of Nine

Totals of Nine

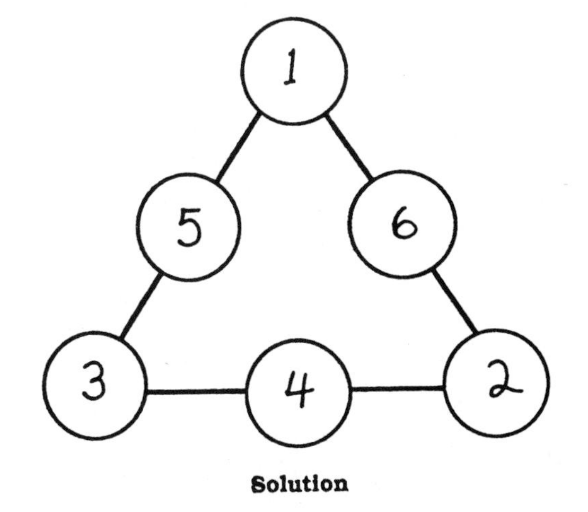

Solution

Solution

THE LADY AND THE TIGER

How many different squares can you count?

How many different triangles can you count?

Answers: 11 squares 19 triangles

YOUR NUMBER

Ask someone to think of a number, but keep it secret. Now tell him to double the number, add eight to the result, divide by two, and subtract the original number. Then have him write down the answer but tell him not to show it to you until you predict the answer. The answer will always be four.

Young children enjoy games. Board, target, and action games offer them opportunities to apply math concepts. Through observation of children playing games, adults can obtain information regarding a child's development of math skills and concepts.

FINGER PLAYS AND SONGS

There are many finger plays and songs that help teach children about math. The rhythm and motions that go along with finger plays and songs can be made up by the children. First, the teacher needs to say the words. Then she can sing the song. Children can repeat the words with the teacher. Next they can sing the song. Now, children can be asked how they would act out the poem or song. Appendix C gives a collection of finger plays that contain math concepts. Below are two examples of action finger poems.

Five Little Ducks

Five little ducks went out for a swim
(5 fingers swimming in a lap "pond")
They swam around and around and ran away *(hand behind back)*
The Mama Duck called, "Quack, quack, quack" *(other arm held straight up from elbow, hand dropped, opens for mouth quacking).*
But only four little ducks came back.
Four little ducks went out for a swim, etc.
(Repeat until "But no little ducks came back")
Then the Mama Duck said, "Quack, quack, quack,"
And five little ducks came back.

Five Little Monkeys
Five little monkeys, jumping on the bed,
(5 fingers, held up high)
One fell off and broke his head. *(Touch head)*
We called for the doctor, and the doctor said,
"No more monkeys jumping on the bed."
Four little monkeys, jumping on the bed. . . . *(4 fingers, held up high)*
(Continue counting backward)

Songs are also used by adults to help children explore math concepts. Three favorite math songs are *Johnny Works With One Hammer, Did You Ever See A Circle*, and *The Numeral Song* (Figures 41–9, 41–10, and 41–11).

Riddles can also teach about math. Children enjoy the guessing game illustrated in Figure 41–12.

Finger plays, songs, and riddles are used for many purposes. They provide a fun way to help children learn math ideas. Through action of the hands and body, rhyme, rhythm, and vocabulary are experienced by the children.

SUMMARY

Math in action means children exploring the environment. This is done through woodworking, block construction, and game playing (Figure 41–13). Math in action means children saying, singing, and acting out math language and concepts. Teachers must incorporate these activities into their early childhood instructional programs to promote math concept development.

FURTHER READING AND RESOURCES

Woodworking

Skeen, P., Garner, A. P., and Cartwright, S. 1984. *Woodworking for young children*. Washington, D. C.: National Association for the Education of Young Children.

Blocks

Hirsch, E. *The Block Book*. Washington, D. C.: National Association for the Education of Young Children.

Games

Arithmetic Teacher. See monthly articles and columns for problems and puzzles to solve.

Dellosa, J., and Carson, P. 1980. *Fluttery Readiness*. Akron, Ohio: Carson-Dellosa.

Games for growing. 1984. *First Teacher*. 5 (2).

Kamii, C. 1985. *Young Children Reinvent Arithmetic*. New York: Teachers College Press.

Kamii, C. 1989. *Young Children Continue to Reinvent Arithmetic*. New York: Teachers College Press.

Kamii, C., and DeVries, R. 1980. *Group Games in Early Education*. Washington, D. C.: National Association for the Education of Young Children.

Platts, M. E. 1964. *Plus*. Educational Service, Inc. Stevensville, Mich.

Schutte, B. 1978. *Readiness Game Boards*. Palos Verdes Penninsula, Calif.: Frank Schaffer.

Warren, J. 1986. *1-2-3 Games: No-Lose Group Games for Young Children*. Totline Press, Warren Publishing House. Everett, Wash.

Worstell, E. V. 1961. *Jump the Rope Jingles*. New York: Collier Books.

Fingerplays and Songs

Bayless, K. M., and Ramsey, M. E. 1987. *Music: A Way of Life for Young Children, 3rd. ed.* Columbus, Ohio: Merrill.

Cromwell, L., and Hibner, D. 1976. *Finger frolics*. Partner Press, Livonia, Mich.

Haines, B. J. E., and Gerber, L. L. 1988. *Leading young children to music, 3rd. ed.* Columbus, Ohio: Merrill. (Continued on page 549)

JOHNNY WORKS WITH ONE HAMMER

Other verses are as follows:

Johnny works with two hammers
Two hammers, two hammers,
Johnny works with two hammers,
Then he works with three.

Johnny works with three hammers,
Three hammers, three hammers,
Johnny works with three hammers,
Then he works with four.

Johnny works with four hammers,
Four hammers, four hammers,
Johnny works with four hammers,
Then he works with five.

Johnny works with five hammers,
Five hammers, five hammers,
Johnny works with five hammers,
Then he goes to sleep.

As each verse is sung the children do the following movements:

1. Pound on one knee with fist.
2. Pound on both knees.
3. Pound with both fists while tapping the floor with one foot.
4. Pound with both fists and tap with both feet.
5. Shake head back and forth along with all other movements.

Figure 41–9 "JOHNNY WORKS WITH ONE HAMMER."

DID YOU EVER SEE A CIRCLE?

1. Children sit in a circle.

2. Each child is given an envelope containing 4 shapes.

3. Everyone removes the shapes from the envelope and places them on the floor in front of him.

4. Dramatize the song by holding up the appropriate shape as it is mentioned in each verse.

   "Did you ever see a circle, a circle, a circle,
   Did you ever see a circle
   Go this way and that?"

   (Choose two appropriate hand movements. Up and down, side to side, circle clockwise and circle counterclockwise, forward and back, clapping hands, tapping parts of the body or the floor, and moving the entire body forward to back.)

5. Continue singing with the motions:

   "Go this way and that way
   Go this way and that way
   Did you ever see a circle
   Go this way and that?"

6. In succeeding verses, substitute square, rectangle or triangle for circle.

Figure 41–10 "DID YOU EVER SEE A CIRCLE?"

THE NUMERAL SONG

1. The children sit in a circle.

2. The teacher keeps one set of 9-inch numerals and places one set out on the floor in the center of the circle.

3. One child is chosen to stand in the center of the circle.

4. The teacher holds up a numeral and everyone sings:

   "Oh, do you know the numeral (one)
   The numeral (one), the numeral (one)?
   Oh, do you know the numeral (one)?
   It looks just like this."

5. The child in the center picks up the matching numeral from the group on the floor and everyone sings:

   "Oh, yes, I know the numeral (one)
   The numeral (one), the numeral (one).
   Oh, yes, I know the numeral (one).
   It looks just like this."

6. Another child is chosen to go to the center and the song is repeated using another numeral. Write the numeral in the air with one hand and sing:

   "Oh, can you write the numeral (one)
   The numeral (one), the numeral (one)?
   Oh, can you write the numeral (one)?
   It looks just like this."

7. Another child is chosen to go to the center of the circle and the game is repeated.

Figure 41–11 "THE NUMERAL SONG."

My voice is quiet
A soft tick-tick
You look at my hand
And you must be quick.

Clock

My color is red
When it's cold I fall
When it's hot I rise
I'm useful to all.

Thermometer

I live on a wall
With dates on my face
I tell you what day
If you're going someplace

1978		November				1978
			1	2	3	4
5	6	7	8	9	10	11
12	13	14	15	16	17	18
19	20	21	22	23	24	25
26	27	28	29	30		

Calendar

I am not a king
But my name sounds strong
I keep lines straight
And tell how long.

Ruler

Figure 41–12 A guessing game

Figure 41–13 Playing with blocks is fun!

Moomaw, S. 1984. *Discovering music in early childhood.* Boston: Allyn & Bacon.

Music for march. 1985. *First Teacher.* 6 (3).

Wiltcher, D. J. 1983. *Lots of wiggles.* Louisiana Association on Children Under Six. Ruston, La.

Records

Math readiness by Hap Palmer. Educational Activities. Freeport, N.Y.

Musical math by Ruth and David White. Rhythms Productions. Los Angeles.

Action Books

Bohning, G. and Radencich, M. C. 1989. Math action books. *Arithmetic Teacher.* 37 (1): 12–13.

SUGGESTED ACTIVITIES

1. Visit a school supply or a toy store and examine the types of blocks available for purchase. Consider how they are like or unlike those discussed in this unit.
2. Examine various early education supply catalogs and check prices and types of blocks available for young children. Make a cost list of blocks you would purchase and tell why you selected them.
3. Observe children playing with blocks. Determine what stage of block construction their play represents. Share your observations in class.
4. Go to a local hardware store and examine woodworking tools. If possible, become familiar with the use of the various pieces of equipment.

5. Learn five math finger plays. Share two with your classmates.
6. Share a different math action song with your classmates. This could be one that you make up.
7. Plan a math action game to use with children. If possible, try it with a group of children. Share with the class what happened during this experience.

REVIEW

A. Indicate the choice which best completes each of the following.

1. The woodworking area is
   a. Not beneficial for teaching math
   b. Well supervised
   c. Not liked by girls
   d. A piece of plywood
2. Woodworking tools should be
   a. Rusty
   b. Nails with small heads
   c. Hard wood
   d. High quality
3. One of the most basic play materials in the preschool and kindergarten is
   a. Swings
   b. Guns
   c. Blocks
   d. Dolls
4. Care of wood blocks includes
   a. Washing daily
   b. Allowing them to become dusty
   c. Throwing them in a pile
   d. Oiling or waxing them occasionally
5. Block accessories do not include
   a. Cars and trucks
   b. Paste
   c. Animals
   d. People
6. The first stage of block construction is
   a. Building vertical rows
   b. Handling blocks
   c. Making enclosures
   d. Making structures for dramatic play
7. Substitutes for building materials do not include
   a. Food
   b. Styrofoam forms
   c. Hollow blocks
   d. Boxes
8. An infant game well liked is
   a. Marbles
   b. Duck duck goose
   c. Musical chairs
   d. Peek-a-boo
9. Which of the following is not a board game?
   a. Candyland
   b. Ten Pin Bowling
   c. Chutes and Ladders
   d. Numberland Counting Game
10. Preschool games
    a. Should be competitive
    b. Should involve many skills
    c. Can be teacher made
    d. Should not include any action

B. Briefly answer each of the following.

1. Describe how woodworking experiences for children can be planned.
2. Describe a way to set up a block area in a school.
3. Describe how to use games for math experiences.
4. Describe how to use math finger plays and songs with children.

UNIT 42 Science in Action

OBJECTIVES

After studying this unit, the student should be able to:

- Plan and use blocks for science experiences
- Describe the benefits of using blocks with primary age children
- Plan and use outdoor activities with primary age children

Children continue to be active learners in the primary grades. This is a fact from research based on Piagetian theory. Unfortunately, the active opportunities provided by blocks and outdoor explorations are not always considered in curriculum plans for primary age children. This is a mistake. Remember, primary age children are still concrete operation thinkers who learn to understand the world around them through actively engaging in explorations.

Block play and outdoor explorations give children many opportunities to investigate, test, and change objects. It is from these interactions that children build their own model of the world. (Refer to concept development in Unit 4 to refresh your memory.)

This unit focuses on the relationship of blocks and outdoor learning to the affective, cognitive, and psychomotor learning needs of the child. The developmental stages of the child are explored, activities and evaluations suggested, and the "hands-on," learn by "doing and acting on" aspects of science are emphasized.

BLOCKS, SCIENCE, AND CHILDREN

Block building and play are not isolated from science. The very nature of building a structure requires that children deal with the processes of science as discussed in Unit 5. As children build they compare, classify, predict, and interpret problems. Scientific thinking is stimulated as children discover and invent new forms, expand experiences, explore major conceptual ideas in science, measure, and work with space, change, and pattern.

Blocks Encourage Thinking

Blocks force children to distinguish, classify, and sort. This can be seen as a group of second graders learns the different properties of blocks by recreating a field trip to the zoo. As they plan and build, they deal with the fact that each block has different qualities. Size, shape, weight, thickness, width, and length are considered. As construction progresses, the blocks become fulcrums and levers. Guiding questions such as, CAN YOU MAKE A RAMP FOR UNLOADING THE RHINOCEROS? WHERE WILL YOU PUT THE ACCESS ROAD FOR DELIVERING FOOD TO THE ANIMALS? will help children focus on an aspect of construction. Some children create zoo animals, workers, and visitors to dramatize a day at the zoo (Figure 42–1).

Allow time for children to verbalize why they are arranging the zoo in a particular way. This will encourage children to share their problem-solving strategy and will help them to clarify thinking. By observing the children at play, you will also gain insight into their thinking.

Figure 42–1 The children create a block zoo.

Balance, Predictions, Interactions, and Movement

One emphasis in science is interactions within systems. In block building, the blocks form a system that is kept in equilibrium through balance. As children build, they work with a cause-and-effect approach to predict stress and to keep the forces of gravity from tumbling their structure. The idea that each part added to the structure contributes to the whole is constantly reinforced as children maintain the stability of their structure.

Balancing blocks is an effective way to explore cause and effect. By seeing the reaction of what they do, children begin to learn cause and effect. The following ideas emphasize action and cause and effect:

1. *Dominoes.* Children develop spatial relationships as they predict what will happen to an arrangement of dominoes (Figure 42–2).

Figure 42–2 "Arrange the dominoes so that they will all be knocked down."

   Ask, CAN YOU ARRANGE THE DOMINOES IN SUCH A WAY THAT THEY WILL ALL BE KNOCKED DOWN?
2. *Construct and roll.* Arrange plastic bottles or blocks in a variety of ways and have the children try to knock them over with a ball. This bowling-like game encourages children to keep score and establish a correspondence between the way blocks are arranged and how the ball is rolled. Ask, WHAT ACTION CAUSED THE BLOCKS TO FALL DOWN? WHAT ACTION STARTED THE BALL MOVING? DID THE BALL KNOCK DOWN EVERY BOTTLE THAT FELL? IF NOT, WHAT MADE THEM FALL?
3. *Pendulum release.* In a pendulum game, a ball moves without being pushed; it is released. Children structure space as they place blocks in a position to be knocked down by the pendulum bob. Have children predict, IF THE BALL PENDULUM IS PULLED BACK AND RELEASED, WILL IT KNOCK OVER THE BLOCK?

Pendulums can be made life-size by securing one end of a length of cotton

Figure 42–3 "How is a swing like a pendulum?"

string to the ceiling and attaching a weighted bob to the other end. Weighted bobs for the pendulum can be made by tying a plastic pill vial filled with sand to the cotton string. However, a smaller model may be more practical. Try this first and then tell them how to expand to a life-size model. One end of a length of cotton string or fishline must be attached to a stable support that allows for a swinging motion. A simple effective pendulum can be constructed by placing an eye screw into a board, suspending the board between the backs of two chairs, and attaching the string with the pendulum bob. Children can sit on the floor and explore the action of the pendulum.

4. *Inclines.* In incline activities, the ball moves when released. Exactly where the ball goes is determined by manipulating the incline and the ball. Have children change the incline in different ways to control what happens to the ball. Children will enjoy creating games such as catching the ball with a cup, racing different-size balls down the incline, and measuring how far the ball travels (Figure 42–4).
5. *Buttons and bobby pins.* The object of this tilt game is to jiggle a button from start to finish without letting it fall off of the board or through a hole. The button, or ball is guided by tilting and wiggling the board. To make the game, cut several holes in cardboard and rub the board with a piece of waxed paper. Position bobby pins with the raised side up on the board, mark a start and finish, and begin the game.

Pinball Wizards

Children can create pinball machines from scraps of plywood, strips of wood, glue, nails, plastic caps, and marbles. The challenge is for children to invent ways to make a marble fall down a series of ramps and make the trip as long as possible. Have the children glue wooden strips to a backboard at different angles. Then have them adjust their obstacles of nails and blocks as they try out the marble. Ask, CAN YOU THINK OF A WAY TO SLOW DOWN THE MARBLE? (You might want to introduce the concept of friction.)

Blocks and Marbles

Balance and action can be seen as children assemble plastic ramps and chutes with commercial toys such as Marbleworks® from Discovery Toys. Children gain a familiarity with concepts such as gravity, acceleration, and momentum when they design and create a maze of movement by fitting pieces together. To further introduce children to the principle of cause and effect, ask, WHAT ACTION STARTS THE MARBLE MOVING? Then have children predict the way in which the marble will move. Creating different pathways and exploring how the marble moves on them can be exciting.

Figure 42–4 "Can you control the speed of the ball?"

Complex block and marble sets seem to fascinate primary age children. In these sets children arrange attractive wooden sections to allow marbles to travel through holes and grooved blocks of different lengths. Children enjoy controlling the movement of the marble down the construction and creating changes that determine direction and speed of the marble (Figure 42–5).

Figure 42–5 Primary age children enjoy controlling the movement of blocks and marbles.

Children can make their own marble runs from decorative molding that is available in paneling supply stores. The track can be nailed onto boards, taped down, or held for observing the movement of marbles (some will move at breakneck speeds). Have your students add a tunnel, try different types of balls, and find ways to use friction to slow down the marbles.

Another Type of Construction

Constructions introduce children to the conditions and limitations of space. They learn to bridge space with appropriate-sized blocks and objects and enclose space in different ways. The following ideas involve creating your own construction set with straws.

Use large straws for straw construction and connect them with string, pipe cleaners, or paper clips. String is the most difficult to use, but makes the most permanent construction. Simply stick the string in one end of the straw and suck on the other end. The string will come through.

You will have to form a triangle with three straws. A triangle is the only shape made with straws that is rigid enough for building. If you are using string as a connector, tie the ends together to form a triangle. Or, thread three straws on one string to form the triangle.

Pipe cleaners as connectors are another method of building with straws. Push a pipe cleaner halfway into the end of one straw, then slip another straw over the other end of the pipe cleaner. Double up the pipe cleaners for a tighter fit. Children can twist and turn this construction in many ways.

Many teachers recommend paper clips as ideal connectors in straw building. Open a paper clip, bend out the two ends, and slip each end into a straw. Paper clips are rigid and allow for complex building. You might have to add as many as three paper clips to give the structure strength. Paper clips may also be chained for a flexible joint between two straws. Challenge children to think and construct. Ask, HOW TALL

A STRUCTURE CAN YOU MAKE? WHY DID YOUR STRUCTURE COLLAPSE? CAN YOU MAKE A BRIDGE?

When the straw frame stands by itself, test it. Ask, CAN YOU THINK OF A WAY TO TEST THE STRENGTH OF YOUR STRUCTURE? Place a paper clip through a paper cup and hang it somewhere on the straw structure. Ask, HOW MANY PAPER CUPS CAN YOUR STRUCTURE SUPPORT? HOW MANY PAPER CLIPS WILL MAKE THE FRAME WORK? (Figure 42–6).

Block City

Blocks in the classroom provide many opportunities to integrate basic reading and writing, science, math skills and concepts, and social studies into the construction process. Opportunities for integration abound as children explore the busy life of a block city.

Mr. Wang's second grade created a city of blocks. Buildings had to be accurate in the city

Figure 42–6 "Can you make a bridge?"

and each child builder represented himself in the daily acting out of city life. The block building sessions were preceded by class discussion as the children planned the daily block activities. Assessories (labeled boxes of food, clothing, typewriters, and the like) were constructed from a variety of materials. Children played the roles of shopkeepers, bankers, and other workers. They made decisions such as where the people in the block city would get their money.

When the children had to put out an imaginary fire, they immediately saw a problem. How would they get water to the blaze? This discovery led to an investigation of how water gets into hydrants, utility covers, and water pipes. The children responded to the emergency by adding plastic tubing to the city as well as wire for electricity and telephones.

Not only was the city becoming more realistic, it was becoming less magical. Children no longer thought that water magically appeared when the water faucet was turned on. They knew that a system of pipes carried the water. In fact, the workings of a city in general became less magical. Many common misconceptions were dispelled, and understandings of how a city functions began to develop (Figure 42–7).

The Edible Village

Mrs. Moore's first grade class integrated the study of their neighborhood with block building. After finding out the different sections of the neighborhood and buildings they needed to create, each child was assigned a building. The class created their neighborhood with blocks made of graham crackers. They used flattened caramels for roadways, and lollipops for streetlights. Coconut spread over white icing gave the illusion of snow (Figure 42–8).

The students mixed yellow and green food coloring into icing to create different colored buildings. Recipes for icing provided opportunities to use measurements and follow directions in

Figure 42–7 A block city begins to take shape

Figure 42–8 Mrs. Moore's first grade class made a Graham Cracker village.

sequence. Writing about the creation of the village and what might be happening within graham cracker walls became a springboard for discussion.

Children made decisions about what should and should not be included in the village. They determined the authenticity of buildings and building size. This activity is especially appropriate for primary age children. Children in this age group are able to incorporate more detail and can be exposed to another's viewpoint. For example, the teacher asked, HOW WILL THE PEOPLE KNOW THAT SCHOOL IS OPEN? Children began asking each other, DO WE NEED A HOSPITAL? WHAT ABOUT A GAS STATION?

If your city or town is located near a river or lake, be sure to include it in construction. Paper straw bridges could be added and the geography of your area explored. You will find that as the children develop questions, they are motivated to find the answers because they need to know something for construction of the city. Thus the block experience also becomes a first research experience.

SCIENCE IN ACTION: OUTDOORS

Virtually all outdoors is science. This is where children can become a part of the natural world. Whether you use the outdoor environment around you to extend and enhance indoor science lessons or design lessons that focus on available outdoor resources, your students will benefit from the experience.

Children will be enthusiastic about exploring the "real" world. After all, "The real thing is worth a thousand pictures." Although this old saying and many of the suggestions for implementing outdoor learning overlap with field trip experiences, many of the learning strategies suggested can also be done in an urban setting. School yards, sidewalks, vacant lots, any strip of ground can be an area for outdoor learning. The important thing is to get your students outdoors and engage them in challenging learning.

Specific plans for outdoor learning will help ensure a successful experience. The following suggestions include teaching strategies that focus on specific science learnings.

Animal Study Activities

Animal homes, habits, and behaviors fascinate children. To begin a successful outdoor experience, assess the previous experiences, skills, and attention spans of your students. Then, review the teacher preparation and control suggestions at the end of this unit and begin.

Animal Homes

Involve children in the study of animal homes. First discuss, WHERE MIGHT AN ANIMAL LIVE? Then plan a field trip to look for animal homes. When planning a trip, keep in mind that most animals make their homes on southern slopes (sunny and warm). When you find a home, examine the area for tracks. Make a cast of the footprints and determine if the home is in use (See Figure 34–7.) Ask, HOW CAN WE TELL IF AN ANIMAL LIVES HERE? (One way is to look for signs such as food scraps and activity around the entrance.) Discuss possible reasons for the selection of this particular location for an animal home and speculate on the possible enemies and living habits of the occupant (Figure 42–9).

Finding Insects

Insects can be hard to find, but signs of their presence are common. The paper nest of a hornet or mud nest of a wasp can be found on buildings, rocks, or some tree branches. Be careful—if the nests are occupied, the owners may sting.

The presence of bark beetles can be seen by the "tunnel" left when they strip bark from logs. Most children are familiar with ant nests and know how to find them. Fallen logs are good locations for observing insects in the winter. The insects are usually sluggish and the stinging types can be more easily observed.

Follow animal home observations with discussion. Ask questions such as, WHAT HAVE WE LEARNED ABOUT THE KINDS OF ANIMALS THAT LIVE AROUND OUR SCHOOL? WHAT ARE THEIR NEEDS? Pick one animal to focus on. Have children write about what one of the animals is thinking as it prepares a home.

A Different Type of Home

The next time you see a swollen, tumor-like bulge on the stem, flower head, stalk, or root of some plants, you might be looking at a unique insect home called a *gall*. The purpose of a gall is to provide an animal home. This happens when some species of insects causes specific kinds of plants to form galls around them. The gall has a hard outer wall and contains a food supply from the plant tissue. Have your students search for galls growing on flowers, bushes, or trees. Lead the children to discover that certain types of galls are found on specific plants. For example, a gall found on a Canada goldenrod (*Solidago canadensis*) is caused by a small, brown-winged fly. This fly (*Eurosta solidaginis*) only forms a gall on the Canada goldenrod (Figure 42–10).

Children will enjoy dramatizing the life of

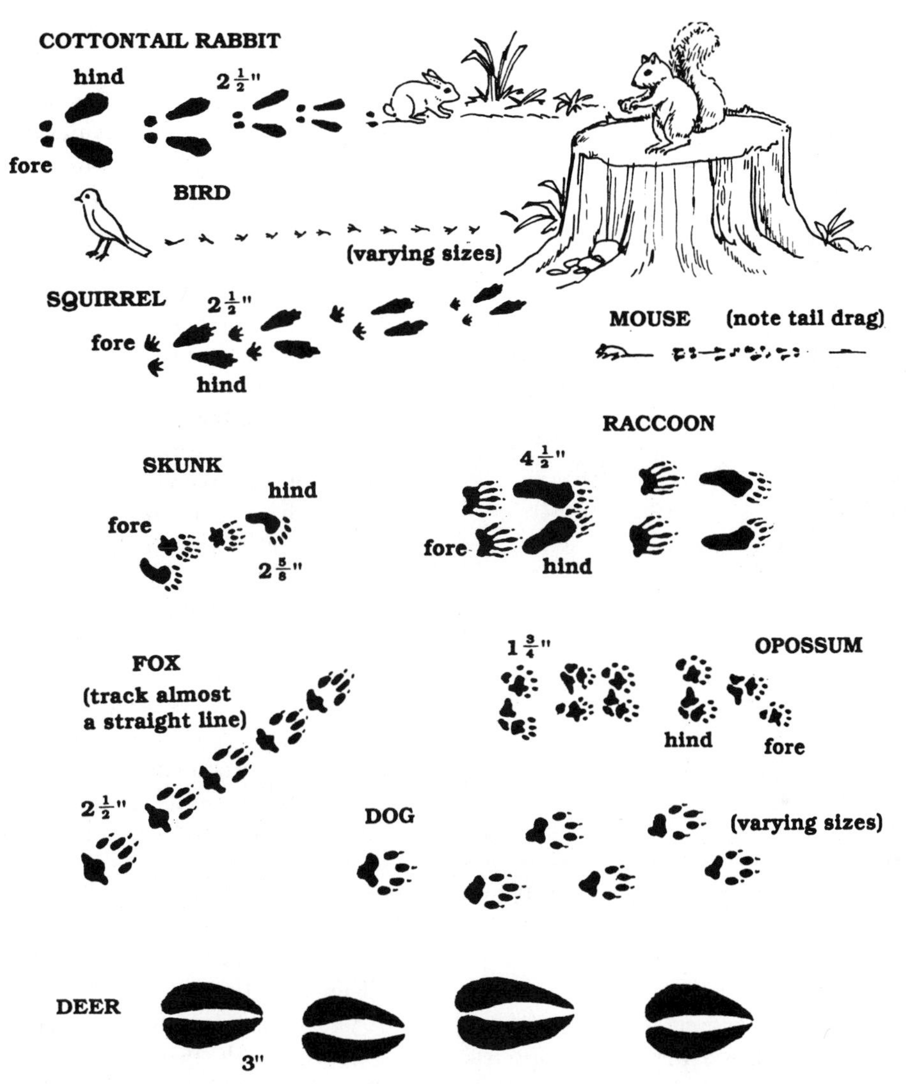

Figure 42–9 You can identify an animal by the tracks that you see.

a gall insect. Say, PRETEND YOU ARE TINY AND HELPLESS. FIND A PLACE WHERE YOU CAN BE SAFE. Have the children pull jackets over their heads to simulate how protected the insect feels. Make a large papier-mâché gall for children to crawl into. Furnish it with a battery-powered light, and children will enjoy crawling in to read, write poems, or turn the light off and simply speculate about what it would be like to be a gall insect inside its home.

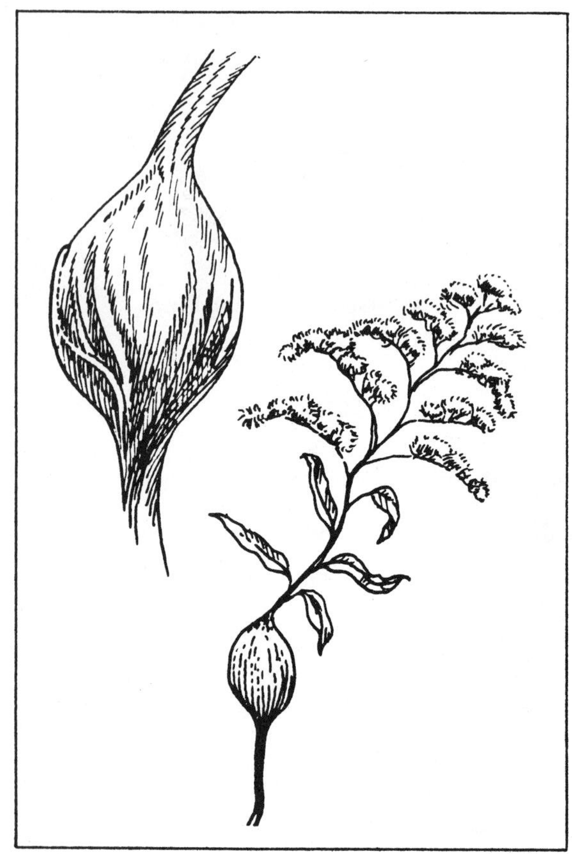

Figure 42–10 A gall is a different type of insect home.

Interview a Spider

Children enjoy becoming reporters and interviewing various wildlife. After discussing what a reporter does and the techniques of interviewing, teams of children can decide on an animal they want to interview.

Birds, Birds, Birds

A bird feeder is a good place to begin observing birds. If you do not have a suitable tree near your school, make your own with limbs or cornstalks tied together and propped up to provide perches and shelter. You will be providing birds with much-needed food, and as the birds come to eat, children will have a change to observe them and their activities at close range.

Children will enjoy making seasonal ornaments and garlands for a holiday tree. Also, use ground feeders or seed dispensers as added attractions for the birds. To decorate a tree, have students string foods that appeal to a wide range of birds. Give them cubes of cheese, popcorn, raisins, and peanuts (in shells). To attract fruit-eating songbirds, add dried fruits to the strings.

Plain peanuts in their shells appeal to both insect- and seed-eating birds, so hang them on fishing line or skewer them on galvanized wire and attach the line to the tree. Then watch the antics of birds such as blue jays as they break open the shells.

Instruct children to keep a notebook of which foods various birds most like to eat. If you prefer, feed birds in aluminum TV dinner trays. Mount the trays on a board and puncture them so that excess moisture can drain out. Fill the different compartments in the trays with cracked corn, sunflower seeds, commercial bird seed, and fruit. Then watch as the birds come to feed and ask questions that help students focus on the differences in the birds' feeding habits. WHICH BIRDS PREFER TO EAT ON THE GROUND? DO ALL OF THE BIRDS EAT SEEDS? WHICH LIKE FRUIT THE BEST?

Have children begin a class (and personal) list of the birds that visit your feeder. With a little practice, children might be beginning a lifelong hobby and interest. You need to be familiar with the birds in your area that will most likely appear at the feeder. Each bird has its own specific habits, food preferences, and actions. Take the children on a field trip and compare birds seen with the birds that visit the school yard feeder (Figure 42–11). (Refer to Unit 44 for additional bird activities.)

Figure 42–11 "Which birds prefer to eat on the ground?"

Figure 42–12 "Describe your plant as it appears today."

Outdoor Plants

My Wild Plant

Observing a wild plant and learning as much as possible through observation makes a good long-term activity for spring. Visit a vacant lot or school parking lot with the children and try to find a spot not likely to be mowed, paved, or interfered with during spring months. Or, contact your school maintenance personnel and ask them to leave a small section of the school yard untouched for a month.

Have each child select one wild plant as their own for close study. A label with a child's name on it can be taped around the stems of the plant they select. Encourage children to begin a plant notebook to make entries about their plant (Figure 42–12). For example:

1. Describe the plant as it appears today.
2. Measure everything you can with a tape measure.
3. Count all plant parts. Does the plant have an odor?
4. What textures did you find on the plant? (You might be able to record these with crayon rubbings.)
5. Does your plant make any sounds?
6. What other plants are the nearest neighbors of yours?
7. Do any animals live on or visit your plant?

During subsequent visits, determine how the plant has grown and changed. Decide what effect the plant has on other plants and animals living nearby, and determine what is good or bad about this plant. Encourage children to make drawings and take photographs so that they will be able to share the story of their plant with others. Then, predict what the plant and growing site will be like in one year.

Hugging a Tree

A variation of selecting a special plant is to have children work in pairs to explore a tree. Blindfold one of the partners. Have the other partner lead the blindfolded one to a tree. Give the children time to touch, smell, and hug the tree. Then, bring the children back to a starting point, take the blindfold off, and ask, CAN YOU FIND THE TREE THAT YOU HUGGED? Children will enjoy finding a special tree to hug; some might even want to whisper a secret to the tree.

Adopt a Rock

Have children select a favorite rock for their collection and examine it closely. Then, put the rock in a bowl with other rocks. Ask, CAN YOU FIND YOUR ROCK? Blindfold students and pass the rocks around the group. Ask, WHICH ROCK FEELS LIKE YOUR ROCK? Then, help children trace the rock to the meal eaten last night. For example, rocks break down into soil, plants grow in soil, and animals live in plants.

What's for Dinner?

After doing adopt a rock, say to the children, GO HOME TONIGHT AND LIST EVERYTHING YOU HAVE FOR DINNER. The next day have children work in groups to discuss the dinner menu and analyze where their food comes from. Help children trace every food back to a plant. For example, Milk to cow to grass. Have reference books available for children to consult. Encourage the children to reach the conclusion that all animals and people need food and that we all depend upon plants for food. Ask, DO THE PLANTS NEED PEOPLE AND ANIMALS? After a lively discussion, point out the decay of animal life that nourishes plants. Then, create a food chain for your bulletin board by using yarn to connect pictures of animals, plants, soil, and rocks. Add the sun and have children create their own food chains to hang from a hanger (Figure 42–13).

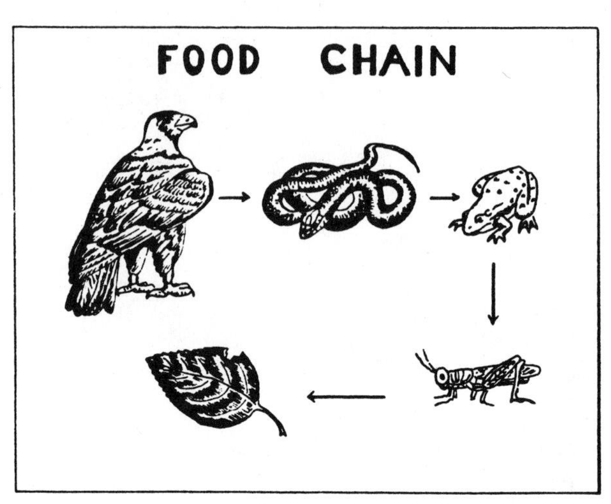

Figure 42–13 A food chain

Scavenger Hunts and Other Ten-Minute Activities

Scavenger hunts are an excellent way to challenge children while focusing their attention on the task at hand. Make up a set of file cards to take with you. Each card should have a challenge on it. For example:

- Find a seed.
- Find three pieces of litter.
- Find something a bird uses for nesting material.
- Find something red.
- Find something a squirrel would eat.
- Find something that makes its own food.
- Find something that shows signs of erosion.
- Find something that shows change.
- Find a bird's feather.

Caution the children to collect only small quantities of the item on their card, or not to collect at all if they will damage something. In this case, write a description of the situation. Pass out the cards and tell the individuals or teams that they have ten minutes to meet the challenge. Discuss the findings back in the classroom.

Circle Game

If the ground is dry, sit in a circle and pass an object such as a rock, leaf, or twig around the circle. As each child touches the object, he or she must say something that is observed about the object. Say, YOU WILL NEED TO LISTEN AND NOT REPEAT AN OBSERVATION MADE BY ANYONE ELSE. Remember, observations are made with the senses. Do not accept inferences or predictions.

While you are still in the circle, move the children apart so that they do not touch each other and ask them to close their eyes and explore the area around them with just their hands (and bare feet if the weather is nice). Then ask them to describe or write the textures that they felt.

Outdoor Learning and Writing Experiences

Writing, drawing, and dictating can be integrated with outdoor learning experiences. Here are a few suggestions.

- Write poems about something that was observed during an outdoor experience.
- Write a story about a living thing in the outdoors that has the power to speak to the humans that come to visit its outdoor area. What might the living thing say? What questions might it ask you?
- Write and draw posters that describe outdoor experiences.
- Keep a written log of outdoor activities.
- Write a short play about the outdoor trip.
- Write letters to someone about one aspect of the outdoor experience.
- Dictate stories about an incident or observation made during the outdoor experience.
- Write how you would feel if you were a plant and it didn't rain for a long time.

Planning for Outdoor Learning

Taking children outdoors can be a challenge if you are not prepared. To ensure the greatest value from the experience, teachers of all age groups should

1. Think about your purpose for including outdoor experiences. How will children benefit? What type of preparation do the children need before they go outdoors?
2. What are the logistics? Will you walk, drive, or ride in cars? What type of clothing is needed? Should you take snacks? Are there people to contact? What are the water and toilet facilities? How much help do you need?
3. Which science concepts will be developed? What do you hope to accomplish?
4. Have you planned what you will be before the experience, during the activity, and after the experience?
5. How much talking do you really need to do?
6. How will you evaluate the experience?
7. What types of follow-up learnings will be provided? What subjects can you integrate into the experience?

Attention Grabbers

Devices for grabbing the attention of a group can be physical, such as pulling out a huge beef thigh bone from your bag when you want to discuss animal bones. The bone will be heavy, but this action is guaranteed to grab the attention of your group.

More subtle attention grabbers include:

1. Look intently at an object to focus group attention on the same object.
2. Have children remind you of tasks, carry various items, assist with the activity, and lead in other tasks. This participation helps keep their attention.
3. Lowering your voice when you want to make a point works well in the classroom and outdoors.
4. Change your position. Sit down with the children when they begin to wander and regroup them.
5. Give the children specific items to look for or match, notes to take, or specific jobs.

Figure 42–14 "How many segments does a worm have?"

Children tend to lose interest if they do not have a task (Figure 42–14).

Additional Control Strategies

Although you might be proficient at controlling children indoors, the outdoors can be quite a different matter. Here are a few tips.

1. Before the outdoor experience, set up a firm set of rules (like indoors). As you know, it is far easier to relax rules than the other way around. However, hurting and frightening animals, crushing plants, and littering should not be tolerated.
2. When children become too active, try an attention grabber or initiate an activity designed to give children a chance to run. For example, RUN TO THE BIG PINE TREE AND BACK TO ME. Relays with rocks as batons will also expend excess energy.
3. Have a prearranged attention signal for activities that require wandering. A whistle, bell, or hand signals work well.
4. When one child is talking and others desperately want your attention, place your hand on their hand to let them know that you recognize them.
5. Let different children enjoy leading the adventure. Occasionally, remove yourself from the line and take a different place in line and go in a different direction. In this way, you lead the group in a new direction with different children directly behind you.
6. Play follow the leader with you as the leader as you guide the line where you want it to go. If there is snow on the ground, have the children walk like wolves: Wolves walk along a trail in single file, putting their feet in the footprints of the wolf ahead of them. If you are at the zoo, try walking to the next exhibit like the animal you have just been observing would walk.
7. Be flexible. If something is not working, just change the activity. Later you can analyze why something was not working the way you had planned.

Exciting outdoor activities do not happen by chance. Begin by planning carefully what you want the children to learn. Then teach the lesson and evaluate the children's learning and your preparation. You will be off to a good start as an outdoor science educator.

SUMMARY

When children make block building decisions, they are thinking like a scientist. They focus on a problem and use thinking skills to arrive at a solution or conclusion. Block building and outdoor explorations give children an opportunity to build their own model of the world. Primary children are still concrete operational thinkers who must learn by manipulating and acting on their environment.

The outdoor environment can be used to

extend and enhance indoor lessons or as a specific place for engaging children in challenging learning. Specific outdoor strategies include the study of animal homes and behavior, wild plants, rocks, food chains, scavenger hunts, and circle games.

Suggestions have been given about how to work directly with outdoor learning as well as strategies for focusing children's attention on outdoor learning and encouraging higher level science thinking. Once children learn how to learn outdoors, they will enjoy the fascinating world around them.

FURTHER READING AND RESOURCES

Cany, S. 1975. *Play Book*. New York: Workman Publishing Company.

Carson, R. 1956. *A sense of wonder*. New York: Harper & Row.

Department of Public Information/Public Affairs. 1988. *Elementary School Notes*. 3 (5). Louisville, Ky.: Jefferson County Public Schools.

Hirsch, E. S. (ed.) 1984. *The block book*. Washington, D.C.: National Association for the Education of Young Children.

Holt, B. G. 1989. *Science with young children*, revised. Washington, D.C.: National Association for the Education of Young Children.

Kamii, S., and DeVries, R. 1976. *Physical knowledge in preschool education*. Englewood Cliffs, N.J.: Prentice-Hall.

Lind, K. K., and Milburn, M. J. 1988. More for the Mechanized Child. *Science and Children*. 25 (6). 39–40.

Lingelbach, J. 1986. *Hands-on Nature*. Woodstock, Vt: Vermont Institute of Natural Science.

McCormack, A. J. 1979. *Outdoor areas as learning laboratories*: Council for Elementary Science International Sourcebook I. Columbus, Ohio: SMEAC Information Reference Center.

Moffitt, M. W. 1977. Children learn about science through block building. *The block book*. Washington, D.C.: National Association for the Education for Young Children.

Provenzo, E. F. Jr., and Brett, A. 1983. *The complete block book*. Syracuse, N.Y.: Syracuse University Press.

Sisson, E. A. 1987. *Nature with children of all ages*. New York: Prentice Hall Press.

Zubrowski, B. 1981. *Messing around with drinking straw construction*. Boston, Mass.: Little, Brown and Company.

SUGGESTED ACTIVITIES

1. Observe children interacting with blocks and other cause-and-effect materials. Record the children's behavior and comments. Do the children control the cause-and-effect reaction? Do they predict what will happen?

2. Prepare a lesson that focuses on blocks, pendulums, inclines, or balls. Teach the lesson to primary age children and evaluate the activity. Share your findings with the class.

3. Design an outdoor activity and teach it to a group of primary age children. Be prepared to teach the lesson to your class and discuss strengths and weaknesses of this approach.

4. Reflect on your own primary education. Were blocks included in your classroom? What types of outdoor activities were used?

5. Secure samples of outdoor education resources. Modify them for primary grades and include them in your file. If possible, write your own teaching lessons around a theme or

science concept suitable for the primary grades.

6. Prepare a list of positive reinforcement statements that you will use when implementing outdoor education. How do the strategies compare with those used in indoor learning? Discuss reasons for any differences.

REVIEW

1. What are the benefits of using blocks with primary age children?
2. Name some ideas that emphasize cause and effect.
3. Why should you include outdoor activities when teaching?

UNIT 43 Math in the Home

OBJECTIVES

After studying this unit, the student should be able to

- Explain the value of parents teaching children at home
- Demonstrate the ability to implement a parent involvement approach
- List guides for parents to effectively teach their child at home
- Share with parents naturalistic, informal, and structured home math teaching experiences

His parents are the child's first teachers (Figure 43–1). The parents and the environment are most influential in the young child's development. Parents can provide the foundation for success or failure. Success is fostered by encouraging a child to be active and to explore. Failure is created by constantly offering "No" each time a child tries to do something new. Praise provides encouragement. Criticism leads to loss of interest. The level of competence reached by the young child is directly related to the type and quality of parent-child interaction.

The competent child has parents who believe that the child wants to learn. They believe the child can learn. They believe learning is fun and natural! These parents recognize that the child imitates them. They recognize that even infants learn through example. They know that the more the child sees and does, the more he learns. These parents also realize that to teach the child to do things for himself is rewarding for the child.

Teachers of young children have contact with parents. Parents often ask for advice about how to help their child do things better. They want to know how their child is doing in school. They want ideas about what they should buy. They want to know how their child gets along with other children his age.

Parents are the child's most important teachers. Teachers need to emphasize this to parents. They need to reassure parents that young children are anxious and eager to learn. Teachers

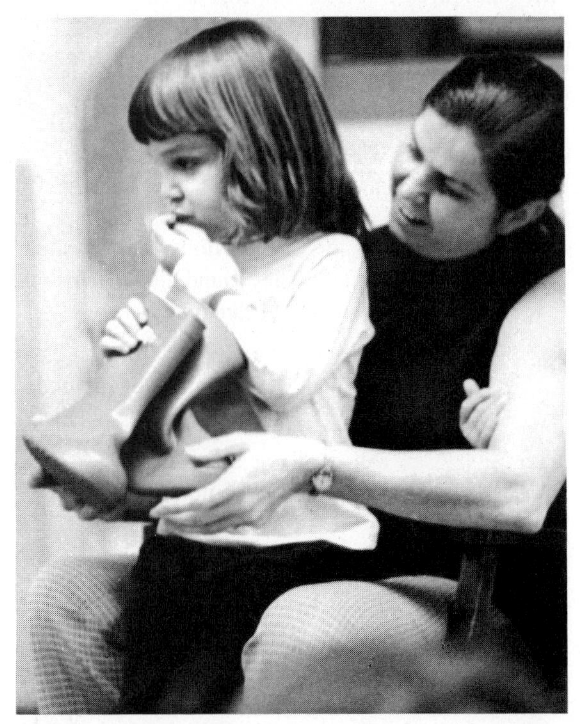

Figure 43–1 As mother helps the young child put on her boots, the young child constructs a concept of one-to-one correspondence.

should help parents understand that their child's natural curiosity drives him to question, explore, and discover more about the world in which he lives. Parents only need to provide the guidance and direction. The child looks to his parents for knowledge and help.

Parents know their child better than anyone else does. Parents can help teachers learn to understand the child better. It is important that parents and teachers communicate on a regular basis.

APPROACHES FOR PARENT INVOLVEMENT IN MATH

There are many ways to help parents understand how to work with their children in specific skill areas. These approaches can be used for math skill development also.

One approach is to send home a newsletter each week or each month (Figure 43–2). In these newsletters, the teacher can list the skills on which the children have been working. She can tell about special experiences the children have had. A clever way to present a newsletter is to write as if the school pet is observing the children. He then informs the parent about the children. This format is enjoyed by parents. Children can also contribute to the newsletter. They can draw pictures and dictate/and or write news stories describing their experiences at school. Two or three children might be asked to contribute to each newsletter.

Some teachers like to have open house for parents. This is another approach for parent involvement. At the open house, the parents can examine the school environment and the teaching

Dear Mr. and Mrs. Foster,

The children in Timmy's class have been learning how to sort things this week. They had all kinds of collections of things out in the classroom. I never thought they could organize all the piles Mrs. Olert put on the table. But guess what – the children found all the things with which to write. They made a separate pile for all the things with which to eat. They found all the items that could be used to tell time. I was really amazed at how much they liked doing this.

We visited a grocery store on Tuesday. When the children came back, Mrs. Olert had pictures of different kinds of foods. The children sorted them according to meats, vegetables, fruits, and beverages. They were excited about visiting the store.

All in all, we had a busy week. Why don't you check and see what items your child can sort for you at home?

As ever,

Freddie the Fish

Figure 43–2 Parent newsletter

materials. Teachers can demonstrate special ways that materials are used to help children learn. Parents particularly like open houses where they can participate by working with the children's materials.

Parents can also be asked to send waste materials to school as needed. The items listed in Figure 43–4 (see page 571) as aids to learning math at home are also useful at school. If the children each bring a pack of small brown paper lunch bags to school at the beginning of the year, when an item is needed the children can draw and/or write the name of the item on the bag, take it home, and ask their parents to put the item in the bag to take to school. Don't be concerned if the younger children write symbols that aren't conventional pictures or words—they will know what it is and can read the symbol to adults.

Parents who have the time may volunteer to assist in the classroom. Parents who are not free during the day or prefer not to be involved in the classroom are often delighted to make games and other materials at home. There should always be an open invitation for parents to visit school.

The parent-teacher conference is another parent involvement approach. This method is used to help parent and teacher learn about each child's needs. Parent concerns about how their child is doing in school can be discussed. Teachers can find out if any special home events might be affecting the child. Specific suggestions for how to best help the child handle any problems can be made. A feeling of mutual concern is needed if the child is to grow and develop to his fullest potential. This can be done by frequent parent-teacher interchanges.

GUIDELINES FOR PARENTS AS TEACHERS

If parents ask for help on how to provide learning experiences for their child, teachers need to be prepared to give them some guidelines. Parents need to be assured that natural ways of learning are best for young children. Parents need to be aware that the child's play is learning. By exploring and discovering through play children construct concepts. However, children do imitate what adults do. Children also model or imitate how adults feel. If the parent is positive about learning, the child will be positive, too.

As the child's first teacher, parents can provide a variety of play experiences. They can provide opportunities to repeat activities. They can provide a lot of conversation. In an atmosphere where children are free to explore and ask questions, they develop self-confidence as they learn many new skills.

There are three basic guidelines that help parents learn how to be effective teachers.

- Patience. The child needs a lot of time to think. If he is pressured to hurry or senses the adult wants him to work faster, he will become upset. When mistakes are made, adults should not scold the child. Learning involves mistakes. The child should be encouraged to try the activity again. Parents can suggest different ways to try the activity. If the child becomes frustrated, parents should stop the activity and try something else.
- Repetition. The young child needs to repeat activities. This is how he learns. He likes repetition. Familiar activities he can do well build his self-confidence. The more successful a child feels, the more he is motivated to try and learn new things (Figure 43–3).
- Concrete experiences. A child is interested in working with real things. He likes a variety of materials to use. The materials can be inexpensive and simple to use. The parents' imagination and the child's ideas can be used to make up concrete math experiences. This should help make these teaching sessions fun for the parent and the child.

Figure 43–3 Grandmother shares memories of the past with this young child.

If a parent asks the teacher for advice on how children learn math, it is important to explain about the naturalistic experiences, the informal experiences, and the more structured ways of learning. The three basic guidelines just listed should be given to parents. A list of inexpensive math materials could also be shared with them. Ideas on specific math activities that can be done at home are included in this unit.

NATURALISTIC

From a very young age, children enjoy counting experiences with their parents. As mother buttons the child's sweater, she counts, "One, two, three buttons on your sweater." As she helps baby put on his shoes, she says, "One shoe, two shoes." As the family sits at the dinner table, older brother may say, "There are three boys in this house and one girl." Mother looks at the clock while carrying baby in her arms and says, "Oh dear, it's almost 2 o'clock. We will be late to the doctor's office." Mother says, "After lunch we are going to grandmother's house. But first, we have to wash the dishes." This type of conversation goes on continually in the home. These experiences provide a natural foundation for later math learnings.

INFORMAL

As the child gets older, mother tells him to put three napkins on the table. Picture books are read to the child that contain math ideas. Birthdays are celebrated, and children count the candles. Mother makes a growth chart. The child weighs himself. Mother and child record this data on the chart. Mother measures the child's height periodically. She uses colored tape to mark changes in height. The child needs new shoes. The shoes sales-person shows the child the instrument he uses to measure feet. Mother teaches the child the proper way to dial the telephone. She teaches the child his telephone number. She helps the child learn his house number and address. As she cooks in the kitchen, she lets the child help her. As she runs errands, the child learns about the post office, the bank, the grocery store, and the gas station. Stamps have different numbers on them. Money is deposited at the bank. Fruits are weighed on a scale at the grocery store. Gas is purchased by the gallon. Experience with an allowance helps young children learn about the value of money. Math is used informally in many activities that people do daily.

STRUCTURED

Structured home math includes providing a specific place for math equipment and a specific time to work on math activities. Structured math activities in the home usually do not begin prior to age three and may be supplemental to school activities. When demonstrating math materials to the child, parents should use few words and movements. For example, the parent who is interested in teaching the child the difference between a circular shape and a square shape should first hold a cutout cardboard circular shape and show it to the child, saying, "This is a circle." The parents should then take their fingers and slowly touch the edges of the circle. Next they should hand the circle to the child and ask, "What is this?" If the child says, "I don't know," then the parent should repeat the word "circle" and put the shape away. The child's refusal indicates he is not interested or he is not ready to participate in a structured shape activity. If the child says, "Circle," then the parents say, "Show me" or "How do you know it's a circle?" The child will touch the outer edges, as he observed the parent doing. Then the activity can continue for the square shape. When the child has demonstrated that he knows a circle and a square, the parent can play games such as "Find me something square in the kitchen," or "Find me a circle shape in the playroom." Preschool children enjoy these kinds of games at home.

When teaching structured activities, parents must be very aware of the importance of being patient. Children need time to do activities. They can be easily distracted. Structured home math activities and materials should be those in which the child is interested and which he is capable of doing. The child should choose the activities. The time spent on structured math activities may be only a few minutes each day. If parents are providing a regular school time for the child at home, they should schedule structured math within that time. The age and interest of the child should determine the length of the formal home school program. A special time may be set aside each week so that the child can anticipate that he will work on math.

Homework becomes an important type of activity in the primary grades. Children can work their way into the more formal homework activities by bringing things requested from home as a part of their prekindergarten and kindergarten experiences. These activities help them to develop responsibility and accustom parents to supporting classroom instruction. Homework should always be an extension of what has been taught at school. It may involve bringing some material to school, doing a simple project, or obtaining some information from a newspaper, magazine, or reference book. Be sure that parents have all the information needed in order to guide the child to completion of the assignment. Assignments for young children should be something that can be easily completed in ten to fifteen minutes.

HOME MATH MATERIALS

The home provides many materials that are suitable for beginning math (Figure 43–4). The button box is full of different shapes and sizes. Ribbons and yarns provide length activities. Pictures, knicknacks, and furniture can be categorized. Sets can be found in clothing, kitchen

Figure 43–4 Children of all ages enjoy repeating stories and acting them out. Predictable stories such as *The Three Billy Goats Gruff* help children develop the concept of sequence; in this case, the story also provides experiences with counting and ordering.

From The Kitchen	From All Around The House
egg cartons	sponges
oatmeal boxes	magazines with pictures
margarine tubs	catalogs (general, seed, etc.)
milk cartons	old crayons
milk jugs	shoe boxes
plastic lids	other cardboard boxes (not corrugated)
egg shells	corrugated cardboard
nuts	cardboard tablet backs
straws	scraps of wallpaper, carpeting
coffee can and lids	Contac paper
other food cans	toothbrushes (for spattering paint)
baby food jars	lumber scraps
potato chip cans	gift wrap (used or scraps)
orange juice cans	gift wrap ribbon and bows
yogurt, cottage cheese, sour cream or dip containers	old greeting cards (pictures)
milk bottle lids	newspapers
individual cereal boxes	jewelry
soft drink bottle caps	wire
plastic holder (soft drink 6-pak carrier)	clothesline
paper plates	clocks
coffee grounds	
plastic bottles or jugs (soap, bleach, etc.)	
plastic or metal tops or lids	
styrofoam meat plates (trays)	
plastic bag twists	
cardboard rolls (paper towels, foil, etc.)	
kitchen scales	
plastic forks, spoons	
From Outside	**From Sewing Scraps**
rocks and stones	buttons
twigs, sticks, bark	snaps
leaves and weeds	fabric
pine cones and nuts	felt
seeds	thread
corn kernels and husks	yarn
soybeans	lace
flowers	ribbon
clay	trim
sea shells	ric rac
	spools

Figure 43–5 Math materials found in the home

items, and father's toolbox. Folding napkins, setting the table, measuring beans, rice, or popcorn include many basic math concepts. A few commercial child math materials should be purchased for use in the home (Figure 43–5). For example, a set of blocks is a basic material. One other type of construction material such as lincoln logs or jumbo legos is also practical. Dominoes and a deck of playing cards can be used for many math activities. Balls of various sizes can be used for math and many other types of activities. Real coins can be used to make comparisons, to count, and to learn coin values. However, most math materials can be homemade. The list of activities that follows uses materials that are available in the home.

MATH ACTIVITIES IN THE HOME

There are many math manipulative activities that parents can do at home with children. The following activities involve counting, matching, measuring, and ordering skills. Math vocabulary can be expanded by these activities. Shape, size, and part-whole relationships can be explored. Numerals are learned, too. Send a suggested activity home in each newsletter.

COUNTING

TITLE: Clap, Bounce, Jump

PURPOSE: To learn how to count

MATERIALS: Child's hands, body

ACTIVITY: Say: LET'S CLAP AND COUNT TO FOUR. EACH TIME WE CLAP WE WILL SAY THE NUMBER. Demonstrate. "Clap-one" "Clap-two." Let the child practice. Increase the range of numbers. (Or you can say Bounce 4 times, Jump 3 times, etc.)

COUNTING

TITLE: Object Counting

PURPOSE: To help the child learn how to count

MATERIALS: Containers of beans, buttons, rocks, coins, beads

ACTIVITY: Let child play with materials. Sit down with child. Say: "COUNT THE BEANS." Then see if the child can count them by himself.

MATCHING

TITLE: Find the Look Alikes

PURPOSE: To help child learn to match two like things

MATERIALS: Make two sets of picture cards. One set of cards should be drawn in one color (orange). The second set should be the same objects but drawn in another color (green) (Figure 43–6). Put the two sets in an envelope.

ACTIVITY: Take out both sets of cards from the envelope. Put all the orange cards in a row on the table. Say: FIND A GREEN CARD LIKE EACH OF THE ORANGE PICTURES. PUT THEM TOGETHER. Show your child what you mean.

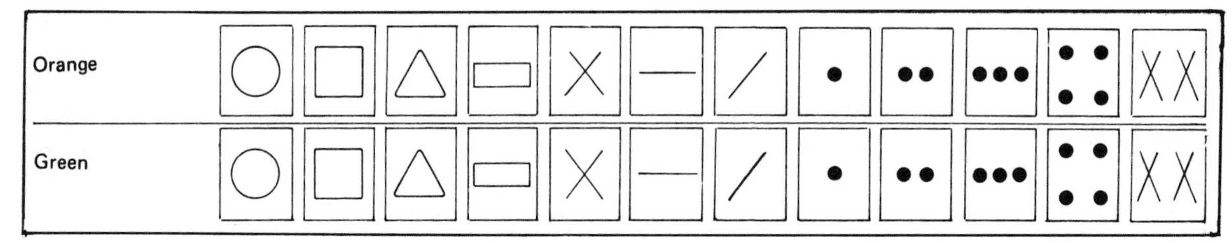

Figure 43–6 Look alike cards

COUNTING/MATCHING SETS TO NUMERALS

TITLE: Egg Container Counting

PURPOSE: To help the child learn how to count and sort

MATERIALS: Egg carton with a numeral written in each cup of the carton (0-11) and the following sets: 11 paperclips, 10 buttons, 9 navy beans, 8 popcorn kernels, 7 beads, 6 sunflower seeds, 5 hairpins, 4 bottle caps, 3 safety pins, 2 poker chips, and 1 small block.

ACTIVITY: First give the child the sets with one to five items. Say, FIND TWO POKER CHIPS AND PUT THEM WHERE IT SAYS TWO. Continue up to the numeral five. The child will think he has been tricked when he is asked "What goes in the zero cup?" It is important to have the zero cup, as too often it is forgotten and later confuses the child. When the child can do these well, add more cups until he can put the right set in each cup.

MEASURING

TITLE: Bathtub Measuring

PURPOSE: To explore liquid volume using containers with different relative dimensions.

MATERIALS: One large, tall, thin container and one short, wide container (both of which hold one liter or one quart of liquid), and a partially filled bathtub

ACTIVITY: Give the child the containers to play with in the bathtub. Observe him pour from one container to the other for several minutes. FILL THE TALL CONTAINER TO THE TOP. NOW POUR THE WATER INTO THE SHORT CONTAINER. DOES THE SHORT CONTAINER HOLD THE SAME AMOUNT OF WATER AS THE TALL CONTAINER?

ENGLISH-METRIC EQUIVALENTS		
Approximate Values		
1 inch	=	25.4 mm
1 inch	=	2.54 cm
1 foot	=	0.305 m
1 yard	=	0.91 m
1 mile	=	1.61 km
1 square inch	=	6.5 cm^2
1 square foot	=	0.09 m^2
1 square yard	=	0.8 m^2
1 acre	=	0.4 hectare
1 cubic inch	=	16.4 cm^3
1 cubic foot	=	0.03 m^3
1 cubic yard	=	0.8 m^3
1 pint	=	0.47 ℓ
1 quart	=	0.95 ℓ
1 gallon	=	3.79 ℓ
1 ounce	=	28.35 g
1 pound	=	0.45 kg
1 U.S. ton	=	0.9 metric ton

SI METRIC BASE UNITS		
Quantity	Unit	Symbol
Length	meter	m
Mass	kilogram	kg
Time	second	s
Temperature	kelvin	K
Electric current	ampere	A
Luminous intensity	candela	cd
Amount of substance	mole	mol
Common Unit	degree Celsius	°C
SUPPLEMENTARY UNITS		
Plane angle	radian	rad
Solid angle	steradian	sr

Figure 43–7 Metric charts

MEASURING: WEIGHT

TITLE: How can it balance?

PURPOSE: To learn about balance and weight

MATERIALS: Hook, flat piece of wood, heavy cord, two plastic bowls

ACTIVITY: Make the balance as shown in Figure 43–8. Hang it in a spot at the child's level. Give the child several cups each half full of a different substance such as dry peas, beans, rice, etc. SAY: PUT SOME RICE IN ONE BOWL. PUT SOME RICE IN THE OTHER BOWL. IS THERE THE SAME AMOUNT IN EACH BOWL? HOW DO YOU KNOW? If they are not the same say: MAKE THEM THE SAME WEIGHT SO THE BOWLS BALANCE.

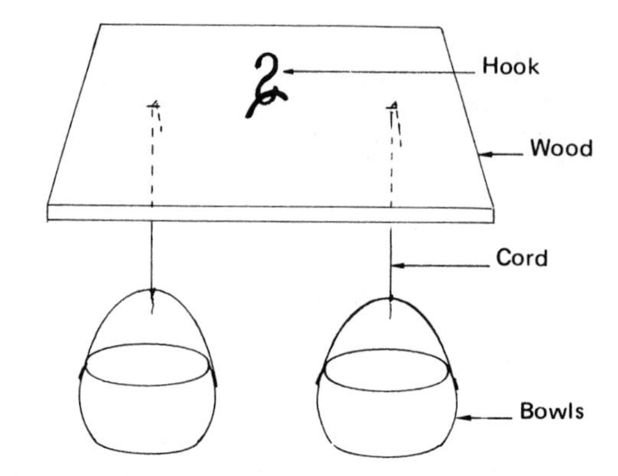

Figure 43–8 A balance to make at home

ORDERING

TITLE: Measuring Cups

PURPOSE: To learn how to arrange items in sequence

MATERIALS: Set of four measuring cups

ACTIVITY: Say to the child, HERE ARE FOUR MEASURING CUPS. EACH IS DIFFERENT IN SIZE. SEE IF YOU CAN ARRANGE THEM IN ORDER OF SIZE, BEGINNING WITH THE BIGGEST AND ENDING WITH THE SMALLEST.

TITLE: Arranging Straws

PURPOSE: To learn how to arrange items in sequence

MATERIALS: Six drinking straws—cut in graduated length. The first can be one inch, the second two inches, etc.

ACTIVITY: Give the graduated straws to the child. Say: ARRANGE THE STRAWS FOR ME, ONE BESIDE THE NEXT, FROM THE SMALLEST TO THE LARGEST.

SIZE/SHAPE

TITLE: Likenesses and Differences

PURPOSE: To help the child notice likenesses and differences in shape (Figure 43–9).

MATERIALS: Containers: *Round* (box, bowl, plate, pie pan, cake pan, clothes basket); *Rectangular* (long cake pan, shoe box, box); *Square* (box, cake pan, clothes basket); *Objects* (magazine, ruler, sheet of paper, jar lid, ball, plates, napkin, handkerchief)

ACTIVITY: Place three containers on a table (one round, one rectangular, and one square). On the table, place many things that have the same shape as the containers. Ask the child to put into each container the things that are shaped most like the container.

TITLE: Size Relationships

PURPOSE: To help the child learn how to order small to large objects

MATERIALS: Cutouts of dogs or any other animal the child likes (Flowers or any objects can be used.)

ACTIVITY: Make several (5 or 6) different sizes of dogs, going from small to large. Give them to the child and say: PUT THE DOGS IN ORDER ACCORDING TO SIZE. GO FROM SMALL TO LARGE.

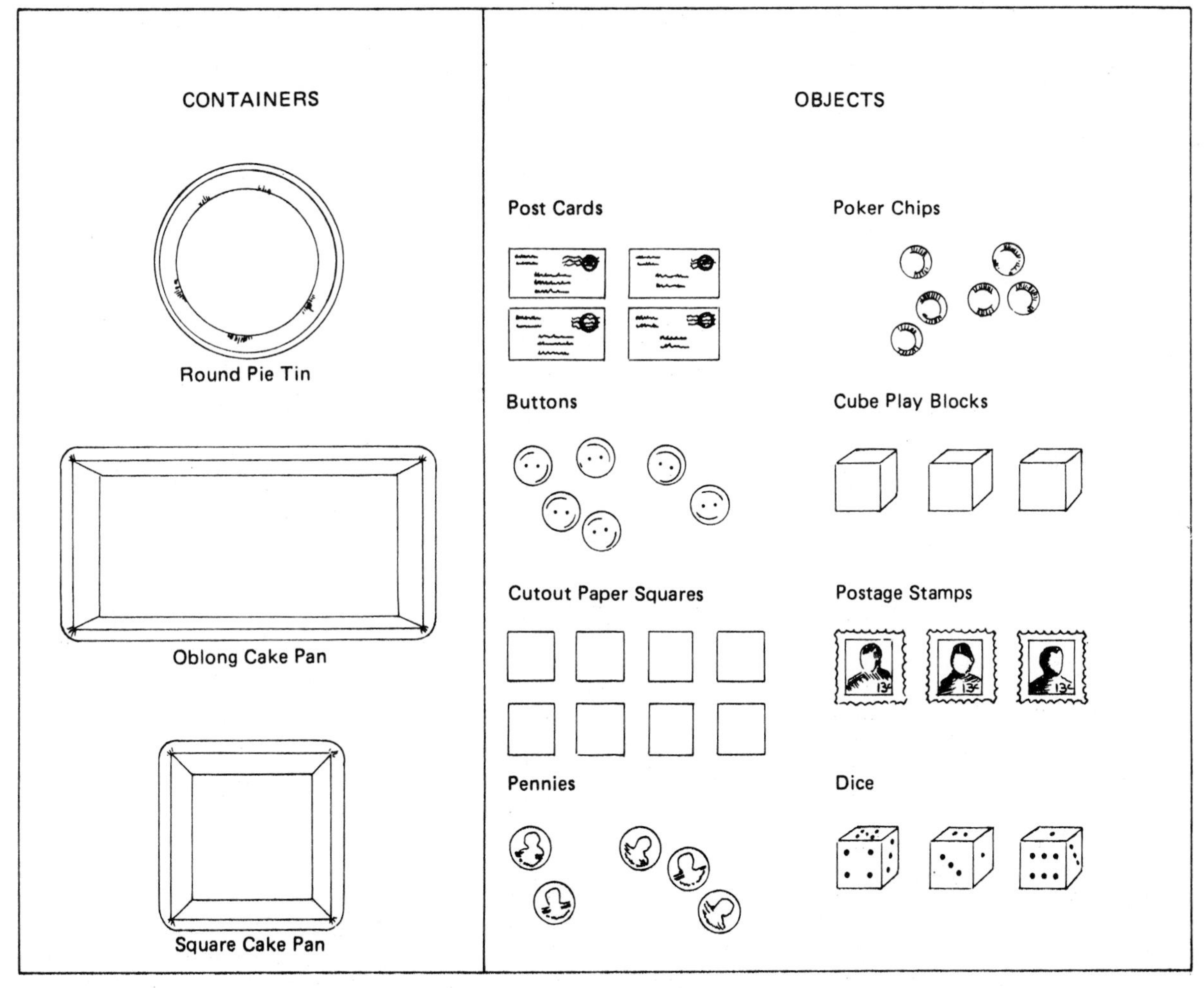

Figure 43–9 Shape—likenesses and differences.

PART/WHOLE

TITLE: Napkin Folding

PURPOSE: To help the child learn the concept of "half" through natural activities

MATERIALS: Napkins

ACTIVITY: Give the child several napkins. Show him how to fold them in half. Then say: FOLD THE NAPKINS IN HALF AND PUT THEM ON THE TABLE. ONE FOR YOU, ONE FOR ME, ONE FOR DAD, ETC.

TITLE: Let's Divide

PURPOSE: To help your child learn parts: Halves, thirds, and quarters

MATERIALS: Pizza, apples, pears, bananas, other food

ACTIVITY: Whenever pizza, apples, pears, bananas, or other suitable foods are served, let the child help slice the food. As the parent and child slice together, talk about the pieces. Say: WE WILL FIRST CUT THE PIZZA IN HALF. NOW WE WILL QUARTER IT. NOW IT IS IN FOUR PARTS OR FOURTHS. Peel a banana—say: LET'S DIVIDE THIS INTO THREE PARTS. THREE PARTS MEAN THAT WE'LL CUT IT INTO THIRDS. Count the portions as you divide the banana.

SHAPES

TITLE: Find the Squares in the Room

PURPOSE: To learn to recognize shapes

MATERIALS: Items of various shapes

ACTIVITY: Say to the child, FIND SOME SQUARE THINGS IN THE ROOM. When the child has found several, you can show him some he may have missed. If he is still interested in playing—ask him to find the round shapes in the room, etc.

TITLE: I'm Thinking About Everything That's Round

PURPOSE: To learn to recognize shapes

MATERIALS: Natural environment

ACTIVITY: Say to the child, LET'S PLAY A GAME. FIRST, I'LL LOOK AROUND THE ROOM AND THINK OF SOMETHING THAT IS ROUND. GUESS WHAT IT IS. Once the child guesses, it is his turn to test the parent. Children love trying to trick their parents.

NUMERALS

TITLE: Sand Printing

PURPOSE: To learn how to write numerals

MATERIALS: Colored sand (purchased at hobby stores or florist shop), aluminum foil pan

ACTIVITY: Pour sand in pan. Say, WATCH ME. Trace a numeral with finger. Have child copy.

VOCABULARY

TITLE: Tonga the Tiger

PURPOSE: To help the child learn about location and space

MATERIALS: A "Tonga the Tiger" puppet (made from paper bag and crayon)

ACTIVITY: Give child verbal directions using Tonga the Tiger. Say, TONGA THE TIGER SAYS, "PUT YOUR HAND ON TOP OF YOUR HEAD." Words to use: Next to your head, at the side of your head, in front of your head, beside your head, behind your head, on your head, off your head, under your head, on top of your head, underneath your head. Other placement words to use: Inside, outside, in, out, by, beside, between, etc. This activity can also be done with objects such as spools and empty margarine containers.

TITLE: Make Yourself Short

PURPOSE: To teach math vocabulary words

MATERIALS: None.

ACTIVITY: Say, LET'S STAND UP AND PLAY A GAME. WATCH ME. SEE IF YOU CAN DO WHAT I ASK YOU. The parent does the body actions with the child at first. Later the parent just tells the child what to do.

1. Make yourself short.
2. Make yourself shorter.
3. Make yourself tall.
4. Make yourself taller.
5. Make yourself fat.
6. Make yourself fatter.
7. Make yourself small.
8. Make yourself smaller.

TITLE: What Did We Do Today?

PURPOSE: To learn time concepts of first, second, and last

MATERIALS: None.

ACTIVITY: When putting the child to bed, the parent talks about what was done first today, what happened second, what happened last. Ask the child what toy he played with *first*, etc. Ask him what TV program he watched *last*. Ask him what his *second* favorite thing was that he did today.

Provide parents of primary children with ideas for practical applications of math such as making a grocery list, picking out the products at the store, and finding out by adding up the prices if the items can be purchased for a predetermined amount. For example, can one quart of milk, two pounds of bananas, and a head of lettuce be purchased for five dollars or less? How much more or less? If a parent is cooking, the child can measure the ingredients and watch the timer. The child who understands standard units can help a parent measure for a carpentry or sewing project. Send homework assignments that relate to these everyday activities. Also explain to parents the importance of having a quiet place and a regular time for homework. If living quarters are crowded and the child doesn't have a private room or a desk, homework supplies can be kept in a bag or a small suitcase and brought out when needed.

SUMMARY

Parents are the child's first teachers. Parents and teachers need to communicate. This helps both adults know more about what the child is learning and doing. This helps adults learn about the child's needs. Approaches for parent involvement include newsletters, open houses, and conferences. To become effective teachers, parents must have patience, make use of repetition, and use concrete experiences. Naturalistic, informal, and structured experiences occur in the home through which children construct math concepts. Most home math equipment and materials are inexpensive. Most home math activities involve manipulative concrete experiences.

FURTHER READING AND RESOURCES

FOR TEACHERS

Berger, E. H. 1981. *Parents as partners in education*. St. Louis, Mo.: C.V. Mosby.

Croft, D. J. 1979. *Parents and teachers: A resource book for home, school and community relations*. Belmont, Calif.: Wadsworth.

Honig, A. 1979. *Parent involvement in early childhood education*. Washington, D. C.: National Association for the Education of Young Children.

McCrea, N. L. 1981. A down-under approach to parent and child food fun. *Childhood Education*. 57 (4): 216–222.

Miller, B. L., and Wilmhurst, A. L. 1980. *Parents and volunteers in the classroom: A handbook for teachers, rev.* Palo Alto, Calif.: R & E Research Associates.

Nedler, S. E., and McAfee, O. D. 1979. *Working with parents. Guidelines for early childhood and elementary teachers*. Belmont, Calif.: Wadsworth.

Parent involvement. 1985. *Dimensions*. 14 (1).

BOOKS FOR PARENTS

Beck, J. 1986. *How to raise a brighter child*. New York: Pocket Books.

Baratta-Lorton, M. 1975. *Workjobs for parents*. Menlo Park, Calif.: Addison-Wesley.

Burtt, K. G. 1981. *Smart toys*. New York: Harper & Row.

Fisher, J. J. 1986. *Toys to grow with infants and toddlers*. New York: Philip Lief Group.

Gesell Institute Series. *Your one year old* up to *Your seven year old*. New York: Delta.

Hagstrom, J. 1986. *Games toddlers play*. New York: Pocket Books.

Jones, S. 1979. *Learning for little kids*. Boston: Houghton-Mifflin.

Marzollo, J., and Lloyd, J. 1973. *How to help your child learn through play.* New York: Scholastic.

Oppenheim, J. 1987. *Buy me! Buy me! the bank-street guide for choosing toys for children.* New York: Pantheon.

Segal, M. 1985. *Your child at play.* New York: Newmarket Press.

Sparling, J., and Lewis, I. 1981. *Learning games for the first three years.* New York: Berkely Books.

White, B. 1985. *The first three years of life, rev.* New York: Prentice-Hall Press.

Wurman, R. S. 1972. *Yellow pages of learning resources.* GEE! Philadelphia, Pa.

PAMPHLET

From the National Association for the Education of Young Children, 1834 Connecticut Avenue, N. W., Washington, D.C. 20009-5786. (800-424-2460). Fifty cents each or 100 for $10. *More than 1, 2, 3: The real basics of mathematics.* Jane Brown McCracken.

SUGGESTED ACTIVITIES

1. Observe a parent-teacher conference. Note how the teacher reports on the child's math concept development and how the parent responds.
2. Write a parent newsletter that focuses on mathematics for the preschool, kindergarten, or primary level. Share the newsletter with the class.
3. Plan a math workshop for parents. Do the workshop with the class. Then do the workshop with a group of parents, if possible.
4. Make a list of guidelines for parents when teaching math at home to their young children.
5. Add fifteen home math activities to your Activities File: five prekindergarten, five kindergarten, and five primary.
6. Volunteer to assist a parent teach his/her child a math concept. Describe what you learned about parents as teachers.
7. Do some of the assessment tasks suggested in the text with a young child. Prepare a parent report. Discuss the report with the class. If possible, share the assessment results with the parent. Give the parent a handout with four or five suggested home math activities.

REVIEW

A. Following is an observation of a parent and child. Read it thoroughly. Then indicate what was appropriate behavior and inappropriate behavior on the part of the parent.

SITUATION: Parent is teaching child about bigger and smaller.

Parent: Deb, Let's see if you can help me find something bigger than this pencil.

Debbie: Oh, that's easy.

Parent: Well, you think you know everything.

Debbie: OK, this book is bigger than the pencil.

Parent: Yeah.

Debbie: And this brush is bigger, but the comb is the same size.

Parent: OK, that's enough, Find anything that is smaller.

Debbie: Oh, this button, key, and penny are smaller.

Parent: You sure are a smartie!

B. Answer each of the following:

1. Why do teachers of young children need to learn how to work with parents?
2. What are five approaches for parent involvement in math?
3. What are some guides to share with parents so they may effectively teach their child at home?
4. Describe two naturalistic, two informal, and two structured home math teaching experiences.
5. Describe the place of homework in the math program.

UNIT 44 Science in the Home

OBJECTIVES

After studying this unit, the student should be able to:

- Explain to parents some strategies for teaching children at home
- Demonstrate the ability to implement a parent involvement approach
- List a variety of science activities that relate science to a child's everyday life

Parents are their children's first teachers. Learning happens on a daily basis in the home: children learn as they cook, observe ants, watch birds, or take a walk in the backyard. Teachers of young children are in a unique position to help parents make good use of these home learning opportunities. Unit 43 provided guidelines for parents as teachers. This unit focuses on specific suggestions for emphasizing science as a vehicle for family learning.

SCIENCE IN THE HOME

Encourage parents to find the science in their home. A large part of a child's time is spent in school, but the majority of time is spent outside of the classroom. Every day at home is filled with opportunities to explore and ask questions that encourage thinking. It should be stressed that family entertainment does not have to be passive such as watching TV. Activities that incorporate daily routines, cooking, and exploring the lives of ants and birds are suggested as opportunities for discovery, science, and family fun.

Daily Routines

Parents should be encouraged to emphasize the skills of science as they go about their daily routines. For example:

1. As laundry is sorted and socks are matched, talk about the differences and similarities in the articles. Then fold the clothes and put them in the correct places. Ask, WHERE DO THE PANTS GO? WHERE SHALL WE PUT THE T-SHIRTS? Even small children will begin the process of classification as they note the differences in characteristics.
2. Children can examine their bodies and compare themselves to animals in a concrete way. They have five toes on each foot, but a horse does not. When a family takes a trip to a farm or a duck pond, the differences and similarities in animal feet can be noted. Ask, HOW MANY TOES DO YOU HAVE? HOW ARE YOUR TOES DIFFERENT FROM A DUCK'S? WHY DO YOU THINK A DUCK HAS A WEB BETWEEN ITS TOES? Activities like this help children be aware of differences in animals and offer opportunities to discuss why an animal is structured in a certain way.
3. When kitchen utensils are returned to drawers after washing or food is put away after a shopping trip, discuss why they go where they do. Say, WHERE SHALL WE PUT THE SPOONS? SHOULD THE

CRACKERS GO IN THE CUPBOARD OR THE REFRIGERATOR? Some parents might want to lay items such as spoons, spatulas, or cups on the table and see how many ways children can device to group items; for example, things you eat with, things you cook with, things you stir with, and so on.

4. Begin a bottle cap collection for classifying, counting, and crafts. Have children sort caps by size, color, and function. Trace around the caps to make designs. Then paste the caps on cardboard and paint it.
5. Collect scraps of wood and make things with the wood. Give children a hammer and some nails and say, LET'S MAKE SOMETHING WITH THIS WOOD. When the wood sculpture is completed, name it and propose a function.
6. As children work with tools such as a hammer, screwdriver, tape measure and various types of screws, bolts, and nails, ask WHAT DO WE DO WITH THIS? HOW IS THIS TOOL USED? If something needs to be fixed, let the children help fix the item. Children enjoy practicing tightening and untightening screws in a board. Simply begin the screws in a board and let children practice the type of motion needed to operate a screwdriver (Figure 44–1).

Figure 44–1 "What do we do with this tool?"

Cooking with Children

Cooking provides many opportunities for parents to provide children practical applications of science. When the parent is cooking, the child can measure the ingredients, observe them as they change form during cooking (or mixing), and taste the final product.

Children should be given as much responsibility for the food preparation as possible. This might include shopping for the food, washing, possibly cutting (carefully supervised, of course), reading and following the recipe, baking, cooking, or freezing, setting the food on the table and cleaning up. The more the parent does, the less the child learns (Figure 44–2).

Try making an easy pizza. You will need muffins, tomato sauce, oregano, and meat and mozzarella cheese slices. Spread one-half of the muffin with a tablespoon of tomato sauce. Add a pinch of oregano. Sprinkle meat on the sauce and add a layer of cheese. Place the little pizza on a cookie sheet and bake for ten minutes at 425 degrees.

Children enjoy getting creative with food. Create "bugs on a log" by spreading peanut butter in pieces of celery. Top off the "log" with raisin "bugs" (Figure 44–3).

Who Invited the Ants?

Encourage families to use their backyard as a resource for teaching science. Many of the activities found in previous units are appropriate

Figure 44–2 Give children responsibility in food preparation.

for use in the home. Select topics that are relevant to family life and suggest them to parents. For example, the following suggestions turn uninvited picnic visitors into a family science exploration:.

1. You will need a spoonful of tuna, a spoonful of honey, and a piece of fruit for this activity. Ask, WHAT WOULD ANTS EAT IF YOU INVITED THEM TO A FAMILY PICNIC? LET'S HAVE A PICNIC FOR THE ANTS IN OUR BACKYARD. Then let the fun begin. However, a few cautions should be noted. Ants belong to the same family as bees and wasps (some ants sting). They have strong jaws and their bites can hurt a lot, so be very careful when dealing with ants.

   To begin explorations, go on an anthill hunt, usually a small pile of dirt with a hole in the middle of it. Or, if you spot any ants, follow them back to their home. If this doesn't work, any bare patch of ground will be fine for observing ants.

   Arrange the food on the ground (one foot apart). Put the honey in a leaf and the meat and fruit directly on the ground. Observe closely. Ask, WHICH KIND OF FOOD DO THE ANTS GO TO FIRST? DO YOU THINK THE ANTS GO TO THEIR FAVORITE FOOD OR TO THE FOOD CLOSEST TO THEM? DO ALL OF THE ANTS CHOOSE THE SAME FOOD?

   Observe ant behavior by asking, DO THE ANTS CARRY THE FOOD BACK TO THE ANT HILL OR EAT IT ON THE SPOT? Sometimes ants act like messenger ants. When they find food, they go back to the nest and tell the others. Then everybody comes to your picnic.

   This activity extends to the sense of smell. Ask, HOW DO YOU THINK THAT THE ANTS KNOW WHERE TO FIND THE FOOD? CAN THEY SEE THE FOOD? CAN THEY HEAR THE FOOD? Explain that as the ant runs to tell about the picnic, it leaves a scent trail for the other ants to follow (Figure 44–4).

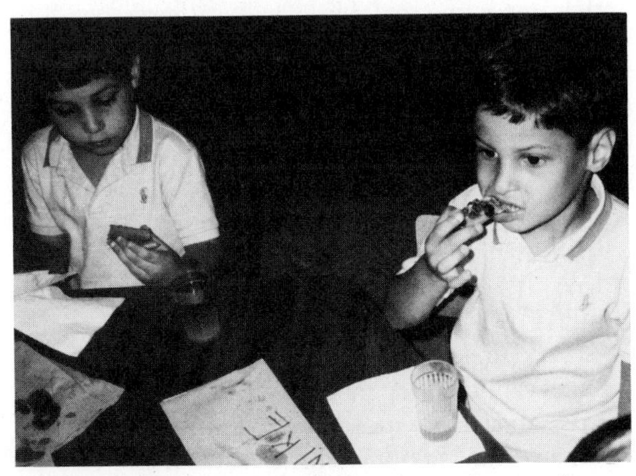

Figure 44–3 Children get creative and eat "Bugs on a Log."

Scent Trails

Which member of the ant family can smell the best? Make a scent trail on your lawn by

Figure 44–4 Watching ants find the food

placing drops of extract on pieces of cutup sponge. Use several distinctive scents such as peppermint, cinnamon, or lemon and create trails. See if family members can find and follow the trail. Mixing extract with water and spraying it with a spray bottle in a trail pattern on the ground also works well.

Families will enjoy learning more about their picnic visitors. Most of the ants that we see are worker ants. Their job is to build and maintain the nest and find food for the colony. Soldier ants live up to their name by defending the anthill against invaders. There is only one queen ant. She rules the nest and lays the eggs that populate the colony. Some queen ants can live fifteen years.

There is a variety of ants with interesting habits and life-styles to discover. For example, some ants even keep tiny insects called aphids to produce a sweet juice for them. The ants "milk" them for the sweet juice in much the same way that cows are milked. Another kind of ant farmer chews up leaves and spreads them out so that an edible fungus will grow. Some worker honey pot ants use their second stomach to store honeydew. They get so fat with honeydew that they hang in their nest like honey pots. Other ants take the honeydew from them when they are hungry.

Feed the Birds

To help the children recognize different kinds of birds and to discriminate differences in birds' sizes and shapes, feeding styles, and food preferences, create your own bird feeding program. In addition to the suggestions for learning about birds found in Unit 42, make beef suet—hard fat from about the kidneys and loins—to help keep up the birds' energy. The suet helps birds maintain their high body temperature. Ask a butcher for suet that is short and not stringy (stringy suet is hard for the birds to eat and does not melt down smoothly).

You can offer the suet to the birds in many ways. Try putting it in a soap dish attached to a tree limb with chicken wire, or hang it in an onion or lobster bait bag. Or make suet ornaments with grapefruit rinds or coconut shells. To do so, chop the suet or put it through a meat grinder and then melt it in a double boiler. Pour the liquid suet into the rinds or shells to which you have already attached wire or string hangers and set the containers aside in a cool place until the suet hardens. Then hang them on your tree.

How about a bottle cap suet log? Simply nail bottle caps to one side of a dead bough. Pour melted suet into the caps and set the bough aside until the suet hardens. Woodpeckers and other medium-sized birds will gather around to eat from the suet log.

Or make attractive suet pinecones. Melt the suet and spoon it over the pinecones (to which you have already attached string or wire). Sprinkle the cones with millet, push sunflower seeds down into the cones' scales, and spoon more warm suet over the cones to secure the seeds. Place the cones on waxed paper and refrigerate until firm. Later, hang the cones from the tree as a snack for small birds like chickadees.

Peanut butter mixture makes great food for birds, too, since peanuts have high nutritional content and mixtures made with them can be spread on tree bark, placed in the holes of a log or a bottle cap feed, or hung from pinecones. But before giving the peanut butter to birds, be sure to mix cornmeal into it (one cup of peanut butter to five cups of cornmeal). This will make the peanut butter mixture easier for the birds to swallow. It is possible for birds to choke on peanut butter when it is not mixed with anything.

Try mounting a whole ear of dried corn in a conspicuous place, perhaps by nailing it to a post. Then have the children predict which birds will be able to eat the corn (Figure 44–5). (Only birds with large beaks will be able to crack the whole kernels of corn.)

A Bird Walk

A bird walk will heighten the observational skills of everyone involved. Here are some things to look for. Families can look for birds that:

- Hop when they move on the ground
- Peck at the ground
- Hold their heads to one side and appear to be listening to something in the ground
- Flap their wings a lot when they fly
- Glide and hardly move their wings
- Climb on the side of trees
- Fly alone
- Fly with many other birds
- Make a lot of noise
- Eat alone
- Blend in well with the grass, trees, or sky

Select a favorite bird and find out as much as you can about it. This can be a family project. Use birdcall audio or videotapes to identify the birds that have been seen and heard on the bird walk. Children will enjoy creating bird stories, art projects, puzzles, and reading more about the birds they have observed (Figure 44–6).

The following books are helpful in identifying birds and creating a backyard habitat:

Burke, Ken 1983. *How to attract birds.* San Francisco: Ortho Books. A well-illustrated book with an informative text about providing food, water, and nest sites for both eastern and western birds.

Cook, Beverly C. 1978. *Invite a bird to dinner.* New York: Lathrop. A children's book about feeding birds with good ideas for making feeders out of everyday materials.

Cosgrove, Irene 1976. *My recipes are for the birds.*

Figure 44–5 "What kind of food do you think birds feed their young?"

Figure 44–6 Children enjoy putting together bird puzzles.

New York: Doubleday. A book full of recipes, featuring treats such as Cardinal Casserole, Finch Fries, and Dove Delight.

Kress, Stephen W. 1985. *The Audubon Society guide to attracting birds.* New York: Scribner's.

SUMMARY

Science provides many opportunities for informal family sharing. Parents can encourage children to explore, ask questions, and think about the world around them. A single guiding question from a parent can turn a daily routine into a learning experience. As children cook, observe, sort, investigate, and construct, they are using the skills needed to learn science. Birds, ants, and other backyard inhabitants make ideal subjects for observation and exploration.

FURTHER READING AND RESOURCES

Lind, K. K. 1986. The bird's Christmas. *Science and Children.* 24 (3): 34–35.

MCCPTA-EPI Hands-on-Science 1987. *Putting together a family science festival.* Washington, D.C.: National Science Teachers Association.

Smithsonian Family Learning Project 1987. *Science activity book.* New York: Galison Books.

Wanamaker, N., Hearn, K., and Richard, S. 1979. *More than graham crackers.* Washington, D.C.: National Association for the Education of Young Children.

SUGGESTED ACTIVITIES

1. Observe a parent-teacher conference. Note how the teacher reports on the child's science concepts development and how the parents respond.
2. Write a parent newsletter that focuses on science in the home. Share the newsletter with the class.
3. Reflect on your own science in the home experiences, either as a child or as a parent. What kinds of opportunities were available to you? Which do you provide?
4. Make a list of guidelines for parents to follow when teaching science to their young children at home.
5. Add fifteen home science activities to your Activities File: five prekindergarten, five kindergarten, and five primary.
6. Plan a science in the home workshop for parents. Present the workshop to the class.
7. Interview at least three parents. Ask them what types of science activities they do with their children. Find out what problems, concerns, or needs they might have in doing science with their children.

REVIEW

1. Why is the home a good place to emphasize science?
2. List three opportunities for learning science in the home.
3. How does cooking relate to science?

APPENDIX A

Developmental Assessment Tasks*

CONTENTS

*Tasks that have been used as samples in the text.

SENSORIMOTOR: LEVEL 1

1A — **Sensorimotor, Age 2 months**

General Development

METHOD: Interview
SKILLS: Perceptual/motor
MATERIALS: Familiar object/toy such as a rattle
PROCEDURES/EVALUATION:

1. Talk to the infant. Notice if he seems to attend and respond (by looking at you, making sounds, and/or changing facial expression).
2. Hold a familiar object within the infant's reach. Note if he reaches out for it.

3. Move the object through the air across the infant's line of vision. He should follow it with his eyes.
4. Hand the small toy to the infant. He should hold it for two to three seconds.

1B

Sensorimotor
Age 4 months

General Development
METHOD: Observation
SKILLS: Perceptual/motor
MATERIALS: Assortment of appropriate infant toys
PROCEDURES/EVALUATION:

1. Note each time you offer the infant a toy. Does she usually grab hold of it?
2. Place the infant where it is possible for her to observe the surroundings (such as in an infant seat) in a situation where there is a lot of activity. Note if her eyes follow the activity and if she seems to be interested and curious.

1C

Sensorimotor
Age 6 months

General Development
METHOD: Interview and observation
SKILLS: Perceptual/motor
MATERIALS: Several nontoxic objects/toys including infant's favorite toy
PROCEDURES/EVALUATION:

1. One by one hand the infant a series of nontoxic objects. Note how many of his senses he uses for exploring the objects. He should be using eyes, mouth, and hands.
2. Place yourself out of the infant's line of vision. Call out to him. Note if he turns his head toward your voice.
3. When the infant drops an object, note whether or not he picks it up again.
4. When the infant is eating, notice if he can hold his bottle in both hands by himself.
5. Show the infant his favorite toy. Slowly move the toy to a hiding place. Note if the infant follows with his eyes as the toy is hidden.

1D

Sensorimotor
Age 12 months

General Development
METHOD: Interview and observation

SKILLS: Perceptual/motor and receptive language
MATERIALS: Two bells or rattles, two blocks or other small objects, two clear plastic cups, pillow or empty box, a cookie, if desired
PROCEDURES/EVALUATIONS:
1. Note if the infant will imitate you when you do the following activities (for each task provide the infant with a duplicate set of materials):
   a. Shake a bell (or a rattle).
   b. Play peek-a-boo by placing your open palms in front of your eyes.
   c. Put a block (or other small object) into a cup; take it out of the cup and place it next to the cup.
2. Partially hide a familiar toy or a cookie under a pillow or a box as the child watches. Note whether the infant searches for it.
3. Note whether the infant is creeping, crawling, pulling up to her feet, trying to walk, or is actually walking.
4. Note whether the infant responds to the following verbal commands:
   a. NO, NO
   b. GIVE ME THE (*Name of object*).

RESOURCES:
Charlesworth, R. 1987. *Understanding Child Development—for Adults Who Work With Young Children. 2nd ed.* Albany, N.Y.: Delmar.

SENSORIMOTOR: LEVEL 2

2A — **Sensorimotor, Ages 12–18 months**

General Development
METHOD: Interview and observation
SKILLS: Perceptual/motor and receptive language
MATERIALS: Several safe containers (i.e., plastic is good) and a supply of safe, nontoxic objects
PROCEDURES/EVALUATION:
1. Give the child several containers and the supply of small objects. Note if he fills the containers with objects and dumps them out repeatedly.
2. Ask the child, POINT TO YOUR NOSE, HEAD, EYES, FOOT, STOMACH.
3. Hide a familiar object completely. Note whether the child searches for it.

2B — **Sensorimotor, Ages 18–24 months**

General Development
METHOD: Interview and observation

SKILLS: Perceptual/motor and receptive and expressive language
MATERIALS: Child's own toys (or other assortment provided by you such as a ball, toy dog, toy car, blocks, baby bottle, doll, and the like)
PROCEDURES/EVALUATION:
1. During playtime observations, note if the child is beginning to organize objects in rows and put similar objects together in groups.
2. Ask the child to point to familiar objects. POINT TO THE BALL (CHAIR, DOLL, CAR).
3. Note whether the child begins to name the parts of her body (usually two parts at 18 months).

RESOURCES:
Charlesworth, R. 1987. *Understanding Child Development—for Adults Who Work With Young Children. 2nd ed.* Albany, N.Y.: Delmar.

PREOPERATIONAL: LEVEL 3

3A* **Preoperational Ages 2–3**

One-to-One Correspondence: Unit 8
METHOD: Observation, individuals or groups
SKILLS: Child demonstrates one-to-one correspondence during play activities
MATERIALS: Play materials that lend themselves to one-to-one activities, such as small blocks and animals, dishes and eating utensils, paint containers and paintbrushes, pegs and pegboards, sticks and stones, and the like
PROCEDURE: Provide the materials and encourage the children to use them.
EVALUATION: Note if the children match items one to one such as putting small peg dolls in each of several margarine containers or on top of each of several blocks that have been lined up in a row.

3B **Preoperational Ages 2–3**

Number and Counting: Unit 9
METHOD: Interview
SKILL: Child understands the concept of "twoness" and can rational count at least two objects
MATERIALS: Ten counters (cube blocks, Unifix® Cubes, or other objects)
PROCEDURES:
1. Ask, HOW OLD ARE YOU?

2. Give the child two objects. HOW MANY (*name of objects*) ARE THERE? If the child succeeds, try three objects. Go on as far as the child can go.

EVALUATION:

1. May hold up appropriate number of fingers or answer "two" or "three."
2. Should be able to rational count two objects (or possibly recognize two without counting).

3C **Preoperational**
Ages 2–3

Sets and Classifying, Informal Sorting: Unit 10

METHOD: Observation and informal interviewing

SKILL: While playing, the child groups toys by various criteria such as color, shape, size, class name, and so on

MATERIALS: Assortment of normal toys for two- to three-year-olds

PROCEDURE: As the child plays, note whether toys are grouped by classification criteria (see Unit 10). Ask, SHOW ME THE RED BLOCKS. WHICH CAR IS THE BIGGEST? FIND SOME SQUARE BLOCKS.

EVALUATION: The child should naturally group by similarities; should be able to group objects by at least one or two colors, and find objects from the same class.

3D **Preoperational**
Ages 2–3

Comparing, Informal Measurement: Unit 11

METHOD: Interview

SKILL: Child can respond to comparison terms applied to familiar objects.

MATERIALS: Pairs of objects that vary on comparative criteria such as:

large-small heavy-light
long-short cold-hot
fat-skinny higher-lower

PROCEDURE: Show the child the pairs of objects one pair at a time. Ask, POINT TO THE BIG (BALL). POINT TO THE SMALL or LITTLE (BALL). Continue with other pairs of objects and object concept words.

EVALUATION: Note how many of the objects the child can identify correctly.

3E

Preoperational
Ages 2–3

Comparing, Number: Unit 11

METHOD: Interview

SKILL: Shown a set of one and six or more, the child can identify which set has more

MATERIALS: Twenty counters (i.e., pennies, Unifix® Cubes, cube blocks)

PROCEDURE: Place two groups of objects in front of the child, one group with a set of one object and one group with a set of six or more. Ask, WHICH HAS MORE (*object name*)? POINT TO THE ONE WITH MORE.

EVALUATION: Note if the child identifies the group that contains more.

3F

Preoperational
Ages 2–3

Shape, Matching: Unit 12

METHOD: Interview

SKILL: Child can match an object or cutout shape to another of the same size and shape.

MATERIALS: Attribute blocks or shape cutouts, one red circle, square, and triangle, one green circle, square, and triangle. All should be the same relative size.

PROCEDURE: Place the three green shapes in front of the child. One at a time, show the child each of the red shapes and ask, FIND A GREEN SHAPE THAT IS THE SAME AS THIS RED ONE.

EVALUATION: The child should be able to make all three matches.

3G*

Preoperational
Ages 2–3

Space, Position: Unit 13

METHOD: Interview

SKILL: Given a spatial relationships word, the child can place objects relative to other objects on the basis of that word.

MATERIALS: A small container such as a box, cup, or bowl and an object such as a coin, checker, or chip

PROCEDURE: PUT THE (*object name*) IN THE BOX (or CUP or BOWL). Repeat using other space words: ON, OFF, OUT OF, IN FRONT OF, NEXT TO, UNDER, OVER.

EVALUATION: Note if the child is able to follow the instructions and place the object correctly relative to the space word used.

3H* — **Preoperational, Ages 2–3**

Parts & Wholes, Missing Parts: Unit 14

METHOD: Interview

SKILL: Child can tell which part(s) of objects and/or pictures of objects are missing.

MATERIALS: Several objects and/or pictures of objects and/or people with parts missing. Some examples are:

Things: Doll with a leg or arm missing
Car with a wheel missing
Cup with a handle broken off
Chair with a leg gone
Face with only one eye
House with no door

Pictures: Mount pictures of common things on poster board. Parts can be cut off before mounting.

PROCEDURE: Show the child each object or picture. LOOK CAREFULLY. WHICH PART IS MISSING FROM THIS (*name of object*)?

EVALUATION: Note if the child is able to tell which parts are missing in both objects and pictures. Does she have the language label for each part? Can she perceive what is missing?

3I — **Preoperational, Ages 2–3**

Ordering, Size: Unit 17

METHOD: Interview

SKILL: Child can order three objects that vary on one size dimension.

MATERIALS: Three objects of the same shape that vary on one size dimension such as diameter:

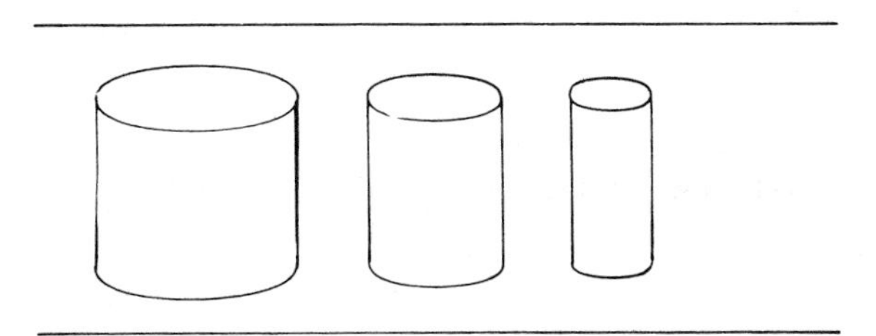

Paper towel rolls can be cut into proportional lengths (heights) for this task. More objects can be available in reserve to be used for more difficult seriation tasks.
PROCEDURE: WATCH WHAT I DO. Line up the objects in order from fattest to thinnest (longest to shortest, tallest to shortest). NOW I'LL MIX THEM UP, (do so) PUT THEM IN A ROW LIKE I DID. If the child does the task with three objects, try it with five.
EVALUATION: Note whether the objects are placed in a correct sequence.

3J — **Preoperational, Ages 2–3**

Measuring, Volume: Unit 18

METHOD: Observation
SKILL: Child evidences an understanding that different containers hold different amounts.
MATERIALS: A large container filled with small objects such as small blocks, paper clips, table tennis balls, or teddy bear counters or with a substance such as water, rice, or legumes. Several different size small containers for pouring.
PROCEDURE: Let the children experiment with filling and pouring. Note any behavior that indicates they recognize that different containers hold different amounts.
EVALUATION: Children should experiment, filling containers and pouring back into the large container, pouring into larger small containers, and into smaller containers. Note behaviors such as if they line up smaller containers and fill each from a larger container or fill a larger container using a smaller one.

PREOPERATIONAL: LEVEL 4

4A — **Preoperational, Ages 3–4**

One-to-One Correspondence, Same Things/Related Things: Unit 8

METHOD: Interview
SKILLS:
1. Child can match, in one-to-one correspondence, pairs of objects that are alike.
2. Child can match, in one-to-one correspondence, pairs of objects that are related but not alike.

MATERIALS:
1. Four different pairs of matching objects (such as two toy cars, two small plastic animals, two coins, two blocks)

2. Two groups of four related objects such as four cups and four saucers, four cowboys and four horses, four flowers and four flowerpots, four hats and four heads.

PROCEDURES:

1. Matching like pairs. Place the objects in front of the child in a random array. FIND THE THINGS THAT BELONG TOGETHER. If there is no response, pick up one. FIND ONE LIKE THIS. When the match is made, FIND SOME OTHER THINGS THAT BELONG TOGETHER. If there is no spontaneous response, continue to select objects and ask the child to find the one like each.
2. Matching related pairs. Place two related groups of four in front of the child in a random array. FIND A CUP FOR EACH SAUCER (or COWBOY FOR EACH HORSE).

EVALUATION:

1. Note if the child matches spontaneously and if he makes an organized pattern (such as placing the pairs side by side or in a row).
2. Note if the child is organized and uses a pattern for placing the objects (such as placing the objects in two matching rows).

4B **Preoperational**
Ages 3–4

Number and Counting, Rote and Rational: Unit 9

METHOD: Interview

SKILL: Child can rote and rational count

MATERIALS: Twenty counters (i.e., cube blocks, pennies, Unifix® Cubes)

PROCEDURE: First have the child rote count. COUNT FOR ME. START WITH ONE AND COUNT. If the child hesitates, ONE, TWO, . . . WHAT COMES NEXT? Next ask, HOW OLD ARE YOU? Finally, place four counters in front of the child. COUNT THE (*objects*). HOW MANY (______) ARE THERE? If the child cannot count four items, try two or three. If she counts four easily, put out more counters and ask her to count as many as she can.

EVALUATION: Note if she can rote count more than five and rational count at least five items. When she rational counts more than four, she should keep track of each item by touching each methodically or moving those counted to the side.

4C **Preoperational**
Ages 3–4

Sets and Classifying, Object Sorting: Unit 10

METHOD: Interview

SKILL: Child can sort objects into groups using logical criteria

MATERIALS: Twelve objects: two red, two blue, two green, two yellow, two orange, two purple. There should be at least five kinds of objects. For example:

Color	*Object 1*	*Object 2*
red	block	car
blue	ball	cup
green	comb	car
yellow	block	bead
orange	comb	cup
purple	bead	ribbon

In addition, you will need six to ten small containers (bowls or boxes).

PROCEDURE: Place the twelve objects in random array in front of the child. Provide him with the containers. PUT THE TOYS THAT BELONG TOGETHER IN A BOWL (BOX). USE AS MANY BOWLS (BOXES) AS YOU NEED.

EVALUATION: Note whether the child uses any specific criteria as he makes his groups.

4D*

Preoperational
Ages 3–4

Comparing, Number: Unit 11

METHOD: Interview

SKILL: Child can compare sets and identify which set has more or less (fewer).

MATERIALS: Two dolls (toy animals or cutout figures) and ten cutout poster board cookies

PROCEDURE: Place the two dolls (toy animals or cutout figures) in front of the child. WATCH, I'M GOING TO GIVE EACH DOLL (OR ______) SOME COOKIES. Put two cookies in front of one doll and six in front of the other. SHOW ME THE DOLL (______) THAT HAS MORE COOKIES. Now pick up the cookies and put one in front of one doll and three in front of the other. SHOW ME THE DOLL (______) THAT HAS FEWER COOKIES. Repeat with different amounts.

EVALUATION: Note whether the child consistently picks the correct amounts. Some children may understand more but not fewer. Some may be able to discriminate if there is a large difference between sets, such as two versus six, but not small differences, such as four versus five.

4E*

Preoperational
Ages 3–4

Shape, Identification: Unit 12

METHOD: Interview

SKILL: When given the name of a shape, the child can point to a drawing of that shape.

MATERIALS: On pieces of white poster board or on 5 1/2 × 8 file cards, draw the following shapes with a black marker (one shape on each card): circle, square, and triangle.

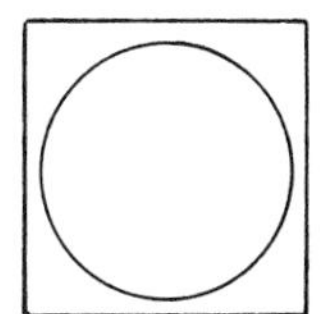

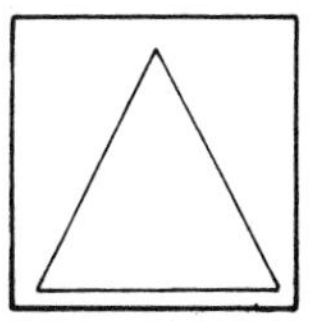

PROCEDURE: Place the cards in front of the child. POINT TO THE SQUARE. POINT TO THE CIRCLE. POINT TO THE TRIANGLE.
EVALUATION: Note which, if any, of the shapes the child can identify.

4F* **Preoperational Ages 3–4**

Space, Position: Unit 13

METHOD: Interview
SKILL: Child can use appropriate spatial relationship words to describe positions in space.
MATERIALS: Several small containers and several small objects; for example, four small plastic glasses and four small toy figures such as a fish, dog, cat, and mouse.
PROCEDURE: Ask the child to name each of the objects so you can use his name for it if it is different from yours. Line up the glasses in a row. Place the animals so that one is *in*, one *on*, one *under*, and one *between* the glasses. Say,

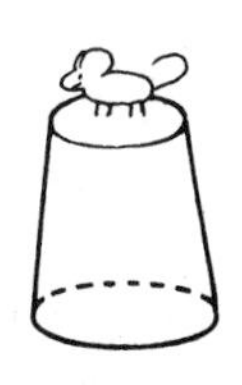

TELL ME WHERE THE FISH IS. Then, TELL ME WHERE THE DOG IS. Then, TELL ME WHERE THE CAT IS. Finally, TELL ME WHERE THE MOUSE IS. Frequently, children will insist on pointing. Then say, DO IT WITHOUT POINTING. TELL ME WITH WORDS.
EVALUATION: Note whether the child responds with position words and whether or not the words used are correct.

4G* **Preoperational**
Ages 3–6

Rote Counting: Unit 9
METHOD: Interview
SKILL: Child can rote count.
MATERIALS: None
PROCEDURE: COUNT FOR ME. COUNT AS FAR AS YOU CAN. If the child hesitates or looks puzzled, ask again. If the child still doesn't respond, say, ONE, TWO, WHAT'S NEXT?
EVALUATION: Note how far the child counts and the accuracy of the counting. Young children often lose track (i.e., "One, two, three, four, five, six, ten, seven, . . .") or miss a number name. Two's and three's may just count their ages, whereas fours usually can count accurately to ten and may try the teens and even beyond. By five or six, children will usually begin to understand the commonalities in the twenties and beyond and move on toward counting to 100. Young children vary a great deal at each age level, so it is important to find where each individual is and move along from there.

4H* **Preoperational**
Ages 3–6

Rational Counting: Unit 9
METHOD: Interview, individual or small group
SKILL: Child can rational count.
MATERIALS: Thirty or more objects such as cube blocks, chips, or Unifix® Cubes
PROCEDURE: Place a pile of objects in front of the child (about ten for a three-year-old, twenty for a four-year-old, and thirty for a five-year-old, and as many as 100 for older children). COUNT THESE FOR ME. HOW MANY CAN YOU COUNT?
EVALUATION: Note how accurately the child counts and how many objects are attempted. In observing the process, note:
1. Does the child just use her eyes or does she actually touch each object as she counts?
2. Is some organizational system used, such as lining the objects up in rows or moving the ones counted to the side, and so on?
3. Compare accuracy on rational counting with rote counting.

4I* **Preoperational**
Ages 3–6

Time, Identify Clock or Watch: Unit 19
METHOD: Interview

SKILL: Child can identify a clock and/or watch and describe its function.
MATERIALS: One or more of the following timepieces: conventional clock and watch, digital clock and watch. Preferably at least one conventional and one digital should be included. If real timepieces are not available, use pictures.
PROCEDURE: Show the child the timepieces or pictures of timepieces. WHAT IS THIS? WHAT DOES IT TELL US? WHAT IS IT FOR? WHAT ARE THE PARTS AND WHAT ARE THEY FOR?
EVALUATION: Note whether the child can label watch(es) and clock(s), and how much he is able to describe about the functions of the parts (long and short hands, second hands, alarms set, time changer, numerals). Note also if the child tries to tell time. Compare knowledge of conventional and digital timepieces.

4J* **Preoperational**
Ages 3–6

Symbols, Recognition: Unit 23
METHOD: Interview
SKILL: Child can recognize numerals zero to ten presented in sequence.
MATERIALS: 5 × 8 cards with one numeral from zero to ten written on each such as:

1	2	3	4

PROCEDURE: Starting with zero, show the child each card in numerical order from zero to ten. WHAT IS THIS? TELL ME THE NAME OF THIS.
EVALUATION: Note if the child uses numeral names (correct or not), indicating she knows the kinds of words associated with the symbols. Note which numerals she can label correctly.

INSTRUCTIONAL RESOURCES:
Charlesworth, R., and Lind, K. 1990, *Math and Science for Young Children*. Albany, N.Y.: Delmar.

PREOPERATIONAL: LEVEL 5

5A **Preoperational**
Ages 4–5

One-to-One Correspondence, Same Things/Related Things: Unit 8
Do tasks in 4A (1 and 2) using more pairs of objects.

5B

Preoperational
Ages 4–5

Number and Counting, Rote and Rational Counting: Unit 9

See 4B, 4G, and 4H.

5C

Preoperational
Ages 4–5

Comparing, Number: Unit 11

METHOD: Interview
SKILL: The child can compare the amounts in groups up to five and label the ones that are more, less, and fewer.
MATERIALS: Ten counters (i.e., chips, inch cubes, Unifix® cubes)
PROCEDURE: Present the following groups for comparison in sequence:

1 versus 5
4 versus 1
2 versus 5
3 versus 2
5 versus 4

Each time a pair of groups is presented, ask, DOES ONE GROUP HAVE MORE? If the answer is yes, POINT TO THE GROUP THAT HAS MORE. Ask, HOW DO YOU KNOW THAT GROUP HAS MORE? If the child responds correctly to *more*, present the pairs again using LESS and FEWER.
EVALUATION: Note for which comparisons the child responds correctly. Can he give a logical reason for his choices (such as "Four is more than one" or "I counted them") or does he place them in one-to-one correspondence?

5D*

Preoperational
Ages 4–5

Comparing, Informal Measurement: Unit 11

SKILL: Child can point to big (large) and small objects.
MATERIALS: A big block and a small block (a big truck and a small truck, a big shell and a small shell), and so on.
PROCEDURE: Present two related objects at a time. Say, FIND (POINT TO) THE BIG BLOCK. FIND (POINT TO) THE SMALL BLOCK. Continue with the rest of the object pairs.
EVALUATION: Note if the child is able to identify big and small for each pair.

5E **Preoperational Ages 4–5**

Shape, Geometric Shape Recognition: Unit 12

METHOD: Interview

SKILL: Child can identify geometric shapes.

MATERIALS: Use the circle, square, and triangle from 4E and add a rectangle, rhombus, and cross.

PROCEDURE: Show the child each shape. TELL ME THE NAME OF EACH OF THESE SHAPES. If there are any she cannot name, you provide the name and ask her to find the shape. POINT TO THE (*shape name*).

EVALUATION: Note how many shapes the child can identify.

5F **Preoperational Ages 4–5**

Part/Whole, Parts of a Whole: Unit 14

METHOD: Interview

SKILL: Child can recognize that a whole divided into parts is still the same amount.

MATERIALS: Apple and knife

PROCEDURE: Show the child the apple. HOW MANY APPLES DO I HAVE? After you are certain the child understands that there is one apple, cut the apple into two equal halves. HOW MANY APPLES DO I HAVE NOW? HOW DO YOU KNOW? If the child says "Two," press the halves together and ask, HOW MANY APPLES DO I HAVE NOW? Cut the apple into fourths, then eighths, following the same procedure.

EVALUATION: If the child can tell you that there is still one apple when it is cut into parts, he is able to mentally reverse the cutting process and may be leaving the preoperational period.

5G* **Preoperational Ages 4–5**

Ordering, Sequence/Ordinal Number: Unit 17

METHOD: Interview

SKILL: Child can order up to five objects relative to physical dimensions and identify the ordinal position of each.

MATERIALS: Five objects or cutouts that vary in equal increments of height, width, length, or overall size dimensions.

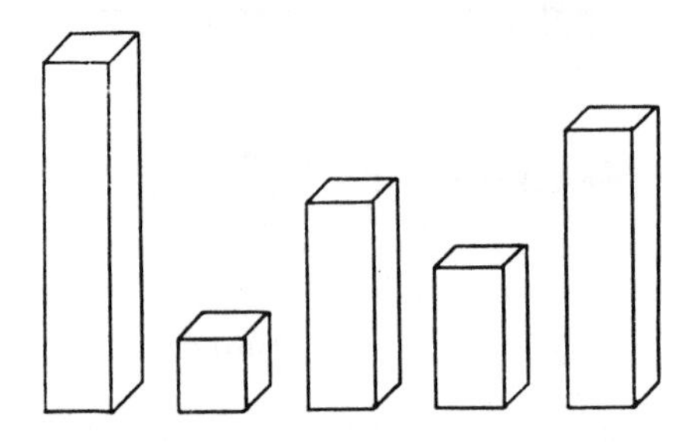

PROCEDURE: Start with five objects (cutouts). If this proves to be difficult, remove the objects (cutouts), then put out three and ask the same questions. FIND THE (TALLEST, BIGGEST, FATTEST) or (SHORTEST, SMALLEST, THINNEST). PUT THEM ALL IN A ROW FROM TALLEST TO SHORTEST (BIGGEST TO LITTLEST, FATTEST TO THINNEST). If the child accomplishes the task, ask, WHICH IS FIRST? WHICH IS LAST? WHICH IS SECOND? WHICH IS THIRD? WHICH IS FOURTH?

EVALUATION: Note whether the children find the extremes, but mix up the three objects (cutouts) that belong in the middle. This is a common approach for preoperational children. Note if children take an organized approach to solving the problem or if they seem to approach it in a disorganized, unplanned way.

5H* **Preoperational Ages 4–5**

Time, Labeling, and Sequence: Unit 19

METHOD: Interview

SKILL: Shown pictures of daily events, the child can use time words to describe the action in each picture and place the pictures in a logical time sequence.

MATERIALS: Pictures of daily activities such as meals, nap, bath, playtime, bedtime

PROCEDURE: Show the child each picture. TELL ME ABOUT THIS PICTURE. WHAT'S HAPPENING? After the child has described each picture, place all the pictures in front of her. PICK OUT (SHOW ME) THE PICTURE OF WHAT HAPPENS FIRST EACH DAY. After a picture is selected, WHAT HAPPENS NEXT? Continue until all the pictures are lined up.

EVALUATION: When describing the pictures, note whether the child uses time words such as breakfast time, lunch time, playtime, morning, night, and so on. Note whether a logical sequence is used when placing the pictures in order.

5I **Preoperational**
Ages 4–5

Practical Activities, Money: Unit 22

METHOD: Observation and interview

SKILL: Child understands that money is exchanged for goods and services and can identify nickel, dime, penny, and dollar bill.

MATERIALS:

1. Play money and store props for dramatic play
2. Nickel, dime, penny, and dollar bill

PROCEDURE:

1. Set up play money and props for dramatic play as described in unit 22. Observe the child and note if he demonstrates some concept of exchanging money for goods and services and of giving and receiving change.
2. Show the child a nickel, dime, penny, and dollar bill. TELL ME THE NAME OF EACH OF THESE.

EVALUATION: Note the child's knowledge of money during dramatic play and note which, if any, of the pieces of money he recognizes.

The following tasks can be presented first between ages four and five and then repeated as the child's concepts and skills grow and expand.

5J* **Preoperational**
Ages 4–6

Sets and Classifying, Free Sort: Unit 10

METHOD: Interview

SKILL: Child can classify and form sets in a free sort.

MATERIALS: Twenty to twenty-five objects (or pictures of objects or cutouts) that can be grouped into several possible sets by criteria such as color, shape, size or category (i.e., animals, plants, furniture, clothing, or toys).

PROCEDURE: Set all the objects out in front of the child in a random arrangement. PUT THE THINGS TOGETHER THAT BELONG TOGETHER. If the child looks puzzled, backtrack to the previous task and hold up one item. FIND SOME THINGS THAT BELONG WITH THIS. When a set is completed, NOW FIND SOME OTHER THINGS THAT BELONG TOGETHER. Keep on until all the items are grouped. Then point to each group. Ask, WHY DO THESE BELONG TOGETHER?

EVALUATION: Note if the child can make logical-looking groups and provide a logical reason for each one. That is, "Because they are cars" ("They are all green," "You can eat with them,").

5K* **Preoperational**
Ages 4–6

Sets and Classifying, Clue Sort: Unit 10

METHOD: Interview

SKILL: Child is able to classify and form sets using verbal and/or object clues.

MATERIALS: Twenty to twenty-five objects (or pictures of objects or cutouts) that can be grouped into several possible sets by criteria such as color, shape, size or category (i.e., animals, plants, furniture, clothing, or toys).

PROCEDURE: Set all the objects in front of the child in a random arrangement. Try the following types of clues:

1. FIND SOME THINGS THAT ARE ________. (Name a specific color, shape, size, material, pattern, function, or class.)
2. Hold up one object, picture, or cutout. FIND SOME THINGS THAT BELONG WITH THIS. After the choices are made, ask, WHY DO THESE THINGS BELONG TOGETHER?

EVALUATION: Note if the child can make a logical-looking group and provides a logical reason for her choices. That is, "Because they are cars" ("They are all green," "You can eat with them.").

5L* **Preoperational**
Ages 4–6

Symbols, Sequencing: Unit 23

METHOD: Interview

SKILL: Child is able to sequence numerals from zero to ten.

MATERIALS: 5 × 8 cards with one numeral from zero to ten written on each.

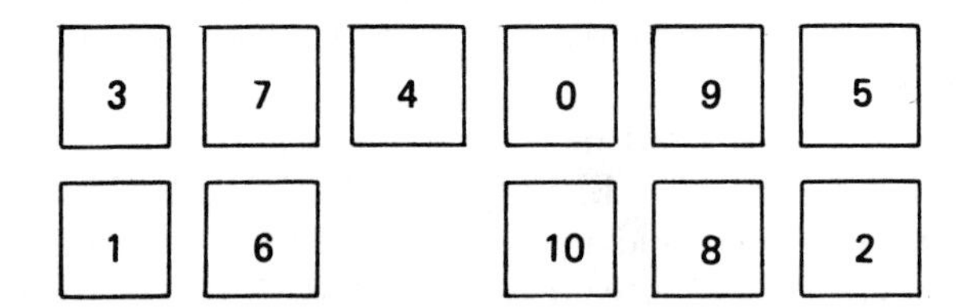

PROCEDURE: Place all the cards in front of the child in random order. PUT THESE IN ORDER. WHICH COMES FIRST? NEXT? NEXT?

EVALUATION: Note whether the child seems to understand that numerals belong in a fixed sequence. Note how many are placed in the correct order and which, if any, are labeled.

5M **Preoperational**
Ages 4–5

Naturalistic and Informal Activities

METHOD: Observation

SKILL: Child can demonstrate a knowledge of math concepts and skills during naturalistic and informal activities.

MATERIALS: Math center (three-dimensional and two-dimensional materials); sand/water/legume pouring table; dramatic play props; unit blocks and accessories; cooking center; math concept books

PROCEDURE: Develop a recording system and keep a record of behaviors such as the following:

- Chooses to work in the math center
- Selects math concept books to look at
- Chooses to work in the cooking center
- Selects working with sand, water, or legumes
- Can give each person one napkin, one glass of juice, and so on
- Spontaneously counts objects or people
- While playing, spontaneously separates objects or pictures into logical groups
- Spontaneously uses comparison words (e.g., This one is *bigger*.)
- Chooses to build with blocks
- Knows the parts of people and objects
- Demonstrates a knowledge of first, biggest, heaviest, and other order concepts
- Does informal measurement such as identifying hot and cold, a bigger container and a smaller container, and so on
- Evidences a concept of time (What do we do next? Is it time for lunch?)
- Points out number symbols in the environment
- Uses the language of math (whether he or she understands the concepts or not)

EVALUATION: Child should show an increase in frequency of these behaviors as the year progresses.

PREOPERATIONAL: LEVEL 6

6A **Preoperational**
Ages 5–6

One-to-One Correspondence: Unit 8

METHOD: Interview

SKILL: Child can place two groups of ten items each in one-to-one correspondence.

MATERIALS: Two groups of objects of different shapes and/or color (such as pennies and cube blocks or red chips and white chips). Have at least ten of each type of object.

PROCEDURE: Place two groups of ten objects in front of the child.

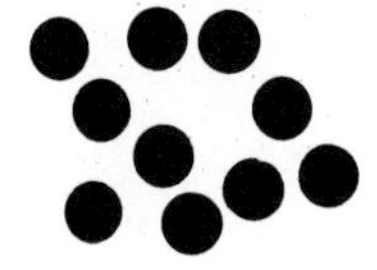

FIND OUT IF THERE IS THE SAME AMOUNT (NUMBER) IN EACH BUNCH (PILE, GROUP, SET). If the child cannot do the task, go back and try it with two groups of five.

EVALUATION: The children should arrange each group so as to match the objects one-to-one or they might count each group to determine equality.

6B

Preoperational
Ages 5–6

Number and Counting, Rote and Rational: Unit 9

METHOD: Interview

SKILL: Child can rote and rational count.

MATERIALS: Fifty counters (i.e., chips, cube blocks, Unifix® Cubes)

PROCEDURES:

1. **Rote counting.** COUNT FOR ME AS FAR AS YOU CAN. If the child hesitates, say ONE, TWO, . . . WHAT'S NEXT?
2. **Rational counting.** Present the child with twenty objects. HOW MANY ________ ARE THERE? COUNT THEM FOR ME.

EVALUATION:

1. **Rote.** By age five the child should be able to count to ten or more; by age six to twenty or more. Note if any number names are missed or repeated.
2. **Rational.** Note the degree of accuracy and organization. Does he place the objects so as to insure that no object is counted more than once or that any object is missed? Note how far he goes without making a mistake. Does he repeat any number names? Skip any? By age six he should be able to go beyond ten objects with accuracy.

6C

Preoperational
Ages 5–6

Shape, Recognition and Reproduction: Unit 12

METHOD: Interview

SKILL #1: Identify shapes, Task 4E.

SKILL #2*: Child can identify shapes in the environment.

MATERIALS: Natural environment

PROCEDURE: LOOK AROUND THE ROOM. FIND AS MANY SHAPES AS YOU CAN. WHICH THINGS ARE SQUARE SHAPES? CIRCLES? RECTANGLES? TRIANGLES?

EVALUATION: Note how observant the child is. Does she note the obvious shapes such as windows, doors, and tables? Does she look beyond the obvious? How many shapes and which shapes is she able to find?

SKILL #3: Child will reproduce shapes by copying.

MATERIALS: Shape cards (4E), plain white paper, a choice of pencils, crayons, and markers

PROCEDURE:

1. COPY THE CIRCLE.
2. COPY THE SQUARE.
3. COPY THE TRIANGLE.

EVALUATION: Note how closely each reproduction resembles its model. Is the circle complete and round? Does the square have four sides and square corners? Does the triangle have three straight sides and pointed corners?

6D* **Preoperational**
Ages 5–6

Part/Whole, Parts of Sets: Unit 14

METHOD: Interview

SKILL: Child can divide a set of objects into smaller groups.

MATERIALS: Have three small dolls (real or paper cutouts) and a box of pennies or other small objects.

PROCEDURE: Have the three dolls arranged in a row. I WANT TO GIVE EACH DOLL SOME PENNIES. SHOW ME HOW TO DO IT SO EACH DOLL WILL HAVE THE SAME AMOUNT.

EVALUATION: Note how the child approaches the problem. Does he give each doll one penny at a time in sequence? Does he count out pennies until there are three groups with the same amount? Does he divide the pennies in a random fashion? Does he have a method for finding out if each has the same amount?

6E **Preoperational**
Ages 5–6

Ordering, Size and Amount: Unit 17

METHOD: Interview

SKILLS: Child can order ten objects that vary on one criteria and five sets with amounts from one to five.

MATERIALS:

1. **Size.** Ten objects or cutouts that vary in size, length, height, or width. An example for length is shown below:

2. **Amount.** Five sets of objects consisting of one, two, three, four, and five objects each.

PROCEDURE:

1. **Size.** Place the ten objects or cutouts in front of the child in a random arrangement. FIND THE (BIGGEST, LONGEST, TALLEST, OR WIDEST). PUT THEM ALL IN A ROW FROM ________ TO ________.
2. **Amount.** Place the five sets in front of the child in a random arrangement. PUT THESE IN ORDER FROM THE SMALLEST BUNCH (GROUP) TO THE LARGEST BUNCH (GROUP).

EVALUATION:

1. **Size.** Preoperational children will usually get the two extremes but may mix up the inbetween sizes. Putting ten in the correct order would be an indication that the child is entering concrete operations.
2. **Amount.** Most fives can order the five sets. If they order them easily, try some larger amounts.

6F **Preoperational Ages 5–6**

Measurement; Length, Weight, and Time: Units 18 and 19

METHOD: Interview

SKILLS: Child can explain the function of a ruler, discriminate larger from heavier, and identify and explain the function of a clock.

MATERIALS:

1. **Length.** A foot ruler
2. **Weight.** A plastic golf ball and a marble or other pair of objects where the larger is the lighter
3. **Time.** A clock (with a conventional face)

PROCEDURES:

1. Show the child the ruler. WHAT IS THIS? WHAT DO WE DO WITH IT? SHOW ME HOW IT IS USED.
2. Give the child the two objects, one in each hand. Ask, WHICH IS BIGGER? WHICH IS HEAVIER? WHY IS THE SMALL ________ HEAVIER?
3. Show the child the clock. WHAT IS THIS? WHY DO WE HAVE IT? TELL ME HOW IT WORKS.

EVALUATION: Note how many details the child can give about each of the measuring instruments. Is she accurate? Can she tell which of the objects is heavier? Can she provide a reason for the lighter being larger and the smaller heavier?

6G **Preoperational**
Ages 5–6

Practical Activities, Money: Unit 22

METHOD: Interview

SKILL: Child can recognize money and tell which pieces of money will buy more.

MATERIALS:

1. Pictures of coins, bills, and other similar looking items
2. Selection of pennies, nickels, dimes, and quarters

PROCEDURE:

1. Show the child the pictures. FIND THE PICTURES OF MONEY. After he has found the pictures of money, ask, WHAT IS THE NAME OF THIS? as you point to each picture of money.
2. Put the coins in front of the child. Ask, WHICH WILL BUY THE MOST? IF YOU HAVE THESE FIVE PENNIES (put five pennies in one pile) AND I WANT TWO CENTS FOR A PIECE OF CANDY. HOW MANY PENNIES WILL YOU HAVE TO GIVE ME FOR THE CANDY?

EVALUATION: Note which picture of money the child can identify. Note if he knows which coins are worth the most. Many young children equate worth and size and thus think a nickel will buy more than a dime.

Check back to 5J, 5K, and 5L, then go on to the next tasks. ***The following tasks can be presented first between ages five and six and then repeated as the child's concepts and skills grow and expand.***

6H* **Transitional Period**
Ages 5–7

Ordering, Double Seriation: Unit 17

METHOD: Interview

SKILL: Child can place two sets of ten items in double seriation.

MATERIALS: Two sets of ten objects, cutouts, or pictures of objects such that they vary in one or more dimensions in equal increments so that there is one item in each set that is the correct size to go with an item in the other set. The sets could be children and baseball bats, children and pets, chairs and tables, bowls and spoons, cars and garages, hats and heads, and so on.

PROCEDURE: Suppose you have decided to use hats and heads. First, place the heads in front of the child in random order.

LINE THESE UP SO THAT THE SMALLEST HEAD IS FIRST AND THE BIGGEST HEAD IS LAST. Help can be given such as, FIND THE SMALLEST. GOOD. NOW WHICH ONE COMES NEXT? AND NEXT? If the child is able to line up the heads correctly, then put out the hats in a random arrangement.

FIND THE HAT THAT FITS EACH HEAD AND PUT IT ON THE HEAD.

EVALUATION: Note how the children approach the problem—in an organized or haphazard fashion. Note whether they get the whole thing correct or partially correct. If they get a close approximation, go through the procedure again with seven items or five to see if they have the concept when fewer items are used. A child going into concrete operations should be able to accomplish the task with two groups of ten. Transitional children may be able to do the task correctly with fewer items in each group.

6I* **Transitional Period**
Ages 5–7

Ordering, Patterning: Unit 17

METHOD: Interview

SKILL: Child can copy, extend, and describe patterns made with concrete objects

MATERIALS: Color cubes, Unifix® Cubes, Teddy Bear Counters, attribute blocks, small toys, or other objects that can be placed in a sequence to develop a pattern

PROCEDURE:

1. Copy patterns. One at a time, make patterns of various levels of complexity (each letter stands for one type of item such as one color of a color cube, one shape of an attribute block, or one type of toy). For example, A-B-A-B could be red block, green block, red

block, green block or big triangle, small triangle, big triangle, small triangle. Using the following series of patterns, tell the child, MAKE A PATTERN JUST LIKE THIS ONE. (If the child hesitates, point to the first item and say, START WITH ONE LIKE THIS):
   a. A-B-A-B
   b. A-A-B-A-A-B
   c. A-B-C-A-B-C
   d. A-A-B-B-C-C-A-A-B-B-C-C
2. Extend patterns. Make patterns as in # 1 but this time say, THIS PATTERN ISN'T FINISHED. MAKE IT LONGER. SHOW ME WHAT COMES NEXT.
3. Describe patterns. Make patterns as in #1 and #2. TELL ME ABOUT THESE PATTERNS (WHAT COMES FIRST? NEXT? NEXT?). IF YOU WANTED TO CONTINUE THE PATTERN, WHAT WOULD COME NEXT? NEXT?
4. If the above tasks are easily accomplished, then try some more difficult patterns such as:
   a. A-B-A-C-A-D-A-B-A-C-A-D
   b. A-B-B-C-D-A-B-B-C-D
   c. A-A-B-A-A-C-A-A-D

EVALUATION: Note which types of patterns are easiest for the children. Are they more successful with the easier patterns? With copying? Extending? Describing?

6J* **Preoperational**
Ages 5 and older

Symbols, One More Than: Unit 23

METHOD: Interview

SKILL: Child can identify numerals that are "one more than."

MATERIALS: 5 × 8 cards with one numeral from zero to ten written on each (see 5L)

PROCEDURE: Place the numeral cards in front of the child in order from zero to ten. TELL ME WHICH NUMERAL MEANS ONE MORE THAN TWO. WHICH NUMERAL MEANS ONE MORE THAN SEVEN? WHICH NUMERAL MEANS ONE MORE THAN FOUR? (If the child answers these, then try LESS THAN.)

EVALUATION: Note whether the child is able to answer correctly.

6K* **Preoperational/Concrete**
Ages 5–7

Sets and Symbols, Reproduce (Write) Numerals: Unit 24

METHOD: Interview

SKILL: Child can reproduce (write) numerals from zero to ten.

MATERIALS: Pencil, pen, black marker, black crayon, white paper, numeral cards zero to ten

PROCEDURE: HERE IS A PIECE OF PAPER. PICK OUT ONE OF THESE (point to writing tools) THAT YOU WOULD LIKE TO USE. NOW, WRITE AS MANY NUMBERS AS YOU CAN. If the child is unable to write from memory, show him the numeral cards. COPY ANY OF THESE THAT YOU CAN.
EVALUATION: Note how many numerals the child can write and if they are in sequence. If the child is not able to write the numerals with ease, this indicates that responding to problems by writing is not at this time an appropriate response. Have him do activities where he can use movable numerals or place markers on the correct answers.

6L* **Preoperational/Concrete**
Ages 5–7

Sets and Symbols, Match Sets to Symbols: Unit 24
METHOD: Interview
SKILL: Child can match sets to symbols using sets of amounts zero to ten and numerals from zero to ten.
MATERIALS: 5 × 8 cards with numerals zero to ten, sixty objects (e.g., chips, cube blocks, coins, buttons)
PROCEDURE: Lay out the numeral cards in front of the child. Place the container of objects within easy reach. MAKE A SET FOR EACH NUMERAL. Let the child decide how to organize the materials.

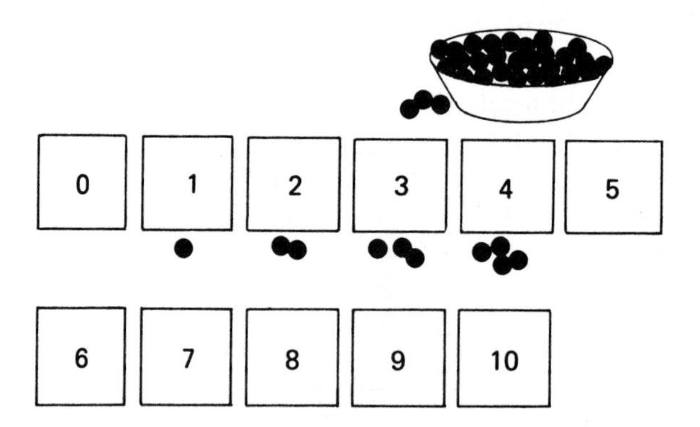

EVALUATION: Note for which numerals the child is able to make sets. Note how the child goes about the task. For example, does she sequence the numerals from zero to ten, does she place the objects in an organized pattern by each numeral, can she recognize some amounts without counting, when she counts, does she do it carefully. Her responses will indicate where instruction should begin.

6M* **Preoperational/Concrete**
Ages 5–7

Sets and Symbols, Match Symbols to Sets: Unit 24

METHOD: Interview

SKILL: Child can match symbols to sets using numerals from zero to ten and sets of amounts zero to ten.

MATERIALS: 5 × 8 cards with numerals zero to ten, ten objects (e.g., chips, cube blocks, buttons)

PROCEDURE: Lay out the cards in front of the child in numerical order. One at a time show the child sets of each amount in this order: 2, 5, 3, 1, 4. For example:

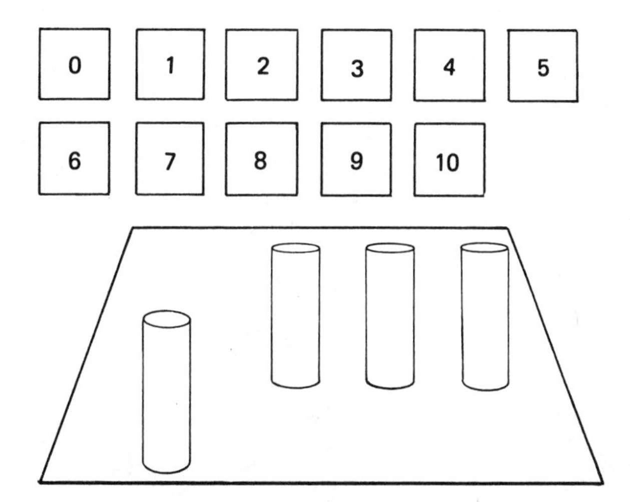

PICK OUT THE NUMERAL THAT TELLS HOW MANY THINGS ARE IN THIS SET. If the child does these correctly, then go on to 7, 9, 6, 10, 8, 0 using the same procedure.

EVALUATION: Note which sets and symbols the child can match. His responses will indicate where instruction can begin.

6N **Preoperational/Concrete**
Ages 5–6

Naturalistic and Informal Activities: Units 1–25

METHOD: Observation

SKILL: Child demonstrates a knowledge of math concepts and skills during naturalistic and informal activities

MATERIALS: See task 5M

PROCEDURE: See task 5M. Add the following behaviors to your list:

- Demonstrates an understanding of *more than*, *the same amount*, and *less than* by responding appropriately to questions such as, "Do we have the same number of children as we have chairs?"
- Can match a set to a symbol and a symbol to a set (for example, if the daily attendance total says 22, he can get 22 napkins for snack).
- Can do applied concrete whole number operations. For example, if four children plan to do drawing and two more children join them, he knows that there are now six children, or if he has three friends and eight cars to play with he figures out that each friend can use two cars.

EVALUATION: Child should show an increase in frequency of these behaviors as the year progresses.

MATH LANGUAGE: LEVEL 7

By the time the child is between 5½ and 6½ years of age, he should be using most of the words listed in Unit 15. The following tasks can be used to find out which words the child uses in an open-ended situation. Show each picture individually. Say, I HAVE SOME PICTURES TO SHOW YOU. HERE IS THE FIRST ONE. TELL ME ABOUT IT. For each picture, tape record or write down the child's responses. Later list all the math words. Compare this with the list of math words he uses in class.

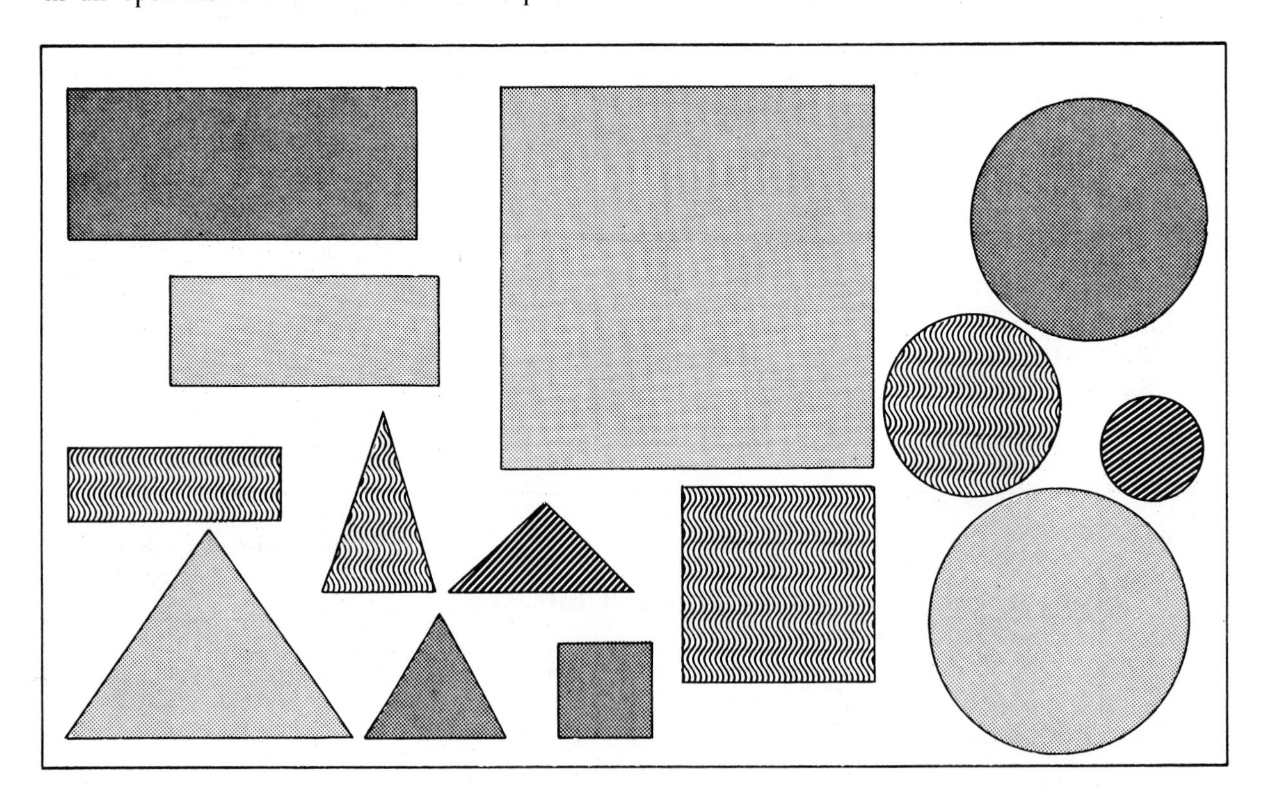

S T L
R

123
125

CONCRETE OPERATIONS: LEVEL 8

The following tasks are all indicators of the child's cognitive developmental level. The child who can accomplish all these tasks should be ready for the primary level instruction described in Section V.

8A **Concrete Operations**
Ages 6–7

Conservation of Number: Unit 1

METHOD: Interview

SKILL: Child can solve the number conservation problem.

MATERIALS: Twenty chips, blocks, or coins, all the same size, shape, and color

PROCEDURE: Set up a row of nine objects. Then proceed through the following four tasks.

1. MAKE A ROW JUST LIKE THIS ONE (point to yours).

   Child □ □ □ □ □ □ □ □ □

   Adult □ □ □ □ □ □ □ □ □

   DOES ONE ROW HAVE MORE BLOCKS (CHIPS, COINS) OR DO THEY BOTH HAVE THE SAME AMOUNT? HOW DO YOU KNOW? If child agrees to equality go on to the next tasks.

2. Task 2

   NOW WATCH WHAT I DO. (Push yours together.)

   Child □ □ □ □ □ □ □ □ □

   Adult □□□□□□□□□

   DOES ONE ROW HAVE MORE BLOCKS OR DO THEY BOTH HAVE THE SAME AMOUNT? WHY? (If the child says one row has more, MAKE THEM HAVE THE SAME AMOUNT AGAIN.) (If the child says they have the same amount, tell him, LINE THEM UP LIKE THEY WERE BEFORE I MOVED THEM.) Go on to task 3 and task 4 following the same steps as above.

3. Task 3

   Child □□□□□□□□□

   Adult □□□□ □□□□□

4. Task 4

   Child □□□□□□□□□

   Adult □□□□□□□□□

EVALUATION: If the child is unable to do Task 1 (one-to-one correspondence), do not proceed any further. He needs to work further on this concept and needs time for development. If he succeeds with Task 1, go on to 2, 3, and 4. Note which of the following categories fit his responses:

Nonconserver 1. Indicates longer rows have more but cannot give a logical reason, (For example, the child may say, "I don't know," "My mother says so," or gives no answer.)

Nonconserver 2. Indicates longer rows have more and gives logical reasons, such as "It's longer," "The long row has more," etc.
Transitional. Says both rows still have the same amount but has to check by counting or placing in one-to-one correspondence.
Conserver. Completely sure that both rows still have the same amount. May say, "You just moved them."

8B

Concrete Operations
Ages 6–7

Symbols and Sets, Matching and Writing: Units 23, 24

METHOD: Interview

SKILL: Child can match sets to symbols and write symbols.

MATERIALS: Cards with numerals zero to twenty, a supply of counters, paper and writing implements

PROCEDURE:

1. Present the child with sets of counters. Start with amounts under ten. If the child can do these, go on to the teens. MATCH THE NUMBERS TO THE SETS.
2. Put the numeral cards and counters away. Give the child a piece of paper and a choice of writing instruments. WRITE AS MANY NUMBERS AS YOU CAN. START WITH ZERO.

EVALUATION: Note how high the child can go in matching sets and symbols and in writing numerals.

8C

Concrete Operations
Ages 6–7

Multiple Classification: Unit 25

METHOD: Interview

SKILL: Child can group shapes by more than one criteria.

MATERIALS: Make thirty-six cardboard shapes

1. Four squares (one each red, yellow, blue, and green)
2. Four triangles (one each red, yellow, blue, and green)
3. Four circles (one each red, yellow, blue, and green)
4. Make three sets of each in three sizes.

PROCEDURE: Place all the shapes in a random array in front of the child. DIVIDE (SORT, PILE) THESE SHAPES INTO GROUPS, ANY WAY YOU WANT TO. After the child has sorted on one attribute (shape, color, or size) say, NOW DIVIDE (SORT, PILE)

THEM ANOTHER WAY. The preoperational child will normally refuse to conceptualize another way of grouping.
EVALUATION: The preoperational child will center on the first sort and will not try another criteria. The concrete operations child will sort by color, shape, and size.

8D **Concrete Operations**
Ages 6–7

Class Inclusion: Unit 25

METHOD: Interview
SKILL: Child can perceive that there are classes within classes.
MATERIALS: Make a set of materials using objects, cutouts, or pictures of objects such as the following:

1. Twelve wooden beads of the same size and shape differing only in color (e.g., four red and eight blue)
2. Twelve pictures of flowers: eight tulips and four daisies
3. Twelve pictures of animals: eight dogs and four cats

PROCEDURE: Place the objects (pictures) in front of the child in random order. PUT THE (object name) TOGETHER THAT ARE THE SAME. Then after they have grouped into two subcategories ask, ARE THERE MORE (WOODEN BEADS, FLOWERS OR ANIMALS) OR MORE (BLUE BEADS, TULIPS OR DOGS)? That is, have them compare the overall class or category with the larger subclass.
EVALUATION: The preoperational child will have difficulty conceptualizing parts and wholes of sets at the same time.

INSTRUCTIONAL RESOURCES:
Charlesworth, R., and Lind, K. 1990. *Math and science for Young Children*. Albany, N.Y.: Delmar.

CONCRETE OPERATIONS: LEVEL 9

9A* **Concrete Operations**
Ages 6–8

Addition, Combining Sets up to Ten: Unit 27

METHOD: Interview
SKILL: Child can combine sets to form new sets up to ten.
MATERIALS: Twenty counters (cube blocks, Unfix® Cubes, chips): ten of one color and ten of another
PROCEDURE: Have the child select two groups of counters from each color so that the total

is ten or less. PUT THREE YELLOW CUBES OVER HERE AND FIVE BLUE CUBES OVER HERE. Child completes task. NOW TELL ME, IF YOU PUT ALL THE CUBES IN ONE BUNCH, HOW MANY CUBES DO YOU HAVE ALTOGETHER? HOW DO YOU KNOW? Do this with combinations that add up to one through ten.
EVALUATION: Note if the child is able to make the requested groups with or without counting. Note the method used by the child to decide on the sum:

1. Does he begin with one and count all the blocks?
2. Does he count on? That is, in the example above, does he put his two small groups together and then say, "Three blocks, four, five, six, seven, eight. I have eight now."
3. Does he just say, "Eight, because I know that three plus five is eight."?

9B*

Concrete Operations
Ages 6–8

Subtraction, Sets of Ten and Less: Unit 27
METHOD: Interview
SKILL: Child can subtract sets to make new sets using groups of ten and smaller.
MATERIALS: Twenty counters (cube blocks, Unifix® Cubes, chips): ten of one color and ten of another and a small box or other small container
PROCEDURE: Pick out a group of ten or fewer counters. I HAVE SEVEN CUBES. I'M GOING TO HIDE SOME IN THE BOX. (Hide three in the box.) NOW, HOW MANY DO I HAVE LEFT? HOW MANY DID I HIDE? If the child cannot answer, give her seven of the other color cubes and ask her to take three away and tell you how many are left. Do this with amounts of ten and less. For the less mature or younger child, start with five and less.
EVALUATION: Note if the child is able to solve the problem without working it out herself. Note whether the child has to count or if she just knows without counting.

9C

Concrete Operations
Ages 6–8

Addition and Subtraction, Understanding Notation: Unit 27
METHOD: Interview or small group
SKILL: Child understands the connection between notation and concrete problems.
MATERIALS: Counters (i.e., chips, Unifix® Cubes, cube blocks) and pencil and paper
PROCEDURE: Each child should have a supply of counters and pencils and paper. TAKE THREE RED (*Name of counter*). NOW PUT TWO GREEN (*Name of counter*) WITH THE THREE RED. WRITE A NUMBER SENTENCE THAT TELLS WHAT YOU DID. When finished, PUT THE RED AND GREEN (*counters*) BACK. TAKE OUT SIX YELLOW (*counters*). SEPARATE THREE OF THE YELLOW (*counters*) FROM THE SIX.

WRITE A NUMBER SENTENCE THAT TELLS WHAT YOU DID. Continue with more addition and subtraction problems. Written story problems could be given to the children who know how to read.
EVALUATION: Note if the children are able to use the correct notation, that is, $3 + 2 = 5$ and $6 - 3 = 3$.

9D **Concrete Operations Ages 6–8**

Addition and Subtraction, Create Problems: Unit 27

METHOD: Interview or small group
SKILL: Given a number sentence, the child can create a problem
MATERIALS: Counters (i.e., chips, Unifix® Cubes, cube blocks) and pencil and paper
PROCEDURE: Give the children the number sentences below. Tell them to make up a story to go with each one using their counters to represent the characters in the story. Nonreaders/writers can dictate their stories; reader/writers can write the stories themselves. Number sentences:
1. $3 + 5 = 8$ 2. $6 - 4 = 2$
EVALUATION: Note if the dictated or written problem relates correctly to the number sentence.

9E **Concrete Operations Ages 6–8**

Addition and Subtraction, Translating Symbols into Concrete Actions: Unit 27

METHOD: Interview or small group
SKILL: The child can translate written problems into concrete actions
MATERIALS: Counters (i.e., chips, Unifix® Cubes, cube blocks), pencil and paper with several addition and subtraction problems

PROBLEMS

1. $9 - 4$	2. $4 + 5$
3. $3 + 2$	4. $8 - 6$
5. $1 + 7$	6. $6 - 3$
7. $4 - 1$	8. $2 + 6$
9. $5 + 3$	10. $7 - 2$

PROCEDURE: Give each child a supply of counters, a pencil, and a paper with one or more written problems like those above. It is best to give the problems one at a time the first time, then give more as the children become more proficient. Point to the first problem if there is

more than one. LOOK AT THIS PROBLEM. SHOW ME THE PROBLEM WITH (*counters*). NOW WRITE THE ANSWER. READ THE PROBLEM AND THE ANSWER TO ME. If they do this one correctly, have them continue on their own. Ask them to show you if there are any problems they can do without the cubes.
EVALUATION: Note whether the children do the problems correctly and especially whether they are accurate in translating the signs. For example, for problem #1, a child might take nine counters and then take four, ignoring the − sign. It is not uncommon for children to omit the = sign. That is, they might write 4 + 5 9. Some children might be able to tell you that 2 + 6 = 8, etc., but not be able to show you with the counters. This behavior indicates the children have learned to use the symbols in a rote fashion but do not understand the concepts that the symbols stand for.

9F* — **Concrete Operations, Ages 7–8**

Multiplication, Readiness: Unit 27

METHOD: Interview
SKILL: Child demonstrates readiness for multiplication by constructing equal groups of different sizes from groups of the same size.
MATERIALS: Twenty counters (cube blocks, Unifix® Cubes chips)
PROCEDURE: Make two groups of six counters each. Ask the child, MAKE THREE GROUPS OF TWO CHIPS (BLOCKS, CUBES) EACH WITH THIS BUNCH OF SIX CHIPS (BLOCKS, CUBES). When the child finishes (right or wrong), point to the other group of counters. NOW MAKE TWO GROUPS OF THREE WITH THESE CHIPS (BLOCKS, CUBES).
EVALUATION: Note if the child is able to make the two different subgroups. Children who are not ready for multiplication will become confused and not see the difference between the two tasks.

9G — **Concrete Operations, Ages 7–8**

Multiplication, The Process: Unit 27

METHOD: Interview
SKILL: Child understands the process of multiplication.
MATERIALS: Counters (i.e., chips, Unifix® Cubes, cube blocks) and pencil and paper
PROCEDURE: Provide the child with counters, pencil, and paper.
1. Show the child patterns of counters such as:

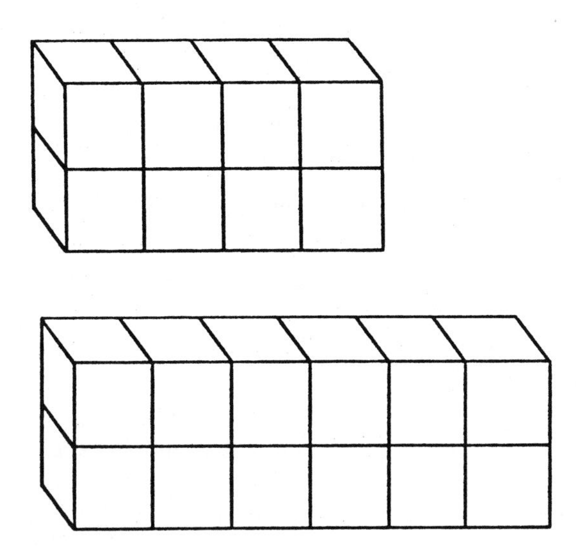

For each pattern, WRITE AN EQUATION (OR NUMBER SENTENCE) THAT TELLS ABOUT THIS PATTERN.

2. WITH YOUR (*counters*), SHOW ME TWO TIMES TWO. (Try 3×4, 5×2, etc.)

EVALUATION:

1. Note if the child writes: a. $2 \times 4 = 8$ or $4 \times 2 = 8$; b. $2 \times 6 = 12$ or $6 \times 2 = 12$. It is not uncommon for a child to write $3 \times 4 = 12$ for the second equation. This response would indicate he has memorized three times four equals twelve without a basic understanding of the multiplication concept.
2. For two times two, the child should make two groups of two; for 3×4, three groups of four; for 5×2, five groups of two. Some children will reverse the numbers such as two groups of five for the last problem. Other children may just make two groups, such as a group of five and a group of two for 5×2.

9H **Concrete Operations**
Ages 7–8

Multiplication, Using Symbols: Unit 27

METHOD: Interview

SKILL: Child can solve written multiplication problems.

MATERIALS: Counters (i.e., chips, Unifix® Cubes, cube blocks) and pencil and paper

PROCEDURE: One at a time give the child written multiplication problems. Tell her,

WORK OUT THE PROBLEM WITH COUNTERS AND WRITE DOWN THE ANSWER.
EVALUATION: Note whether the child makes the appropriate number of sets of the right amount. Note whether the child does the correct operation. Sometimes children will forget and add instead of multiply. That is, a child might write $3 \times 2 = 5$.

9I* **Concrete Operations**
Ages 7–8

Division, Basic Concept: Unit 27

METHOD: Interview
SKILL: Child demonstrates an understanding that division consists of grouping or sharing objects.
MATERIALS: Thirty counters (cube blocks, Unifix® Cubes, chips) and five small containers (such as clear plastic glasses)
PROCEDURE: Put out eight chips and four containers. DIVIDE UP THE CHIPS SO THAT EACH CUP HAS THE SAME AMOUNT. When the chips are divided ask, HOW MANY CUBES DO YOU HAVE IN EACH CUP? The child should respond "Two in each cup" rather than "I have two, two, and two." Try the same procedure with more cups and larger amounts to divide. Then try it with uneven amounts. Note if the child becomes confused or can recognize that there are more than needed. Also do some sharing problems. That is, for example, put out sixteen chips.
I WANT TO GIVE THREE FRIENDS EACH THE SAME AMOUNT OF CHIPS. HOW MANY WILL EACH ONE RECEIVE? ARE THERE ANY LEFTOVER?
EVALUATION: Note how the children handle the problem. Do they proceed in an organized fashion? Can they deal with the remainders?

9J **Concrete Operations**
Ages 7–8

Division, Symbols: Unit 27

METHOD: Interview
SKILL: Child understands the use of symbols in division problems.
MATERIALS: Counters (i.e., chips, Unifix® Cubes, cube blocks) and pencil and paper
PROCEDURE: Give the child a division story problem such as: Chan has eight bean seeds. He wants to plant two in each small pot. How many pots will he need? READ THE PROBLEM AND SOLVE IT. WRITE THE NUMBER SENTENCE FOR THE PROBLEM ON YOUR PAPER.

EVALUATION: Note if the child can write the correct number sentence. If she can't do the problem from memory, does she figure out that she can use her counters or draw a picture to assist in finding the answer?

9K* **Concrete Operations**
Ages 6–8

Patterns, Extension in Three Dimensions: Unit 28

METHOD: Interview

SKILL: Child can extend complex patterns in three dimensions by predicting what will come next.

MATERIALS: Inch or centimeter cubes, Unifix® Cubes, or other counters, that can be stacked

PROCEDURE: Present the child with various patterns made of stacked counters. Ask the child to describe the pattern and to continue it as far as he can. Stack the blocks as follows one pattern at a time:

a.

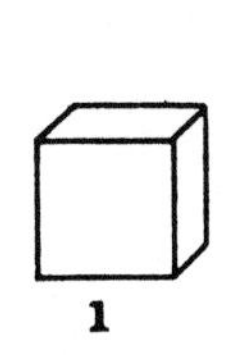

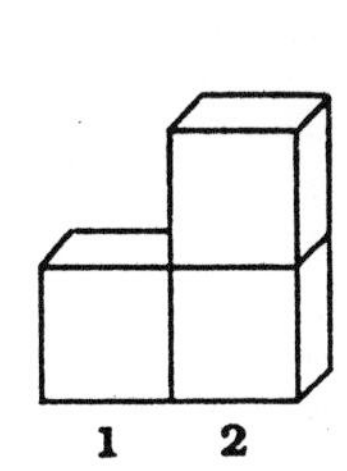

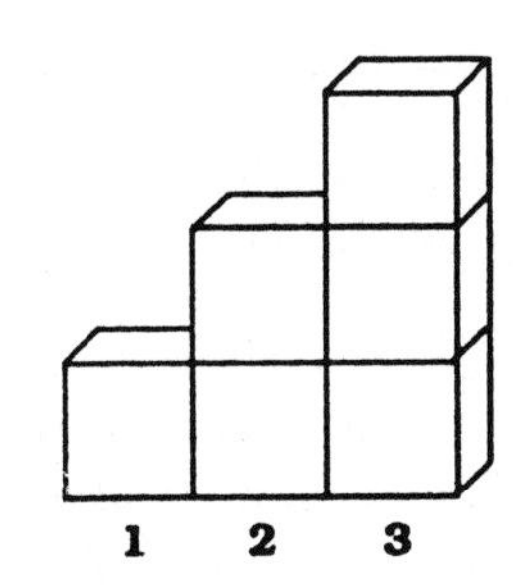

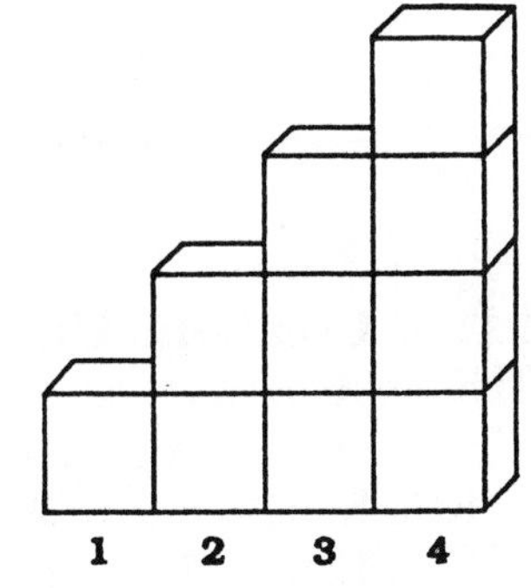

b.

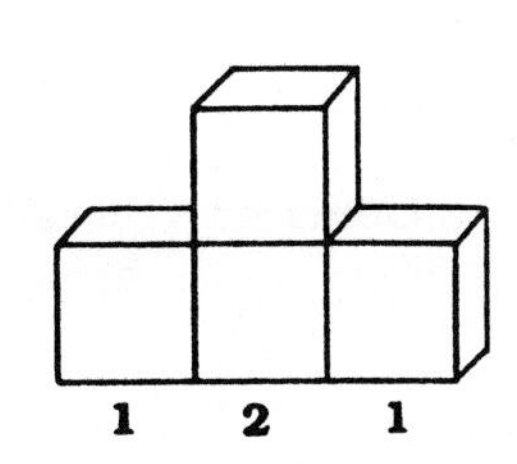

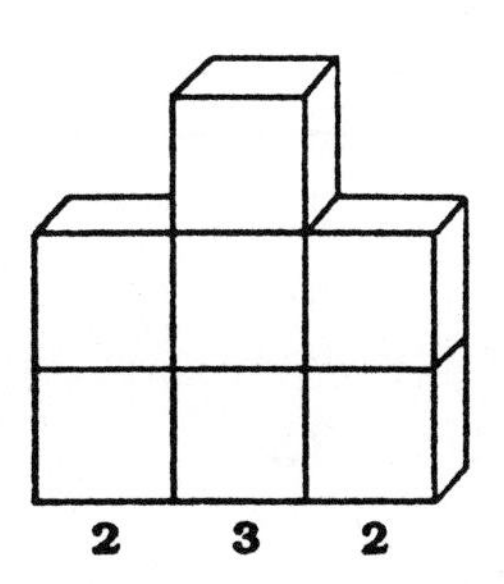

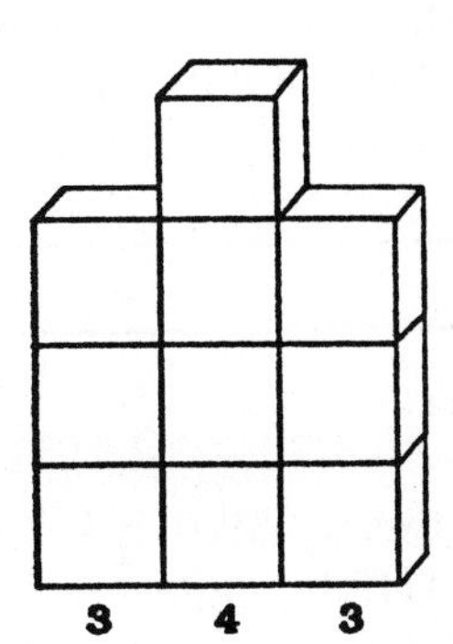

c.

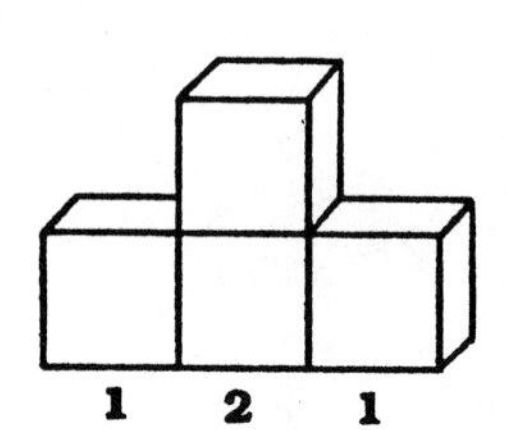

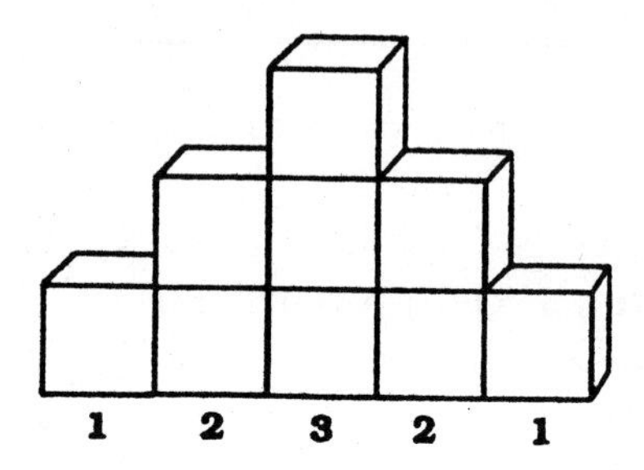

For each pattern ask, TELL ME ABOUT THIS PATTERN. WHAT COMES NEXT? HOW DO YOU KNOW? CONTINUE THE PATTERN FOR ME.
EVALUATION: Note if the child can continue each pattern and state his rationale. Note where the child may need further help and practice.

9L

Concrete Operations
Ages 7–8

Patterns, Creation: Unit 28
METHOD: Interview
SKILL: Child can create patterns using discrete objects.
MATERIALS: Concrete objects such as chips, Unifix® Cubes, or cube blocks
PROCEDURE: USING YOUR (*counters*), MAKE YOUR OWN PATTERN AS YOU HAVE DONE WITH PATTERN STARTERS I HAVE GIVEN YOU. When the child is finished, TELL ME ABOUT YOUR PATTERN.
EVALUATION: Note if the child has actually developed a repeated pattern and if she is able to tell you the pattern in words.

9M*

Concrete Operations
Ages 7–9

Patterns, Multiple Number: Unit 28
METHOD: Interview
SKILL: Child can use a 00 to 99 chart to discover and predict number multiple patterns.

MATERIALS: Inch or centimeter cubes, Unifix® Cubes, or other counters, and a 00 to 99 chart (see Figure 28–1)
PROCEDURE: Start a pattern using multiples of two blocks. Ask the child, CIRCLE OR MARK THE AMOUNT IN MY GROUP ON THE CHART. If the child has a problem, show her the 2 and circle it, if necessary. Next to the group of two blocks construct a group of four blocks. Use the same procedure as above. Continue up to ten. Then ask, SHOW ME WHICH NUMBERS YOU WOULD CIRCLE IF I KEPT CONTINUING WITH THIS PATTERN. When children can predict accurately with multiples of two, try 3's, 4's, 5's, etc.
EVALUATION: Note whether the children can connect the numbers in the pattern to the numerals on the chart and whether they can predict what comes next. If they cannot accomplish these tasks, note where their errors are. Do they need more help with basic pattern construction? With counting? With connecting sets to symbols? With finding numbers on the chart?

9N* **Concrete Operations**
Ages 6–8

Fractions Equivalent Parts: Unit 29
METHOD: Interview
SKILL: Child can divide a rectangle into smaller equal parts.
MATERIALS: A supply of paper rectangles of equal size (8½″ × 2¾″) in four different colors and a pair of scissors
PROCEDURE: Show the child a paper rectangle. THIS IS A RECTANGLE. Place two more rectangles (color # 2) below the first one. HERE ARE TWO MORE RECTANGLES. ARE ALL THREE THE SAME SIZE? Be sure the child agrees. Let him compare them to be sure. NOW I'M GOING TO FOLD ONE OF THE RECTANGLES (color #2) SO BOTH PARTS ARE THE SAME. Fold the rectangle. NOW YOU FOLD THIS OTHER ONE (also color # 2) JUST LIKE I DID. Offer assistance if necessary. The three rectangles should look like this:

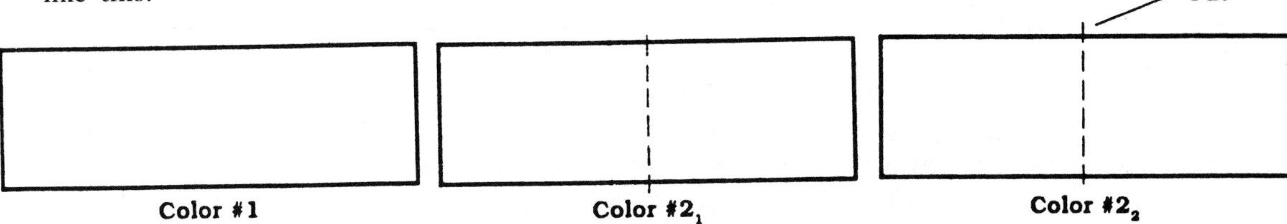

ARE THE PARTS OF (color #2) RECTANGLE THE SAME SIZE AS THE PARTS OF THIS ONE? (also color #2) SHOW ME HOW YOU KNOW. I'M GOING TO CUT THIS ONE (second color #2) ON THE FOLD. HOW MANY PARTS DO I HAVE NOW? IF I PUT THEM BACK TOGETHER, WILL THEY BE THE SAME SIZE AS THIS WHOLE RECTANGLE? (color #1) AS YOUR RECTANGLE? WHAT IS A SPECIAL NAME FOR THIS AMOUNT OF THE WHOLE RECTANGLE? Point to the half. If the

response is one-half, go through the procedure again with one third and one-fourth using colors #3 and #4 respectively.
EVALUATION: Note whether the child has to check on the equivalency of the three rectangles. Can he keep in mind that the parts still equal the whole, even when cut into two or more parts? Does he know the terms one-half, one-third, and/or one-fourth?

9O*

Concrete Operations
Ages 6–8

Fractions, One-half of a Group: Unit 29
METHOD: Interview
SKILL: Child can divide a set of objects into smaller groups when given directions using the term *one-half*.
MATERIALS: Ten counters (cube blocks, chips, Unifix® Cubes, or other concrete objects).
PROCEDURE: Place the counters in front of the child. I HAVE SOME (*name of counters*). DIVIDE THESE SO THAT WE EACH HAVE ONE-HALF OF THE GROUP. If the child completes this task easily, go on to nine counters and ask him to divide the group into thirds and eight counters, then ask him to divide the group into fourths.
EVALUATION: Note the method used by the child. Does he use counting or does he pass the counters out: "One for you and one for me . . . ." Does he really seem to understand the terms one-half, one-fourth, and one-third?

9P*

Concrete Operational
Ages 7–8

Place Value, Groups of Ten: Unit 30
METHOD: Interview
SKILL: Child can count groups of eleven or more objects and tell how many tens are in the groups.
MATERIALS: Container of 100 counters (e.g., chips, cubes, or sticks)
PROCEDURE: Place the container of counters in front of the child. HERE ARE A BUNCH OF COUNTERS. COUNT OUT AS MANY OF THEM AS YOU CAN. If the child counts out eleven or more, ask, HOW MANY TENS DO YOU THINK YOU HAVE? HOW MANY ONES?
EVALUATION: If the child answers correctly, then she probably has the concept of place value for tens. If she answers incorrectly, this indicates that although she might be able to rational count groups of objects greater than ten that she does not yet understand the meaning of each of the numerals in her response.

9Q* **Concrete Operational**
Ages 7–8

Place Value, Grouping to Identify an Amount: Unit 30

METHOD: Interview

SKILL: Child can form two or more subgroups of ten objects each with some remaining from the original group and tell how many he has without counting each individual object.

MATERIALS: Container of 100 counters (e.g., chips, cubes, or sticks)

PROCEDURE: Place a pile of counters (start with about thirty-five) in front of the child. MAKE AS MANY GROUPS OF TEN AS YOU CAN. HOW MANY (counters) DO YOU HAVE ALTOGETHER?

EVALUATION: Note if the child can come up with the answer by counting the number of groups of ten and adding on the number of ones, or if he has to count each object to be sure of the total. If the child can determine the answer without counting by ones, this is an indicator that he is developing the concept of place value.

9R **Concrete Operations**
Ages 7–8

Place Value, Symbols to Concrete Representations: Unit 30

METHOD: Interview or small group

SKILL: Child can translate from written numerals to concrete representations.

MATERIALS: Base ten blocks or similar material (i.e., Unifix® Cubes or sticks or straws and rubber bands for bundling), eight or ten cards, each with a double-digit number written on it; for example, 38, 72, 45, 83, 27, 96, 51, 50.

PROCEDURE: USING YOUR BASE TEN BLOCKS, MAKE GROUPS FOR EACH NUMERAL. After each group has been constructed, TELL ME HOW YOU KNOW THAT YOU HAVE (*number*).

EVALUATION: Note if the child constructs the correct amount of tens and units and can explain accurately how the construction is represented by the numeral.

9S **Concrete Operations**

Geometry, Graphs, Charts, and Tables: Unit 31

See the prerequisite concepts and skills in Units 12, 13, 20, and 25 and Assessment Tasks 3F, 3G, 4E, 5E, and 6C.

9T **Concrete Operations**

Measurement with Standard Units: Unit 32

See prerequisite concepts in Units 18 and 19, and Assessment Tasks 3J, 4I, 5H, 5I, Level 7 (2), 8A, and 8B.

9U **Concrete Operations**

Problem Solving: Unit 33

Observe each child as he works on problems. Use the list of behaviors suggested in Unit 33:

1. The child reads the problems carefully.
2. The child has a way of attacking problems.
3. The child applies a strategy.
4. If one strategy fails the child tries another.
5. The child is persistent and consistent in applying strategies.
6. The child avoids making careless errors. If he does make careless errors are there patterns to them?
7. Note how long the child persists in trying to solve a problem.
8. How long does it take before the child will ask for help when needed?
9. What strategies does the child use most frequently?
10. The child uses manipulative materials when needed.
11. Note facial and physical indicators of interest and involvement.

REFERENCES

Baroody, A. J. 1988. *Children's mathematical thinking*. New York: Teachers College Press. Includes many examples of children's common mistakes and misconceptions.

Engelhardt, J. M., Ashlock, R. B, and Wiebe, J. H. 1994. *Helping children understand and use numerals*. Boston: Allyn & Bacon. Chapter 4 contains many diagnostic tasks.

Labinowicz, E. 1985. *Learning from children: New beginnings for teaching numerical thinking*. Menlo Park, Calif.: Addison-Wesley. Chapter 2 describes the basics of the interview method, Appendix B gives interview hints, and Appendix F provides some starting points for interviews.

Richardson, K. 1984. *Developing number concepts using Unifix® Cubes*. Menlo Park, Calif.: Addison-Wesley. At the end of each chapter there is a section on analyzing and assessing children's needs.

APPENDIX B

Children's Books with Math and Science Concepts

CONTENTS*

*Numbers indicating relevant units follow each entry.

FUNDAMENTAL CONCEPTS

One-to-One Correspondence

Gag, W. 1941. *Nothing at all.* New York: Coward-McCann. Ages 3–5; **8, 9, 17**.

Slobodkina, E. 1976. *Caps for sale.* New York: Scholastic. Ages 3–6; **8, 15**.

The three bears. 1973. New York: Golden Press. Ages 2–5; **8, 9, 17**.

The Three Billy Goats Gruff. 1968. New York: Grosset & Dunlap. Ages 2–5; **8, 9**.

Number and Counting

Allen, R. 1968. *Numbers: A first counting book.* New York: Platt & Munk. Ages 3–7; **9**.

Anno, M. 1982. *Anno's counting house.* New York: Philomel. Ages 4–7; **9**.

Bang, M. 1983. *Ten, nine, eight.* New York: Greenwillow. Ages 3–5; **9**.

Becker, J. 1973. *Seven little rabbits.* New York: Scholastic. Ages 3–6; **9, 25**.

Boynton, S. 1978. *Hippos go berserk.* Chicago, Ill.: Recycled Paper Press. Ages 3–6; **9, 11, 25, 27**.

Budney, B. 1962. *A cat can't count.* New York: Lothrop. Ages 5–8; **9**.

Carle, E. 1971. *1, 2, 3 to the zoo.* Mountain View, Calif.: Collins and World. Ages 3–5; **9**.

______. 1971. *The rooster who set out to see the world.* New York: Franklin Watts. Ages 3–5; **9**

______. 1969. *The very hungry caterpillar.* Mountain View, Calif.: Collins and World. Ages 3–5; **9, 17, 19, 21, 26**.

______. 1972. *The very long train: A folding book.* New York: Crowell. Ages 3–5; **9, 18**.

Chwast, S. 1971. *Still another number book.* New York: McGraw-Hill. Ages 3–5; **9**.

Cleveland, D. 1978. *April rabbits.* New York: Scholastic. Ages 3–5; **9, 19**.

Craig, H. 1983. *The little mouse 123.* New York: Little Simon. Ages 3–5; **9**.

Crews, D. 1985. *Ten black dots, revised.* New York: Greenwillow. Ages 4–6; **9**.

Crowther, R. 1981. *The most amazing hide and seek counting book.* New York: Viking. Ages 3–6; **9**.

Davis, B. S. 1972. *Forest hotel—A counting story.* Racine, Wis.: Western Publishing. Ages 3–6; **9**.

Dodd, L. 1978. *The nickle nackle tree.* New York: Macmillan. Ages 3–6; **9, 11, 15, 27**.

Duke, K. 1985. *Seven froggies went to school.* New York: Dutton. Ages 2–6; **9**.

Eichenberg, F. 1955. *Dancing in the moon.* New York: Harcourt, Brace. Ages 3–5; **9, 23, 24**.

Elkin, B. 1968, 1971. *Six foolish fishermen.* New York: Scholastic. Ages 3–6; **9, 33**.

Ernst, L. C. 1986. *Up to ten and down again.* New York: Lothrop, Lee, & Shepard. Ages 2–7; **9**.

Feelings, M. 1976. *Moja means one: Swahili counting book.* New York: Dial. Ages 3–6; **9**.

Frances, M. 1972. *Mr. Mac-a-doodle.* Mahwah, N.J.: Troll Associates. Ages 3–5; **9**.

Friskey, M. 1946. *Chicken little count to ten.* New York: Harcourt, Brace. Ages 3–8; **9, 23, 24**.

Frith, M. 1973. *I'll teach my dog 100 words.* New York: Random House, Ages 3–6; **9**.

Gag, W. 1928, 1956, 1977. *Millions of cats.* New York: Coward-McCann. Ages 3–5; **9, 15, 11**.

Gerstien, M. 1984. *Roll over.* New York: Crown. Ages 3–6; **9**.

Gordon, M. 1986. *Counting.* Morristown, N.J.: Silver-Burdett. Ages 3–6; **9**.

Gretz, S. 1969. *Teddy bears 1 to 10.* Chicago: Follett. Ages 3–6; **9**.

Hughes, S. 1985. *When we went to the park.* New York: Lothrop. Ages 1–2; **9**.

Hoban, T. 1972. *Count and see.* New York: Macmillan. Ages 2–5; **9**.

Howe, C. 1983. *Counting penguins.* New York: Harper. Ages 3–5; **9, 21, 26**.

Hynard, J. 1981. *Percival's partu.* Winchester, England: Hambleside. Ages 3–5; **9**.

Ipcar, D. 1958. *Ten big farms.* New York: Alfred A. Knopf. Ages 3–5; **9**.

Keats. E. J. 1972. *Over in the meadow.* New York: Scholastic. Ages 3–5; **9**.

Kinkaid, L., and Kinkaid, E. 1976. *The skittles.* Great Britain: Brimax Books, Ages 3–5; **9**.

Kitamura, S. 1986. *When sheep cannot sleep? The counting book.* New York: Farrar, Straus, & Giroux. Ages 2–5; **9**.

Kulas, J. E. 1978. *Puppy's 1 2 3 book.* Racine, Wis.: Western Publishing Co. Ages 3–5; **8, 9**.

Leedy, L. 1985. *A number of dragons.* New York: Holiday House. Ages 1–3; **9**.

Little, M. E. 1974. *1 2 3 for the library.* New York: Atheneum. Ages 2–5; **9**.

Mack, S. 1974. *Ten bears in my bed: A goodnight countdown.* New York: Pantheon Books. Ages 2–6; **9**.

Maestro, B. 1977. *Harriet goes to the circus: A number concept book.* New York: Crown. Ages 2–6; **9**.

Matthews, L. 1980. *Bunches and bunches of bunnies.* New York: Scholastic. Ages 2–5; **9**.

McMillan, C. 1986. *Counting wild flowers.* New York: Lothrop, Lee, & Shepard. Ages 2–7; **9, 21, 26, 35**.

Miller, J. 1983. *The farm counting book.* Englewood Cliffs, N.J.: Prentice-Hall. Ages 3–5; **9, 21, 26**.

Nolan, D. 1976. *Monster bubbles.* Englewood Cliffs, N.J.: Prentice-Hall. Ages 3–5; **9**.

Noll, S. 1984. *Off and counting.* New York: Green Willow. Ages 3–5; **9**.

Parish, P. 1974. *Too many rabbits.* New York: Scholastic. Ages 3–6; **9, 11, 15**.

Patience, J. 1985. *The fancy dress party counting book.* New York: Outlet Book Co. Ages 4–7; **9**.

Pavey, P. 1979. *One dragon's dream.* Scarsdale, N.Y.: Bradbury Press. Ages 3–6; **9**.

Peppe, R. 1969. *Circus numbers.* New York: Delacorte. Ages 3–8; **9**.

Petie, H. 1975. *Billions of bugs.* Englewood Cliffs, N.J.: Prentice-Hall. Ages 3–7; **9**.

Pomerantz, C. 1984. *One duck, another duck.* New York: Greenwillow. Ages 3–7; **9**.

Presland, J. 1975. *How many.* Restrop Manor, Purton Wilts, England: Child's Play (International) Ltd. Ages 3–7; **9**.

Quackenbush, R. 1975. *Too many lollipops.* New York: Parent's Magazine Press. Ages 3–6; **9, 11**.

Ross, H. L. 1978. *Not counting monsters.* New York: Platt and Munk. Ages 3–5; **9**.

Scarry, R. 1975. *Best counting book ever.* New York: Random House. Ages 2–8; **9, 19**.

Schwartz, D. M. 1985. *How much is a million?* New York: Lothrop. Ages 5–9; **9**.

Seuss, Dr. 1960. *One fish, two fish, red fish, blue fish.* New York: Random House. Ages 3–7; **9, 11, 15, 18**.

______. (1938) *The 500 hats of Bartholomew Cubbins.* Eau Claire, Wis.: Hale and Co. Ages 3–7; **9**.

Sitomer, M., and Sitomer, H. 1976. *How did numbers begin?* New York: Harper and Row. Ages 4–6; **9**.

Slobodkin, L. 1955. *Millions and millions.* New York: Vanguard. Ages 3–5; **9**.

Steiner, C. 1960. *Ten in a family.* New York: Alfred A. Knopf. Ages 3–5; **9**.

Stoddart G., and Baker, M. 1982. *One, two number zoo.* London: Hodder & Stoughton. Ages 5–7; **9, 21, 36, 35**.

Sussman, S. 1982. *Hippo thunder.* Niles, Ill.: Whelman. Ages 2–5; **9, 21, 26**.

Tafuri, N. 1986. *Who's counting?* New York: Greenwillow. Ages 2–5; **9**.

Thompson, S. L. 1980. *One more thing, Dad.* Chicago: Whitman. Ages, 3–6; **9**.

Ungerer, T. 1962. *The three robbers.* New York: Antheum. Ages 3–5; **9**.

Wildsmith, B. 1965. *Brian Wildsmith's 1, 2, 3's.*

New York: Franklin Watts. Ages 3–5; **9**.
Zolotow, C. 1955. *One step, two step.* New York: Lothrop, Lee, & Shepard. Ages 2–5; **9**.
The Sesame Street 1, 2, 3 Book. 1973. New York: Random House. Ages 2–6; **9**.

Classification

Gordon, M. 1986. *Colors.* Morristown, N.J.: Silver-Burdett. Ages 2–6; **10**.
Hill, E. 1982. *What does what?* Los Angeles: Price/Stern/Sloan. Ages 2–4; **10**.
Hoban, T. 1978. *Is it red? Is it yellow? Is it blue?* New York: Greenwillow. Ages 2–5; **10**.
Hughes, S. 1986. *Colors.* New York: Lothrop. Ages 2–4; **10**.
Johnson, J. 1985. *Firefighters A–Z.* New York: Walker. Ages 5–8; **10, 22**.
Wandro, M., and Blank, J. 1981. *My daddy is a nurse.* Reading, Mass.: Addison-Wesley. Ages 5–8; **10, 22**.
White, R., and Rehwald, M. 1976. *Mix and match: Activities for classification.* Los Angeles: Rhythms Productions. Ages 3–5; **10**.
Wildsmith, B. 1967. *Brian Wildsmith's wild animals.* New York: Franklin Watts. Ages 3–8; **10, 21, 26, 35**.
Wildsmith, B. 1968. *Brian Wildsmith's fishes.* New York: Franklin Watts. Ages 3–8; **10, 21, 26, 35**.
Winthrop, E. 1986. *Shoes.* New York: Harper & Row. Ages 3–7; **10, 22**.

Comparing

Brenner, B. 1966. *Mr. Tall and Mr. Small.* Menlo Park, Calif.: Addison-Wesley. Ages 4–7; **11, 21, 26, 35**.
Broger, A., and Kalow, G. 1977. *Good morning whale.* New York: Macmillan. Ages 3–6; **11, 21, 26, 35**.
Carle, E. 1977. *The grouchy ladybug.* New York: Crowell. Ages 3–5; **11, 21, 26**.
Eastman, P. D. 1973. *Big dog, Little dog.* New York: Random House. Ages 3–3; **11, 21, 26**.
Gordon, M. 1986. *Opposites.* Morristown, N.J.: Silver-Burdett. Ages 3–5; **11**.
Grender, I. 1975. *Playing with shapes and sizes.* New York: Knopf/Pinwheel Books. Ages 3–6; **11, 12**.
Heide, F. P. 1970. *Benjamin Budge and Barnaby Ball.* New York: Scholastic. Ages 3–5; **11**.
Hoban, T. 1972. *Push pull, empty full.* New York: Macmillan. Ages 3–5; **11**.
Holl, A., and Reit, S. 1970. *Learning about sizes.* Indianapolis: Bobbs-Merrill. Ages 3–5; **11, 18**.
Horn, A. 1974. *You can be taller.* Boston: Little, Brown. Ages 3–5; **11**.
Hughes, S. 1985. *Bathwater's hot.* New York: Lothrop. Ages 1–2; **11, 21, 26**.
———. 1985. *Noises.* New York: Lothrop. Ages 1–2; **11, 21, 26**.
Lewis, J. 1963. *The tortoise and the hare.* Chicago, Ill.: Whitman. Ages 4–8; **11, 18**.
Lionni, L. 1968. *The biggest house in the world.* New York: Pantheon. Ages 2–6; **11, 21, 26**.
McMillan, B. 1986. *Becca backward, Becca forward.* New York: Lothrop. Ages 3–6; **11**.
Presland, J. 1975. *Same and different.* Purton Wilts, England: Child's Play (International), Ltd. Ages 4–7; **11**.
Scarry, R. 1976. *Short and tall.* New York: Golden Press. Ages 2–7; **11**.
Scarry, R. 1986. *Big and little: A book of opposites.* Racine, Wis.: Western. Ages 3–7; **11**.
Shapiro, L. 1978. *Pop-up opposites.* Los Angeles: Price/Stern/Sloan. Ages 3–5; **11**.

Shape

Budney, B. 1954. *A kiss is round.* New York: Lothrop, Lee, & Shepard. Ages 2–6; **12**.
Carle, E. 1974. *My very first book of shapes.* New York: Crowell. Ages 3–6; **12**.
Emberley, E. 1972. *Ed Emberley's drawing book:*

Make a world. Boston: Little, Brown. Ages 6–8; **12**.

______. 1970. *Ed Emberley's drawing book of animals.* Boston: Little, Brown. Ages 6–8; **12**.

______. 1961. *A wing on a flea: A book about shapes.* Boston: Little, Brown. Ages 5–8; **12**.

Gordon, M. 1986. *Shapes.* Morristown, N.J.: Silver-Burdett. Ages 3–5; **12**.

Hefter, R. 1976. *The strawberry book of shapes.* New York: Weekly Reader Books. Ages 3–7; **12**.

Hoban, T. 1974. *Circles, triangles, and squares.* New York: Macmillan. Ages 5–8.

Hoban, T. 1983. *Round and round and round.* New York: Greenwillow. Ages 3–6; **12**.

Hoban, T. 1986. *Shapes, shapes, shapes.* New York: Greenwillow. Ages 3–7; **12**.

Kessler, E., and Kessler, L. 1966. *Are you square?* Garden City, N.Y.: Doubleday. Ages 5–7; **12**.

Salazar, V. 1967. *Squares are not bad.* New York: Golden. Ages 5–8; **12**.

Schlein, M. 1952. *Shapes.* Glenview, Ill.: Scott, Foresman. Ages 2–6; **12**.

Sullivan, J. 1963. *Round is a pancake.* New York: Holt, Rinehart, and Winston. Ages 3–5; **12**.

Supraner, R. 1975. *Draw me a square, Draw me a triangle, & Draw me a circle.* New York: Simon and Schuster/Nutmeg. Ages 3–6; **12**.

Space

Barton, B. 1981. *Building a house.* New York: Greenwillow. Ages 4–7; **13, 14, 19**.

Berenstain, S., and Berenstain, J. 1968. *Inside, outside, upside down.* New York: Random House. Ages 3–7; **13**.

Brown, M. 1949. *Two little trains.* New York: Scott, Foresman. Ages 2–4; **13, 15**.

Carle, E. 1972. *The secret birthday message.* New York: Crowell. Ages 3–7; **13**.

Dunrea, O. 1985. *Fergus and the bridey.* New York: Holiday. Ages 4–7; **13**.

Hill, E. 1980. *Where's Spot?* New York: Putnam's Sons. Ages 2–4; **13**.

Lionni, L. 1983. *Where?* New York: Pantheon. Ages 2–3; **13**.

Maestro, B., and Maestro, G. 1976. *Where is my friend?* New York: Crown. Ages 2–4; **13**.

Martin, B., Jr. 1971. *Going up, going down.* New York: Holt, Rinehart, & Winston. Ages 6–8; **13, 9**.

Russo, M. 1986. *The line up book.* New York: Greenwillow. Ages 3–5; **13**.

Parts and Wholes

Dubov, C. S. 1986. *Alexsandra, where is your nose?* New York: St. Martin's Press. Ages 1 1/2–3; **14, 21, 26**.

______. 1986. *Alexsandra, where are your toes?* New York: St. Martin's Press. Ages 1 1/2–3; **14, 21, 26**.

Le Tord, B. 1985. *Good wood bear.* New York: Bradbury. Ages 4–7; **14, 32**.

Mathews, L. 1979. *Gator pie.* New York: Scholastic. Ages 4–7; **14, 25**.

Language: Books That Contain Several Concepts

Bemelmans, L. 1969. *Madeline.* New York: Viking. Ages 4–7; **15, 8, 9, 19**.

Duvoisin, R. 1974. *Petunia takes a trip.* New York: Knopf/Pinwheel. Ages 4–7; **15, 11, 18, 21, 26, 35**.

Hoff, S. 1959. *Julius.* New York: Harper & Row. Ages 4–7; **15, 21, 26, 35**.

Mathematics in the kitchen, Mathematics at the farm, Mathematics in buildings, Mathematics on the playground, Mathematics in the circus ring. (1978) Milwaukee, Wis.: MacDonald-Raintree. Ages 3–7; **15**.

APPLICATION OF FUNDAMENTAL CONCEPTS

Ordering

Asbjörsen, P. C., and Moe, J. E. 1957. *The Three Billy Goats Gruff.* New York: Harcourt, Brace, Jovanovich. Ages 2–5; **17, 9, 11, 15**.

Bishop, C. H., and Wiese, K. 1938. *The five Chinese brothers.* New York: Coward, McCann, & Geoghegan. Ages 5–8; **17**.

Brett, J. 1987. *Goldilocks and the three bears.* New York: Dodd, Mead. Ages 3–5; **17, 21, 26**.

Ipcar, C. 1972. *The biggest fish in the sea.* New York: Viking. Ages 3–6; **17, 21, 26, 35**.

Macauly, D. 1987. *Why the chicken crossed the road.* Boston: Houghton-Mifflin. Ages 4–8; **17**.

Maestro, B., and Maestro, G. 1977. *Harriet goes to the circus.* New York: Crown. Ages 5–8; **17, 21, 26, 35**.

Mahy, M. 1987. *17 kings and 42 elephants.* New York: Dial. Ages 2–6; **17, 9, 21, 26, 35**.

Martin, B., Jr. 1963. *One, two, three, four.* New York: Holt, Rinehart, and Winston. Ages 5–7; **17, 9**.

______. 1970. *Monday, Monday, I like Monday.* New York: Holt, Rinehart, and Winston. Ages 5–8; **17, 19**.

*Note: Many of the books listed in the section on *Number and counting* include ordinal numbers.

Measurement: Volume, Weight, and Length

Allen, P. 1983. *Who sank the boat?* New York: Coward. Ages 3–5; **18**.

Anderson, L. C. 1983. *The wonderful shrinking shirt.* Niles, Ill.: Whitman. Ages 3–5; **18**.

Battles, E. 1973. *One to teeter-totter.* Niles, Ill.: Whitman. Ages 3–7; **18, 32**.

Bennett, V. 1975. *My measure it book.* New York: Grosset & Dunlap. Ages 3–5; **18**.

Briggs, R. 1970. *Jim and the beanstalk.* New York: Coward, McCann, & Geoghegan. Ages 4–6; **18**.

Charles, D. 1977. *Fat, fat calico cat.* Chicago: Children's Press. Ages 3–6; **17, 38**.

Grender, I. 1975. *Measuring things.* New York: Knopf/Pantheon Pinwheel. Ages 4–6; **18**.

Holl, A., and Reit, S. 1970. *Time and measuring.* Indianapolis: Bobbs-Merrill. Ages 3–6; **18, 19**.

Larranaga, R. D. 1970. *The king's shadow.* Minneapolis: Carolrhoda. Ages 5–8; **18, 32**.

Linn, C. 1970. *Estimation.* New York: Crowell. Ages 8–9; **18, 32, 33**.

Lionni, L. 1960. *Inch by inch.* New York: Astor-Honor. Ages 3–5; **17, 21, 26**.

McMillan, B. 1987. *Step by step.* New York: Lothrop. Ages 3–6; **18, 19**.

Myller, R. 1972. *How big is a foot?* New York: Atheneum. Ages 6–8; **18, 32**.

Parkinson, K. 1986. *The enormous turnip.* Niles, Ill.: Whitman. Ages 4–7; **18, 32, 35**.

Russo, M. 1986. *The lineup book.* New York: Greenwillow. Ages 2–4; **18**.

Schlein, M. 1954. *Heavy is a hippopotamus.* New York: Scott. Ages 3–6; **18, 21, 26**.

Scrivastava, J. J. 1980. *Spaces, shapes, and sizes.* New York: Crowell. Ages 5–8; **18, 32**.

Shapp, M., and Shapp, C. 1975. *Let's find out about what's light and what's heavy.* New York: Franklin Watts. Ages 6–8; **18, 32**.

Väes, A. 1985. *The wild hamster.* Boston: Little, Brown. Ages 5–8; **18, 21, 26, 32, 35**.

Ward, L. 1952. *The biggest bear.* Boston: Houghton-Mifflin. Ages 3–5; **18, 19**.

Zion, G. 1959. *The plant sitter.* New York: Harper & Row. Ages 3–6; **18, 19**.

Measurement: Time

Aiken, C. 1966. *Tom, Sue, and the clock.* New York: Collier Books. Ages 5–8; **19, 32**.

Bancroft, H., and Van Gelde, R. G. 1963. *Ani-*

mals in winter. New York: Scholastic. Ages 3–6; **19, 32**.

Barrett, J. 1976. *Benjamin's 365 birthdays.* New York: Atheneum. Ages 3–6; **19**.

Berenstain, S., and Berenstain, J. 1973. *The bear's almanac.* New York: Random House. Ages 3–6; **19, 32**.

Bonne, R. 1961. *I know an old lady.* New York: Scholastic. Ages 3–5; **19**.

Brenner, B. 1984. *The snow parade.* New York: Crown. Ages 4–7; **19, 32**.

Brown, M. W. 1947. *Goodnight moon.* New York: Harper & Row. Ages 3–6; **19**.

______. 1950. *A child's goodnight book.* New York: W. R. Scott. Ages 3–6; **19**.

Brown, M. 1984. *Arthur's Christmas.* Boston: Little, Brown. Ages 6–8; **32**.

Carle, E. 1977. *The very hungry caterpillar.* New York: Collins & World. Ages 3–5; **19, 21, 26**.

Castle, C. 1985. *The hare and the tortoise.* New York: Dial. Ages 5–8; **19, 32**.

Chalmers, M. 1988. *Easter parade.* New York: Harper. Ages 3–6; **19, 32**.

DeArmand, F. U. 1963. *A very, very special day.* New York: Parents' Magazine Press. Ages 3–6; **19, 32**.

DePaola, T. 1986. *Merry Christmas, Strega Nona.* San Diego, Calif.: Harcourt, Brace. Ages 3–6; **19, 32**.

Downer, M. N. 1945. *The flower.* New York: W. R. Scott. Ages 3–6; **19, 21, 26, 32**.

Duvoisin, R. 1956. *The house of four seasons.* New York: Lothrop, Lee, & Shepard. Ages 3–6; **19, 32**.

Flournoy, V. 1978. *The best time of day.* New York: Random House. Ages 3–5; **19**.

______. 1985. *Patchwork quilt.* New York: Dial. Ages 4–8; **19, 32**.

Gibbons, G. 1983. *Thanksgiving day.* New York: Holiday House. Ages 4–8; **19, 32**.

Hauge, C., and Hauge, M. 1974. *Gingerbread man.* New York: Golden Press. Ages 2–5; **19, 17**.

Hall, B. 1973. *What ever happens to baby animals?* New York: Golden Press. Ages 2–5; **19, 17**.

Hayes, S. 1986. *Happy Christmas Gemma.* New York: Lothrop. Ages 2–5; **19**.

Hooper, M. 1985. *Seven eggs.* New York: Harper & Row. Ages 3–5; **19, 17, 8**.

Kelleritti, H. 1985. *Henry's Fourth of July.* New York: Greenwillow. Ages 3–6; **19**.

Krementz, J. 1986. *Zachary goes to the zoo.* New York: Random House. Ages 2–8; **19, 21, 26, 32, 35**.

Kraus, R. 1972. *Milton the early riser.* New York: Prentice-Hall. Ages 2–5; **19**.

Leslie, S. 1977. *Seasons.* New York: Platt & Munk. Ages 2–5; **19**.

Lester, A. 1986. *Clive eats alligators.* Boston: Houghton-Mifflin. Ages 3–6; **19, 21, 26**.

McCully, E. A. 1985. *First snow.* New York: Warner. Ages 3–5; **19, 21, 26**.

Miles, B. 1973. *A day of Autumn.* New York: Random House. Ages 3–5; **19**.

Ormerodi, J. 1981. *Sunshine.* New York: Lothrop, Lee, & Shepard. Ages 2–6; **19**.

Pearson, S. 1988. *My favorite time of year.* New York: Harper & Row. Ages 3–7; **19, 32, 35**.

Porter Productions. 1975. *My tell time book.* New York: Grosset & Dunlap. Ages 5–7; **19, 32**.

Prelutsky, J. 1984. *It's snowing! It's snowing!.* New York: Greenwillow. Ages 4–7; **19, 32, 21, 26, 37.**.

Provensen, A., and Provensen, M. 1976. *A book of seasons.* New York: Random House. Ages 3–5; **19**.

Robison, A. 1973. *Pamela Jane's week.* Racine, Wis.: Whitman Books, Western Publishing. Ages 2–5; **19**.

Rockwell, A. 1985. *First comes Spring.* New York: Crowell. Ages 2–6; **19, 32, 21, 26, 37**.

Rutland, J. 1976. *Time.* New York: Grosset & Dunlap. Ages 2–7; **19, 32**.

Scarry, R. 1976. *All day long.* New York: Golden

Press. Ages 3–6; **19, 32.**
Schlein, M. 1955. *It's about time.* New York: Young Scott. Ages 3–7; **19, 32.**
Schwerin, D. 1984. *The tommorrow book.* New York: Pantheon. Ages 3–6; **19.**
Stein, S. B. 1985. *Mouse.* San Diego, Calif.: Harcourt, Brace, Jovanovich. Ages 4–8; **19, 32, 21, 26, 35.**
Todd, K. 1982. *Snow.* Reading, Mass.: Addison-Wesley. Ages 3–8; **19, 21, 26, 32, 37.**
Tudor, T. 1957. *Around the year.* New York: Henry Z. Walck. Ages 3–5; **19.**
———. 1977. *A time to keep: The Tasha Tudor book of holidays.* New York: Rand McNally. Ages 3–6; **19.**
Vincent, G. 1984. *Merry Christmas, Ernest & Celestine.* New York: Greenwillow. Ages 4–8; **19, 32.**
Wolff, A. 1984. *A year of birds.* New York: Dodd, Mead. Ages 3–6; **19, 21, 26.**
Zolotow, C. 1984. *I know an old lady.* New York: Greenwillow. Ages 4–8; **19, 32, 21, 26.**

Practical Activities

Money

Asch, F. 1976. *Good lemonade.* Ontario, Canada: Nelson, Foster, & Scott. Ages 6–8; **32.**
Brenner, B. 1963. *The five pennies.* New York: Random House. Ages 6–7; **22, 32.**
Credle, E. 1969. *Little pest Pico.* Ontario, Canada: Nelson, Foster, & Scott. Ages 6–8; **32.**
Hoban, L. 1981. *Arthur's funny money.* New York: Harper & Row. Ages 4–7; **22, 32.**
Kirn, A. 1969. *Two pesos for Catalina.* New York: Scholastic. Ages 6–8; **22, 32.**
Martin, B., Jr. 1963. *Ten pennies for candy.* New York: Holt, Rinehart, and Winston. Ages 5–7; **22, 32.**
Rockwell, A. 1984. *Our garage sale.* New York: Greenwillow. Ages 3–5; **22.**

Food (see also Unit 22)

Brown, M. 1947. *Stone soup.* New York: Charles Scribner's. Ages 3–5; **22, 32, 38.**
Carle, E. 1970. *Pancakes, pancakes.* New York: Knopf. Ages 3–5; **22.**
Hoban, R. 1964. *Bread and jam for Frances.* New York: Scholastic. Ages 3–7; **22, 38.**
McCloskey, R. 1948. *Blueberries for Sal.* New York: Viking. Ages 3–6; **22, 38.**
Norquist, S. 1985. *Pancake pie.* New York: Morrow. Ages 4–8; **22, 32, 38.**
Sendak, M. 1962. *Chicken soup with rice.* New York: Harper & Row. Ages 3–5; **22.**
———. 1970. *In the night kitchen.* New York: Harper & Row. Ages 4–6; **22, 32, 38.**
Seymour, P. 1981. *Food.* Los Angeles: Intervisual Communications. Ages 2–5; **22.**
Thayer, J. 1961. *The bluberry pie elf.* Edinburgh, Scotland: Oliver & Boyd. Ages 4–7; **22, 32, 38.**

Cookbooks (see also Unit 22)

Ault, R. 1974. *Kids are natural cooks.* Boston: Houghton-Mifflin. **22, 32, 38.**
Better Homes and Gardens new junior cookbook. 1979. Des Moines: Meredith. **22, 32, 38.**
Blanchet, F., and Kaup, D. 1979 *What to do with an egg.* Woodbury, N.Y.: Barrons Educational Series, Inc. **22, 32, 38.**
Cooking is fun. 1970. New York: Dell. **22, 32, 38.**
Holly Hobbie's cookbook. 1979. New York: House. **22, 32, 38.**
Kementz, J. 1985. *The fun of cooking.* New York: Knopf. **22, 32, 38.**
Sesame Street cookbook. 1978. New York: Platt & Munk, **22, 32, 38.**
Walt Disney's Mickey Mouse cookbook. 1975. New York: Golden Press. **22, 32, 38.**

SYMBOLS AND HIGHER-LEVEL ACTIVITIES

Sets and Symbols

Most of the books listed in the section on *Number and counting* include number symbols.

The following are some of the better sets and symbols books and a few additional titles.

Alain (Bruslein, Alain). 1964. *One, two, three going to sea.* New York: Scholastic. Ages 5–7; **23, 24, 25, 27**.

Anno, M. 1977. *Anno's counting book.* New York: Crowell. Ages 5–7; **23, 24, 25**.

Balet, J. B. 1959. *The five Rollatinis.* Philadelphia: Lippincott. Ages 4–7; **23, 24, 25**.

Brown, M. 1976. *One, two, three an animal counting book.* New York: Atlantic Monthly Press. Ages 5–7; **23, 24, 25, 21, 26, 35**.

Daly, E. 1974. *1 is red.* Racine, Wis.: Western Publishing. Ages 4–7; **23, 24, 25**.

Duvoisin, R. 1955. *Two lonely ducks.* New York: Knopf. Ages 4–7; **23, 24, 25, 21, 26, 35**.

______. 1955. *1000 Christmas beards.* New York: Knopf. Ages 3–7; **23, 24, 25**.

Federico, H. 1963. *The golden happy book of numbers.* New York: Golden Press. Ages 3–7; **23, 24, 25**.

Francoise (Seignobosc, Francoise). 1951. *Jean-Marie counts her sheep.* New York: Charles Scribner's Sons. Ages 3–6; **23, 24, 25**.

Friskey, M. 1940. *Seven diving ducks.* New York: McKay. Ages 4–6; **23, 24, 25**.

Hoban, T. 1987. *Letters & 99 cents.* New York: Greenwillow. Ages 4–8; **23, 24, 25, 32**.

Keats, E. J. 1971. *Over in the meadow.* New York: Scholastic. Ages 3–5; **23, 24**.

LeSeig, T. 1974. *Whacky Wednesday.* New York: Random House. Ages 5–8; **25**.

McNutt, D. 1979. *There was an old lady who lived in a 1.* Palo Alto, Calif.: Creative Publications. **9, 23, 24**.

Numbers: Match-up flip book. 1984. St. Paul, Minn.: Trend. Ages 4–8; **23, 24**.

The count's number parade. 1973 and *The count's poem* 1978. Racine, Wis.: Western Publishing. Ages 3–6; **23, 24**.

MATHEMATICS CONCEPTS AND ACTIVITIES FOR THE PRIMARY GRADES

As already noted, many of the books listed are appropriate for preprimary and primary children. Many books that are read-along books for the younger children become books for individual reading for older children. A few additional titles are included here.

Abisch, R. 1968. *Do you know what time it is?* Englewood Cliffs, N.J.: Prentice-Hall. Ages 6–8; **32**.

Allen, J. 1975. *Mary Alice, Operator No. 9.* New York: Scholastic. Ages 6–8; **32**.

Anderson, L. 1971. *Two hundred rabbits.* New York: Penguin Books. Ages 7–9; **27**.

Belov, R. 1971. *Money, money, money.* New York: Scholastic. Ages 6–8; **32**.

Dennis, J. R. 1971. *Fractions are parts of things.* New York: Crowell. Ages 7–8; **29**.

Friskey, M. 1963. *Mystery of the farmer's three fives.* Chicago: Children's Press. Ages 6–8; **27**.

Hawkins, C. 1984. *Take away monsters.* New York: Putnam's Sons. Ages 3–5; **25, 27**.

Martin, B. Jr. 1964. *Delight in number.* New York: Holt, Rinehart, & Winston. Ages 6–8; **9, 27, 32**.

______. 1963. *Five is five.* New York: Holt, Rinehart, & Winston. Ages 6–8; **25, 27**.

______. 1964. *Four threes are twelve.* New York: Holt, Rinehart, & Winston. Ages 6–8; **25, 27**.

______. 1964. *If you can count to ten.* New York: Holt, Rinehart, & Winston. Ages 6–8; **25, 27**.

______. 1971. *Number patterns make sense.* New York: Holt, Rinehart, & Winston. Ages 8–9; **28, 33**.

Schertle, A. 1987. *Jeremy Bean's St. Patrick's Day.* New York: Morrow. Ages 5–8; **19, 32**.

Schleim, M. 1972. *Moon months and sun days.* Reading, Mass.: Young Scott. Ages 6–8; **19, 32**.

BOOKS THAT SUPPORT SCIENCE INVESTIGATIONS

Life Science

Animals

Arnold, C. 1987. *Kangaroo/Koala.* New York: Morrow. Ages 7–10; **35**.

Arnosky, J. 1986. *Deer at the Brook.* New York: Lothrop. Ages 1–6; **21, 26, 35**.

———. 1987. *Raccoons and Ripe Corn.* New York: Lothrop. Ages 3–6; **21, 26, 35**.

Berger, M. 1983. *Why I Cough, Sneeze, Shiver, Hiccup, and Yawn.* New York: Crowell. Ages 5–8; **21, 26, 35**.

Bond, F. 1987. *Wake up, Vladimir.* New York: Harper & Row. Ages 3–5; **21, 26**.

Crow, S. L. 1985. *Penguins and Polar Bears.* Washington, D.C.: National Geographic. All ages; **21, 26, 35**.

Cole, J. 1984. *How You Were Born.* New York: Morrow. Ages 4–8; **21, 26, 35**.

Cole, S. 1985. *When the Tide is Low.* New York: Lothrop. Ages 3–9; **21, 26, 35**.

Fischer-Nagel, H. and Fischer-Nagel, A. 1986. *Inside the Burrow: The Life of the Golden Hamster.* Minneapolis: Carolrhoda. Ages 7–10; **35**.

Flack, M. 1930. *Angus and the Ducks.* New York: Doubleday. Ages 3–6; **21, 26, 35**.

———. 1937. *The Restless Robin.* New York: Houghton Mifflin. Ages 4–6; **21, 26, 35**.

———. 1933. *The Story about Ping.* New York: Viking. [(1977) Penguin] Ages 4–6; **21, 26, 35**.

Freeman, D. 1968. *Corduroy.* New York: Viking. [(1976) Penguin] Ages 3–6; **21, 26, 35**.

George, L. 1987. *William and Boomer.* New York: Greenwillow. Ages 3–7; **21, 26, 35**.

Girard, L. W. 1983. *You Were Born on Your Very First Birthday.* Niles, Ill.: Whitman. Ages 3–8; **21, 26, 35**.

Greenberg, P. 1983. *Birds of the World.* New York: Platt and Munk. Adult information, children pictures, **21, 26, 35**.

Hill, E. 1982. *Animals.* Los Angeles: Price/Stern/Sloan. Ages 3–5; **21, 26**.

Hillert, M. 1975. *The Sleepytime Book.* Racine, Wis.: Western. Ages 3–5; **21, 26**.

Hoban, T. 1985. *A Children's Zoo.* New York: Greenwillow. Ages 3–6; **21, 26**.

King, T. 1983. *The Moving Animal Book.* New York: Putnam. Ages 3–5; **21, 26**.

Koelling, C. 1978. *Whose House is This?* Los Angeles: Price/Stern/Sloan. Ages 3–5; **21, 26**.

Kramer, S. P. 1986. *Getting Oxygen: What to Do if You're Cell 22.* New York: Crowell. Ages 4–8; **21, 26, 35**.

Krementz, J. 1986. *Holly's Farm Animals.* New York: Random House. Ages 3–8; **21, 26, 35**.

Lionni, L. 1963. *Swimmy.* New York: Pantheon. Ages 3–6; **21, 26, 35**.

McCloskey, R. 1941. *Make Way for Ducklings.* New York: Viking. [(1976) Penguin] Ages 4–6; **21, 26, 35**.

McGrath, S. 1985. *Your World of Pets.* Washington, D.C.: National Geographic Society. Ages 5–10; **35**.

Mellonie, B. and Ingpen, R. 1983. *Lifetimes.* New York: Bantam Books. Ages 4–6; **21, 26**.

National Geographic Society (ed.). 1985. *Books for Young Explorers—Set XII.* Washington, D.C.: Editor. Ages 3–8; **21, 26, 35**.

Nicholson, D. 1987. *Wild Boars.* Minneapolis: Carolrhoda. Ages 6–10; **35**.

Nockels, D. 1981. *Animal Acrobats.* New York: Dial. Ages 4–6; **21, 26, 35**.

Oppenheim, J. 1986. *Have You Seen Birds?* New York: Scholastic. Ages 4–7; **21, 26, 35**.

Patent, D. H. 1987. *All about Whales*. New York: Holiday House. Ages 6–9; **35**.

Potter, B. 1902. *The Tale of Peter Rabbit*. New York: Warne. Ages 4–6; **21, 26, 35**.

Powzyk, J. 1985. *Wallaby Creek*. New York: Lothrop. Ages 6–9; **35**.

Rankin, C. 1985. *How Life Begins: A Look at Birth and Care in the Animal World*. New York: Putnam. Ages 7–11; **35**.

Roy, R. 1982. *What Has Ten Legs and Eats Cornflakes? A Pet Book*. New York: Clarion/Houghton. Ages 5–8; **35**.

Seymour, P. 1985. *Animals in Disguise*. New York: Macmillan. Ages 4–6; **21, 26, 35**.

Sussman, S. and Sussman, R. J. 1987. *Lies (People Believe) about Animals*. Niles, Ill.: Whitman. Ages 7–12; **35**.

Tarrant, G. 1983. *Frogs*. New York: Putnam. Ages 4–6; **21, 26, 35**.

Watson, J. W. 1958. *Birds*. New York: Golden Press. Ages 3–5; **21, 26**.

Wildsmith, B. 1983. *The Owl and the Woodpecker*. New York: Oxford University Press. Ages 4–7; **21, 26, 35**.

Bugs, Spiders, and Bees

Bason, L. 1974. *Spiders*. Washington, D.C.: National Geographic Society. Ages 3–8; **21, 26, 35**.

Berenstain, S. and Berenstain, J. 1962. *The Big Honey Hunt*. New York: Random House. Ages 3–8; **21, 26, 35**.

Carle, E. 1969. *The Very Hungry Caterpillar*. New York: Philomel. Ages 3–6; **21, 26, 35**.

______. 1977. *The Grouchy Ladybug*. New York: Crowell. Ages 3–6; **21, 26, 35**.

______. 1981. *The Honeybee and the Robber*. New York: Philomel. Ages 3–6; **21, 26, 35**.

______. 1984. *The Very Busy Spider*. New York: Philomel. Ages 3–6; **21, 26, 35**.

Clay, P. and Clay, H. 1984. *Ants*. London: A & C Black. Adult resource, good pictures for children; **21, 26, 35**.

Fischer-Nagel, H. and Fischer-Nagel, A. 1986. *Life of the Ladybug*. Minneapolis: Carolrhoda. Ages 7–10; **35**.

Fisher, A. 1986. *When it Comes to Bugs*. New York: Harper & Row. Ages 4–8; **21, 26, 35**.

Hooker, Y. 1981. *The Little Green Caterpillar*. New York: Grosset & Dunlap. Ages 3–6; **21, 26, 35**.

______. 1984. *The Little Red Ant*. New York: Grosset & Dunlap. Ages 3–6; **21, 26, 35**.

Johnson, S. A. 1983. *Ladybugs*. Minneapolis: Lerner. Adult resource, good photos; **21, 26, 35**.

Kaufman, J. 1985. *Joe Kaufman's Slimy, Creepy, Crawly Creatures*. New York: Golden.

McNulty, F. 1986. *The Lady and the Spider*. New York: Harper & Row. Ages 5–8; **35**.

Overbeck, C. 1982. *Ants*. Minneapolis: Lerner. Adult resource, good photos for children; **21, 26, 35**.

Parker, N. W. 1987. *Bugs*. New York: Greenwillow. Ages 8–10; **35**.

Rood, R. N. 1960. *Insects We Know*. New York: Wonder Books. Ages 3–8; **21, 26, 35**.

Seymour, P. 1984. *Insects: A Close-Up Look*. New York: Macmillan. Ages 3–6; **21, 26, 35**.

Selsam, M. E. and Goor, R. 1981. *Backyard Insects*. New York: Fourwinds. Ages 4–8; **21, 26, 35**.

Tarrant, G. 1983. *Butterflies*. New York: Putnam. Ages 3–6; **21, 26, 35**.

______. 1984. *Honeybees*. New York: Putnam. Ages 3–6; **21, 26, 35**.

Yabuuchi, M. 1983. *Animals Sleeping*. New York: Philomel. Ages 5–8; **21, 26, 35**.

______. 1985. *Whose Footprints*. New York: Philomel. Ages 3–4; **21, 26**.

Plants

Gibbons, G. 1984. *The Seasons of Arnold's Apple*

Tree. San Diego, Calif.: Harcourt Brace. Ages 3–9; **21, 26, 35**.

Johnson, S. A. 1986. *How Leaves Change.* Minneapolis: Lerner. Ages 7–13; **35**.

Krauss, R. 1945. *The Carrot Seed.* New York: Harper & Row. Ages 3–5; **21, 22, 26, 28**.

Mitgutsch, A. 1986. *From Wood to Paper.* Minneapolis: Carolrhoda. Ages 4–7; **21, 26, 35**.

Oechsli, H. and Oechsli, K. 1985. *In My Garden: A Child's Gardening Book.* New York: Macmillan. Ages 5–9; **21, 26, 35**.

Romanova, N. 1985. *Once There Was a Tree.* New York: Dial. Ages 3–9; **21, 26, 35**.

Schnieper, C. 1987. *An Apple Tree Through the Year.* Minneapolis: Carolrhoda. Ages 7–10; **35**.

Schweitzer, I. 1982. *Hilda's Restful Chair.* New York: Atheneum. Ages 3–6; **21, 26, 35**.

Silverstein, S. 1964. *The Giving Tree.* New York: Harper & Row. Ages 3–8; **21, 26, 35**.

Physical Science

Ardrizzone, E. 1960. *Johnny the Clock Maker.* New York: Walck. Ages 3–5; **21, 26**.

Brandt, K. 1985. *Sound.* Mahwah, N.J.: Troll Associates. Ages 3–5; **21, 26**.

Burton, V. L. 1939. *Mike Mulligan and his Steam Shovel.* Boston: Houghton Mifflin. Ages 3–5; **21, 26**.

Bushey, J. 1985. *Monster Trucks and Other Giant Machines on Wheels.* Minneapolis: Carolrhoda. Ages 5–9; **21, 26, 36**.

Cobb, V. 1983. *Gobs of Goo.* Philadelphia: Lippincott. Ages 6–8; **36**.

Crampton, G. 1986. *Scuffy the Tugboat.* New York: Western. Ages 3–5; **21, 26**.

Crews, D. 1981. *Light.* New York: Greenwillow. Ages 3–7; **21, 26, 36**.

Cole, J. 1983. *Cars and How They Go.* New York: Crowell. Ages 7–11; **36**.

Fowler, R. 1986. *Mr. Little's Noisy Boat.* New York: Grosset & Dunlap. Ages 0–9; **21, 26, 36**.

Gabb, M. 1980. *The Question and Answer Books: Everyday Science.* Minneapolis: Lerner. All ages; **21, 26, 36**.

Gibbons, G. 1982. *The tool book.* New York: Holiday House. Ages 3–5; **21, 26**.

———. 1983. *New Road!* New York: Crowell. Ages 5–8; **36**.

Gramatky, H. 1959. *Little Toot.* New York: Putnam. Ages 3–5; **21, 26**.

Iveson-Iveson, J. 1986. *Your Nose and Ears.* New York: The Bookwright Press. Ages 3–6; **21, 26, 36**.

Isadora, R. 1985. *I Touch.* New York: Greenwillow. Ages 0–2; **21**.

Macaulay, D. 1988. *The Way Things Work.* Boston: Houghton Mifflin. Ages 8–adult; **36**.

McNaught, H. 1978. *The Truck Book.* New York: Random House. Ages 3–5; **21, 26**.

Peet, B. 1971. *The Caboose Who Got Loose.* Boston: Houghton Mifflin. Ages 3–5; **21, 26**.

Piper, W. 1984. *The Little Engine that Could.* New York: Putnam. Ages 3–5; **21, 26**.

Pluckrose, H. 1986. *Think about Hearing.* New York: Franklin Watts. Ages 4–8; **21, 26, 36**.

Rockwell, A. 1986. *Things That Go.* New York: Dutton. Ages 3–5; **21, 26**.

Rockwell, A. and Rockwell, H. 1971. *The toolbox.* New York: Harper & Row. Ages 3–6; **21, 26, 36**.

———. 1971. *Machines.* New York: Harper & Row. Ages 3–6; **21, 26, 36**.

Scarry, R. 1986. *Splish-Splash Sounds.* Racine, Wis.: Western. Ages 3–7; **21, 26, 36**.

Simon, S. 1985. *Soap Bubble Magic.* New York: Lothrop. Ages 6–9; **36**.

Swift, H. 1942. *Little Red Lighthouse and the Great Gray Bridge.* New York: Harcourt Brace Jovanovich. Ages 3–6; **21, 26, 36**.

Wyler, R. 1986. *Science Fun with Toy Boats and Planes.* New York: Julian Messner. Ages 5–9; **36**.

Earth and Space Science

Arnold, C. 1987. *Trapped in Tar: Fossils Form the Ice Age.* New York: Clarion. Ages 7–10; **37**.

Bauer, C. F. 1987. *Midnight Snowman.* New York: Atheneum. Ages 4–7; **19, 21, 26, 37**.

Brandt, K. 1985. *Air.* Mahwah, N.J.: Troll Associates. Ages 3–6; **21, 26, 37**.

Branley, F. M. 1987. *The Moon Seems to Change/ The Planets in Our Solar System/Rockets and Satellites.* New York: Crowell. Ages 5–8; **37**.

______. 1986. *Air is All Around Us.* New York: Crowell. Ages 3–6; **21, 26, 37**.

______. 1986. *Journey into a Black Hole.* New York: Crowell. Ages 8–10; **37**.

______. 1985. *Volcanoes.* New York: Crowell. Ages 6–8; **37**.

______. 1985. *Flash, Crash, Rumble, and Roll.* New York: Crowell. Ages 5–7; **37**.

______. 1983. *Rain and Hail.* New York: Crowell. Ages 5–7; **37**.

______. 1982. *Water for the World.* New York: Crowell. Ages 7–11; **37**.

Carrick, C. 1983. *Patrick's Dinosaurs.* New York: Clarion. Ages 4–8; **21, 36, 35, 37**.

Cole, J. 1987. *The Magic School Bus Inside the Earth.* New York: Scholastic. **21, 26, 37**.

Cole, J. 1987. *Evolution.* New York: Crowell. Ages 5–8; **35, 37**.

Elting, M. 1984. *Dinosaurs and Other Prehistoric Creatures.* New York: Macmillan. Ages 4+; **21, 26, 37**.

Giesen, R. 1977. *Where am I, Revised.* Minneapolis: Lerner. Ages 5–8; **37**.

Gibbons, G. 1987. *Weather Forecasting.* New York: Macmillan. Ages 5–8; **31, 32, 21, 26, 37**.

Keats, E. J. 1981. *Regards to the Man in the Moon.* New York: Four Winds. Ages 3–6; **21, 26, 37**.

Knowlton, J. 1985. *Maps and Globes.* New York: Harper & Row. Ages 5–8; **37**.

Lye, K. 1987. *Deserts.* Morristown, N.J.: Silver Burdett. Ages 8–14; **37**.

Malnig, A. 1985. *Where the Waves Break: Life at the Edge of the Sea.* Minneapolis: Carolrhoda. Ages 7–10; **37**.

Markle, S. 1987. *Digging Deeper.* New York: Lothrop. Ages 8–12; **37**.

Ride, S., with S. Oakie. 1986. *To Space and Back.* New York: Lothrop. Ages 8–12; **37**.

Rocks and Minerals. 1988. London: Natural History Museum. Ages 7–12; **37**.

Selberg, I. 1982. *Our Changing World.* New York: Philomel. Ages 4–6; **21, 26, 37**.

Sattle, H. R. 1985. *Pterosaurs: The Flying Reptiles.* New York: Lathrop. Ages 5–10; **37**.

Simon, S. 1985. *Jupiter.* New York: Morrow. Ages 5–9; **37**.

______. 1985. *Saturn.* New York: Morrow. Ages 5–9; **37**.

______. 1987. *Mars/Uranus.* New York: Morrow. Ages 8–11; **37**.

Szilagyi, M. 1985. *Thunderstorms.* New York: Bradbury. Ages 3–9; **21, 26, 37**.

Wade, H. 1977. *Sand.* Millwaukee: Raintree. Ages 4–8; **21, 26, 37**.

Health Science

Bayle, L. 1987. *Picture Books for Preschool Nutrition Education: A Selected Annotated Bibliography.* Order from author, 72 Meriam St., Lexington, Mass. 02173. **21, 26, 38**.

Brown, M. 1947. *Stone soup.* New York: Charles Scribner's Sons. Ages 3–6; **21, 26, 38**.

Brandenburg, A. 1976. *Corn is Maize: The Gift of the Indians.* New York: Crowell. Ages 3–5; **21, 22, 26, 38**.

Carle, E. 1970. *Pancakes, Pancakes.* New York: Knopf. Ages 3–5; **21, 22, 26, 38**.

Ontario Science Center. 1987. *Foodworks.* Reading, Pa.: Addison-Wesley. Ages 8–12; **38**.

Pomerantz, C. 1984. *Whiff, Whiff, Nibble, and Chew.* New York: Greenwillow. Ages 4–8; **21, 26, 36**.

Sendak, M. 1962. *Chicken Soup with Rice.* New York: Harper & Row. Ages 3–5; **21, 22, 26, 28.**

______. 1970. *In the Night Kitchen.* New York: Harper & Row. Ages 3–5; **21, 26, 28.**

Shaw, D. 1983. *Germs!* New York: Holiday House. Ages 8–12; **38.**

Skeleton. 1988. London: Natural History Museum. Ages 7–12; **38.**

CHILDREN'S PERIODICALS THAT EMPHASIZE MATH AND SCIENCE CONCEPTS

Chickadee: The Canadian Magazine for Children. Young Naturalist Foundation, 56 the Esplanade, Suite 304, Toronto, Ontario, Canada M5E 1A7. Ages 3–9.

Child Life. P.O. Box 10681, Des Moines, Iowa 50381. Ages 7–9.

Children's Playmate Magazine. Children's Better Health Institute, 1100 Waterway Blvd., P.O. Box 567, Indianapolis, Ind. 46206. Ages 4–8.

Koala Club News. Zoological Society of San Diego, Inc., P.O. Box 551, San Diego, Calif. 92212. Ages 6-15.

National Geographic News. P.O. Box 2330, Washington, D.C. 20009. Ages 5–12.

Ranger Rick's Nature Magazine. The National Wildlife Federation, 1412 16th. St., N. W., Washington, D. C. 20036. Ages 5–11.
D. C. 20036. Ages 5–11.

Scienceland, Inc. 501 5th Ave., Suite 2102, New York, N.Y. 10017. Ages 3–12.

Science Weekly. P.O. Box 70154, Washington, D.C. 20088. Ages 4–12.

Scholastic Let's Find Out. Scholastic Magazines, 1290 Wall Street, W., Lyndhurst, N.J. 07071. Age 5.

Sesame Street. Children's Television Workshop, P.O. Box 2896, Boulder, Col. 80322. Ages 3–8.

3 2 1 Contact. P.O. Box 2933, Boulder, Col. 80322. Ages 6–14.

Your Big Back Yard. National Wildlife Federation, 1412 16th Street, NW, Washington, D.C. 20036. Ages 3–5.

SOFTWARE PUBLISHERS USED IN THIS TEXT

Broderbund Software
17 Paul Dr.
San Rafael, Calif. 94903-2101

D.C. Heath and Company
125 Spring Street
Lexington, Mass. 02173

DLM
25115 Ave. Standford
Suite 130
Valencia, Calif. 91355

Lawrence Hall of Science
University of California
Berkeley, Calif. 94720

Learning Company
4370 Alpine Rd.
Portola Valley, Calif. 94015

Learning Well
200 S. Service Rd.
Roslyn Heights, N.Y. 11577

MECC
3490 Lexington Ave. N.
St. Paul, Minn. 55126

Milliken Publishing Co.
1100 Research Blvd.
St. Louis, Mo. 63132-0579

Scholastic, Inc.
730 Broadway
New York, N.Y. 10003

Sunburst Communications
39 Washington Ave.
Pleasantville, N.Y. 10570

APPENDIX C

Finger Plays and Rhymes

The Monkey

One little monkey was looking at you,
He was joined by another and then there were two.
Two little monkeys playing in a tree,
Were joined by another and then there were three.
Three little monkeys saw one more,
She came to play with them,
And then there were four.
Four little monkeys happy to be alive,
Were joined by another,
And then there were five.

Two Apples

Way up high in the apple tree,
Two little apples smiled at me.
I shook the tree as hard as I could.
Down fell the apples–
m-m-m-m-m-m! Were they good!

1, 2, 3 Clap

Your hand on your shoulder,
And then on your toes.
Your right hand on your knee,
Then on your lap it goes.
Place the left hand on your nose.
Touch your left hand to the floor.
Clap your hands 1, 2, 3
Would you like to do that some more?

1, 2, 3 Clap

See the birds in the tree,
Now let's count them: 1, 2, 3
Now do it backwards: 3, 2, 1
Away fly the birds, every one.

Feet and Hands

Two little feet go jump, jump, jump,
Two little hands go thump, thump, thump.
One little body turns round and round.
One little child sits quietly down.
Two little feet go tap, tap, tap.
Two little hands go clap, clap, clap.
A quick little leap up from the chair.
Two little arms reach high in the air.

Two Little Blackbirds

Two little blackbirds standing on a hill.
This one is Jack, and this one is Jill.
Fly away Jack; Fly away Jill.
Come back, Jack; come back, Jill.
Now there are two little blackbirds
standing on the hill.

I See Three

I see three: one, two, three
Three little bunnies reading the funnies.
I see three: one, two, three
Three little kittens, all wearing mittens.
I see three: one, two, three
Three little bears climbing the stairs.
I see three: one, two, three
Three little ducks riding in trucks.
I see three: one, two, three
Three little frogs sitting on logs.
I see three: one, two, three
Three little bees, buzzing in trees.
(Children may suggest more rhyming verses.)

Four Little Snowmen

I built four snowmen
All in a line.
They looked so strong,
They looked so fine.
Along came a child
As silly as can be.
Who packed down one snowman,
And that left three.
Down came the rain
From a sky of blue,
Smush went one snowman,
And that left two.
The clouds passed by
And out came the sun.
It melted one snowman,
And that left one.
One little snowman
Alone once more,
I built three more again,
And that made four.

Five Soldiers

Here are five soldiers standing by the door,
One marches away and then there are four.
Four little soldiers looking at me,
One falls down and then there are three.
Three little soldiers going to the zoo,
One goes home and then there are two.
Two little soldiers left all alone,
One says, "Good-bye" and then there is one.
One little soldier lifts up his gun,
He marches away and then there are none.

Here is Thumbkin

Here is thumbkin, Number 1
He is big, but see him run.
Here is pointer, Number 2
He can show the way for you.
This is long man, Number 3
He's the tallest one you see.
This is lazy, Number 4
Try to lift him just once more.
Here's the baby, Number 5
Tiny, yes, but quite alive.
Now move your fingers, do a dance!
Lift them high, like horses prance.
Now into their houses they all creep,
Make a fist–they've gone to sleep!

Five Little Kittens

Five little kittens sitting on the floor,
One ran away and then there were four.
Four little kittens playing round a tree,
One went to sleep and then there were three.
Three little kittens beginning to "Mew,"
One climbed the tree and then there were two.
Two little kittens playing in the sun,
One went home and then there was one.
One little kitten left all alone.
He chased a mouse and then there was none.

Five Little Chickadees

Five little chickadees sitting by the door (5 fingers)
One flew away and then there were four.
(Chorus) Chickadee, chickadee, happy and gay
Chickadee, chickadee, fly away.
Four little chickadees sitting in a tree
One flew away and then there were three.
(Chorus)
Three little chickadees sitting just like you,
One flew away and then there were two.
(Chorus)
Two little chickadees having lots of fun,
One flew away and then there was one.
(Chorus)
One little chickadee sitting all alone.
It flew away and there were none.
(Chorus)

Here Is the Beehive

Here is the Beehive.
Where are the Bees?
Hidden away where nobody sees.
Soon they come creeping out of the hive:
One, Two, Three, Four, Five.

Five Little Ducks

Five little ducks standing in a row,
They walk to the water, going to and fro.
This duck cries, "Quack, quack, quack."
This one says, "Let's go back."
This one yells, "Stay in line."
This one says, "The water's fine."
The last little duck just pecked at the ground,
Eating the bugs that crawled around.

Birthday Years

This is ______________________________ .
It's his birthday today.
He's __________ years old, so they say.
Let's count his age as we clap our hands,
To let him know we think he's grand.
1, 2, 3, 4, 5.

Add One More

This is one cat,
Now two cats I see.
Another cat will make it three.
I add another,
Then there are four.
Mother cat joins them.
To make one more.
Now let's count them: 1, 2, 3, 4, 5.

Addition

One and one are two. That I always knew.
Two and two are four. They could be no more.
Three and three are six. Whether stones or sticks.
Four and four are eight, if I can keep them straight.
Five and five are ten. Let's try it all over again.

Ten Fingers

I have ten little fingers.
And they all belong to me.
I can make them do things.
Would you like to see?
I can shut them up tight
Or open them wide.
I can put them together
Or make them all hide.
I can make them jump high,
I can make them jump low,
I can fold them up quietly,
And fold them just so.

USE OF ORDINAL NUMBERS

Little Dogs

The first little dog barked very loud.
The second little dog ran after a crowd.
The third little dog said, "Let's eat, let's eat!"
The fourth little dog said, "Let's have meat"
The fifth little dog said, "I think I will stay
in my own backyard and sleep all day!"

Five Little Snowmen

Five little men all made of snow,
Five little snowmen, all in a row,
Out came the sun and stayed all day.
The first little snowman melted away.
(repeat with second, third, fourth, and fifth.)

Five Little Pumpkins

Five little pumpkins sitting on a gate
The first one said, "Oh, my, it's getting late."
The second one said, "There are witches in the air."
The third one said, "But we don't care."
The fourth one said, "Let's run, and run, and run."
The fifth one said, "I'm ready for some fun."
"Ooo" went the wind, and out went the light.
And the five little pumpkins rolled out of sight.

Ten Little Snowmen

Ten little snowmen standing in a line;
The first one melted and then there were nine.
Nine little snowmen standing tall and straight;
The second one melted and then there were eight.
Eight little snowmen, white as in heaven;
The third one melted and then there were seven.
Seven little snowmen with arms made of sticks;
The fourth one melted and then there were six.
Six little snowmen looking alive;
The fifth one melted and then there were five.
Five little snowmen with mittens from the store;
The sixth one melted and then there were four.
Four little snowmen beneath a green pine tree;
The seventh one melted and then there were three.
Three little snowmen with pipes and mufflers, too;
The eighth one melted and then there were two.
Two little snownen standing in the sun;
The ninth one melted and then there was one.
One little snowman standing all alone;
He started to run, run, run, and then there was none.

POEMS TO TEACH SIZE AND SHAPE

Big and Fat

Santa Claus is big and fat.
He wears black boots
and a bright red hat.
His nose is red,
just like a rose,
And he Ho-Ho-Ho's
From his head to his toes.

Big Round Cookie

I looked in the cookie jar, and what did I see?
A big round cookie Mother put there for me.
Mother looked in the cookie jar,
But she didn't see
The big round cookie she put there for me.

The Ball

A little ball,
A larger ball,
The largest ball I see.
Now let's count them:
One, Two, Three.

Fingers

My fingers make a circle,
One that's very small.
Now I make a big one,
Just like a rubber ball.
This is my ring finger,
This one is the smallest.
I put all my fingers together,
To see which is the tallest.

ADDITIONAL RESOURCES (see also Unit 41)

Bayley, N. 1977. *One Old Oxford Ox*. New York: Atheneum. **9**.

Bouton, J. 1967. *For the Children's Hour.* New York: Platt & Munk. "Playgrounds." **11, 15, 18**; "The Little Elf Man." **11, 15, 18**.

Brewerton, S., and Brewerton, J. E. 1960. *Birthday Candles Burning Bright: A Treasury of Birthday Poetry.* New York: Macmillan. "The Difference." **11, 15, 19**; "Comparison." **11, 15, 17, 18, 19**.

Cole, W. 1964. *Beastly Boys and Ghastly girls*. New York: Collins-World. "The Sweet Tooth." **11, 15, 18, 38**; "Little Thomas." **11, 15, 18, 38**.

Ferris, H. 1957. *Favorite Poems, Old and New.*

New York: Doubleday, "People" by Lois Lenski. **11, 15, 18**.

Knight, H. 1973. *Hillary Knight's Mother Goose* New York: Golden Press. "Elsie Marley," "Deedle deedle dumpling," "Cocks crow in the morn," "A diller, a dollar," "Hickory, dickory, dock," "Solomon Grundy," "Wee Willie Winkie," "The man in the moon." **19, 32;** "The house that Jack built." **22**; "Two black birds," "The dove," "St. Ives," "Sing a song of sixpence." **9**.

McDonald, G. D. 1959. *A Way of Knowing: A Collection of Poems for Boys*. New York: Crowell. "Measurement." **11, 15, 18, 19, 37**.

Milne, A. A. 1982. *Pooh's Counting Book*. New York: Dutton. **9**.

Opie, I., and Opie, P. 1973. *The Oxford Book of Children's Verse*. New York: Oxford University Press. "The dream of a boy who lived at Nine Elms." **9**; "The dream of a girl who lived at Sevenoaks." **9**.

Silverstein, S. 1974. *Where the Sidewalk Ends*. New York: Harper & Row. "Smart." **22, 32**; "One inch tall." **18, 32**; "Me and my giant." **9, 11**; "Skinny." **11**.

Untermeyer, L. 1959. *The Golden Treasury of Poetry*. New York: Golden. "Old Joe Brown." **18**; "There was an old man" by E. Lear. **11, 15**.

Wyndam, R. 1968. *Chinese Mother Goose Rhymes*. New York: Collins-World. "Little." **11**.

APPENDIX D

Code of Practice on Use of Animals in Schools

This code of practice is recommended by the National Science Teachers Association for use throughout the United States by elementary, middle/junior high, and high school teachers and students. It applies to educational projects conducted and lessons taught, involving live organisms in schools or in school-related activities such as science fairs, science clubs, and science competitions.

The purpose of these guidelines is to enrich education by encouraging students to observe living organisms and to learn proper respect for life. The study of living organisms is essential for an understanding of living processes. This study must be coupled with the observance of humane animal care and treatment.

I. CARE AND RESPONSIBILITY FOR ANIMALS IN THE CLASSROOM

A. A teacher must have a clear understanding of and a strong commitment to the responsible care of living animals before making any decision to use live animals for educational study. Preparation for the use of live animals should include acquisition of knowledge on care appropriate to the species being used including housing, food, exercise, and the appropriate placement of the animals at the conclusion of the study.

B. Teachers should try to assure that living animals entering the classroom are healthy and free of transmissible disease or other problems that may endanger human health. Not all species are appropriate. Wild animals are not appropriate because they may carry parasites or serious diseases.

C. Maintaining good health and providing optimal care based on an understanding of the life habits of each species used is of primary importance. Animal quarters shall be spacious, shall avoid overcrowding, and shall be sanitary. Handling shall be gentle. Food shall be appropriate to the animal's normal diet and of sufficient quantity and balance to maintain a good standard of nutrition at all times. No animal shall be allowed less than the optimum maintenance level of nutrition. Clean drinking water shall always be available. Adequate provision for care shall be made at all times including vacation times.

D. All aspects of animal care and treatment shall be supervised by a qualified ADULT WHO IS KNOWLEDGEABLE ABOUT RESEARCH METHODS, BIOLOGY, CARE, AND HUSBANDRY OF THE SPECIES BEING STUDIED.

E. Supervisors and students should be familiar with *literature on care and handling* of living organisms. Practical training in these techniques is encouraged.

F. Adequate plans should be made to *control possible unwanted breedings* of the species during the project period.

G. Appropriate plans should be made for future care of animals at the conclusion of the study.

H. As a general rule, laboratory-bred animals should not be released into the wild as they may disturb the natural ecology of the environment.

I. On rare occasions it may be necessary to sacrifice an animal for educational purposes. This shall be done only in a manner accepted and approved by the American Veterinary Association, by a person experienced in these techniques, and at the discretion of the teacher. It should not be done in the presence of immature or young students who may be upset by witnessing such a procedure. Maximum efforts should be made to study as many biological principles as possible from a single animal.

J. The procurement, care, and use of animals must comply with existing local, state, and federal regulations.

II. EXPERIMENTAL STUDIES OF ANIMALS IN THE CLASSROOM

A. When biological procedures involving living organisms are called for, every effort should be made to use plants or invertebrate animals when possible.

B. No experimental procedure shall be attempted on mammals, birds, reptiles, amphibians, or fish that causes the animal unnecessary pain or discomfort.

C. It is recommended that preserved vertebrate specimens be used for dissections.

D. Students shall not perform dissection surgery on vertebrate animals except under direct supervision of a qualified biomedical scientist or trained adult supervisor.

E. *Experimental procedures* including the use of pathogens, ionizing radiation, toxic chemicals, and chemicals producing birth defects must be under the supervision of a biomedical scientist or an adult trained in the specific techniques. Such procedures should be done in appropriate laboratory facilities that adhere to safety guidelines.

F. *Behavior studies should use only reward* (such as providing food) and not punishment in training programs. When food is used as a reward, it should not be withheld for more than 12 hours.

G. If embryos are subjected to invasive or potentially damaging manipulation, the embryo must be destroyed prior to hatching. If normal embryos are hatched, provisions must be made for their care and maintenance.

III. RESEARCH INVESTIGATIONS INVOLVING VERTEBRATE ANIMALS

The National Science Teachers Association recognizes that an exceptionally talented student may wish to conduct research in the biological or medical sciences and endorses procedures for student research as follows:

A. Protocols of extracurricular projects involving animals should be reviewed in advance of the start of the work by a qualified adult supervisor.

B. Preferably, extracurricular projects should be carried out in an approved area of the school or research facility.

C. The project should be carried out with the utmost regard for the humane care and treatment of the animals involved in the project.

—Adopted by the NSTA Board of Directors in July 1985.

Hampton, C.H., Hampton, C.D., & Kammer, D.C. (1988) *Classroom creature culture*. Washington, D.C.: National Science Association.

1. NSTA Care of Animals Position Statement
2. Guidelines for Proper Care of Living Organisms
3. Sources for Teaching About Plants, Animals, and the Environment

Food and Water	Rabbits	Guinea Pigs	Hamsters	Mice	Rats
Daily					
pellets or	rabbit pellets: keep dish half full		large dog pellets: one or two		
grain		corn, wheat, or oats		canary seeds or oats	
green or leafy vegetables, lettuce, cabbage, and celery tops or	keep dish half full 4–5 leaves	2 leaves	1 1/2 tablespoon 1 leaf	2 teaspoons 1/8–1/4 leaf	3–4 teaspoons 1/4 leaf
grass, plantain, lambs' quarters, clover, alfalfa or	2 handfuls	1 handful	1/2 handful	—	—
hay, if water is also given					
carrots	2 medium	1 medium			
Twice a week					
apple (medium)	1/2 apple	1/4 apple	1/8 apple	1/2 core and seeds	1 core
iodized salt (if not contained in pellets)	or salt block	sprinkle over lettuce or greens			
corn, canned or fresh, once or twice a week	1/2 ear	1/4 ear	1 tablespoon 1/3 ear	1/4 tablespoon or end of ear	1/2 tablespoon or end of ear
water	should always be available		necessary only if lettuce or greens are not provided		

Food and Water	Water Turtles	Land Turtles	Small Turtles
Daily			
worms or night crawlers or	1 or 2	1 or 2	1/4 inch of tiny earthworm
tubifex or blood worms and/or			enough to cover 1/2 area of a dime
raw chopped beef or meat and fish-flavored dog or cat food	1/2 teaspoon	1/2 teaspoon	
fresh fruit and vegetables		1/4 leaf lettuce or 6–10 berries or 1–2 slices peach, apple, tomato, melon or 1 tablespoon corn, peas, beans	
dry ant eggs, insects, or other commercial turtle food			1 small pinch
water	always available at room temperature; should be ample for swimming and submersion		
	3/4 of container	large enough for shell	half to 3/4 of container

Food and Water Plants (for Fish)	Goldfish	Guppies
Daily		
dry commercial food	1 small pinch	1 very small pinch; medium size food for adults; fine size food for babies
Twice a week		
shrimp—dry—or another kind of dry fish food	4 shrimp pellets or 1 small pinch	dry shrimp food or other dry food: 1 very small pinch
Two or three times a week		
tubifex worms	enough to cover 1/2 area of a dime	enough to cover 1/8 area of a dime
Add enough "conditioned" water to keep tank at required level	allow one gallon per inch of fish; add water of same temperature as that in tank—at least 65°F	all 1/4–1/2 gallon per adult fish; add water of same temperature as that in tank—70°–80°F
Plants: cabomba, anarcharis, etc.	should always be available	

	Newts	Frogs
Daily		
small earthworms or mealworms	1–2 worms	2–3 worms
or		
tubifex worms	enough to cover 1/2 area of a dime	enough to cover 3/4 area of a dime
or		
raw chopped beef	enough to cover a dime	enough to cover a dime
water	should always be available at same temperature as that in tank or at room temperature	

(Reprinted from Science and Children *(1965) with permission from the National Science Teachers Association, 1742 Connecticut Avenue, Washington, D.C. 20009, Pratt, G.K. "How to . . . care for living things in the classroom.")*